The Routledge History of Global War and Society

The Routledge History of Global War and Society offers a sweeping introduction to the most significant research on the causes, experiences, and impacts of war throughout history. This collection of twenty-seven essays by leading historians demonstrates how war and society studies have dramatically expanded the chronological, geographic, and thematic breadth of the field of military history. Each chapter addresses the ways in which recent scholarship has integrated cultural, ethical, environmental, medical, and ideological factors to explain both conventional conflicts and genocide, terrorism, and other forms of mass violence. The broad scope of the collection makes it the perfect primer for scholars and students seeking to understand the complex interactions of warfare and those affecting and affected by conflict.

Matthew S. Muehlbauer is Assistant Professor of History at the United States Military Academy.

David J. Ulbrich is Associate Professor and Program Director of the Master of Arts in History and Military History degrees at Norwich University.

The Routledge Histories

For a full list of titles in this series, please visit www.routledge.com

The Routledge Histories is a series of landmark books surveying some of the most important topics and themes in history today. Edited and written by an international team of world-renowned experts, they are the works against which all future books on their subjects will be judged.

The Routledge History of Global War and Society

Edited by
Matthew S. Muehlbauer
and David J. Ulbrich

Routledge
Taylor & Francis Group

LONDON AND NEW YORK

First published 2018 by Routledge

2 Park Square, Milton Park, Abingdon, Oxon, OX14 4RN
605 Third Avenue, New York, NY 10017

Routledge is an imprint of the Taylor & Francis Group, an informa business

First issued in paperback 2020

Library of Congress Cataloging-in-Publication Data
A catalog record for this book has been requested

ISBN: 978-1-138-84980-8 (hbk)
ISBN: 978-0-367-73517-3 (pbk)

Typeset in Bembo
by Apex CoVantage, LLC

The opinions expressed by the contributing authors in this work do not necessarily reflect official policies or positions of any institution or agency, including the U.S. Army, the U.S. Department of Defense, or the U.S. Government or of the Australian government.

Contents

Contributors

Bianka J. Adams is a Historian in the Office of History, U.S. Army Corps of Engineers. She received her master's degree in Political Science from the Christian-Albrechts University in Kiel, Germany, and earned her doctorate in U.S. Diplomatic History from the Catholic University of America in Washington, DC. Adams is the author of *From Crusade to Hazard: The Denazification of Bremen Germany*.

Alexander M. Bielakowski is Assistant Professor of History at University of Houston-Downtown. He edited *Ethnic and Racial Minorities in the U.S. Military: An Encyclopedia*. He has also published on African American military history, the cavalry in the interwar U.S. Army, and Dwight D. Eisenhower as NATO's first commander. Bielakowski is Editor-in-Chief of the peer-reviewed online journal *U.S. Military History Review*.

Stephen Connor is Assistant Professor of History at Nipissing University, North Bay, Ontario. His research and teaching focus on terrorism, insurgencies, genocide, comparative revolutions, and twentieth-century war and society. Among his publications is "Bringing in the Harvest: The German Civil Administration and Occupation Management in Brest-Litovsk, 1942–1944," which was published in *Global War Studies*.

Peter J. Dean currently serves on the faculty of University of Western Australia. He is author of several books on Australian military affairs and military history, including most recently *MacArthur's Coalition: U.S. and Australian Military Operations in SWPA, 1942–1945*. Dean has also served as Senior Fellow at the Australian and New Zealand Centre, Edmund A. Walsh School of Foreign Service, Georgetown University, Washington, DC.

Michael Dolski is a Historian in the Defense POW/MIA Accounting Agency. He is author of *D-Day Remembered: The Normandy Landings in American Collective Memory* and co-editor of *D-Day in History and Memory: The Normandy Landing in International Remembrance and Commemoration*. With Eric Klinek, Dolski is co-authoring a book on the U.S. Army's ways of war in the twentieth and twenty-first centuries to be published by the University of Oklahoma Press.

Hilary Earl is Associate Professor of European History at Nipissing University, North Bay, Ontario. Her research and teaching interests include war crimes trials, perpetrator testimony, and the cultural impact of the Holocaust and genocide in the twenty-first century. Earl is co-editing the forthcoming *Wiley-Blackwell Handbook on the Holocaust*, and she is making a documentary film on Nazi perpetrators.

Antulio J. Echevarria II is currently Editor of the *U.S. Army War College Quarterly, Parameters*. Previously, he served as Director of Research at the U.S. Army War College. Echevarria has written several books, including *Reconsidering the American Way of War: US Military Practice From the Revolution to Afghanistan* and *Military Strategy: A Very Short Introduction*.

Aaron Graham is a Leverhulme Early Career Fellow at University College London. His research examines politics, finance, and warfare in Britain and its empire between 1660 and 1840. He is author of *Corruption, Party, and Government in Britain, 1702–1713*. With Patrick Walsh, Graham also coedited *The British Fiscal-Military States, 1660–1783*.

Eric W. Klinek is a Historian in the Defense POW/MIA Accounting Agency. His doctoral dissertation from Temple University is titled "The Army's Orphans: The United States Army Replacement System in the European Campaign, 1944–1945." With Michael Dolski, Klinek is co-authoring a book on the U.S. Army's ways of war in the twentieth century and twenty-first centuries to be published by the University of Oklahoma Press.

Jay Lockenour is Associate Professor and Chair of the Department of History at Temple University. He is author of *Soldiers as Citizens: Former Wehrmacht Officers in the Federal Republic of Germany, 1945–1955*. Lockenour is also completing a biography of Erich Ludendorff that focuses on his military career and anti-Semitic politics. Lockenour is also co-host of the podcast series "New Books in Military History."

Tristan Moss is an Adjunct Lecturer at the University of New South Wales, and has published his book *Guarding the Periphery: the Australian Army in Papua New Guinea, 1951–75*. He is also a researcher on the Official Histories of Australian Operations in Iraq, Afghanistan, and East Timor.

Matthew S. Muehlbauer teaches military history at the United States Military Academy at West Point. He is author of "Holy War and Just War in Early New England, 1630–1655," which has been published in *Journal of Military History*. With David J. Ulbrich, Muehlbauer has coauthored *Ways of War: American Military History from the Colonial Era to the Twenty-First Century*, now in its second edition. His next project is a book for Naval Institute Press to be titled *Ares and His Sons: War, Terror, and Terrorism in History*, with an anticipated publication date of 2020.

Kaushik Roy is Guru Nanak Chair Professor in the Department of History, Jadavpur University, Kolkata, India, and Global Fellow at Peace Research Institute Oslo (PRIO), Norway. He has published many books and edited anthologies for Ashgate, Bloomsbury, Brill, Cambridge University Press, Oxford University Press, Routledge, and Pickering & Chatto.

Richard A. Ruth is Associate Professor and Chair of the Department of History at the U.S. Naval Academy. He is a specialist in Southeast Asian history, with an emphasis on Thailand. Ruth has published the book *In Buddha's Company: Thai Soldiers in the Vietnam War*.

Eileen Ryan is Assistant Professor of History at Temple University. She is author of the recent book *Religion and Resistance: Negotiating Authority in Italian Libya*. Among Ryan's other publications is her article "Violence and the Politics of Prestige: The Fascist Turn in Colonial Libya," appearing in *Modern Italy*.

Margaret Sankey is Director of Research at the U.S. Air War College. Her publications include the book *Jacobite Prisoners of the 1715: Preventing and Punishing Insurrection in Early Hanoverian Britain* and the award-winning article "Elite Culture and the Decline of Scottish Jacobitism 1716–1745" with Dr. Daniel Szechi in *Past and Present*.

Adam R. Seipp is a Professor of History at Texas A&M University. His research and teaching focus on war and social change in modern Germany. Seipp has published the books *Strangers in the Wild Place: Refugees, Americans, and a German Town, 1945–1952* and *The Ordeal of Peace: Demobilization and the Urban Experience in Britain and Germany, 1917–1921*.

Jason W. Smith is an Assistant Professor of History at Southern Connecticut State University. Previously Smith has published articles in the *Journal of Military History* and *Environmental History*. Smith is the author of the recent book *To Master the Boundless Sea: The U.S. Navy, the Marine Environment, and the Cartography of Empire*.

Paul J. Springer is Professor of History at the U.S. Air Command and Staff College and Senior Fellow of the Foreign Policy Research Institute. He has written several books, including *America's Captives: Treatment of POWs From the Revolutionary War to the War on Terror* and *Outsourcing War to Machines: The Military Robotics Revolution*.

Heather Marie Stur is Associate Professor and the current General Buford Blount Professor of Military History at the University of Southern Mississippi. She is also a fellow at the Dale Center for the Study of War and Society. Stur is author of *Beyond Combat: Women and Gender in the Vietnam Era* and, more recently, co-editor of *Integrating the U.S. Military: Race, Gender, and Sexuality Since World War II*.

Kenneth M. Swope is Professor of History and Senior Fellow at the Dale Center for the Study of War & Society at University of Southern Mississippi. From 2015 to 2017, Swope also held the General Buford Blount Professorship of Military History at USM. During 2015, Swope was a member of the School of Historical Studies, Institute for Advanced Study at Princeton University. Among his many publications, Swope authored *On the Trail of the Yellow Tiger: War, Trauma, and Social Dislocation in Southwest China During the Ming-Qing Transition*.

Ellen D. Tillman is Associate Professor of History at Texas State University. She is author of *Dollar Diplomacy by Force: Nation-Building and Resistance in the Dominican Republic*. Tillman is currently writing two monographs: one on the role of U.S. expansionism and interventionism in the formation of the state of Panama, and another on the role of Central America in the American Civil War.

Ingo Trauschweizer is an Associate Professor of History and Director of the Contemporary History Institute at Ohio University. He has published widely on the Cold War, including his award-winning book *The Cold War U.S. Army: Building Deterrence for Limited War*. Trauschweizer is currently completing a biography of General Maxwell Taylor.

Matthew Trundle is Professor and Chair of Classics and Ancient History at the University of Auckland, New Zealand. He is author of *Greek Mercenaries From the Late Archaic Period to Alexander*, as well as editor of volumes titled *New Perspectives on Ancient Warfare* and *Beyond the Gates of Fire: New Perspectives on the Battle of Thermopylae*.

David J. Ulbrich is Associate Professor and Program Director of Norwich University's MA in Military History and History degree programs. Ulbrich is the author of the award-winning *Preparing for Victory: Thomas Holcomb and the Making of the Modern Marine Corps, 1936–1943*. Together with Matthew S. Muehlbauer, Ulbrich coauthored *Ways of War: American Military History From the Colonial Era to the Twenty-First Century*, now in its second edition. Ulbrich and Bobby Wintermute are completing a book titled *Introducing Race and Gender as Tools of Analysis in Military History*, to be published by DeGruyter Press.

Bruce Vandervort is Emeritus Professor of History at the Virginia Military Institute. He is currently the Editor of *The Journal of Military History*. Vandervort is the author of *Wars of Imperial Conquest in Africa, 1830–1914* and of *Verso la Quarta Sponda: La Guerra italiana per la Libia (1911–1912)* [*To the Fourth Shore: Italy's War for Libya (1911–1912)*].

Jacqueline E. Whitt is an Associate Professor of Strategy at the U.S. Army War College. She teaches courses on strategy, national security policy, American studies, religion and war, and the social and cultural history of the U.S. military. She is the author of *Bringing God to Men: American Military Chaplains and the Vietnam War.*

Bobby A. Wintermute is Associate Professor of History at Queens College-CUNY. He is the author of *Public Health and the U.S. Military: A History of The Army Medical Department, 1818–1917.* Wintermute is also co-host of the podcast series "New Books in Military History." Together with David Ulbrich, Wintermute is completing a book titled *Introducing Race and Gender as Tools of Analysis in Military History,* to be published by DeGruyter Press.

Marilyn Zilli is a retired attorney from Harrisburg, Pennsylvania, who also earned advanced degrees in French language and literature. She made several trips to Francophone West and Central Africa as a participant in the "experts program" then run by the State Department.

Acknowledgments

This anthology had its genesis over lunch at the 2014 Society for Military History Conference with Kimberly Guinta, then in her role as acquisitions editor at Routledge. We had already worked with her to publish the successful textbook *Ways of War: U.S. Military History From the Colonial Era to the Twenty-First Century* in 2013. Kim believed that a niche existed for a reference work on the historiography of global war and society. We agreed that this project had great merit.

Over the next three years, we worked with three other supportive editors at Routledge: Genevieve Aoki, Margo Irvin, and most recently Eve Mayer. We are grateful for their flexibility and enthusiasm in seeing this anthology through to publication. Theodore Meyer also deserves all our thanks for providing support at key points in the lengthy process. Several contractors also helped with the copyediting and indexing tasks.

We are indebted to all the contributors who willingly gave of their time and expertise to write the chapters appearing herein. They accepted suggestions for revisions with grace, made timely submissions, and produced what we believe will be the premier reference of its kind for years to come. Among the contributors, Bobby Wintermute provided essential advice throughout the project.

Lastly, the greatest of appreciation must go to our families: Elizabeth Schoetz for her enduring patience, good humor, and support; Samantha and Jacob Muehlbauer for occasionally getting their father away from the computer and providing perspective; and Tom and Pat Ulbrich for always being there with good advice and enjoyable vacations.

Matthew S. Muehlbauer
New City, New York

David J. Ulbrich
Barre, Vermont
November 2017

Introduction

Matthew S. Muehlbauer and David J. Ulbrich

What was once known as the "new" military history has now been a distinct field for over a generation, and is better known as war and society studies. Its initial innovation was that in considering the historical phenomenon of war, practitioners went beyond accounts of battles and campaigns—ones that privileged the role and decisions of top commanders. Academics began to scrutinize the experience of war in itself, while also examining the broader interaction societies and polities have had with respect to preparing for and engaging in military conflict. As these scholars have pursued these methodologies and applied them to a growing range of topics and issues, new ones have emerged, particularly studies of how culture has shaped perceptions of military service, conduct, and sacrifice (for an overview, see Morillo with Pavkovic, 2014).

These expanding perspectives on military history followed the advent of social, feminist, Marxist, the *Annales*, postmodern, and other conceptual frameworks in the mid-twentieth century, albeit sometimes a decade or more after they had gained footholds in the academy as a whole. Historical studies also began incorporating interdisciplinary approaches in this era, drawing on research in culture, psychology, sociology, literary criticism, religious studies, ethnography, biology, ecology, and medicine. The interactions of these fields and disciplines have allowed scholars of war and society to ask questions, discover evidence, analyze factors, and draw conclusions that have greatly enriched our understanding of military history (for an overview, see Hughes and Philpott, 2006).

In the mid-twentieth century, Bell Wiley broke ground as one of the first academics to consider the experiences of soldiers on their own terms, looking specifically at those who had fought in the American Civil War (1943, 1952; both updated 2008). This "bottom-up" approach became more established with the publication of John Keegan's *Face of Battle* (1976), which compared the nature of combat for three famous battles involving English and British troops: Agincourt, Waterloo, and the Somme. Since then, similar studies have considered conflict in various eras, including Greek hoplite warfare (Hanson, 2000). But such bottom-up studies of combatants also intersected powerfully with methodologies evolving from the 1960s that highlighted groups previously ignored or marginalized in broad historical narratives, leading to examinations of noncombatant and particularly women's experiences in war (Campbell, 1984), as well as those of minorities (McPherson, 1965).

While some scholars began scrutinizing the experiences of particular social groups, others considered the relationship between war and society in the most expansive of terms. In the 1950s, for example, George Clark published a series of essays (1958) that explored topics such as the relationship between war and economic interdependence among countries in the seventeenth century, and how war helped define the "community" of Europe. However, the most famous concept derived from this period is Michael Roberts' idea of the "military revolution," wherein he argued that in early modern Europe, changes in weapons technology produced larger armies, which in turn led to the development of stronger centralized states that could better extract from their societies the resources needed to conduct war. Geoffrey Parker later reworked and revamped this concept, while also broadening it to include naval developments and European expansion into other regions of the world (Rogers, 1995; Parker, 1996).

Since the mid-1900s, scholars have employed many lenses to scrutinize how the state has meditated the relationship between war and society. These range from how governments developed more efficient means to raise and spend tax revenue (Brewer, 1988) to the role conscription has played in fostering national identity (Weber, 1976). As the last point illustrates, numerous works incorporate bottom-up approaches, noting how the state interacted with social groups in order to generate resources and organize forces and labor for military ends. Many others focus on a specific armed force—or even a group within such an organization, such as the officer corps—to scrutinize both its relationships with broader society and the state and its internal organizational structure and dynamics (Craig, 1955). Research that considers American cases have an added wrinkle to these approaches, in that early government and military institutions were notably weaker compared to Europe (Kohn, 1975; Karsten, 1986).

In more recent years, another group of methodologies has increasingly made its mark on the study of war. These fall within the general rubric of "culture," though they include a number of distinct approaches. One looks at how media—including print, visual, and other forms—shapes attitudes toward war, as well as reflects the experience of conflict (Fussell, 1975). Others examine the factors that shape widespread memories and interpretations of historical episodes in popular culture (Reardon, 1997), and how societies have memorialized and commemorated military service and sacrifice (Winter, 1995). Some works specifically explore the factors that shape military cultures (Hull, 2005). Though scholars such as Rob Citino (2007) distinguish between "war and society" and "memory and culture" studies, we see the latter as the most recent set of research methods within the former. Culture and its various manifestations shape how societies—and distinct groups within them—both impute meaning to and derive meaning from military experience, obligation, and loss, and should not be considered apart from them (for a recent example in a survey text, see Lee, 2016).

The foregoing touches on just the broadest swathes of methodological approaches in the contemporary historical study of war. But before noting how subsequent chapters consider them, one other historiographical development requires commentary. Until the late twentieth century, the vast bulk of academic research addressed Western experiences. Recent scholarship has begun to rectify this issue, and has produced a rich and engaging literature that addresses war and military service among non-Western peoples from their own perspectives (Black, 2004).

The Routledge History of Global War and Society is designed to serve as a primer for students and scholars entering the field of war and society. Readers should use it to begin their research, cognizant of the fact that there are many other academic books and articles that limited space prevents from including in these essays. What follows this introduction is a series of historiographical essays that demonstrate the ideas and conceptual frameworks that scholars of war and society have employed to explore various historical topics and themes—emphasizing those that have been prominent or emerged since the last decades of the twentieth century. (Readers in need of an introduction to historiography can consider Arnold, 2000.) Each chapter ends with a bibliography to facilitate further research.

The volume contains two broad sections. The first addresses distinct geographical regions. Because the West has dominated the historiography, two chapters address modern Europe and North America each, with separate discussions for the ancient Mediterranean and the British Isles. The section begins, however, with the non-Western world, with chapters exploring three distinct regions of Asia as well as Australasia, the Middle East and North Africa, sub-Saharan Africa, and Latin America.

In the second section, chapters focus on particular methodologies and frameworks used within the war and society field. A number of these have been prominent within the broader academic discipline of history for some time. These essays address gender, race, and ethnicity, and explore the intersections of war with religion and ethics, and nationalism and militarism. Others explore topics that are more specific to the study of war, such as combatant experiences, treatment of prisoners, and studies of "homefronts." Two chapters examine scholarship that considers the application of violence beyond combat, particularly the phenomena of terror and atrocities. The remaining essays in this section illustrate research involving newer approaches, such as the interaction of war with media, health, the environment, memory and memorialization, and culture.

While striving to balance prominent and emerging methodologies in the war and society field, the second section cannot devote distinct chapters to every possible topic. Many, however, are addressed within several of this volume's essays, such as the question of technology. Similarly, approaches that are the subject of particular chapters are also discussed in others. To note one example, many authors in the first section note the notion of "martial races" among Western powers that cultivated minority populations to enforce imperial rule in Africa and Asia. Readers are thus advised to read broadly across this volume's offerings.

Doing so will reveal connections among historiographical themes that demonstrate why and how war and society studies has become a vibrant academic field. This work demonstrates the numerous impacts war and armed forces can have on a society: exacerbating preexisting tensions within it, enabling some groups to augment their status and influence at the expense of others, or as a means to enhance political cohesion. But it also illustrates how social dynamics and cultural ideas shape the planning, training for, and prosecution of war. Whatever their particular interests, our hope is that *The Routledge History of Global War and Society* will enable students and scholars to begin intellectually rich and stimulating research.

Bibliography

Arnold, John H. *History: A Very Short Introduction*. New York: Oxford, 2000.

Black, Jeremy. *Rethinking Military History*. New York: Routledge, 2004.

Brewer, John. *The Sinews of Power: War, Money, and the English State, 1688–1783*. Cambridge: Harvard University Press, 1988.

Campbell, D'Ann. *Woman at War With America: Private Lives in a Patriotic Era*. Cambridge: Harvard University Press, 1984.

Citino, Robert M. "Military Histories Old and New: A Reintroduction," *American Historical Review* 112 (2007): 1070–1090.

Clark, George. *War and Society in the Seventeenth Century*. New York: Cambridge University Press, 1958.

Craig, Gordon A. *The Politics of the Prussian Army, 1640–1945*. New York: Oxford University Press, 1955.

Fussell, Paul. *The Great War and Modern Memory*. New York: Oxford, 1975.

Hanson, Victor Davis. *The Western Way of War: Infantry Battle in Classical Greece*. Second Edition. Berkeley: University of California Press, 2000.

Hughes, Matthew, and William J. Philpott, eds. *Modern Military History*. Houndsmills, UK: Palgrave Macmillan, 2006.

Hull, Isabell V. *Absolute Destruction: Military Culture and the Practices of War in Imperial Germany*. Ithaca: Cornell University Press, 2005.

Karsten, Peter, ed. *The Military in America: From Colonial Times to the Present*. New, Revised ed. New York: The Free Press, 1986.

Keegan, John. *The Face of Battle*. New York: Penguin, 1976.

Kohn, Richard H. *Eagle and Sword: The Federalists and the Creation of the Military Establishment in America, 1783–1802*. New York: The Free Press, 1975.

Lee, Wayne E. *Waging War: Conflict, Culture, and Innovation in World History*. New York: Oxford University Press, 2016.

McPherson, James M. *The Negro's Civil War: How American Blacks Felt and Acted During the War for the Union.* New York: Pantheon Books, 1965; New York: Knopf, 2003.

Morillo, Stephen with Michael F. Pavkovic. *What Is Military History?* Second ed. Malden, MA: Polity Press, 2014.

Paret, Peter. *Imagined Battles: Reflections of War in European Art.* Chapel Hill: The University of North Carolina Press, 1997.

Parker, Geoffrey. *The Military Revolution: Military Innovation and the Rise of the West 1500–1800.* 2nd ed. New York: Cambridge University Press, 1996.

Reardon, Carol. *Pickett's Charge in History and Memory.* Chapel Hill: The University of North Carolina Press, 1997.

Rogers, Clifford, ed. *The Military Revolution Debate: Readings on the Military Transformation of Early Modern Europe.* Boulder, CO: Westview Press, 1995.

Weber, Eugen. *Peasants into Frenchmen: The Modernization of Rural France.* Stanford: Stanford University Press, 1976.

Wiley, Bell Irvin. *The Life of Billy Yank: The Common Soldier of the Union.* Baton Rouge: Louisiana State University Press, 1952; updated edition 2008.

———. *The Life of Johnny Reb: The Common Soldier of the Confederacy.* Baton Rouge: Louisiana State University Press, 1943; updated edition 2008.

Winter, Jay. *Sites of Memory, Sites of Mourning: The Great War in European Cultural History.* Cambridge: Cambridge University Press, 1995.

Part I

Regional and Temporal Approaches

1 War and Society in East Asia

Kenneth M. Swope

While the transformation of traditional operational military history to the field more broadly known today as "war and society" might be roughly dated to the 1970s with respect to the study of the West, in East Asian circles it might be argued that the "war and society" approach was the norm from ancient times, as official histories dating from the second century BCE and earlier sought to put military affairs and warriors themselves within a broader social and political context. Thus, the study of battles themselves was often glossed over in favor of ruminations on the importance of good government or measures to "rescue the people from fire and water." Later primary sources stressed the negative social ramifications of prolonged conflict and the need for restoring order. While descriptions of the horrors of war are rife in such sources, specific tactical details are often lacking. Upon their first serious academic exploration of East Asian (primarily Chinese) military topics in the nineteenth and twentieth centuries, this often led Western scholars to wrongly conclude that these societies were in fact passive and/or unconcerned with military matters. In the era of high imperialism this dovetailed with notions of Social Darwinism to perpetuate the idea that such areas were ripe for the taking by more vigorous societies.

Therefore, the study of war and society in East Asia (in the West) has followed a rather curious trajectory, conditioned by assumptions emanating from nineteenth-century imperialist clashes and then colored by the massive military conflicts of the twentieth century, most notably the Second World War. But of course even this conflict is highly contested in East Asian memories and designated differently depending upon one's perspective. The Japanese generally call it the "Fifteen Years' War," or the "Greater East Asian and Pacific War," designations that highlight both the chronological and geographic scope of the conflict and also somewhat rightly downplay the centrality of American involvement, a perspective that might be surprising to students of American history. The Chinese, on the other hand, refer to this conflict as the "War of Resistance Against Japan," whereas for Koreans, this was merely the final stage of their colonial nightmare under Japanese domination and the lead-in to Korea's own massively destructive civil war. Incidentally that war itself has several designations. In North Korea it is known as the "Fatherland Liberation War"; South Koreans refer to it as "Six Two-Five" in reference to the date it began in 1950; the Chinese, echoing designations for their war to assist Korea against the Japanese in the 1590s, refer to it as "The War to Resist America and Aid Korea." In the twenty-first century, such self-serving rhetoric can be seen playing out in hyper-nationalist debates among Chinese, Japanese, and Korean scholars, as well as entire populations, regarding the blame, conduct, and memory of their nation's wartime actions.

But while these wars have certainly been important in highlighting the complex connections between war and society in modern Asia, they also draw more attention to the rich past of East Asian military history as unbeknownst to many outside the region. These conflicts were all waged within the context of deep-seeded historical rivalries and often painful memories, passed down for generations. While space precludes going into great detail, this chapter highlights the most salient developments in the field of war and society in East Asia, which encompasses China, Japan, and Korea. It will first review major trends in the field, and then highlight some key works for each of these three places in chronological fashion, noting that the political boundaries and identities of said regions evolved considerably over time. This essay will focus on works from the past three decades because they provide the clearest view of the state of the field and represent by far the most fruitful period of scholarship.

In several ways the field of war and society in East Asia parallels its developments elsewhere, on a somewhat delayed time frame. The earliest studies in the West tended to focus upon topics of intersection with Western topics, such as the Opium War of 1839–1841, the Taiping Rebellion of 1850–1864, or popular rebellions connected to Western incursions (Wakeman, 1966, and Kuhn, 1970). In the case of Chinese history this was because of two major factors. First, the earliest avenues of access and interest to Chinese history and culture occurred via missionary and commercial efforts, and the first generation of Western scholars on China was drawn from these ranks. Second, the victory of the Chinese communists in the Chinese Civil War of 1946–1949 meant that Chinese archives (outside of Taiwan) remained largely closed to outsiders. Since most of the archives in Taiwan pertained to the Qing (1644–1911) Dynasty, and because the Chinese themselves stressed the importance of imperialism in their modern history, these topics became the most prominent and served as the starting point for explorations in other eras and fields.

The first edited volume, in English, devoted to Chinese ways of warfare noted that China's distinctive military record had been "too little studied" and suggested that further study of China's military past could help ease the rest of the world's adjustment to China's participation in the new transnational order (Kierman and Fairbank, 1974, 1–2). And while laying out what he believed to be certain underlying characteristics of traditional Chinese warfare, such as the need for balancing *wen* (civil) and *wu* (military) authority, as embodied in the military and philosophical classics of ancient China, editor John King Fairbank admitted that China's rich military past had barely been examined and future scholarship would likely reach new conclusions. To cite just two examples, Fairbank posited that bureaucratic control of the military and a constant preoccupation with military threats from Central Asia were hallmarks of premodern Chinese states. He also perpetuated the old notion that Central Asians were inherently more inclined to martial pursuits than the Han Chinese, who supposedly preferred to let "barbarians" do their dirty work of fighting whenever feasible. Therefore one could presume that non-Chinese dynasties like those of the Mongols and Manchus were inherently more aggressive, a characterization that continues to find support among some Chinese scholars.

But more recent scholarship (Di Cosmo, 2009) has revealed a far more nuanced understanding of the dynamics between the civil and military bureaucracies and a far more complicated understanding of Central Asians in Chinese military affairs (Robinson, 2013) and vice versa. The contributors to Di Cosmo's study on military culture in Chinese history show how the military permeated virtually all segments of Chinese society, including popular culture, drama, and even painting. David Robinson's work problematizes the simple Han-barbarian dichotomy by showing how Mongol culture and martial values influenced the Han rulers of the Ming dynasty. In addition to widely employing Mongols in their armies, the Ming rulers patronized such distinctively Central Asian institutions as royal hunts, mounted archery contests, and "willow shoots," whereby mounted archers had to strike single branches of willow trees while loosing arrows from horseback. Utilizing heretofore largely ignored unofficial histories and poems among other sources, these scholars paint an increasingly textured picture of the multifarious links between war and society in East Asia.

In fact perhaps the most striking development of the past three decades in the field of East Asian war and society has been the emergence of the "New Qing History" (Waley-Cohen, 2006), wherein scholars make use of Manchu language materials to study Qing military and political institutions (Elliott, 2001). Elliott argues for the pervasive role of the Eight Banners in Manchu life, even beyond military operations. Recasting the Qing from a supine enervated state that practically invited Western exploitation in the nineteenth century, these scholars (Perdue, 2005, and Dai, 2009) draw attention to the dynamism of the high Qing military and situate its accomplishments more broadly in comparison to its contemporaries, such as the Russian Empire. Some (Kim, 2004) have even conducted complete studies of Muslim revolts in nineteenth-century Central Eurasia that embroiled the Chinese, Russian, Ottoman, and British empires, shifting the focus away from mere Western imperialist exploitation. Others (Larsen, 2008) have demonstrated how the Qing used the rhetoric and tools of nineteenth-century imperialism to assert its interests in its traditional tributary, Korea, even after disastrous defeat at the hands of Japan. While produced in a postcolonial atmosphere and perhaps influenced by such studies, these works bring the empire back into late imperial history and draw parallels between the Qing and other early modern empires.

Still others have returned to older topics in new ways—for example, highlighting the surprising impact of the Taiping Rebellion on the American Civil War (Platt, 2012) or looking at its social implications for China in terms of reconstructing shattered communities (Meyer-Fong, 2013). In fact neither of these historians is really interested in the military aspects of what was the bloodiest civil war in world history. Platt wants to show how China was important economically to the United States even in the nineteenth century, while also exploring the unintended ramifications of missionary activity. Meyer-Fong focuses on reconstruction and memory, drawing extensively on literary and trauma studies to inform her analysis of diaries, memoirs, and other unofficial histories. Ironically, she seems rather ignorant of the broader war and society literature that might further inform her study. This shortcoming plagues many historians of East Asia who continue to marginalize themselves by not engaging more widely with literature and debates outside their immediate fields.

Nonetheless, others have sought to examine China's military past in explicit comparison with the West, especially engaging debates concerning the "military revolution" proposed by Geoffrey Parker (1996) and others. Tonio Andrade (2011) has studied the only full-scale clash between Chinese and Western parties, between the pirate lord Koxinga and the Dutch over Taiwan in the early 1660s, and determined that the Chinese armies were in fact superior to those of the Dutch in many respects, though they remained confounded by certain technologies, such as the artillery fortress. In the process, Andrade also engages issues such as climate change and cultural predilections concerning the development and deployment of new technologies, having been influenced by his former mentor, Geoffrey Parker. Most notably, Andrade notes how weather patterns affected ship construction and development and challenges the idea that Westerners were more adept at drilling and training than their Eastern counterparts.

Andrade takes up the comparison more explicitly in a subsequent work (2016) that examines military innovation and the broader question of the rise of the West in world history despite China's early edge in gunpowder technologies. He concludes that, because of alternate phases of innovation and stagnation, European technology did not surge ahead of China until the Industrial Revolution. Moreover, this surge was aided by constant warfare between European states roughly in parity, a situation not present in East Asia for centuries except for the especially violent era of 1550–1683, as noted by Geoffrey Parker in his recent book on global crisis in the seventeenth century (2014). Indeed, Parker finds that China was the area of the globe most profoundly affected by the seventeenth-century crisis, losing perhaps 40 percent of its total population between 1600–1680, owing to incessant warfare, natural disasters, and epidemics, not to mention attacks by wild animals (Swope, 2018). The Chinese even developed a new term, the "soldier calamity" (*bing huo*), to refer to hardships caused by the depredations of roving soldiers and bandits.

During the past three decades, the study of modern Chinese warfare has been an especially fruitful area of study. The field has expanded exponentially with the opening of Chinese archives to the West. In addition to general surveys covering all or part of the twentieth century, a plethora of monographs have appeared. These include studies of such phenomena as warlordism (McCord, 1993), the War of Resistance against Japan (Lary, 2010a; Lary and MacKinnon, 2001; Peattie et al., 2011), and the Chinese Civil War, though most studies of the latter have still been of the operational variety. Surprisingly, only Diana Lary (2015), building upon her fine earlier work on Chinese warlord soldiers and on her study of the Japanese invasion, has embarked upon a serious study of the societal impact of the Chinese Civil War. She concludes that the social leveling begun during the Japanese invasion accelerated during the subsequent civil war, and culminated in the disasters of the Cultural Revolution during the 1960s. Because declassification of Chinese documents offers greater access to archives, China's involvement in the Cold War and Korean and Vietnam Wars has attracted increased attention among scholars (Chen, 2001). But, to date, the bulk of these works focuses more on traditional diplomatic and military history and does not explicitly address topics addressed in the war and society field, such as the social implications of mobilization for war.

In marked contrast to China, premodern Japan and Korea have not fared as well in Western scholarship despite their rich military pasts. While there remains tremendous popular interest in topics such as samurai and ninja, few academics in the West have deigned to examine these topics in any depth. The creation of the Japanese *bakufu*, or shogunate, government was the subject of many earlier studies (Mass and Hauser, 1993), and the institutional history of the samurai remains quite popular, though such studies tended to be remarkably free of the messy details that actually precipitated military rule in Japan. This has continued with studies of institutions such as *sankin kotai*, the "alternate attendance" system of the Tokugawa (Vaporis, 2009), which has been identified as one of the key sources for Japan's rapid modernization as it created a truly "national" culture and established the infrastructure needed for rapid industrial development.

Other scholars have examined the origins of the samurai themselves, coming to rather different conclusions about their origins as either "hired swords" (Friday, 1992) or "heavenly warriors" (Farris, 1992), but making important connections between the Japanese state and society. In particular they note how the samurai emerged out of the noble class itself and gradually arrogated more authority to themselves by virtue of their ability to collect rents and maintain order in the countryside, with Friday highlighting warrior agency and Farris positing a more evolutionary model. This was a long process as the old authorities resisted their loss of influence and the rising elements of society were forced to work with those they sought to replace. Such conclusions led to the more intensive study of the effects of warfare itself on medieval Japanese society (Conlan, 2003, and Friday, 2003). These authors demonstrate how central authority devolved from the court to the localities and how Japan became increasingly militarized. These processes culminated in the era of Warring States (Sengoku, 1467–1603), which finally ended in the Tokugawa defeat of its rivals in part because earlier military leaders had effectively crushed militant religious sects and other sources of strife.

Modern topics have received much more extensive treatment, no doubt due to Japan's prominence in world affairs in the first half of the twentieth century and its key role in the Greater East Asian and Pacific War. Earlier studies tended to focus more specifically on colonial conflicts (Duus, 1989, 1995) or on relations with the United States. But more comprehensive surveys of military institutions, drawing upon both primary sources and the latest Japanese scholarly research (Drea, 2009), have filled important gaps and highlighted the importance of Meiji-created institutions for subsequent military developments. Such studies have demonstrated that, while conditions certainly existed to facilitate the rise of a fascist, hyper-military state, it was by no means a foregone conclusion.

Recent research on the manufacturing of the imperial concept of *bushido*, or "Way of the Warrior," as a motivating tool for Japanese soldiers (Benesch, 2014) attests to the powerful role the mythologization of the country's military past continues to play. Benesch finds that the Japanese government, aided by popular publications, such as Inazo Nitobe's work on *bushido*, manufactured an idealized value system that helped Japanese distinguish themselves from the rest of the world and fueled their sense of victimization at the hands of the West and sense of entitlement vis-à-vis the rest of Asia. Benesch and others note that popular culture was so bombarded by distorted renditions of an idealized samurai past that the general populace more readily embraced the dangerous military adventures embarked upon by its military leaders.

The almost total neglect of Korea's military past prior to the twentieth century can be attributed to several factors, some of which are related to those that have retarded the study of war and society in China. Korea's traditional bureaucrats, the *yangban*, though technically divided into both civil and military branches, tended to disesteem military accomplishments, and military matters and affairs were glossed over in standard histories, with a few notable exceptions. Moreover, Korea was situated between the two powers of China and Japan throughout its long history, and although it warred and traded with both (Batten, 2006), its own military activities and accomplishments have been overshadowed by those of its neighbors until recently. Only in the past twenty years have scholars highlighted Korea's role in northeast Asia with respect to the Mongol Empire (Shultz, 2000; Robinson, 2010; Van de Ven, 2000), surprisingly noting that certain Korean nobles prided themselves as being selected for the Mongol royal bodyguards and marrying into the Yuan Dynasty royal family. Of course it was in the interest of the Japanese colonizers to downplay Korea's martial traditions and military past. Subsequently, memories of the Korean War and the realities of twenty-first-century political rhetoric have conspired to make Korea's military past less attractive for scholars.

Having briefly addressed the field's major recent trends, this essay next turns to a chronological discussion of the field, starting with the ancient period and moving to the present. Much of the early discussion of war and society in ancient China was centered upon statecraft issues as revealed in the classic philosophical texts, including Sunzi's (Sun Tzu) *Art of War*, most of which were purportedly compiled during the Spring and Autumn and Warring States Periods (ca. 722–221 BCE). Thus, there was much emphasis upon topics such as social mobility (Hsu, 1965) and state formation (Sage, 1992), a trend that has continued in recent comparative work (Hui, 2005). But with Ralph Sawyer's translation of the so-called Seven Military Classics (which include *The Art of War*) in 1993, there has been a new interest in revisiting military texts and using them to study the relationship between war, society, and statecraft in early China. New archaeological discoveries have enabled a much more thorough reconstruction of warfare as early as the Shang Dynasty (ca. 1600–1122 BCE) and also allowed historians to more clearly delineate the origins of the long-standing rivalry between sedentary Chinese states and their Central Asian competitors (Di Cosmo, 2002).

By contrast, the middle period (ca. 300–900 CE) of Chinese military history has received comparatively short shrift. David Graff (2002) made an important contribution with his richly detailed survey of medieval Chinese warfare that highlights the relationship between settlement, population growth, and the need for a military that was both functional and cost-effective. Graff shows how the Chinese imperial state tried to create a military that was both affordable and effective. The Tang initially devised a system of farmer militia (*fubing*), but it proved ultimately untenable, leading to the creation of a more professional albeit more expensive military that ultimately turned against the dynasty, leading to still further administrative adjustments under succeeding dynasties. In the process Graff highlights the social implications of these restructurings, explaining how Chinese military institutions related to social structures and how the military could function as an important avenue of social mobility, a point also raised by Kenneth Swope in his work on the Ming.

Graff continues to extend the field in his recent (2016) work that posits a Eurasian "way of war" by comparing Chinese and Byzantine practices, noting how both devised similar strategies for managing a multiplicity of mobile steppe threats by creating sophisticated bureaucracies for resource mobilization and employing divide-and-conquer measures. He contends that the comparable geostrategic situations of these empires and their connections to earlier imperial states prompted their similar responses to analogous military problems. Importantly, this book moves the discussion away from the simple Western "way of war" thesis posited by Victor Davis Hanson (1989), and toward a more productive comparison of states that had much in common. A fine recent compilation of essays (Wyatt, 2008) bridges the gap from middle to late imperial China, drawing attention to the diversity of power brokers and interest groups on the Chinese political scene while eschewing simple civil-military dichotomies, a theme also found in a collection edited by Kai Filipiak (2015) that spans the entirety of Chinese history.

Among the more interesting developments of the past two decades has been the rehabilitation of the image of the Song Dynasty's (960–1279) military. Long decried as the nadir of Chinese military power in the face of more powerful neighboring states from Central and Northeast Asia, recent scholarship emphasizes the technological adaptability and diplomatic guile of the Song in achieving policy aims. Peter Lorge (2015) has led this charge, noting how the Tang collapse compelled early Song rulers to seek military victories to justify their position, and exploring the genesis of trends such as the constant jockeying for imperial favor between civil and military officials that played out in later dynasties.

In relation to this there has been a minor revival of interest in the military tactics and exploits of the Mongols, including studies of the failed invasions of Japan (Conlan, 2001). These indicate a shift of sorts away from the more benign aspects of the *Pax Mongolica* that permeated much of the scholarship around the turn of the century. Likewise, the ongoing influence of the Mongols upon the Ming court has been highlighted in the work of David Robinson (2008, 2013). These recent works highlight the influence of Mongol culture upon the military culture and institutions of its neighbors, but also move away from the socially destructive aspects of said culture that were such a feature of traditional studies of the Mongols.

Turning back to medieval Japan (ca. 1185–1600 CE), many studies tend to focus on subjects such as social relations (Adolphson, 2000), government and military institutions (Souyri, 2001), and even the "culture" of civil war (Berry, 1994), while glossing over the messy details of war itself. Following developments in the field of Chinese history, scholars of Japan are now more explicitly exploring the relationship between war and state building in Japan (Frerejohn and Rosenbluth, 2010), while also examining specific Japanese phenomena, such as monastic warriors (Adolphson, 2007) and militant Buddhist sects (Tsang, 2007), that were well known but understudied in the West. While no work comparing such figures to Western Crusaders or jihadists has been attempted, links between religion and warfare might prove profitable areas for future exploration.

Again mirroring developments in the field of Chinese history, scholars of Japan are investigating the international implications of piracy and its relationship to land warfare in early modern East Asia (Shapinsky, 2014). In particular, this emphasis upon the maritime realm shows how commerce and violence were frequently intertwined in early modern East Asia, and how piracy was often the glue that held different elements of society together. In a situation perhaps not unlike privateering in the West, powerful lords in Japan and officials in China directly or indirectly sponsored pirates and profited from smuggling, plunder, and simple protection rackets. In Japan, China, and Korea the legitimacy of the ruling house was often connected to its ability to curtail or control piracy, and "sea bandits" were a recurrent source of friction between the three polities. Pirates were employed by land-based warriors in Japan and civil officials in China to advance their own aims, often to the detriment of local coastal populations, who could be robbed, impressed into service, or even dislocated by governmental pirate eradication measures.

The Ming Dynasty (1368–1644 CE) was also traditionally overlooked in terms of its military dimensions, despite the fact that more military texts and manuals were produced during the Ming than in any other era in premodern Chinese history. Building upon classic studies of peasant rebellion and piracy, newer scholarship has examined such topics as the presence of "men of force" and the militarization of local society in the mid-Ming (Robinson, 2001) and the military factors behind the fall of the Ming (Swope, 2014) despite its military achievements of the 1590s (Swope, 2008). Ming colonial efforts on the southwest frontier have also received considerable attention (Shin, 2006 and Herman, 2007). The emerging picture is one of a dynamically expansive Ming state that had the wherewithal to combat a bewildering array of foes but was also beset with structural problems, such as factionalism and widespread corruption, that hampered its effectiveness. Likewise, the Qing has been effectively recast as an expansionist colonial power (Crossley et al., 2006; Weinstein, 2014), much to the dismay of contemporary Chinese nationalists.

While earlier studies of the Ming-Qing transition focused on issues of civil-military relations and the Manchu ability to reconstruct order, the latest scholarship examines the impact of war on society as a whole, particularly in terms of local militarization and its effects on the social order (Swope, 2017). Relying on newly available primary sources, scholars can now get at the ground level of the Ming-Qing transition, exploring such topics as the creation of private military installations across China, the spread of feral animals, and the psychological effects on survivors who were unable to intern deceased relatives or perform the proper ceremonies that were considered integral to the survival of family lines.

The Great East Asian War of 1592–1598 has become a particularly fertile area of study, in part because of its rich source base and its continuing resonance in the popular mind, especially in Korea. Until recently it was known in the West primarily through a few translated texts from Korea's famous admiral Yi Sunsin (1592–1598) and prime minister Yu Songnyong (1542–1607). While the early books about the conflict were directed at a more general audience (Turnbull, 2002; Hawley, 2005), Kenneth Swope (2009) produced the first monograph in English on the war, albeit focusing primarily on the Ming. Since its publication, an international collection of essays looking at the war from all three sides has been published (Lewis, 2015), as has yet another translated primary source—the writings of a Korean war captive in Japan (Haboush and Robinson, 2013). Most recently, the work of the Korean scholar Jahyun Kim Haboush was published posthumously (2016), highlighting this war as the origin of the modern Korean nation by virtue of the Korean king's efforts to rally his populace using exhortations directed at them in vernacular Korean, rather than in the standard literary Chinese. Grounded in literary theory and informed by deep textual analysis, works like this demonstrate the degree to which war and society topics are entering the mainstream in scholarship about East Asia.

Studies of secret societies and religiously motivated uprisings in the Qing dynasty have also been prominent. Indeed, they preceded the evolution of the New Qing History and were often used to make connections between traditional uprisings and modern rebels and revolutionaries (Perry, 1980). In later eras these secret societies sometimes became conflated with proto-nationalist groups (Wakeman, 1966), such as the famous Boxer rebels (Esherick,

1987; Bickers and Tiedemann, 2007)—though it is now clear that the myths about the Boxers have largely obscured the reality of their mission and their message, and become subsumed into somewhat romanticized discourses about the exploitation of China by imperialist powers (Cohen, 1997). Still, the study of secret societies and heterodox organizations is important in how the current Chinese government still regards alternative sources of loyalty and seeks to impose its own brand of orthodoxy upon the populace. Finding common threads with China's imperial past can help better explain the context in which the current government operates. It also illuminates areas of revolutionary potential and helps fill in pieces of the accepted narrative of modern Chinese history.

Unsurprisingly, studies of warlordism (McCord, 1993), and the colonial and imperialist relationship between Japan, Korea, and China, respectively, have dominated the field. In addition to studies of specific warlords and topics such as the Northern Expedition of 1926–1928, comprehensive treatments of warlord soldiers (Lary, 2010b) paint a great picture of the effects of militarization on local Chinese society, as do studies of banditry. There has also been a concentrated effort to bring interdisciplinary fields, such as trauma studies, into the analysis of warfare in twentieth-century China (Flath and Smith, 2011; Lary, 2010a). Lary uses sources such as cartoons and patriotic songs to show how the Chinese government used new media to mobilize resistance against Japan. Other works (Schoppa, 2011) offer more detailed treatments and case studies grounded in primary sources that deal with such sensitive topics as collaboration and wartime rape. Earlier studies, while often visceral and filled with statistics, lack the comparative dimension of these recent books.

Given the sheer size of East Asia and the scope of the Greater East Asian War and the ongoing struggles between the Communists and the Nationalists in China, many studies have opted for the regional, or even metropolitan (Zhu, 2015; MacKinnon, 2008) approach, examining conflict in a specific environment. This is fruitful in that it allows historians to see how warfare affected different cities and populations differently, and how the effects varied over time. For example, MacKinnon stresses the unity of wartime Wuhan in 1938, while Zhu's analysis of wartime Guilin shows how the city became increasingly independent and divorced from the Guomindang government as locals rallied around the semi-independent warlords who controlled the city and permitted greater intellectual freedom.

Also especially fascinating are new studies of Japan's imperial aspirations highlighting propaganda (Earhart, 2008), use of new technologies to demonstrate modernity, and celebrations of the empire's uniqueness (Ruoff, 2014). These works build upon the fine work of previous scholars such as John Dower (1986), who emphasized the racial dimensions of both American and Japanese propaganda. Indeed, Dower has influenced subsequent Asia-Pacific war studies by placing racism firmly into war and society historiography. Taken together, one gets a far more nuanced view of the Japanese state and its efforts to mobilize its populace for war. Additionally, the Japanese people themselves are humanized and can be recognized as patriots in their own right.

While the involvement of China in the Korean War has been referenced earlier, it is still rather surprising how little has been published about this conflict using primarily Korean sources. The leading voice has certainly been Bruce Cumings (1981–1983), whose work is meticulously documented, though more concerned with political machinations than battles, and has been challenged for its leftist political slant. Chen Jian (1996) has outlined the reasons for China's intervention in Korea by highlighting internal political concerns, showing how Mao manipulated both the members of the party elite and the general public into pursuing ends that he believed kept both the United States and the Soviet Union on the defensive with respect to China. Recent scholarship from the Chinese side makes impressive use of primary sources but remains based in traditional diplomatic and military history realms.

Leaving the past few decades aside as the province of the political scientists or international relations scholars, it is time to consider where the field might be headed in the near future. Judging from the recent provenance of the vast majority of the publications cited herein, it is reasonable to presume that the future of war and society in East Asia is bright. The first serious wave of scholarship was produced in the 1980s–90s, and as those who finished graduate school around the turn of the millennium publish their monographs, they are attracting new generations of scholars. The sheer volume of surviving source materials, many of which have been largely overlooked by scholars but are increasingly available via both hard copy compilations and online databases, should help facilitate further research into East Asia's military past. The main obstacle in this respect might be the fact that more and more universities, at least in the United States, are cutting foreign language programs or even eliminating foreign language requirements. Such shortsighted actions do a disservice to future generations of would-be scholars.

In terms of specifics, early modern China and Korea are intriguing areas for future research. On the one hand, there are numerous sources still waiting to be examined. And, as seen with the florescence of the New Qing History, the influence of the field of world history toward finding parallels across cultures and trying to move away from teleological history has had a very positive effect on Chinese studies. Important comparative work should continue to come out, and one only hopes that Western scholars will follow the lead of their Asianist colleagues and begin to use the work of their counterparts to illuminate aspects of war and society in other parts of the world. The trend toward interdisciplinary studies will also most likely continue, as evidenced by the spate of recent sociocultural histories of the Greater East Asian and Pacific War and the Manchurian colonial experience. Recent studies of Imperial Japan have demonstrated the applicability of examining Japanese imperial projects and rhetoric within a broader context, and that should likely continue.

Increasing access to Asian archives and primary sources should help fill in important gaps in the study of twentieth-century conflicts. As noted earlier, while we now know much more about China's reasons for intervening in Korea and Vietnam, the actual effects of these wars on the Chinese population as a whole or even certain elements (e.g., munitions workers) are far less clear. Forthcoming studies promise to shed more light—for example, upon the conditions faced by Chinese support units in Vietnam and what they thought of their mission there. The role of gender has finally received some attention (Edwards, 2016), and one should expect future studies along these lines. We might also see studies examining the role of China's modern Chinese military in fomenting nationalism, something becoming increasingly obvious amid contemporary Chinese furor over territorial disputes in Southeast Asian waters.

As the Cold War recedes into the past and America's economic ties with East Asia continue to grow, greater understanding of these relatively recent (and still very much in the East Asian news) events should aid in maintaining positive relationships and providing areas of common ground for future endeavors. Students remain fascinated by these conflicts; they just need direction to look at them in different ways and consider the historical backdrops. For example, it is more than coincidental that America's two major military clashes in the Cold War both involved China, and both were fought in states formerly considered tributaries of the Chinese Empire. In fact, both places were long considered to be part of China's defense "umbrella," and the current leadership remains committed to retaining its traditional primacy in East Asia, as evidenced by recent dustups along China's borders on both land and sea. And both Vietnam and the Koreas remain ambivalent about China's aims and interests, welcoming friendship, but remaining cognizant of past conflicts. They have no desire to become tributaries of China again and are very suspicious of China's "peaceful rise."

On the more negative side, it is difficult to foresee much of a revival of premodern history in the United States, and this bodes ill for the field of East Asian war and society. Despite the increasing adoption of world history sequences at the college level, departments are consistently replacing ancient and premodern historians with scholars of more recent eras. This will certainly impact research decisions by graduate students and will likely close off promising avenues of research. So while the works of Sunzi and Musashi have become virtually cottage industries, too little attention has been devoted to the contexts in which they produced their works or to the way their teachings were understood and transmitted (Tokitsu, 2005), and it is difficult to see this trend reversing. A detailed examination of the intellectual milieu and values of the samurai, as opposed to facile references to romanticized *bushido* codes, is long overdue.

International and regional studies will likely become more common, as seen with the proliferation of recent works on World War II and the Cold War in Asia, and books on the Great East Asian War of 1592–1598. These will be useful in establishing the coherence of East Asia as a field and in examining the transnational impact of war and (often contested) memories of conflict. Finally, more attention needs to be paid to the actual conduct of military operations throughout premodern East Asia and the effects these had on local populations, perhaps following the example of Ulrich Theobald's work on the logistics of the Second Jinchuan War of 1771–1776 (2013). Such efforts would be valuable for scholars of war and society everywhere, and judging from the popular interest in topics such as samurai warfare, they could even draw more people into the field in the future.

Bibliography

Adolphson, Mikael S. *The Gates of Power: Monks, Courtiers, and Warriors in Premodern Japan*. Honolulu: University of Hawaii Press, 2000.
———. *The Teeth and Claws of the Buddha: Monastic Warriors and Sōhei in Japanese History*. Honolulu: University of Hawaii Press, 2007.
Andrade, Tonio. *The Gunpowder Age: China, Military Innovation and the Rise of the West in World History*. Princeton: Princeton University Press, 2016.
———. *Lost Colony: The Untold Story of China's First Great Victory Over the West*. Princeton: Princeton University Press, 2011.
Batten, Bruce L. *Gateway to Japan: Hakata in War and Peace, 500–1300*. Honolulu: University of Hawaii Press, 2006.
Benesch, Oleg. *Inventing the Way of the Samurai: Nationalism, Internationalism, and Bushido in Modern Japan*. Oxford: Oxford University Press, 2014.
Bey, Mary Elizabeth. *The Culture of Civil War in Kyoto*. Berkeley: University of California Press, 1994.
Bickers, Robert, and R. G. Tiedemann, eds. *The Boxers, China, and the World*. Lanham: Rowman & Littlefield, 2007.
Cohen, Paul A. *History in Three Keys: The Boxers as Event, Experience, and Myth*. New York: Columbia University Press, 1997.
Conlan, Thomas D. *In Little Need of Divine Intervention: Takezaki Suenaga' Scrolls of the Mongol Invasions of Japan*. Ithaca: Cornell University Press, 2001.
———. *State of War: The Violent Order of Fourteenth-Century Japan*. Ann Arbor: Center for Japanese Studies, 2003.
Crossley, Pamela Kyle, et al., eds. *Empire at the Margins: Culture, Ethnicity and Frontier in Early Modern China*. Berkeley: University of California Press, 2006.
Cumings, Bruce. *The Origins of the Korean War, Vol. 1: Liberation and the Emergence of Separate Regimes, 1945–1947*. Princeton: Princeton University Press, 1981.
———. *The Origins of the Korean War, Vol. 2: The Roaring of the Cataract, 1947–1950*. Princeton: Princeton University Press, 1992.
Dai, Yingcong. *The Sichuan Frontier and Tibet: Imperial Strategy in the Early Qing*. Seattle: University of Washington Press, 2009.

Di Cosmo, Nicola. *Ancient China and Its Enemies: The Rise of Nomadic Power in East Asian History*. Cambridge: Cambridge University Press, 2002.

———, ed. *Military Culture in Imperial China*. Cambridge, MA: Harvard University Press, 2009.

Dower, John W. *War Without Mercy: Race & Power in the Pacific War*. New York: Pantheon Books, 1986.

Drea, Edward J. *Japan's Imperial Army: Its Rise and Fall, 1853–1945*. Lawrence: University Press of Kansas, 2009.

Duus, Peter, et al., eds. *The Japanese Informal Empire in China, 1895–1937*. Princeton: Princeton University Press, 1989.

———, et al., eds. *The Abacus and the Sword: The Japanese Penetration of Korea, 1895–1910*. Berkeley: University of California Press, 1995.

———, et al., eds. *The Japanese Wartime Empire, 1931–1945*. Princeton: Princeton University Press, 1996.

Earhart, David C. *Certain Victory: Images of World War II in the Japanese Media*. Armonk: M.E. Sharpe, 2008.

Edwards, Louise. *Women Warriors and Wartime Spies of China*. Cambridge: Cambridge University Press, 2016.

Elliott, Mark C. *The Manchu Way: The Eight Banners and Ethnic Identity in Late Imperial China*. Stanford: Stanford University Press, 2001.

Esherick, Joseph W. *The Origins of the Boxer Uprising*. Berkeley: University of California Press, 1987.

Farris, William Wayne. *Heavenly Warriors: The Evolution of Japan's Military 500–1300*. Cambridge, MA: Harvard University Press, 1992.

Ferejohn, John, and Frances McCall Rosenbluth, eds. *War and State Building in Medieval Japan*. Stanford: Stanford University Press, 2010.

Filipiak, Kai, ed. *Civil-Military Relations in Chinese History: From Ancient China to the Communist Takeover*. London: Routledge, 2015.

Flath, James, and Norman Smith, eds. *Beyond Suffering: Recounting War in Modern China*. Vancouver: UBC Press, 2011.

Friday, Karl F. *Hired Swords: The Rise of Private Warrior Power in Early Japan*. Stanford: Stanford University Press, 1992.

———. *Samurai, Warfare and the State in Medieval Japan*. London: Routledge, 2003.

Graff, David A. *The Eurasian Way of War: Military Practice in Seventh Century China and Byzantium*. London: Routledge, 2016.

———. *Medieval Chinese Warfare, 300–900*. London: Routledge, 2002.

Haboush, Jahyun Kim, and Kenneth R. Robinson, eds. *A Korean War Captive in Japan, 1597–1600: The Writings of Kang Hang*. New York: Columbia University Press, 2013.

———, eds. *The Great East Asian War and the Birth of the Korean Nation*. New York: Columbia University Press, 2016.

Hanson, Victor Davis. *The Western Way of War: Infantry Battle in Classical Greece*. Berkeley: University of California Press, 1989.

Hawley, Samuel. *The Imjin War: Japan's Sixteenth-Century Invasion of Korea and Attempt to Conquer China*. Seoul: Royal Asiatic Society, 2005.

Herman, John E. *Amid the Clouds and Mist: China's Colonization of Guizhou, 1200–1700*. Cambridge, MA: Harvard University Press, 2007.

Hsu, Cho-yun. *Ancient China in Transition: An Analysis of Social Mobility*. Stanford: Stanford University Press, 1965.

Jansen, Marius B. *Warrior Rule in Japan*. Cambridge: Cambridge University Press, 1995.

Jian, Chen. *Mao's China and the Cold War*. Durham: University of North Carolina Press, 2001.

Kierman, Frank A., Jr., and John King Fairbank, eds. *Chinese Ways in Warfare*. Cambridge, MA: Harvard University Press, 1974.

Kim, Hodong. *Holy War in China: The Muslim Rebellion and State in Chinese Central Asia, 1864–1877*. Stanford: Stanford University Press, 2004.

Kuhn, Philip. *Rebellion and Its Enemies in Late Imperial China: Militarization and Social Structure, 1796-1864*. Cambridge, MA: Harvard University Press, 1970.

Larsen, Kirk W. *Tradition, Treaties, and Trade: Qing Imperialism and Chosôn Korea, 1850–1910*. Cambridge, MA: Harvard University Press, 2008.

Lary, Diana. *China's Civil War: A Social History, 1945–1949*. Cambridge: Cambridge University Press, 2015.

———. *The Chinese People at War: Human Suffering and Social Transformation, 1937–1945*. Cambridge: Cambridge University Press, 2010a.

————. *Warlord Soldiers: Chinese Common Soldiers*. Cambridge: Cambridge University Press, 2010b.

Lary, Diana, and Stephen MacKinnon, eds. *Scars of War: The Impact of Warfare on Modern China*. Vancouver: UBC Press, 2001.

Lewis, James B. *The East Asian War, 1592–1598: International Relations, Violence, and Memory*. London: Routledge, 2015.

Lewis, Mark Edward. *Sanctioned Violence in Early China*. Albany: SUNY Press, 1990.

Li, Xiaobing. *China's Battle for Korea: The 1951 Spring Offensive*. Bloomington: Indiana University Press, 2015.

Lorge, Peter, ed. *Debating War in Chinese History*. London: Routledge, 2013.

————. *The Reunification of China: Peace Through War Under the Song Dynasty*. Cambridge: Cambridge University Press, 2015.

————. *War, Politics, and Society in Early Modern China, 900–1795*. London: Routledge, 2005a.

————, ed. *Warfare in China to 1600*. Aldershot, UK: Ashgate, 2005b.

MacKinnon, Stephen R. *Wuhan, 1938: War, Refugees, and the Making of Modern China*. Berkeley: University of California Press, 2008.

Mass, Jeffrey, and William Hauser, eds. *The Bakufu in Japanese History*. Stanford: Stanford University Press, 1993.

McCord, Edward. *The Power of the Gun: The Emergence of Modern Chinese Warlordism*. Berkeley: University of California Press, 1993.

Meyer-Fong, Tobie. *What Remains: Coming to Terms With Civil War in 19th-Century China*. Stanford: Stanford University Press, 2013.

Naquin, Susan. *Shantung Rebellion: The Wang Lun Uprising of 1774*. New Haven: Yale University Press, 1981.

Parker, Geoffrey. *Global Crisis: Climate Change and Catastrophe in the Seventeenth Century*. New Haven: Yale University Press, 2014.

————. *The Military Revolution: Military Innovation and the Rise of the West*. 2nd ed. Cambridge: Cambridge University Press, 1996.

Peattie, Mark, et al., eds. *The Battle for China*. Stanford: Stanford University Press, 2011.

Perdue, Peter C. *China Marches West: The Qing Conquest of Central Eurasia*. Cambridge, MA: Belknap Press, 2005.

Perry, Elizabeth J. *Rebels and Revolutionaries in North China 1845–1945*. Stanford: Stanford University Press, 1980.

Platt, Stephen R. *Autumn in the Heavenly Kingdom: China, the West, and the Epic Story of the Taiping Civil War*. New York: Vintage, 2012.

Robinson, David M. *Bandits, Eunuchs and the Son of Heaven: The Economy of Violence in Mid-Ming China*. Honolulu: University of Hawaii Press, 2001.

————, ed. *Culture, Courtiers, and Competition: The Ming Court (1368–1644)*. Cambridge, MA: Harvard University Press, 2008.

————. *Empire's Twilight: Northeast Asia Under the Mongols*. Cambridge, MA: Harvard University Press, 2010.

————. *Martial Spectacles of the Ming Court*. Cambridge, MA: Harvard University Press, 2013.

Ruoff, Kenneth J. *Imperial Japan at Its Zenith: The Wartime Celebration of the Empire's 2600th Anniversary*. Ithaca: Cornell University Press, 2014.

Sage, Steven F. *Ancient Sichuan and the Unification of China*. Albany: SUNY Press, 1992.

Sawyer, Ralph D. *The Seven Military Classics of Ancient China*. Boulder: Westview Press, 1993.

Schoppa, R. Keith. *In a Sea of Bitterness: Refugees During the Sino-Japanese War*. Cambridge, MA: Harvard University Press. 2011.

Shapinsky, Peter D. *Lords of the Sea: Pirates, Violence, and Commerce in Late Medieval Japan*. Ann Arbor: Center for Japanese Studies, 2014.

Shin, Leo K. *The Making of the Chinese State: Ethnicity and Expansion on the Ming Borderlands*. Cambridge: Cambridge University Press, 2006.

Shultz, Edward J. *Generals and Scholars: Military Rule in Medieval Korea*. Honolulu: University of Hawaii Press, 2000.

Souyri, Pierre Francois. *The World Turned Upside Down: Medieval Japanese Society*. New York: Columbia University Press, 2001.

Swope, Kenneth M. "Bestowing the Double-Edged Sword: Wanli as Supreme Military Commander." In *Culture, Courtiers and Competition*. Ed. David M. Robinson. Cambridge: MA: Harvard University Press, 2008, 61–115.

———. *A Dragon's Head and a Serpent's Tail: Ming China and the First Great East Asian War, 1592–1598*. Norman: University of Oklahoma Press, 2009.

———. *The Military Collapse of China's Ming Dynasty, 1618–1644*. London: Routledge, 2014.

———. *On the Trail of the Yellow Tiger: War, Trauma, and Social Dislocation in Southwest China During the Ming-Qing Transition*. Lincoln: University of Nebraska Press, 2018.

———, ed. *Warfare in China Since 1600*. Aldershot, UK: Ashgate, 2005.

Theobald, Ulrich. *War, Finance and Logistics in Late Imperial China: A Study of the Second Jinchuan Campaign, 1771–1776*. Leiden: Brill, 2013.

Tokitsu, Kenji. *Miyamoto Musashi: His Life and Writings*. Boston: Weatherhill, 2005.

Tsang, Carol Richmond. *War and Faith: Ikko Ikki in Late Muromachi Japan*. Cambridge, MA: Harvard University Press, 2007.

Turnbull, Stephen. *Samurai Invasion: Japan's Korean War, 1592–1598*. London: Cassell, 2002.

Van de Ven, Hans, ed. *Warfare in Chinese History*. Leiden: Brill, 2000.

Vaporis, Constantine Nomikos. *Tour of Duty: Samurai, Military Service in Edo, and the Culture of Early Modern Japan*. Honolulu: University of Hawaii Press, 2008.

Wakeman, Jr., Frederic. *Strangers at the Gate: Social Disorder in South China, 1839–1861*. Berkeley: University of California Press, 1966.

Waley-Cohen, Joanna. *The Culture of War in China: Empire and Military Under the Qing Dynasty*. London: I.B. Tauris, 2006.

Wang, Wensheng. *White Lotus Rebels and South China Pirates: Crisis and Reform in the Qing Empire*. Cambridge, MA: Harvard University Press, 2014.

Weinsten, Jodi L. *Empire and Identity in Guizhou: Local Resistance to Qing Expansion*. Seattle: University of Washington Press, 2014.

Wyatt, Don J., ed. *Battlefronts Real and Imagined: War, Border, and Identity in the Chinese Middle Period*. New York: Palgrave Macmillan, 2008.

Zhu, Pingchao. *Wartime Culture in Guilin, 1938–1944: A City at War*. Lanham, MD: Lexington Books, 2015.

2 War and Society in Southeast Asia

Richard A. Ruth

Historians of Southeast Asia writing about warfare in the region have been addressing what came to be known as "war and society" since long before the term gained currency among academic military historians. They have addressed the interrelationship between war and the societies affected by it in three important sets of questions spanning the seventy years or so that Southeast Asian studies has been in existence. And to a certain extent, the questions being considered can be labeled as existential to the field.

The first set concentrates on whether systems of warfare in premodern Southeast Asia can be used to justify the geographic category "Southeast Asia." Do similarities in the methods, rituals, and objectives of warfare in the region help render the category coherent? Do the complex processes of war making pursued by the disparate and varied peoples within this sprawling region share more commonalities than differences? And do these commonalities differ significantly from those practiced in the regions that adjoin it, particularly China and India?

The second major set of questions examines whether Western scholars are capable of writing histories of Western imperialism (c. 1819–1954) that can transcend a Western-centric perspective. Does the focus on the construction, maintenance, and exploits of Western-led colonial regimes operating inside Southeast Asia overshadow the experiences of indigenous historical actors affected by those enterprises? Does the heavy reliance on Western colonial archives ensure that European colonialists emerge in sharp focus while their Asian subjects remain blurred? Does the reliance on Western colonial archives generate historical studies that implicitly downplay or even sanction colonial excesses?

Finally, the third set of questions asks whether the Vietnam War can be studied as a Southeast Asian historical event centered on the experiences of the Vietnamese, Cambodian, Laotian, and Thai societies affected by it. How do scholars develop historical narratives of the Vietnam War that transcend old but enduring America-dominated military and political perspectives to include historical experiences of these Southeast Asian peoples?

Most historians of war and society in Southeast Asia have strong backgrounds in regional studies. They usually have facility in at least one Asian language and, if studying the colonial period, the language of one or more colonizing power. As regionalists, they often have profound understandings of the Southeast Asian cultures they study. They are expected to know the religions, literature, and social structures of their area(s) of expertise. Most were trained to consider the influence of earlier phases of history upon events in the modern period. They usually conduct some (and sometimes all) of their research in the Southeast Asian countries whose wars they study. Generally speaking, they do not study war as a topic in and of itself. They do not produce conventional operational histories of Western militaries fighting in Asia. Most pursue topics concerning the relationship between war making and the societies affected by it. And some produce studies in diplomatic, social, or political history set during wartime rather than studying the military operations of that conflict.

There are historians who combine regional expertise with the skills of a military historian to produce substantive studies of war and society in Southeast Asia. Some of these scholars have had experience serving in Western militaries before pursuing academic degrees in Southeast Asian studies. Others acquired the skills of a military historian through study and time spent with Asian armed forces. This historiographical essay focuses on those historians whose work combines the best of a regionalist approach with that of a conventional military historical approach. It evaluates the strength and standings of these studies as they apply to three developments in the field of Southeast Asian history.

Warfare in premodern Southeast Asia was influenced strongly by demographics. The region's sparse population shaped the objectives, modes of combat, and levels of intensity of military campaigns. In mainland Southeast Asia, rulers fought not for territory but for control of human labor. The victor seized the vanquished's people, bringing them back to the areas he controlled. All enslaved war prisoners were valued because they could work in productive ventures, such as growing food, clearing forests, digging canals, and, of course, making war. Especially prized were those with special skills; artisans, scribes, musicians, and religious figures received priority over peasants. Thus a successful military campaign not only enriched a state's productive base but also greatly enriched its cultural reserves. A ruler whose state was so enriched could compound his battlefield successes by attracting more settlers to areas under his authority. Southeast Asia's "endemic slave-gathering warfare," as historian Bryce Beemer (2009) has illustrated, helped transmit culture throughout the region.

Historians working in the 1920s and 1930s saw the region as an extension—almost an ontological colony—of either Indic or Sinic civilizations. They sought examples of Chinese or Indian influence on Southeast Asian warfare, but glossed over the profound differences they found in the social structures of Southeast Asian cultures compared to those of China and India. The French historian and archaeologist George Coedès (1968) defined Southeast Asia through an Indic idiom as embraced by the rulers of the earliest identifiable kingdoms. He produced the first edition of his landmark study while working in Hanoi during World War II—in a complex social milieu delineated by a nominal French colonial authority, an increasingly voracious Japanese military occupation, and a desperate Vietnamese population suffering hunger, shortages, and bombing raids. But despite living and writing from within this complex wartime society, Coedès barely lowered his focus from the lofty realms of royal courts and godly icons to examine the role of the anonymous human societies who carried out these wars of long ago. He drew upon royal chronicles and engraved bas-reliefs to describe conflicts that enriched, debilitated, and destroyed storied kingdoms. In describing the "Indianization" of Southeast Asia, however, he did not account for the absence of such obvious "Hindu" social institutions as caste (*varna*) in any of the cultures he studied. His analysis of Southeast Asia's upper strata supported his thesis of "Indianization." But a more thorough study of those who filled out the ranks of pike-bearing infantry would have undercut his principal conclusion.

H. G. Quaritch Wales (1952) was among the first to focus exclusively on what might now be classified as war and society in the early Southeast Asian polities. A graduate of Queen's College, Cambridge, Quaritch Wales had served Siam's royal court as a foreign advisor before turning to historical scholarship. In its day, his volume was an authoritative overview of premodern Southeast Asian war on par with the more numerous studies of ancient warfare in China and India. It has endured as an indispensable text for those studying early warfare in the period before the arrival of European soldiers. Not confined to weapons and battlefield tactics, Quaritch Wales explored social, cultural, and ritual aspects of warfare in numerous Southeast Asian cultures from both mainland and maritime subregions. But like his contemporary Coedès, Quaritch Wales was a product of academic programs that saw Southeast Asian culture as the semi-legitimate love child of its better understood neighbors.

Many scholars working in the 1960s and 1970s ultimately rejected George Coedès' and H. G. Quaritch Wales' theories of "Indianization" or "Hinduization" that stressed the absorption of Brahmanist culture by Southeast Asian rulers during the region's "classical age." Similarly, they were not content to study the kingdoms of Vietnam as "lesser dragons" that imitated China at the expense of their own sociopolitical systems, as many early Sinologist-trained scholars had. Instead, they sought evidence of unique systems in the study of what we now call "war and society."

O. W. Wolters led the way for this change with his highly original *Early Indonesian Commerce: A Study of the Origins of Srivijaya* (1967). In it he posited many of the concepts that still guide studies of Southeast Asia's premodern era. Among the most influential was his description of Southeast Asian rulers as "men of prowess" that stressed charismatic leadership, especially in times of war. In Wolters' thesis the bond between a warrior-ruler and the peoples he led was fundamental in establishing legitimacy. This dynamic applied to polities of all sizes and could be measured in leaders of all ranks and circumstances. Battlefield success at the city walls or the village gates hinged on his powers to motivate the populations he gathered around him. Beyond the obvious role that victory played in conquest and security, a successful military campaign demonstrated that the ruler had the necessary reserves of spiritual power—what we might call charisma in some circumstances—to hold on to a potentially wayward population. A ruler's moral power was measured by his ability to attract talented subjects, sire numerous children, and, most importantly, defeat opposing armies. War offered an excellent opportunity for these "men of prowess" to acquire the "soul stuff" necessary for uncontested rule. Their success in rallying and directing armies inspired confidence in their leadership, and their killing fed their spirit reserves with the souls of the dead. Warfare gave them opportunities to construct stable polities that, in turn, facilitated peace and, when need be, launched successful wars.

Wolters made a winning case for the study of Southeast Asia as a whole while acknowledging that specific arguments developed in subregional studies were probably more intellectually defendable and useful than any theory rendered in broad strokes—including the compelling frameworks he himself offered. His "men of prowess" thesis has endured over the last half century, and has settled into the intellectual framework of scholars of Southeast Asian studies working in numerous disciplines and approaches. This notion of proto-state leaders who gathered followers through charisma, talent, and efficacy persists in contemporary studies of revolutionaries and military officers who became leaders of Southeast Asian states in the postcolonial era. Conventional military historians, however, have not acknowledged or adopted the argument in their work.

Southeast Asia's historically sparse population also affected the intensity of battle. Some historians, such as Anthony Reid (1988), have suggested that the importance of capturing manpower inhibited the ferocity of warfare. They argued that commanders tasked with capturing war slaves were hesitant to spill blood at the point where victory appeared guaranteed. Other scholars have cast doubt on this perception of restraint (Charney, 2004; Knaap, 2003). They argue that such rational considerations amid the chaos and bloodshed of life-and-death contests are difficult to imagine.

Low-density populations also opened up opportunities for women in the making of war. Historians such as Barbara Watson Andaya (2006) have explored the role that women played in such war-related practice as headhunting in the island states of maritime Southeast Asia. Women presided over the most important rituals in transforming a taken head into a prized totem by removing the flesh and tissue that were consumed in the cannibalistic rites common to headhunting. They could influence the fate of their men on the battlefield through correct behavior at home.

The weapons of the premodern period convey social complexity. They were largely dual-purpose, used for hunting, security, and farming when not employed on the battlefield.

Soldiers carried individual weapons, such as spears, javelins, tridents, halberds, daggers, swords, and bows and arrows. They carried shields of wood and leather for protection, and some had helmets of leather or iron. Few beyond rulers and top generals wore armor. The common soldiers had other devices to ensure their survival. Southeast Asian warriors relied on charms, amulets, and magic spells to confer protection and even invulnerability in battle (Craig J. Reynolds, 2011, and Ruth, 2012). In many areas, soldiers tattooed mystical patterns and formulae, often the *yantra* schemes adopted from Hinduism's Tantric traditions, on their bodies. Warriors donned white cotton vests imprinted with similar mystical schemes. Throughout the Buddhist mainland, many soldiers wore Buddhist amulets derived from votive tablets.

The second major question addressed by historians of war and society in Southeast Asia concerns debates over what constitutes "Europe-centric" versus "Asia-centric" approaches. In the 1950s and 1960s, these conflicts reflected the unease between the sometimes unrealistic and utilitarian histories of colonial-era Western historians—some of which provided tacit support for Western imperialism—and the emerging corpus of often condemnatory and politicized anti-colonial histories emerging in the immediate postcolonial period by scholars in both the West and in Asia. These "-centric" labels accusatorily dismissed or discounted works produced during the colonial era and its aftermath. John R. W. Smail's landmark address to the 1st International Conference of Southeast Asian Historians in January 1961 in Singapore has endured as the best articulation of the parameters of these debates.

Smail used Dutch historian Jacob Cornelis van Leur's (1967) observation of Western historians' tendency to present Southeast Asia as "observed from the deck of the ship, the ramparts of the fortress, [and] the high gallery of the trading house" (261) in his articulation of the moral and perspectival limitations of what has been called "Europe-centric" approaches to the region's history. He recognized the importance of history that did not privilege Western experience nor mitigate the wrongs of Western colonial regimes. He also recognized that an "Asia-centric" view of history built upon traditional Southeast Asian notions of historical observation, preservation, and articulation was no longer possible in an age in which Southeast Asian historians were now writing history in the Western method. In his identification of independent "thought-worlds" occupied by historians who had rejected the narrow and damning "-centric" labels, Smail suggested the possibility of an approach that would allow historians to transcend the limitations of cultural identification and moral posturing. Smail used Vlekke's history of the Dutch war in Aceh (1959) as an example of a Western historian capable of escaping the moral bias of the Europe-centric approach without necessarily escaping its perspective. Smail argued that for a Westerner to write an autonomous history of Southeast Asia, he or she would adopt a methodological approach that engaged social and cultural norms of the region without applying those borrowed from Western histories. It was an autonomy of the mind. Smail's articulation of "autonomous histories" of Southeast Asia influenced subsequent generations of scholars seeking to write the history of the region—especially the colonial and immediate postcolonial period—without privileging or pardoning Western historical actors.

Writing Smail's style of "autonomous history" is not as easily achieved as many might assume. Historians studying war and society in colonial Southeast Asia examine complex sociopolitical circumstances in European and American colonial regimes that had been constructed upon the former independent states of the region. The societies they examine—at war and peace—existed within colonial-administered territories. The colonial armies that fought upon these historical landscapes were diverse units that may have had only the smallest Western representation in the ranks. Historians of the mid-nineteenth to mid-twentieth century must rely on Western archives in The Hague, Aix-en-Provence, Kew, and College Park, Maryland. Likewise, the memoirs and newspapers used to research this Southeast Asian history are largely products of Western intellects and sensibilities. Even today many ask whether works drawn so heavily from Western archives can accurately convey non-Western perspectives on war and society, given that much early twentieth-century scholarship—the foundation of Southeast Asian historical studies—elevated Western colonial actors to the leading roles in all colonial-era dramas. (The principal theme in many of these studies was "modernization," which meant "Westernization.") To rely heavily on Western sources and themes, especially the effect of external stimuli on indigenous systems, was to risk reducing Southeast Asians to what historian David Joel Steinberg (1971) characterized as "passive receptors, mere bit players, too weak to do more than reflect the brilliance of other civilizations" (3).

Heavy reliance on Western sources is not necessarily, however, a sign of ethnic chauvinism on the part of Western scholars or a rejection of Smail's "autonomous histories." It is not because Western historians have limited themselves to European-language sources that are more easily read and understood than those produced in Southeast Asian languages. The Western-focused approach has a simple and rather dispiriting cause: many national archives in Southeast Asia prohibit scholars from accessing documents related to military affairs. The governmental institutions overseeing archived collections are wary of allowing scholars, foreign or local, open access to their materials—which this author experienced firsthand while conducting research in Thailand and Vietnam in the early 2000s.

Many of the more influential historians of Southeast Asia of the mid- and late twentieth century were former colonial administrators, soldiers, or missionaries. J. S. Furnivall (1939), O. W. Wolters (1967), Victor Purcell (1962), John F. Cady (1960), J. C. van Leur (1955), H. G. Quaritch Wales (1952), and others carried with them experience of serving Southeast Asian colonial regimes when they set out to write the first generation of scholarship that bore the label "Southeast Asia." Furnivall, the former British colonial servant in Burma, gave the field a brilliant case study (1939) for understanding how European colonial domination of Southeast Asian indigenous regimes began. He also contributed frameworks for analyzing the systems by which ethnic groups within a colonial society could be pitted against each other to prevent resistance. Furnivall identified a "plural society" in which diverse ethnic groups came together solely for economic exchanges. With all groups focused narrowly on their own economic progress, Furnivall argued, these competing groups could be played off against each other to prevent anti-colonial resistance to the Western colonial regimes.

Historians of the colonial era, however, have given less attention to conflicts of the modern period that do not include Western colonial armies, such as the Siamese-Vietnamese wars of the early and mid-nineteenth century. The need for fluency in multiple Southeast Asian languages makes the pursuit of such research difficult. The few Thai books on these conflicts are written almost exclusively from the Siamese perspective using Thai sources and betraying Thai sympathies, though Vietnamese scholarship on the conflict is equally chauvinistic in its arguments and sources. Understandably, French colonial control of Cambodia, Laos, and Vietnam receives far more attention in English-language studies, while the period that preceded it is understudied. There are exceptions (Ngaosyvathn and Ngaosyvathn, 1998), but they are rare.

Western military involvement in a Southeast Asian conflict, however, does not guarantee scholarly attention. The Philippine-American War remains understudied by both Southeast Asianists and military historians. The former tend to favor the revolution (Agoncillo, 1956, 1960) rather than the war with American forces that followed it. For this reason, Milagros Guerrero's study (1977) of the war and its impact on Philippine society stands out. His use of a broad range of Philippine sources provides a sharp contrast to the works by American-based military historians who tend to rely almost exclusively on American archives.

Military historians give the Philippine-American War less attention than almost any other twentieth-century conflict, as it falls outside dominant American historical narratives. If acknowledged, the war is subsumed under triumphalist presentations of the Spanish-American War that preceded, and is not studied as a Southeast Asian event. American military historians without regional expertise tend to highlight the successful but destructive military campaigns against Philippine insurgents as a messy crucible in which American imperialism was forged. Others use them as a comparative foreshadowing to later U.S. operations in Vietnam, highlighting American atrocities to suggest national character flaws that link the Philippines to the My Lai massacre. Those who write about the military aspects of the war tend to devote most of their narrative to the American side while depicting Filipino fighters as opaque "others." Some U.S. military historians have recently broadened their studies to encompass Philippine military forces in their analysis, particularly Brian McAllister Linn (2000) and David Silbey (2007). Linn was among the first to apply a sophisticated post–Vietnam War analysis to American counterinsurgency efforts in the Philippines. Downplaying American racism as a principal cause of the high Filipino death toll, he presents the miscalculations and inexperience of the insurgents as the main cause for their disastrous failure.

The Southeast Asian theater in World War II garners less attention than does almost any other theater in the global conflict. The modest attention it does get is inadequate relative to its importance. Military historians have long studied the anti-Japanese campaigns in the Philippines, Singapore, and Burma. But most use American or British military and government sources while paying almost no attention to Japan's various historical archives, despite their relative openness to foreign scholars. Conversely, only a small group of regionalist scholars considers World War II. The result is that there are two World War IIs in Southeast Asian history. The first, and by far the largest in terms of volume and attention, studies the experiences of the Allies—Britain and the United States—as they fought the Imperial Japanese Army in naval and land battles in the Philippines, Malaya, Singapore, and Burma (Willmott, 2005, Cruickshank, 1983). The second, much smaller corpus examines Southeast Asians—their cultures, politics, and society—during World War II and its immediate aftermath. It is this second category that deserves to be called Southeast Asian military history rather than Western (or American and British) military history.

E. Bruce Reynolds' works (1994 and 2005) on Thailand during World War II is indispensable for understanding the war's impact on Southeast Asian societies. In addition to using a full range of diplomatic sources from Washington, Bangkok, and Tokyo, Reynolds employed a remarkable set of Thai-language sources from interviews (many that he conducted), memoirs, contemporary periodicals, and Thai-language histories. Shane Strate's study (2015) of Thailand in the early war period and of the long shadow that the "lost territories" have cast over Thailand's history is a valuable contribution in the tradition of Reynolds' work. Another valuable source for examining war and society during World War II is Alfred W. McCoy's edited volume (1980). McCoy gathered several important studies of the region by leading scholars of Southeast Asia, with an emphasis on Indonesia and the Philippines. Other scholars, such as Benedict R. O'G. Anderson (1961 and 1972), have addressed the war as a preamble to the wars of liberation ignited after 1945. Studies of revolution in colonial societies—many done by students of Anderson and McCoy—paved the way for studies of the Vietnam War that transcended an American-centric perspective.

Southeast Asianists with backgrounds in colonial studies have produced several studies of revolutionary movements that are necessary reading for any historian interested in understanding the three Indochina wars of the late twentieth century. These scholars study the various revolutionary figures—their anti-colonial struggles and the nationalist movements they fostered—who would later lead national armies in the French Indochina War, the Vietnam War, and the Cambodian-Vietnamese War. Christopher E. Goscha (1999 and 2009) has produced several indispensable works examining the creation of Vietnamese revolutionary networks across Southeast Asia that opposed the French (later the Americans), sought to impose a Moscow-directed communist ideology upon the left-leaning postcolonial nationalist movements across the region, and attempted to muster these disparate nationalist movements behind Ho Chi Minh's Indochinese Communist Party. Goscha is one of a small number of Western scholars who has language facility in French, Vietnamese, and Thai, and thus is capable of accessing the broadest range of sources. Not only does he trace the early days of the most prominent Vietnamese communist leaders in the 1920s through 1940s, but also he writes about how their organizations affected the communities in which they lived, worked, and hid. The geographic range of Goscha's studies extends beyond the former French Indochina to include Siam, Malaya, and the Dutch East Indies. Historians David G. Marr (1981), Hue-Tam Ho Tai (1992), and Peter Zinoman (2001) have likewise contributed important scholarship on French colonialism in Southeast Asia that should be read by those seeking to understand the origins of the Vietnam War.

Military historians and Southeast Asianists alike initially approached the French Indochina War as a bilateral conflict that pitted Vietnamese communists against French colonial armies. In English-language studies, Bernard Fall led the way. By the mid-1960s—with American combat troops fighting in South Vietnam—Fall had produced two classic studies of the French war that provided compelling frameworks for understanding the earlier conflict, and offered subsequent historians a compelling model of the historian as war observer in the Second Indochina War. Fall had both a strong understanding of military affairs and an ongoing interest in war's impact on the political sphere. But Fall was not a regional specialist, and did not speak Vietnamese. His perspective is firmly embedded within the French colonial military apparatus that he once traveled with. While Fall is still read, historians in recent years have come to see the First Indochina War less as an anti-colonial struggle involving two competing sides than as something akin to a civil war that pitted numerous competing factions against each other (Logevall, 2014).

The third major issue surrounding studies of war and society in Southeast Asia concerns competing and contested approaches to the Vietnam War. In the last couple decades, a new generation of Southeast Asianist historians have pursued studies of the Vietnam War that differ dramatically from the huge body of work on the conflict produced by military historians. American-centric approaches have long dominated histories produced by the latter, privileging American perspectives at the cost of local experiences, particularly Vietnamese ones. The lively industry in military histories of the Vietnam War appears nearly as strong as ever, and there is little indication that its vigor will subside. But for regionalists considering war and society in the Indochina Wars, there have been several major changes since Southeast Asianist historians have been studying the conflict.

Regional studies scholars pioneered the field of Vietnam War studies in the mid-1960s in the midst of the conflict. Their impetus came less from an interest in traditional military history than from a concern for the populations debased by the war. George Kahin (1969), the influential American political scientist working in Indonesian studies, was among the first to generate critical interest in the conflict through his involvement in the earliest anti-war "teach-ins" in 1965. Kahin was a U.S. Army veteran who had trained as a paratrooper during World War II before becoming an academic in the postwar period. In the 1950s, he helped establish one the earliest and most influential Southeast Asian studies programs at Cornell University, which during the Vietnam War years inspired graduate students from many disciplines (Chaloemtiarana, 1979, and Elliott, 2000) to research the effects of war, revolution, and violence on regional societies. At the heart of their studies were examinations of the social disruptions caused by the postcolonial wars of liberation. Similarly, the Committee of Concerned Asian Scholars, formed in 1969, used opposition to the war to focus scholarly attention on these issues. The committee encouraged radical criticisms of American beliefs regarding Southeast Asia that brought the United States into the war, and in doing so brought still more young scholars into Southeast Asian studies and Vietnam War studies. Historians associated with this movement (Young, 1992; Ngo Vinh Long, 1973) produced some of the earliest and most enduring scholarship on the Second Indochina War. Emphasizing the destructive consequences of American military intervention, these historians of Asia nonetheless highlighted American historical figures in their studies.

Some of the best-known studies of war and society during what the Vietnamese call the American War occurred while the United States still had combat troops in South Vietnam. The journalist-historian Francis FitzGerald attempted to put the Vietnam War in a broader historical context in her landmark study (1972). The volume won her the Pulitzer Prize. But despite FitzGerald's attempts at using premodern history to bolster her claims about Vietnam's rural and urban societies, many scholars—including Marine Corps veteran-turned-historian of Vietnam David G. Marr (1995)—judged her work to be superficial and unpersuasive (564–565).

Others were more successful. Jeffrey Race's study (1972) of a single province overcome by the turmoil of revolution and warfare resulted in one of the classics of Vietnam War studies. A former U.S. Army officer who served as a district advisor to the South Vietnamese forces in the Mekong River delta province of Long An, Race demonstrated how the rural communities in Long An could be simultaneously both victimized and a coveted source of support. Race traced the evolution of exogenous political movements as they were changed by the peasants they targeted—with weapons and aid programs—in the struggle between the Saigon government and the southern communists. Race put village society at the forefront of his study, and in doing so gave these communities a kind of collective agency that transcended and complicated their victimhood. Race's work is still read today by both military historians and regional studies specialists.

Following the end of the Second Indochina War in 1975, interest in military history among Southeast Asianists waned. Other topics emerged to challenge their focus. Imperialism and related colonial studies became the principal focus of their work. In the 1980s and 1990s, it was the military historians without backgrounds in regional studies who produced the plethora of books about the Vietnam War that now fill libraries and book collections around the world. The market for memoirs by journalists, military veterans, diplomats, and other participants flourished in this period as well. It was not until the early 2000s that a new generation of historians of Southeast Asia refocused new scholarly attention on the war.

Since the turning of the new millennium, the biggest change in Vietnam War studies has been the rejection of a U.S.-dominant narrative that places American historical actors at the center of every event and development. Challenging that approach are scholars interested in internationalizing the conflict and who seek, to use a deliciously loaded phrase, "the Vietnamization of Vietnam War studies" (Spector, 5). This movement has sought to break away from any framework that casts the war as merely a geopolitical contest between Washington and Hanoi over the spread of communism into Southeast Asia. Instead, this new generation of scholars has sought to apply extensive foreign language skills, global perspectives, and new forms of evidence to study the war from multiple viewpoints in several disciplines. Their research has contributed much-needed nuance to oft-stated and overly simplistic descriptions of North Vietnam's interaction with the People's Republic of China. Most significantly, they have brought Vietnamese-language skills (as well as time living and working in Vietnam) to study the Vietnam War as a Vietnamese war.

Among the leading figures in this scholarly movement are Edward Miller (2013) and Lien-Hang T. Nguyen (2012). Miller used Vietnamese sources extensively to conjure a portrait of Ngo Dinh Diem's regime that looks less like the creation of American advisors than that of an independent Vietnamese political class. Miller is not an apologist for Diem or for his brother Ngo Dinh Nhu, nor does he level all the blame at the United States for the brothers' downfall. Rather, he gives agency to a Vietnamese historical figure who has long borne quotation-marked American epithets to his name ("America's mandarin," "America's miracle man," "America's puppet," etc.). Lien-Hang T. Nguyen's study breaks with the enduring framework that puts Ho Chi Minh, Vo Nguyen Giap, and Pham Van Dong at the center of Hanoi's decision making. Instead, Nguyen shifts her focus to Le Duan and Le Duc Tho, and shows how the pair outmaneuvered the old leadership to wrest control of the Vietnamese Workers' Party, the prosecution of the war against the United States, and, ultimately, North Vietnam itself. Nguyen also vivifies the complex international dimensions of the war by tracing Hanoi's delicate diplomacy with Beijing and Moscow, as well as with Washington, Saigon, and Paris. What sets apart Nguyen's book from William J. Duiker's earlier study of the Hanoi leadership are the sources. Nguyen is one of a very small number of scholars to gain access to wartime documents in Vietnam's national archives.

The People's Liberation Armed Forces (PLAF)—the southern guerrillas and their supporters known to much of the world as the Viet Cong—have remained a disconcertingly under-examined topic in Southeast Asian studies. Douglas Pike's (1966) and Paul Berman's (1974) wartime studies carried the insights of researchers with a strong understanding of Vietnamese society. But in the years since those studies were published, the southern communist forces have received little attention from regional specialists. Some military historians working in Vietnam War studies have written about the PLAF, but their focus has almost always been operational. For many years these military historians relied on Truong Nhu Tang's memoir (1985) for insight into the organization, although his account shows the upper echelon without much insight into the peasant soldiers. Alternatively, they cited uncritically the translation of Nguyen Thi Dinh's memoir (1976) despite its shortcomings as a historical source. Recent attempts at studying the PLAF have been hindered by Vietnam's postwar politics. To this author's knowledge, Hanoi has not allowed Western scholars to use Vietnam's national archives to study the National Liberation Front or its affiliated soldiers. Scholars who attempt to study the PLAF in Vietnam risk being expelled and banned from the country. Meanwhile non-specialists, most with no Vietnamese-language skills, continue to produce studies of the Vietcong using American archives. Recently declassified CIA sources—however welcome—offer an American-mediated view that does not add as much to our understanding of these guerrillas as Vietnamese sources would.

The trend toward "Vietnamizing" Vietnam War studies has not yet included new or groundbreaking studies of anti-communist Vietnamese soldiers. Studies of the Army of the Republic of Vietnam (ARVN) by military historians have focused on the units' combat failures and institutional obstacles, while giving less attention to social and economic circumstances that compelled these young men—willingly and not—into uniform. Andrew Wiest's study (2008) acknowledges the martial shortcomings of the South Vietnamese soldiers, but places the blame for their inadequacy on their corrupt and conniving leadership. He underscores his points by highlighting several exemplary South Vietnamese soldiers rather than dwelling on the cowardly and ineffectual stereotype known as "Marvin the ARVN." Robert K. Brigham's short and sympathetic work (2006) delves more deeply into the field of social history, but its focus remains firmly fixed on the ARVN's lopsided relationship with the U.S. military. Similarly, with a few exceptions (Ang, 2010, Ruth, 2011), historians of Southeast Asia have not pursued studies of war and society in the surrounding "Free World" countries of Southeast Asia despite the regional breadth of the Second Indochina War. Nguyen Thi Dieu (1999) takes a broad regional approach in her study of the Mekong River during the Second Indochina War. Her scholarly scope transcends borders to include all major countries of the Mekong basin, and combines aspects of environmental, economic, military, and diplomatic history.

Historical studies of the Vietnam War with a focus on gender are scant. The major studies that have been produced have come primarily from scholars with backgrounds in regional studies. Karen Turner's study of female combatants and support troops in North Vietnam (1998) remains one of the few histories of this important topic. More recently, Heather Marie Stur produced a study of gender relations on U.S. military bases in South Vietnam (2011). The military historian Karen Vuic has applied gender studies paradigms to her scholarship on female nurses in the Vietnam War (2009). Gender studies focusing on masculinity have largely come from literary scholars rather than historians.

More so than any other development, the Vietnamization of Vietnam War studies has added greater sophistication to studies of war and society. But there are shortcomings evident in this approach. In elevating the profile of Vietnamese figures, scholars run the risk of obscuring or diminishing the prominence of American figures in the conflict. There is also the problem of ignoring the military aspects of war. An overly regionalist approach can leave the erroneous impression that the fighting was little more than a backdrop to social, political, cultural, and religious developments that occurred throughout Southeast Asia during this period. And scholars pursuing studies of America's soldiers in South Vietnam should not be dismissed as irrelevant or outdated. Several American military historians have published studies in the last decade that have explored—and sometimes exploded—prevalent myths about American military involvement. Among the best are Meredith Lair's study of American noncombatant support troops (2011) and George Lepre's work (2011) on the wartime fratricide called "fragging."

The three issues examined in this essay are relevant to what will certainly be a major concern in our day: the South China Sea territorial disputes. Efforts to define Southeast Asia through an examination of war and society will help us understand efforts by China and most of the Southeast Asian countries to delineate the geographical boundaries of the Spratly, Paracel, and other island groups in the South China Sea. Likewise, the debates Smail identified over what constitutes "Asia-centric" and "Western-centric" approaches to historical production help us understand China's use of premodern history to make its territorial claims, and the limitations of those claims in the twenty-first century. Regionalists writing about Asian ideas of war can help Western military historians to contextualize the actions of present-day Asian leaders through invocations of ideas such as "men of prowess" and others. And efforts to "Vietnamize" studies of the Vietnam War come at a time when Vietnam and the United States are in the midst of a historically tinged rapprochement over shared concerns with an expansive China.

Bibliography

Agoncillo, Teodoro. *Malolos: The Crisis of the Republic.* Quezon City: University of the Philippines, 1960.

———. *The Revolt of the Masses: The Story of Bonifacio and the Katipunan.* Quezon City: University of the Philippines, 1956.

Andaya, Barbara Watson. *The Flaming Womb: Repositioning Women in Early Modern Southeast Asia.* Honolulu: University of Hawai'i Press, 2006.

Anderson, Benedict R. O.' G. *Java in a Time of Revolution: Occupation and Resistance, 1944–1946.* Ithaca, NY: Cornell University Press, 1972.

———. *Some Aspects of Indonesian Politics Under the Japanese Occupation: 1944–1945.* Ithaca: Modern Indonesia Project, Southeast Asia Program, Department of Far Eastern Studies, Cornell University, 1961.

Ang Cheng Guan. *Southeast Asia and the Vietnam War.* New York: Routledge, 2010.

Beemer, Bryce. "Southeast Asian Slavery and Slave-Gathering Warfare as a Vector for Cultural Transmission: The Case of Burma and Thailand," *The Historian* 71 (Fall 2009): 481–506.

Berman, Paul. *Revolutionary Organization: Institution-Building Within the People's Liberation Armed Forces.* Lexington, MA: Lexington Press, 1974.

Brigham, Robert K. *ARVN: Life and Death in the South Vietnamese Army*. Lawrence: University Press of Kansas, 2006.

Cady, John F. *A History of Modern Burma*. Ithaca, NY: Cornell University Press, 1960.

Charney, Michael W. *Southeast Asian Warfare, 1300–1900*. Boston: Brill, 2004.

Coedès, George. *The Indianized States of Southeast Asia*. Honolulu: East-West Center Press, 1968.

Cruickshank, Charles. *SOE in the Far East*. New York: Oxford University Press, 1983.

Elliott, David. *The Vietnamese War: Revolution and Social Change in the Mekong Delta*. Armonk, NY: M.E. Sharpe, 2000.

Fall, Bernard B. *Street Without Joy: Indochina at War, 1946–1954*. Harrisburg, PA: Stackpole Press, 1961.

———. *The Two Viet-Nams: A Political and Military Analysis*. New York: Frederick A. Praeger, 1967.

FitzGerald, Frances. *Fire in the Lake: The Vietnamese and the Americans in Vietnam*. Boston: Little Brown, 1972.

Furnivall, J. S. *The Fashioning of Leviathan: The Beginnings of British Rule in Burma*. Rangoon: Burma Research Society, 1939.

Goscha, Christopher E. *Thailand and the Southeast Asian Networks of the Vietnamese Revolution, 1885–1954*. Richmond: Curzon Press, 1999.

Goscha, Christopher E., and Christian F. Ostermann, ed. *Connecting Histories: Decolonization and the Cold War in Southeast Asia, 1945–1962*. Stanford: Stanford University Press, 2009.

Guerrero, Milagros C. "Luzon at War: Contradictions in Philippine Society." PhD dissertation, University of Michigan, 1977.

Hack, Karl, and Tobias Rettig, ed. *Colonial Armies of Southeast Asia*. New York: Routledge, 2006.

Kahin, George McTurnan, and John Wilson Lewis. *The United States in Vietnam*. 2nd ed. New York: Dial Press, 1969.

Knaap, Gerrit. "Headhunting, Carnage and Armed Peace in Amboina, 1500–1700," *Journal of the Economic and Social History of the Orient* 46, no. 2 (2003): 165–192.

Lair, Meredith. *Armed With Abundance: Consumerism & Soldiering in the Vietnam War*. Chapel Hill: University of North Carolina Press, 2011.

Lepre, George. *Fragging: Why U.S. Soldiers Assaulted Their Officers in Vietnam*. Lubbock: Texas Tech University Press, 2011.

Linn, Brian McAllister. *The Philippine War*, 1899–1902. Lawrence: University Press of Kansas, 2000.

Logevall, Fredrik. *Embers of War: The Fall of an Empire and the Making of America's Vietnam*. New York: Random House, 2014.

Luce, Don, and John Somner. *Viet Nam: The Unheard Voices*. Ithaca, NY: Cornell University Press, 1969.

Marr, David G. *1945: The Quest for Power*. Berkeley: University of California Press, 1995.

———. Review of *Fire in the Lake: The Vietnamese and the Americans in the Vietnam War. Journal of Asian Studies* 32, no. 3 (May 1973): 564–565.

———. *Vietnamese Tradition on Trial, 1920–1945*. Berkeley: University of California Press, 1981.

McCoy, Alfred W. *Southeast Asia Under Japanese Occupation*. New Haven: Yale University Southeast Asia Studies, 1980.

Miller, Edward Garvey. *Misalliance: Ngo Dinh Diem, the United States, and the Fate of South Vietnam*. Cambridge: Harvard University Press, 2013.

Ngaosivat, Mayuri, and Pheuiphanh Ngaosyvathn. *Paths to Conflagration: Fifty Years of Diplomacy and Warfare in Laos, Thailand, and Vietnam, 1778–1828*. Ithaca, NY: Southeast Asia Program, Cornell University, 1998.

Nguyen, Lien-Hang T. *Hanoi's War: An International History of the War for Peace in Vietnam*. Chapel Hill: University of North Carolina Press, 2012.

Pike, Douglas. *Viet Cong: The Organization and Techniques of the National Liberation Front of South Vietnam*. Cambridge, MA: MIT Press, 1966.

Purcell, Victor. *The Revolution in Southeast Asia*. London: Thames & Hudson, 1961.

Race, Jeffrey. *War Comes to Long An: Revolutionary Conflict in a Vietnamese Province*. Berkeley: University of California Press, 1972.

Reid, Anthony. *Southeast Asia in the Age of Commerce, 1450–1680. Vol 1. The Land Below the Winds*. New Haven: Yale University Press, 1988.

Reynolds, Craig J. "Rural Male Leadership, Religion and the Environment in Thailand's Mid-South, 1920s–1960s," *Journal of Asian Studies* 42 (February 2011): 39–57.

Reynolds, E. Bruce. *Thailand and Japan's Southern Advance, 1940–1945*. New York: St. Martin's Press, 1994.

————. *Thailand's Secret War: The Free Thai, OSS, and SOE During World War II*. Cambridge: Cambridge University Press, 2005.

Ruth, Richard A. "Dressing for Modern War in Old-fashioned Magic: Traditional Protective Charms of Thailand's Forces in the Vietnam War." In *The Spirit of Things: Materiality and Religious Diversity in Southeast Asia*. Ed. Julius Bautista Ithaca: Southeast Asia Program, Cornell University, 2012.

————. *In Buddha's Company: Thai Soldiers in the Vietnam War*. Honolulu: University of Hawai'i Press, 2011.

Silbey, David. *A War of Frontier and Empire: The Philippine-American War, 1899–1902*. New York: Hill and Wang, 2007.

Spector, Ronald. Review of *Journal of Vietnamese Studies*, Volume 4, Number 3 (Fall 2009). *H-Diplo Roundtable Reviews* XI, no. 12 (2010). Accessed February 31, 2017. https://issforum.org/roundtables/PDF/Roundtable-XI-12.pdf.

Steinberg, David Joel, ed. *In Search of Southeast Asia: A Modern History*. New York: Praeger, 1971.

Strate, Shane. *The Lost Territories: Thailand's History of National Humiliation*. Honolulu: University of Hawai'i Press, 2015.

Stur, Heather Marie. *Beyond Combat: Women and Gender in the Vietnam War Era*. New York: Cambridge University Press, 2011.

Tai, Hue-Tam Ho. *Radicalism and the Origins of the Vietnamese Revolution*. Cambridge: Harvard University Press, 1992.

Thak Chaloemtiarana. *The Politics of Despotic Paternalism*. Bangkok: Social Science Association of Thailand, 1979.

Thi Dieu, Nguyen. *The Mekong River and the Struggle for Indochina: Water, War, and Peace*. Westport: Prager, 1999.

————. *No Other Road to Take: Memoir of Mrs. Nguyen Thi Dinh*. Trans. Mai V. Elliot. Ithaca, NY: Cornell Southeast Asia Program, 1976.

Truong Nhu Tang. *A Viet Cong Memoir*. San Diego: Harcourt Brace, 1985.

Turner, Karen Gottschang. *Even the Women Must Fight: Memories of War From North Vietnam*. New York: Wiley, 1998.

Van Leur, J. C. *Indonesian Trade and Society: Essays in Asian Social Economic History*. 2nd ed. The Hague: W. Van Hoeve, 1967.

Vinh Long, Ngo. *Before the Revolution: The Vietnamese Peasants Under the French*. Cambridge: MIT Press, 1973.

Vlekke, Bernard H. M. *Nusantara: A History of Indonesia*. The Hague: W. van Hoeve, 1959.

Vuic, Kara Dixon. *Officer, Nurse, Woman: The Army Nurse Corps in the Vietnam War*. Baltimore: Johns Hopkins University Press, 2009.

Wales, H. G. Quaritch. *Ancient Southeast Asian Warfare*. London: B. Quaritch, 1952.

Wiest, Andrew A. *Vietnam's Forgotten Army: Heroism and Betrayal in the ARVN*. New York: New York University Press, 2008.

Willmott, H. P. *The Battle of Leyte Gulf: Last Fleet Action*. Bloomington: Indiana University Press, 2005.

Wolters, O. W. *Early Indonesian Commerce: A Study of the Origins of Srivijaya*. Ithaca: Cornell University Press, 1967.

Young, Marilyn Bratt. *The Vietnam Wars 1945–1990*. New York: Harper Perennial, 1992.

Zinoman, Peter. *The Colonial Bastille: A History of Imprisonment in Vietnam, 1862–1940*. Berkeley: University of California Press, 2001.

3 War and Society in Australia, New Zealand, and Oceania

Peter J. Dean and Tristan Moss

Introduction

Warfare in Australia, New Zealand, and Oceania covers thousands of years and countless generations. However, the long inhabitation of these island states largely escapes written records, and much of the oral history has dissipated with passing generations. Only since the time when European empires transgressed the region has evidence been more systematically gathered and analyzed to form a more complete understanding of the effects of war on these societies.

Though separate entities, these three areas are linked by geography and the impacts of colonization and empire. With a common history of settlement by the British Empire, New Zealand military history and Australian military history are invariably compared. In many respects, given the wars they fought for, and their corresponding places within the British Empire and Commonwealth, there is much to be gained by looking at these two countries in tandem. However, significant differences exist between New Zealand's and Australia's military experiences, and in particular how they have remembered and memorialized these conflicts. In contrast, while they are often perceived as conforming to similar patterns, experiences, scholarship, and interpretation, that of Oceania is seen as occupying the separate historical sphere of "Pacific history," despite the deep involvement of both New Zealand and Australia in this region. This chapter charts the separate military histories of these three areas while identifying elements of commonality between them.

Australia

Within the broader Oceania region Australia is the largest country in terms of its size, population, and relative power. As a nation it has a short but remarkably eventful military history. While warfare on the Continent dates back thousands of years before European settlement in 1788, over the last century military events have been especially prominent. The Boer War, two world wars, Korea, Vietnam, the Gulf War, Iraq, and Afghanistan, plus scores of peace-keeping and peace enforcement deployments and missions, have placed military conflict toward the center of Australia's historical experience.

The centerpiece of Australian military history is the storming of the beaches at Anzac Cove on April 25, 1915. (ANZAC being the acronym for the Australian and New Zealand Army Corps. In modern parlance "ANZAC" refers to the military formation and "Anzac" to soldiers, a place, a national day, a myth, and a legend.) This event is seen as one of the defining points in the national story, and for many it is regarded as the birthplace of the nation. Debate over how the death and wounding of thousands of young Australian males gave "birth" to the country has, however, never been reconciled. But in the absence of a truly national day that unites the country, Anzac Day (part Remembrance Day, part 4th of July foundation day) has, in recent decades, evolved into a spiritually important day of national significance focused on remembrance and sacrifice—one that evokes what some have called the civic religion of Anzac. Key works on the notion of "Anzac" and its influence on Australian society can be found in two excellent studies by Graham Seal (2004) and Carolyn Holbrook (2014).

Before the events at Gallipoli seared themselves into the Australian consciousness, the nation already had a long experience at war. One of the most controversial aspects of Australian history is the notion of the conflict on the "frontiers" between European settlers and Aboriginal Australia. This debate has been mainly fought out between historians of Indigenous Australia, most prominently Henry Reynolds (1987), Richard Broome (1994), Lyndall Ryan (1997), and the controversial Keith Windschuttle (2002); Attwood and Foster (2003) offer the best summary of the differing views and core debates. This discourse was also a microcosm of the so-called history wars in Australia, which, at the time, saw a highly politicized use of the discipline centered on the debate between "black armband" (focused on the history of indigenous society and minorities) and "white blindfold" (focused on the white Anglo-Saxon history of Australia and often rejecting the notion of dispossession and violence against indigenous communities) versions of Australian history. Among more strategic and operationally focused "traditional" military historians, the notion of war on the Australian colonial frontier has been widely accepted for some time, with the best account being John Connor's excellent 2002 study. This period is also very well covered in Jeffery Grey's chapters on the frontier conflicts in his widely read and cited *A Military History of Australia* (2008b), now in its third edition.

The British colonies in Australia next experienced war through a series of small colonial conflicts across the British Empire, including the Māori Wars, Sudan War, Boxer Rebellion, and the Boer War. These conflicts are well covered in Craig Stockings and John Connor's edited collection (2013). The story of the Australian officer Harry "Breaker" Morant and the controversy surrounding his execution by the British military in South Africa in 1902 is the most controversial element of this period, and it has dominated much of the scholarship on Australia's involvement in the Boer War. The case in support of Morant has been put by popular historian Nick Bleszynski (2003) and countered (among other authors) by Craig Wilcox (2002). Wilcox's volume, a product of an Australian War Memorial project, provides a detailed coverage of Australia's involvement in the war and acts as a pseudo-official history in the absence of one being commissioned by the Australian government.

Despite numerous military operations before 1914, the First World War established the role and place of military history in Australia, as well as historiographical elements that still dominate the field today. As noted, the landing at Anzac Cove by the I ANZAC Corps (made up of the 1st Australian Division and the New Zealand Division) was the pivotal event in Australian military history from which all subsequent history has evolved. The Anzac legend was largely fostered by C.E.W. Bean, chosen in 1914 as the official war correspondent and later appointed as the official war historian. Bean edited the fifteen volumes of the official history, *Australia in the War of 1914–1918* (1921–1942), and wrote the six volumes covering the operations at Gallipoli and on the Western Front. In a somewhat radical move for the time, Bean's history was written largely from the viewpoint of the regimental experience as opposed to a focus on the events at headquarters and on strategy. This focus on the soldiers fostered what has been called a "democratic" war history tradition, which has had a profound effect on the writing of military history in Australia. This approach continues to dominate most of the works that have been published on Australians' war experience up to and including today.

Bean inspired a long tradition of journalists writing Australia's military history. Heavily focused on the "diggers" (slang for an Australian soldier that evolved on the Western Front), it has evolved into a form of writing that the historian Robert Stevenson has called "diggerography": one that encapsulates a populist approach reflecting the notion that a good story is more important than good history. Such works, covering all of Australia's major conflicts, are often large in size and heavily reliant on oral histories of aging veterans. They also tend to be short on archival evidence and analysis. Another key feature is their focus on the supposed incompetence of the high command, especially if these officers are provided by Australia's "great and powerful friends" Great Britain or the United States.

Diggerography-styled books remain highly popular among the general public. They are also often the favorite of nationalist politicians, with Les Carlyon's (an award-winning sport journalist) *The Great War* being the co-inaugural winner of the Prime Minister's Award for Australian History in 2007. Unfortunately, the quality of these popular military histories remains mixed, with most falling well short of demonstrating solid historical knowledge and use of evidence. While this trend is replicated in many other countries, such as Great Britain and the United States, Australia lacks high-quality journalist-historians, like Anthony Beevor, Max Hastings, and Rick Atkinson. Furthermore, few Australian scholars have produced works of a similar focus, contributing to a chasm between the small number of historians in the academy and the larger numbers writing from outside of it.

This separation also reflects the academy's rejection of "traditional" military history (focused on strategy, campaigns, operations, tactics, leadership and command, doctrine, etc.) that began with the mass expansion of university education in Australia coinciding with the Vietnam War. As Professor Joan Beaumont, one of the country's leading historians of war, argues, the majority of academics came to see war as a "morally suspect activity." As a result of this factor, along with the corresponding culturalist turn in the humanities, and the "leftist traditions of . . . Australian intellectual life" (Beaumont, 2003, 166), when the academy has looked at war it has done so from the perspective of war and society, with a focus on the social effects of war, including the rise of memory studies, commemoration, and pilgrimage.

This war and society approach has provided fertile ground in Australian military history. One of the key factors has been the prevalence in twentieth-century Australia of largely fighting wars in distant places (mostly in Europe and the Middle East and remote parts of Asia-Pacific). This factor, coupled with the decision to not repatriate the remains of Australian soldiers, sailors, airmen, and nurses from the two world wars, coupled with the strong notions around the Anzac mythology, has given rise to a strong collection of work on war, memory, grief, and pilgrimage. In this the First World War has continued to dominate, with works written by Bart Ziino (2007), Tanja Lukins (2004), and Bruce Scates (2006) serving as key texts. The Second World War has also been well served in this field by the likes of Liz Reed (2004) and, again, Bruce Scates (2013). Operating at the intersection of military history, the Anzac legend, and war and memory has been the importance of the prisoner of war (POW) experience in Australia. Driven largely by the thousands of Australians who went into captivity at the hands of the Japanese during the Pacific war, this field has seen some excellent studies, notably by Hank Nelson (1985) and Joan Beaumont (1988). The best overview of the POW experience across Australia's wars can be found in Joan Beaumont et al. (2015). Overall the losses that occurred during the world wars had a profound effect on Australian society, and this is captured most prominently in the claim that Australia has the highest number (per capita) of war memorials of any country on earth. This is exemplified by the national Australian War Memorial in Canberra established by Charles Bean that serves as a shrine, museum, memorial, and research center. The importance of these memorials is captured brilliantly in Ken Inglis' multi-award-winning *Sacred Places: War Memorials in the Australian Landscape* (1998).

The debate over Bean's influence and the role of military history in Australian society, as well as within the discipline, continues to court controversy. For many, the Anzac legend has been co-opted and exploited by the state in particular, but also by nationalist elements and right-wing politics. In 2010 four prominent academic historians produced *What's Wrong With Anzac? The Militarisation of Australian History* (2010). The book represents a long over-due critique of the legend, and it displayed considerable courage given the sacredness that Anzac had achieved among large sections of Australian society. Its basic premise is that Anzac has overshadowed other (more) important elements of the national story, such as federation and labor history, and is particularly exclusive of women and the multicultural nature of Australian society. However, this legitimate argument was weakened by some poor sections of the work surrounding the publishing of military history in Australia and overseas, a lack of knowledge on strategy and operations, and a misinterpretation of Anzac as an extension of militarism and imperialism. James Brown has written a much more effective, recent critique (2014). A defense analyst, former army officer, and veteran of Iraq and Afghanistan, Brown deconstructs how Anzac has been traded on by companies and associations and how it limits criticism of the Australian Defence Force.

Another of Bean's legacies has been the role and importance of official histories to Aus-tralia. Following up on his epic, another journalist, Gavin Long, was appointed and produced a twenty-two-volume history of *Australia in the War of 1939–1945*. In the post–World War II era, however, the official historians have all come from the academy, and a number of them, includ-ing Professors Robert O'Neill, David Horner, and Craig Stockings, are also former military officers and veterans. Each of these series, covering the conflicts in Korea, Vietnam, peacekeep-ing and Cold War operations, and a newly commissioned (2016) series on East Timor, Iraq, and Afghanistan have evolved considerably from Bean's time to include more detailed coverage of diplomacy and strategy as well as operations, services, and the home front. The enduring foundations for the Australian tradition in official histories is that they have access to official records, that they tell the national story, and that they are written independent of government with control only over the preservation of classified technical secrets. They remain the standard reference volumes for their conflicts and the starting points for any serious researcher.

Besides official histories there have been a number of government- or service-sponsored publications, the two most important being the seven-volume *Australian Centenary History of Defence* (2001). This series has a volume on the history of all three services and is especially notable for the first history of the Department of Defence, Professor Joan Beaumont's excel-lent *Australians Defence: Sources and Statistics*, and Lieutenant General John Coates' superb *An Atlas of Australia's Wars*, now in its second edition. Most recently Professor Jeffery Grey (2015–2016) was series editor of the five-volume *The Centenary History of Australia and the Great War*. This series brings to light the best of recent scholarship in Australia on the First World War, with volumes covering the air war, the war with the Germans, the home front, and a history of the Australian Imperial Force. In tandem with this production is the Royal Australian Navy's comprehensive century volume, *In All Respects Ready*, produced by the nation's leading naval historian, David Stevens. Of note here is also the excellent and broad sweeping *Oxford Companion to Australian Military History* (2008).

Biography has also proven fertile ground for military history, particularly in the last three decades. In a comprehensive overview of this genre, Peter Dean (2011) notes that virtually all of Australia's major army commanders have been the recipient of at least one biography, as have many of the Victoria Cross winners. Most of the more recent biographies have been produced by professionally trained historians. However, there has been a plethora of populist studies on General Sir John Monash, commander of the Australian Corps in 1918, all of which offer little if any new historical insight and most of which delve the depths of hagi-ography and "diggerography." The best biographies remain two of the originals by Geoffrey Serle (1982) and Peter Pederson (1985). Australia's most senior military officer (and one of its most controversial), Field Marshall Thomas Blamey, has also been the subject of a number of studies, the most significant being David Horner's excellent and balanced account (1998).

The Australian Army has dominated the genre of Australian military history, both in the number of volumes and the focus on its battles. These studies have also been particularly well served by investments from the institution in the Australian Army History Series. This includes an excellent academic series, first with Oxford, and then with Cambridge University Press, that accounts for most of the high-quality biographies and battle, campaign, and other studies produced over the last twenty years. The other two services have also been well supported through the navy's Sea Power Centre and the Air Power Centre, whose publications mostly appear in electronic form on their websites.

In terms of specific conflicts, the history of the First World War dominates. Among the overwhelming number of titles a few volumes stand out, including the *Centenary History of the Great War* and Jean Bou's two important works (2009 and 2016). The Second World War is the next most prominent, given its size and scope. Besides the texts already mentioned a number stand out for their approach or coverage, or the quality of the analysis. The absence of a volume on strategy in the Second World War official history is striking, especially given Australia's role in the Pacific war; however, David Horner, Australia's preeminent and most prolific military historian, produced a comprehensive study in 1982, filling this gap with *High Command: Australia and Allied Strategy 1939–1945*. Peter Stanley's three works (1997; 2002, with Mark Johnston; and 2008) demonstrate some of the best analysis, whereas Peter Dean's series of edited books on the Pacific war (2013; 2014; 2016) have been praised for their insight and coverage, combining battlefront analysis across all three services along with the home front, the POW experience, and perspectives from the United States and the Japanese. In the postwar period no texts have as yet gone far beyond the records and synthesis of the official histories, and few texts have yet to appear covering Australia's more recent conflicts in the Middle East—although recent notable editions include Peter Edward's synthesis of the Vietnam War official histories (2014).

Overall, while there has been much written on Australian military history, the gaps in the literature are still large. Across thematic approaches and conflicts, even with the much historically fought-over First World War, particularly the Gallipoli campaign, there is still plenty left to be covered. Other battles, operations, and campaigns still lack any or adequate coverage, and across the scope of conflict technical military history and logistics are very rarely addressed. Australian military history tends to be empirical and narrative-driven and has, more often than not, eschewed theory. This leaves fertile ground for current and future generations of historians to explore.

Furthermore, there still remains a large divide between those who work in more traditional military areas and those covering the "new military history" or war and society approaches. A few studies, such as Joan Beaumont's award-winning First World War history (2013) and Michael McKernan's integrated coverage of the Second World War (2008), bridge this divide, but it is all too rare in the historiography. As the late great Australian military historian Jeffery Grey has noted, "Perhaps it's time Australian military historians on both sides of the divide [traditional military and social history of war] take a leaf from practice elsewhere" and started to work together to advance the field of study (Grey, 2008a, 468).

New Zealand

Military history in New Zealand has not occupied the same central place as it has in Australian historiography. While participating in similar conflicts, including war with indigenous peoples, the wars of Britain and the United States, peacekeeping, and the War on Terror, myriad gaps exist in New Zealand's military history. Within the most comprehensive overview of New Zealand's military history, *The Oxford Companion to New Zealand Military History* (2000), Jeffrey Grey argues that relative to Māori history in particular, military history has been "a minority interest" in New Zealand. Similarly, Deborah Montgomerie (2003, 62) argues that the history of New Zealand's wars has been quarantined from social history, and connections between the two "are made only in passing, or treated as statements of historical verities rather than theses to be proved." Montgomerie then explains that, although war is acknowledged as important in New Zealand's history, "the details of the process . . . remain indistinct" (62).

With the exception of the history of the Māori people, and perhaps the First World War, the empirical and foundational detail of New Zealand's wars is provided by an extensive official history program, which has received the greatest historiographical attention. Robert Rabel (2001) and McGibbon (2003), both official historians, argue that the "dominance" of official history is the result of a combination of factors, including the limited number of historians in a country of around four million, a more limited sense of nationalism than that found in Australia, and the focus of academic historical work on issues such as gender, culture, and race rather than more "traditional" areas of study, such as diplomatic and military history. Moreover, as Montgomerie argues, the nature of official histories has shaped the body of New Zealand war and social history. Funded by the government rather than universities, and writing for a general audience, authors are less likely to engage explicitly with historical debates, concentrating instead on narrative detail. While Rabel (2001, 64) decries the "unremitting empiricism" of the official histories, like in Australia, they do provide a solid foundation on which New Zealand military history rests.

Māori history and culture are central to New Zealand's national identity, which marks a stark contrast with the place of indigenous peoples in Australia. In contrast to the literature on Australian frontier wars, conflict between and with Māori has received a significant proportion of military historians' attentions. The reasons for this are complex, but reflect the higher proportion of the population made up by Māori, compared with Aboriginal Australians (around 15 and 2 percent respectively); a far stronger Māori political voice; and an early acceptance of Māori warfare as similar to Western conceptions of conflict.

The Musket Wars offer a prime example of the latter point. Fought among Māori peoples between 1807 and 1845, their name incorporates the long-standing assumption that Western weapons, introduced by traders before the establishment of European colonies in New Zealand, changed the nature of Māori warfare. Angela Ballara (2003) has challenged this interpretation, arguing that these wars were an extension of previous modes of warfare, with muskets acting as a complementary tool rather than a catalyst for change.

European colonization in New Zealand saw the most sustained regular campaign against an indigenous people by imperial troops in the Pacific region. The most significant period of conflict occurred during the 1860s and 1870s, in what was known as the Māori Wars, now termed the New Zealand Wars. Early scholarship claimed they stemmed from land disputes in the new colony, and saw them as one of a series of colonial "land grabs." James Cowen's older two-volume work (1922–1923) remains a foundational text in its operational detail. Revisionist historians, of which James Belich (1987) is the most well-known, have argued that the conflict between Europeans and Māori was instead over sovereignty. While others have continued to revisit these conflicts, Belich's work, which was also presented in a television documentary, sparked greater historical attention.

As throughout much of the British Commonwealth, the First World War has attracted the lion's share of historical attention in New Zealand. While many nations lay claim to the title of "greatest sacrifice," the sheer number of men sent overseas from New Zealand, relative to the population, was among the highest in the world. Following their return, there was a rush of publications on New Zealand's war experience, particularly memoirs. The First World War also saw four volumes of official history published between 1921 and 1923, which covered the campaigns in Palestine, Turkey, and France. As McGibbon (2003) points out in his introduction to the official histories of the First and Second World Wars, these volumes were managed within the military, written by officers for the general public. Although the volumes were commissioned without a great deal of coordination or general official enthusiasm, they have nonetheless dominated the historiography of New Zealand's First World War. Less formally, unit histories produced by veterans groups were overseen by a government committee.

With the centrality of the First World War to New Zealand history, and surges in scholarly and public interest during the 1980s and around the centenary, the First World War has been subject to the most rigorous reexamination by scholars of all New Zealand's wars. Christopher Pugsley's study of New Zealand's Gallipoli campaign (1984) marked one of the first since the official history. The depth of scholarship now being produced on New Zealand and the First World War is epitomized in John Crawford and Ian McGibbon's *New Zealand's Great War: New Zealand, the Allies and the First World War* (2007), in which an array of authors examine issues ranging from New Zealand and empire to repatriation, operations, gender, and community.

New Zealand's Second World War, despite the proximity of the Japanese advance into the Pacific, was focused on the Middle East and Europe. While Australia withdrew its divisions from the fight against Germany, the New Zealand 2nd Division remained in North Africa, and fought in Italy until the war's end. Although the 3rd Division fought limited campaigns in the Pacific, aided by air and naval units, the Mediterranean remained its principal effort. Writing on the Second World War continues to be dominated by the mammoth forty-eight volume official history (1949–1986). In some respects, the scale of this project was a reaction to the haphazard and uneven First World War official histories, and the project covers the entire breadth of New Zealand's wartime experience, including a volume devoted to prisoners of war. Rabel (2001) argues that the sheer scope of these histories has "daunted" historians, but there undoubtedly remain many more gaps to fill than for the First World War.

New Zealand's post-1945 conflicts were marked by small, niche contributions in support of the British, the United States, and, most recently, New Zealand's own independent strategic path. Each deployment is explored by official histories, but has attracted few other studies. McGibbon (1991, 1996) explores the experiences of the around 4,700 soldiers and sailors sent to Korea, while Pugsley (2003) documents New Zealand's decades-long contribution to the defense of Malaya. Both McGibbon (2010) and Rabel (2005) have covered New Zealand's role in Vietnam, which saw the country integrate its forces with that of Australia in the Phuoc Tuy province. After Vietnam, New Zealand's defense strategy departed from that of Australia, as the country embraced a smaller and more pacific defense posture, notably barring U.S. nuclear powered–ships from its ports. New Zealand has also had a long involvement in peacekeeping operations, examined by Crawford (1996). However, in most cases these histories form the bulk of scholarly work on New Zealand's military operations in the second half of the twentieth century.

Beyond the battlefield, New Zealand's historians have examined the way in which war has forged their national identity. Unlike in Australia, Anzac has not occupied the central place in New Zealand society. Indeed, despite supplying the key consonants to the term "Anzac," New Zealand tends to be forgotten in Australian commemoration in the war. The role of war in New Zealanders' identities has been only superficially examined, with the exception of Jock Phillips' (1989) exploration of Pakeha (white New Zealander) male identity, and the place of war within it.

By contrast, the study of Māori soldiers and their martial identity has been a significant subset of New Zealand military history. The 28th (Māori) battalion, for instance, which served in the Mediterranean during the Second World War, received its own volume in the official history of that conflict. Māori soldiers were, and continue to be, lionized for their martial qualities, and Māori traditions, most famously the Haka war dance, have been incorporated into the New Zealand Defence Force's ceremonies. Recently, the idea of Māori military identity has been examined by Francesca Walker (2012) through the prism of "martial race," itself a resilient myth among minority and colonial soldiers throughout the colonial world. The central place and unique involvement of Māori soldiers in New Zealand's wars represent a rich area of study that intersects social, military, and racial histories. Yet, no detailed scholarly work on Māori in the defense force has been attempted.

New Zealand military history will continue to be shaped by the relatively small size of the historical academy, which, combined with the absence of a national obsession with its wartime past in the vein of Australia, ensures that there is still much work to be done in the field. Examining the state of New Zealand war history, Montgomerie (2003, 74) argues that the rise in social history has reduced military history to "mere background." Yet, she notes that military and social historians are "on common, though substantially uncharted, ground," which bodes well for the future of the study of war and society in New Zealand.

Oceania

If there are gaps in the history of Australia and New Zealand's past conflicts, historical writing on conflict in Oceania is exceptionally meager, conditioned by the region's isolation and its tangential relationship to centers of geopolitical power and historical study. Compounding this is the difficulty in defining the Pacific, although it can be said to encompass the broad linguistic, ethnic, and cultural groups of Melanesia, Micronesia, and Polynesia, with the smaller Oceania centered on the tropical Pacific south of the equator. With small populations separated by the vast Pacific Ocean or, in the case of the largest landmass of New Guinea, by jungle and mountains, the region has a profoundly diverse range of languages, political systems, and cultures. Indeed, Papua New Guinea alone is home to more than one-quarter of the world's languages. Despite, or because of this complexity, the region is understudied by historians, and any engagement with Oceania's conflicts and military necessarily embraces the fields of anthropology, archaeology, and political science.

As a result, the Pacific's historiography is as disparate as its geography and, as David Armitage and Alison Bashford point out in their edited collection (2014), is set apart from histories of other world regions. Within the "kaleidoscopic" Pacific historiography, the study of war and society is sparse, and occupies two broad categories. First, it focuses on the structures and periodization of external, usually Western powers, particularly in relation to imperialism and colonialism. Although colonized later than the rest of the world, largely during the late nineteenth century, the Pacific was subject to the similar waves of European contact, annexation, involvement in the two world wars, and decolonization. Broadly, initial imagining of the effect of colonialism in the Pacific focused on the way in which the arrival of Europeans constituted a "fatal impact," while postwar historical work sought to acknowledge the ways in which islanders negotiated their place with the colonial powers, in what is termed "islander agency." The second category is composed of history that moves beyond a Western periodization, and looks to local, Pacific histories. This approach, however, tends to ignore military history as representative of a Western-oriented approach to the islands' histories, and instead focuses on culture, race, and the environment. Armitage and Bashford's book, for instance, contains chapters on these three themes, but none on conflict.

Moreover, the fragmentary nature of oral and documentary records has resulted in a tendency for precontact warfare to be presented in a "generalized" manner by historians and anthropologists. As Knuaft (1990) argues, warfare was a common occurrence throughout Melanesia, although early European accounts marginalize the ferocity and magnitude of conflicts in the region. The small Western presence in the Pacific during the first decades of colonialism—the result of disinterest from colonial metropoles rather than benevolence—meant that "traditional" warfare continued into the twentieth century. Indeed, Australian colonial officials entered the Papua New Guinean highlands region only during the 1930s, where they "discovered" hundreds of thousands of people. Oceania, therefore, offers a rich field for the anthropological and archaeological study of warfare; a significant proportion of Otto, Thrane, and Vankilde's edited collection on this topic (2006) draws its material from Oceania and Papua New Guinea (PNG) in particular.

Colonialism influenced indigenous ways of war by changing the composition, governance, and interrelations of subject societies. Chris Gosden (2006) argues that as the colonial government extended its monopoly of violence—often through the use of violence itself—the governed peoples of PNG to varying degrees changed their methods of exchange, social relations, rituals, and concepts of maleness. Reflecting the broad church of academic scholarship in Oceania, in Gosden's case the study of war and society has been used as "a crucial diagnostic aspect of social forms, their core values and history" (208).

Essentially, the First World War in the Pacific was an extension of the previous fifty years of the conflict between imperial powers in the region, albeit more overtly. By the end of the first year of the war, German colonies in New Guinea, Samoa, and the Marianas had been seized by Allied powers. Only a handful of land battles were fought, primarily in New Guinea. For some Pacific peoples, one colonial master was swapped for another, while for the majority, such as those in the New Guinean hinterland, the war meant little. Pacific peoples, marginalized in a region that was, at best, peripheral to the conflict, have attracted little scholarly interest. An exception is found in those histories that sought to reorientate away the Western periodization of the Pacific. Adrian Muckle's study of 1917 in New Caledonia (2012) is an example of the way in which recent historians have reimagined the period away from the Western Front to localize the history of conflict. Examining the last of Kanak resistance in the French colony, Muckle here explores not just how war "happened" to indigenous peoples but also how local conflicts, politics, and life interacted with the larger war, or existed outside of it.

The Second World War, by contrast, caused profound upheaval throughout the Pacific, but until recently, there was also little scholarship on the Pacific experience of war. While much of the region remained the possession of Western nations, official histories and other writing conceptualized the indigenous populations as passive participants in the war, often willingly aiding their colonial masters but otherwise tangential to the conflict. Australia historian Hank Nelson led the field in exploring the racially based relationships between Allied troops and Papua New Guineans, who formed by far the largest group of indigenous participants in Oceania, with over fifty thousand men serving as laborers and three thousand as soldiers. Nelson (1980) also shed light on the racialized language used in histories of indigenous troops. Writing on the Pacific Islands Regiment, Nelson wrote that "while the praise is fulsome . . . there was tendency to see Papua and New Guinean talents in the bush and unsteadiness under artillery fire as fixed, as though they are as inherent as the curly hair and black skin" (203). Even in the twenty-first century, words such as "native" abound in campaign histories, and the indigenous troops are described as part of the backdrop of jungle, beach, and ocean. An exception is Alan Powell's *The Third Force* (2003), which examines the military government of PNG during the war and the place of Papua New Guineans within it, placing the colonial nature of Australia's place in PNG at the forefront of the study of military history.

A handful of studies have been written on the military contributions of the two major military forces provided by islanders in the Pacific, from PNG and Fiji. However, these tend to be narrative and nationalist, and focus on the fact that these men did serve, rather than offering any historical analysis of how, why, and in what context they did so. Only recently have scholars attempted to examine indigenous service in an analytical rather than anecdotal manner, such as Noah Riseman (2012), who compares the experiences of Yonglu soldiers from Northern Australia, Papua New Guineans, and Navajo "code talkers" in order to explore the "hidden histories" of indigenous soldiers.

After the Second World War, scholars overwhelmingly saw conflict in the Pacific through the lens of decolonization and national development. While the bloody wars of independence of the scale seen in Asia or Africa did not occur in Oceania, the region has become a byword for national instability and widespread, low-level violence. From the wave of Pacific decolonization in the late 1960s, the region has seen coups in Fiji in 1987, 2001, and 2006, continued ethnic and community violence in PNG and the Solomon Islands, a small insurgency in New Caledonia in 1980, and a decade-long civil war on the Papua New Guinean island of Bougainville from 1988. Invariably, these conflicts are studied as contemporary problems to be solved. As a result the majority of scholarship can be categorized as contemporary political science rather than history. Indeed, older articles by political scientists have simply been taken as "history" by those examining the latest crisis, and in-depth archival and oral work remains to be attempted for much of the period. The strongest of these studies is Ron May's *The Changing Role of the Military in Papua New Guinea* (1993), which places the development of the PNG Defence Force (PNGDF) in its historical context while examining the interaction between the Australian-developed military with the nascent Papua New Guinean state.

The wealth of archival material on the colonial period and the relatively recent decolonization of Pacific nations leave a rich field of study of "new" military history in the Pacific. Tristan Moss' *Guarding the Periphery: The Australian Army in Papua New Guinea, 1951–75* (2017) explores the transformation of the Pacific Islands Regiment (PIR) from a colonial-style unit to the basis for the independent PNGDF. By integrating the study of the role and structure of the PIR with the experience of Papua New Guineans and the progression to independence, Moss bridges the traditional focus on outside influence in the Pacific with the local context. Such an approach also examines Australia as a Pacific nation, given the presence of Papua New Guineans in its army, and reveals the way in which the Pacific could reach back to touch the institutions of the colonial powers.

Like Pacific history more broadly, war and society approaches to the subject remain a niche area, with relatively few historians plying their trade among the islands and jungle. Moreover, the relatively recent ascent to nationhood among Pacific nations has meant that the region has only recently been seen in a historical light. Future work, therefore, will see historians shift their work away from the (admittedly pressing) needs of the present, to a close and dispassionate examination of the past. Moving beyond narrow definitions of what constitutes "war" to include ethnic and non-state conflict, as well as an effort not to see Pacific nations as "developing" and therefore unworthy of the often state-oriented discipline of military history, is key to this process.

Bibliography

Australia

Attwood, Bain, and Foster, S. G., eds. *Frontier Conflict: The Australian Experience*. Canberra: National Museum of Australia, 2003.
Australian Centenary History of Defence. 7 vols. Melbourne: Oxford University Press, 2001.
Beaumont, Joan. *Broken Nation: Australians in the Great War*. Crows Nest: Allen & Unwin, 2013.
———. *Gull Force, Survival and Leadership in Captivity 1941–1945*. Sydney: Allen & Unwin, 1988.
———. "The State of Australian History of War," *Australian Historical Studies*, 35, no. 21 (2003): 165–168.

Beaumont, Joan, et al., eds. *Beyond Surrender: Australian Prisoners of War in the Twentieth Century*. Melbourne: Melbourne University Press, 2015.

Bleszynski, Nick. *Shoot Straight, You Bastards!: The Truth Behind the Killing of Breaker Morant*. Sydney: Random House Australia, 2003.

Bou, Jean, ed. *The AIF in Battle: How the Australian Imperial Force fought 1914–1918*. Melbourne: Melbourne University Press, 2016.

———. *The Light Horse: A History of Australia's Mounted Army*. Melbourne: Cambridge University Press, 2009.

Broome, Richard. *Aboriginal Australians: Black Responses to White Dominance, 1788–1994*. St. Leonards: Allen and Unwin, 1994.

Brown, James. *Anzac's Long Shadow: The Cost of Our National Obsession*. Melbourne: Black, 2014.

Connor, John. *The Australian Frontier Wars 1788–1838*. Sydney: UNSW Press, 2002.

Dean, Peter J. *The Architect of Victory: The Military Career of Lieutenant General Sir Frank Horton Berryman*. Melbourne: Cambridge University Press, 2011.

———. *Australia 1942: In the Shadow of War*. Melbourne: Cambridge University Press, 2013.

———. *Australia 1943: The Liberation of New Guinea*. Melbourne: Cambridge University Press, 2014.

———. *Australia 1944–45: Victory in the Pacific*. Melbourne: Cambridge University Press, 2016.

Dennis, Peter, and Jeffrey Grey. *Oxford Companion to Australian Military History*. 2nd ed. Melbourne: Oxford University Press, 2008.

Edwards, Peter. *Australia and the Vietnam War*. Sydney: NewSouth, 2014.

Grey, Jeffery, ed. *The Centenary History of Australia and the Great War*. 5 vols. Melbourne: Oxford University Press, 2015–2016.

———. "Cuckoo in the Nest? Australian Military Historiography: The State of the Field," *History Compass* 6, no. 2 (2008a): 455–468.

———. *A Military History of Australia*. Melbourne: Cambridge University Press, 2008b.

Holbrook, Carolyn. *Anzac: The Unauthorized Biography*. Sydney: NewSouth, 2014.

Horner, David. *Blamey: The Commander-in-Chief*. Sydney: Allen & Unwin, 1998.

———. *High Command: Australia and Allied Strategy 1939–1945*. Sydney: Allen & Unwin, 1982.

Inglis, Ken. *Sacred Places: War Memorials in the Australian Landscape*. Melbourne: Melbourne University Press, 1998.

Lake, Marilyn, et al. *What's Wrong With Anzac? The Militarisation of Australian History*. Sydney: University of New South Wales Press, 2010.

Lukins, Tanja. *The Gates of Memory: Australia's People's Experiences and Memories of Loss and the Great War*. Fremantle: Curtin University Press, 2004.

McKernan, Michael. *Strength of a Nation Six Years of Australians Fighting for the Nation and Defending the Home-front in World War II*. Sydney: Allen & Unwin, 2008.

Nelson, Hank. *P.O.W. Prisoners of War: Australians Under Nippon*. Sydney: ABC Books, 1985.

Pederson, Peter. *Monash as a Military Commander*. Melbourne: Melbourne University Press, 1985.

Reed, Liz. *Bigger Than Gallipoli: War, History and Memory in Australia*. Perth: University of Western Australia Press, 2004.

Reynolds, Henry. *Frontier: Aborigines, Settlers and Land*. St Leonards: Allen and Unwin, 1987.

Scates, Bruce. *Anzac Journeys: Returning to the Battlefields of World War II*. Melbourne: Cambridge University Press, 2013.

———. *Return to Gallipoli: Walking the Battlefields of the Great War*. Melbourne: Cambridge University Press, 2006.

Seal, Graham. *Inventing Anzac: The Digger and National Mythology*. St. Lucia: University of Queensland Press, 2004.

Serle, Geoffrey. *John Monash: A Biography*. Melbourne: Melbourne University Press, 1982.

Stanley, Peter. *Invading Australia: Japan and the battle for Australia, 1942*. Camberwell: Viking, 2008.

———. *Tarakan: An Australian Tragedy*. St. Leonard's, Allen & Unwin, 1997.

Stanley, Peter, and Mark Johnston. *Alamein: The Australian Story*. Melbourne: Oxford University Press, 2002.

Stockings, Craig, and John Connor, eds. *Before the Anzac Dawn: A Military History of Australia Before 1915*. Sydney: New South, 2013.

Wilcox, Craig. *Australia's Boer War: The War in South Africa, 1899–1902*. Melbourne: Oxford University Press, 2002.

Windschuttle, Keith. *The Fabrication of Aboriginal History: Volume 1, Van Diemen's Land 1803–1847*. Sydney: McLeay, 2002.

Ziino, Bart. *A Distant Grief: Australians, War Graves and the Great War*. Perth: University of Western Australia Press, 2007.

New Zealand

Ballara, Angela. *Taua: "Musket Wars," "Land Wars" Or Tikanga?: Warfare in Māori Society in the Early Nineteenth Century*. Auckland: Penguin Books, 2003.

Belich, James. *The New Zealand Wars and the Victorian Interpretation of Racial Conflict*. Auckland: Auckland University Press, 1987.

Cowan, James. *New Zealand Wars: A History of the Māori Campaigns and the Pioneering Period*. 2 vols. Wellington: WAG Skinner, Government Printer, 1922–1923.

Crawford, John. *In the Field of Peace: New Zealand's Contribution to International Peace-Support Operations: 1950–1995*. Wellington: NZDF, 1996.

Crawford, John, and Ian McGibbon. *New Zealand's Great War: New Zealand, the Allies and the First World War*. Auckland: Exisle, 2007.

McGibbon, Ian, C. *New Zealand and the Korean War*. 2 vols. Auckland: Oxford University Press, 1991, 1996.

———. *New Zealand's Vietnam War: A History of Combat, Commitment and Controversy*. Exisle, Auckland, 2010.

———. *"Something of Them Is Here Recorded": Official History in New Zealand*. Westport, CT: Praeger, 2003, New Zealand Electronic Text Centre.

McGibbon, Ian C., and Paul Goldstone, eds. *The Oxford Companion to New Zealand Military History*. Auckland: Oxford University Press, 2000.

Montgomerie, Deborah. "Reconnaissance: Twentieth-Century New Zealand War History at Century's Turn," *New Zealand Journal of History* 37, no. 1 (2003): 62–79.

New Zealand in the First World War 1914–1918. 4 vols. Wellington: Whitcombe and Tombs, 1921–1923.

Official History of New Zealand in the Second World War 1939–45. 48 vols. Wellington: War History Branch, Department of Internal Affairs, 1949–1986.

Phillips, Jock. *The Pakeha Male: A History*. Auckland: Penguin, 1989.

Pugsley, Christopher. *From Emergency to Confrontation: The New Zealand Armed Forces in Malaya and Borneo 1949–66*. South Melbourne: Oxford University Press, 2003.

Rabel, Robert. *New Zealand and the Vietnam War: Politics and Diplomacy*. Auckland: Auckland University Press, 2005.

———. "War History as Public History: Past and Future." In *Going Public: The Changing Face of New Zealand History*. Eds. Bronwyn Dalley and Jock Phillips. Auckland: Auckland University Press, 2001.

Templeton, Malcolm. *Ties of Blood and Empire: New Zealand's Involvement in Middle East Defence and the Suez Crisis*, 1994.

Walker, Franchesca. "'Descendants of a Warrior Race': The Māori Contingent, New Zealand Pioneer Battalion, and Martial Race Myth, 1914–19," *War & Society* 31 (March 2012): 1–21.

Pacific and Oceania

Armitage, David, and Alison Bashford, eds. *Pacific Histories: Ocean, Land, People*. London: Palgrave Macmillan, 2014.

Gosden, Chris. "Warfare and Colonialism in the Bismarck Archipelago Papua New Guinea." In *Warfare and Society: Archaeological and Social Anthropological Perspectives*. Eds. Tom Otto, Henrik Thrane, and Helle Vankilde. Langelangade: Aarhus University Press, 2006.

Knauft, Bruce M. "Melanesian Warfare: A Theoretical Theory," *Oceania* 60, no. 4 (June 1990): 250–311.

May, Ronald. *The Changing Role of the Military in Papua New Guinea*. Canberra: Strategic and Defence Studies Centre, 1993

Moss, Tristan. *Guarding the Periphery: The Australian Army in Papua New Guinea, 1951–75*. Cambridge: Cambridge University Press, 2017.

Muckle, Adrian. *Specters of Violence in a Colonial Context*. Honolulu: University of Hawai'i Press, 2012.

Nelson, Hank. "Hold the Good Name of the Soldier: Discipline of Papuan and New Guinea Infantry Battalions, 1940–1946," *The Journal of Pacific History* 15, no. 4 (1980): 202–216.

Powell, Alan. *The Third Force: ANGAU's New Guinea War, 1942–46*. South Melbourne: Oxford University Press, 2003

Riseman, Noah. *Defending Whose Country?: Indigenous Soldiers in the Pacific War*. Lincoln: University of Nebraska Press, 2012.

Ryan, Lyndall. *The Aboriginal Tasmanians*. Reprint. Crow's Nest, New South Wales: Allen and Unwin, 1997.

4 War and Society in South Asia

Kaushik Roy

This essay concentrates on the linkages between war and society with a strong historiographic bent. The bulk of this essay will deal with India for two reasons. Firstly, there was no Pakistan before 1947 and no Bangladesh before 1971. Nepal, Bhutan, and Sikkim play marginal roles in the evolution of warfare and society in the subcontinent. Secondly, the sheer size and demographic resources of India overwhelm the other countries of South Asia put together. Within the rubric of the war and society approach, this essay will also include technological developments and evolution of polities until the present era. The focus will mostly be on land warfare, as naval warfare played a marginal role in South Asia's history and the air force came into play only in the 1930s. The central issue of Indian military history is repeated conquests of the subcontinent, despite its huge demographic and economic resources, by a small number of foreigners until the early modern era. Chariot-borne nomads who came through the north-west frontier passes were the first wave; mounted nomads along the same route composed the second wave; the third and last wave was the coming of the British from the sea. The theme of British conquest is related with the nature of professionalism of the armed forces in postcolonial South Asia.

Ancient India produced eulogies of kings. In medieval India, Persian chroniclers chronicled the political, diplomatic, and military activities of the rulers in an attempt to glorify their reigns. Modern military history writing on India started in the second half of the nineteenth century by the British officers. They composed the imperialist school. The first half of the twentieth century witnessed the rise of the nationalist school in opposition to the imperialist school. The proponents of the nationalist school are all Indian scholars. The 1960s witnessed the rise of Marxist historical scholarship and the simultaneous fruition of the organizational school. The next decade saw the flowering of social historians and the prominence of the technology determinist school. The 1980s saw the emergence of the subaltern school as a variant of Marxist historiography. The next decade saw the coming of the maritime school of historians. As regards the writings on the post-colonial military, the field is dominated by the Indian military officers and political scientists. And from the first decade of the new millennium, history has taken a culturalist turn and a focus on environmental perspectives.

The imperialist school was concerned with the rise of British power in India. The proponents of this school argue that before the advent of the British, India lacked standing armies and bureaucratic state structures. Pre-British India had weak polities which maintained rabbles that were incapable of waging decisive battles. Pre-British warfare in India, according to this interpretation, was unorganized melees. Further, Indian warfare was always in a state of stasis. The principal advocates of this interpretation were Colonel G. B. Malleson (1883) and William Irvine (1896), who asserted that the Indians were racially incapable of constructing professional armies capable of conducting sustained sieges and decisive battles. Malleson in his *Decisive Battles of India* focuses on thirteen great battles which enabled the British to conquer India. Malleson's framework is shaped by Edward Creasy. Creasy in his *Fifteen*

Decisive Battles of the World (1851) portrays how the West (especially the British) established supremacy over the rest. Malleson writes that the British won all these battles against the Indians due to their moral superiority. Irvine moves away from battle-centric study and focuses on the armies of the later Mughals and the Mughal successor states of the early eighteenth century. Irvine concludes that the Indians failed to build up a professional army due to their racial inferiority.

Imperialist scholars began a debate about who were the original inhabitants of India. Lieutenant General G. F. MacMunn (1911) introduced the Aryan invasion theory to explain the rise of British power in India. Employing the works of William Jones (1788) and others, MacMunn argues that the invasion of the Aryans resulted in the destruction of the Bronze Age Indus Valley Civilization around 1200 BCE. While the north and north-west parts of the subcontinent were inhabited by the Aryan invaders, the Dravidians were pushed into south and east India and the latter gradually became non-martial. MacMunn continues that the British maintained their power within the subcontinent by providing their own officers for the Indian Army and recruiting soldiers only from the martial Aryan communities.

Currently, most Marxist Indian scholars like D.N. Jha (1977) and Romila Thapar (2002) claim that there was no Aryan invasion. Rather, they note, the Aryans were not a race or an ethnic group but a language group which moved peacefully and gradually into the subcontinent between 1500 BCE and circa 800 BCE. Whether a language group can dominate another language group peacefully is a problematic question.

Thomas R. Trautmann (2005), following techniques of comparative linguistics, suggests that the Indus Valley people were proto-Indo-Europeans, who were followed by Indo-European Aryans. Of the Indo-Europeans who moved out from south Russia, some branched off in Iran, known as Indo-Iranians. Of these, another group migrated to Afghanistan, known as the Indo-Aryans, who then entered the subcontinent and introduced horses, chariots, and iron weapons. The use of iron weapons (swords, spears, and iron-tipped arrows) and horse-drawn chariots gave the Aryans military superiority over the Indus people, who fought on foot. Bronze weapons of the Indus civilization proved ineffective against the iron weapons of the Aryans. Bronze was softer and more costly than iron, which is available in greater quantities. Moreover, wheeled chariots provided the Indo-Aryans mobility against the Indus people. Thus, the destruction of the Bronze Indus Valley civilization by the Indo-Aryans fits more or less within Robert Drews' (1993) theory that the Indo-Europeans, bursting out from southern Russia, destroyed all the Bronze Age river valley–based civilizations of Eurasia. However, in Drews' framework the invaders equipped with iron swords fought as infantry skirmishers against the chariot elite of the bronze civilizations of Eurasia. In the case of South Asia, the Indo-European invaders with iron swords and bows fought from the chariots against the Chalcolithic Indus Valley infantry.

The nationalist school accepts the argument of an Aryan invasion of India. But its advocates, such as P. C. Chakravarti (1941) and V.R.R. Dikshitar (1944), use Kautilya's *Arthasastra* (composed around 300 BCE) to assert that disciplined standing armies and decisive battles were present in India long before the arrival of the British. They explicitly reject imperialist interpretations, and instead promote pride in India's ancient military heritage. They note the Indo-Aryan settlement of Punjab around 1200 BCE, followed by their migration onto the Ganga-Jamuna doab about 800 BCE. The Indo-Aryans at that time carried out raids for collecting booty, with cows regarded as a particularly valuable form of plunder. The power of a tribal chief (*rajan*) depended on the success of the plundering raids. Gradually the pastoral society of the Indo-Aryans was transformed into an agricultural society. The two epics *Ramayana* and *Mahabharata*, composed circa 400 BCE, depict the gradual Aryan intrusion against the Dravidians and a civil war amongst the Aryans in north India. Like the Nationalist historians, advocates of the organizational school (this group focuses on the internal organizational structure of the armed forces), like U. P. Thapliyal (2010), note the soldiers of ancient India were organized in *vyuhas* (battle formations) and they fought decisive battles from 1000 BCE onwards. Thus, unlike V. D. Hanson's (1994) argument, the Classical Greeks were not the only ones who discovered the concept of decisive infantry battles.

By 500 BCE the spread of agriculture had transformed the pastoral society in the Gangetic plain and begun the emergence of the caste system. For social determinists like S. P. Rosen (1996) following a *longue durée* approach, the caste system shaped the structure of the Indian armies throughout its history. The Indo-Aryans with their iron ploughs were able to clear the forest of the Ganga-Jamuna doab and the fertile soil yielded bumper crops. The peasant society in turn generated stratification. At the top of the fourfold hierarchy were the Brahmins, who carried out the Hindu ritual activities. Then came the Kshatriyas, who provided the warriors and the ruling class. Thirdly came the Vaishyas, who engaged in trade and commerce, and finally the Shudras, who carried out agricultural and menial tasks. Indian armies throughout history mirrored the social divisions inherent in the caste system. These divisions, concludes Rosen, debilitated Indian armies, and prevented the emergence of an effective combat force, which in turn resulted in repeated foreign invasions of India. But Rosen's reductionist model cannot explain why the armies of the Muslim polities—which obviously did not employ the *varna* system—collapsed against the British in the eighteenth century.

Gradually villages developed, and about 400 BCE they coalesced into big settlements, with sixteen large states (*mahajanapadas*) emerging in north India. With their rise, plundering raids were replaced by systematic campaigning for territorial annexations. Among the *mahajanapadas*, Magadha (south Bihar) emerged supreme and established the first empire in India, known as the Nanda Empire (344–322 BCE).

One of the characteristics of the ancient Indian armies was the large number of war elephants maintained by them. The size of the Nanda Army's elephant corps (2,000) spread terror amongst Alexander's army, so much so that the Greeks soldiers refused to advance east of the river Beas. Alexander confined his activities to Punjab, which was outside the Nanda Empire. Alexander defeated a small king of Punjab named Paurava (Porus) in the Battle of Jhelum/Hydaspes (May 326 BCE). Porus' *chaturanga bala* (fourfold army), comprising infantry, cavalry, chariots, and elephants, was defeated by the phalangites-cavalry combination of Alexander. Though Porus was able to deploy only eighty-five elephants, Alexander's generals and the Greek writers chronicling the invasion of Alexander were impressed. The elephants caused most of the Greek casualties, which exceeded the number of casualties suffered by Alexander's force at Gaugemela. J. W. M'Crindle (1896), the imperialist historian, emphasizes the stasis of an ancient Indian military system vis-à-vis Alexander's dynamic and efficient military machine. In contrast, Indian historians belonging to both the nationalist (e.g., Chakravarti) and organizational camps (e.g., B. K. Majumdar, 1960, G. S. Sandhu, 2000, Thapliyal) accept that change was inherent in the Indian military tradition. They note that due to Alexander's invasion, war chariots vanished from the Indian order of battle and the importance of elephants increased in the Indian armies. Trautmann (2015) in his *longue durée* environmental history of elephants harps on the importance of these beasts for Indian royalty till the arrival of the British.

Post-Alexander India was dominated by the Maurya Empire (322–185 BCE), which was succeeded by the Sungas and the Kanvas. The latter was defeated by the Greek rulers of Bactria, who had declared independence from the Seleucid Empire. Back in India, the big empire to emerge after the Kanvas was the Gupta (240–543 CE). The Greeks of Bactria and the Guptas were extirpated by a wave of Central Asian nomadic invaders, which marked the beginning of a new epoch in Eurasian history. While the nationalist and organizational historians, such as Jadunath Sarkar (1960), S.A. Bhakari (1981), and J.N. Sarkar (1984), argue that these empires were centralized bureaucratic entities which maintained large standing armies, practitioners of social history, like Burton Stein (1998), assert that all Indian empires before British arrival were loose realms.

The Hindu kingdoms of north India displayed a learning curve. The Guptas realized that to counter the Huns they would require horse archers. From the numismatic evidence and Kalidasa's *Raghuvamsam* (composed around 600 CE), it is clear that the Gupta emperors adopted mounted archery within their force. The Gupta archers, equipped with composite bows and mounted on horses clad in armour, were able to stop the Huns for the time being. However, mounted archery never took firm root in the Indian soil. Sandhu (2000) and Bhakari (1981) emphasize internal organizational factors for the decline of mounted archery in the later Gupta and post-Gupta Indian armies of the Rajputs.

The rise of the Rajputs constituted a landmark in Indian military history. Imperial historians identified the Rajputs as one of the Aryan martial races. More recently, Brajadulal Chattopadhyay (1994) has offered a social perspective about their rise to power. He writes that from around 700 CE large Kshatriya landowners known as *thakurs* dominated the political and military landscape of north India. Eventually known as Rajputs, some had descended from the pastoral tribes and Central Asian invaders who had merged with the settled society of north India. Jadunath Sarkar (1960) and Bhakari (1981) emphasize the atavistic feudal system and clan structure for the shortcomings of the Rajput way of warfare. Regarding war as a form of sport, the numerous small Rajput principalities engaged in internecine struggle with each other. Aristocratic Rajputs fought on horseback with *khandas* (straight swords) and looked down upon infantry. Clan chiefs and their kinsmen composed the core of their forces. Their chivalric ethos also precluded nocturnal attacks or tactical retreats. Rather, Rajputs favoured headlong charges during the battle. Their aim was to destroy the enemy or to be killed in combat and achieve *viragati* (ascent to heaven after death). Such warfare was predominant when Central Asian nomadic Turks who had adopted Islam burst upon the Indian landscape.

Between the tenth and the thirteenth centuries, the Central Asian Islamic Turks pushed back the Hindu frontier from eastern Afghanistan into the interior of India, culminating in Mahmud of Ghori's lieutenant Qutub ud din Aibak establishing the Delhi Sultanate (1206–1526). For technological determinists, such as Simon Digby (1971) and Ali Athar (2006), repeated victories of the small number of Turks over their Rajput opponents stemmed from better horses, bows, and stirrups. The Central Asian mounts and composite bows were better than the Rajputs' Indian-bred horses and simple bamboo bows. Finally, the Turks' iron stirrups enabled them to stand and shoot while their mounts were moving, which the Rajputs could not do with their wooden stirrups. In contrast to these interpretations, social historian Andre Wink (1997) takes a global perspective, writing that the ideology of jihad and the Islamic concept of *qaum*—with its sense of equality among all the members—acted as a force multiplier against the caste-ridden hierarchical *varna* society of the Hindus. Long before Wink, in the 1930s, M. Habib (2001) had noted that Islam's slogan of social equality toppled the top-heavy Rajput polities, which had exploited the lower castes.

The Delhi Sultanate checked the Mongol onslaught on the subcontinent. The Turkish sultans were able to maintain large number of *mamluks* (elite slaves mounted on the horses and equipped with composite bows) in their payroll thanks to the *iqta* system, which assigned land revenue in return for maintaining cavalry contingents. Peter Jackson's sociopolitical analysis (1999) shows the linkages between the *mamluks* and the *iqta* system, which had spread from Egypt in the west to India in the east. To enable Turkish warlords to establish roots in the agrarian society of Hind, the sultans granted them *iqtas*, producing a fusion of pastoral nomadic mobile society with settled sedentary society.

The weakness of the nationalist school, composed mostly of Hindu writers, is its discomfort with the Muslim medieval "past" in India. In this perspective, the Muslim invasions resulted in a sort of "Dark Age" in India. The rise of materialist historiography from the 1960s provided an alternative in the work of Marxist and primarily Indian Muslim historians, such as Irfan Habib (1963), S. Nurul Hasan (2005), and M. Athar Ali (2006). Known as the Aligarh school, they focus on the issue of financing the Mughal Army, and regard Akbar (r. 1556–1605) as the true founder of the Mughal Empire (1526–1857) in India. Emphasizing *Akbar Nama* and *Ain-i-akbari*, they argue that Akbar elaborated and systematized the *iqta* system as the *jagirdari* system. Unlike the *iqtas* and the medieval European fiefs, the *jagirs* were not hereditary. The Mughal nobles were granted *jagirs*—that is, assignment of land revenue for a temporary period—in return for the obligation of maintaining cavalry troopers. Moreover, all the *jagirdars* (holders of *jagirs*) were included in the hierarchical Mughal official ranking system, known as the *mansabdari* system. Akbar broadened the foundation of Mughal rule by granting *mansab* (office) ranks to the defeated Rajput chiefs who became vassals, and also to the Hindu *zamindars* (large landowners). The Mughals under Aurangzeb (d. 1707) siphoned off 50 percent of the gross agricultural produce, 80 percent of which was used for maintaining the army.

The Mughal Army of about one million soldiers was drawn from a population of 150 million. For the Aligarh historians, the very fact that the Mughals appropriated half of the agricultural produce from their pan-Indian empire, as well as the large size of their army, proves that the Mughal Empire was a centralized state. They contend that the *mansabdari* system gradually gave rise to feudal tendencies and in the long run resulted in the decline of the empire. Overexploitation of the peasantry destabilized the *mansabdari* system when the gap between estimated revenue and actual revenue collected grew, producing the early-eighteenth-century peasant rebellions that caused the ruin of the great Mughals.

Jadunath Sarkar (1930) differs from the Aligarh historians and argues that Aurangzeb's policy resulted in a Hindu reaction. Aurangzeb's Islamic policies, emphasizes Sarkar, alienated the Rajput *mansabdars* and encouraged the Marathas to revolt under their Hindu king Shivaji. The Maratha-Rajput axis led to the crisis of the Mughal Empire. Non-Marxist materialist historians M. Alam (1986) and John F. Richards (1993) challenge the concept of Mughal overexploitation of the peasantry. Rather, they claim peasants were underexploited. Moreover, the middle peasantry in particular made rapid economic gains that encouraged and enabled them to challenge the Mughal Empire. In fact, Richards asserts that the empire was not facing a "cash crisis" at the beginning of the eighteenth century, but rather external invasion under the Persian ruler Nadir Shah.

Social historians like Jos Gommans (2002), Dirk Kolff (Gommans and Kolff, 2001), and Douglas Streusand (2011) attempt to fuse their social perspective with ecological underpinnings by arguing that India's unique ecosystem made the establishment of centralized political entities in the pre-modern era impossible. They regard the terrain as having produced an amalgam of sedentary and pastoral, nomadic societies interspersed with each other. Open frontiers among them, along with the presence of numerous armed peasants, made demilitarization of the countryside (as in early modern Europe) impossible. Gommans notes that state power in pre-modern India was weaker than the authority exerted by pre-modern Chinese polities. Attempting a comparative analysis of the Muslim empires of Asia, Streusand writes that while a poor agrarian economy prevented the Safavids from building up a centralized bureaucratic entity, economic resources in the hands of Indian peasantry thwarted Mughal centralization policy. These historians conclude that practically every peasant in India had access to swords and horses, which in turn severely limited the power of central government. In their view the Mughal Empire consisted of a complex mosaic of semi-autonomous polities under the loose control of the emperor at Delhi.

The prominence of gunpowder weapons in Nadir's victory at the Battle of Karnal (24 February 1739) provides the entrée for technology-oriented historians. The crucial development in the field of military technology during the sixteenth century was the use of guns in Indian field and siege warfare. Unlike the social historians, like Gommans, Kolff, and Streusand, who assert that Mughal warfare was based on theatrics and political co-option, the technological determinists, like Zaman (1983) and I. A. Khan (2004), as well as organizational historians, like Roy (2015), assert that the Mughal Army was capable of conducting decisive gunpowder battles. Zaman, Khan, and Roy argue that the tactical combination of field guns and mounted archers enabled Zahir-ud-din Babur (r. 1526–1530) to win the First Battle of Panipat (1526) with only twelve thousand troops against the Delhi Sultanate's army of about sixty thousand, led by Ibrahim Lodhi. With its monopoly of guns, the Mughal government thereafter established its control over the countryside. Khan notes that gunpowder was introduced in north India by the Mongols and after the seventeenth century the Indians failed to innovate in casting guns. In contrast R. Eaton and Phillip B. Wagoner (2014) assert that gunpowder was first introduced in Deccan by the Ottomans and then the Portuguese. The Deccani Sultanates during the sixteenth century, building upon this technology, manufactured guns that were better than the Portuguese and European ones. Scholarly disagreement regarding the issue of gunpowder in late medieval India will likely continue for the foreseeable future.

The repeated victories enjoyed by the British East India Company (EIC) in India during the eighteenth and nineteenth centuries encourage historians to use the concepts of military revolution and the Western way of war. Several social historians, like Douglas M. Peers (1995) and R.G.S. Cooper (2005), argue that the British introduced the Western way of warfare, which had emerged in early modern Europe as part of the so-called military revolution (Parker, 1988), which featured drilled and disciplined infantry supported by field artillery. These scholars also utilize John Brewer's concept of the fiscal-military state, and accept Stein's contention that all pre-British Indian states were characterized by divisible sovereignty. In their perspective, the forces deployed by, and the underlying financial and logistical strength of, the EIC proved supreme against the feudal cavalry forces of the Indian princes ruling over fluid realms.

Arguing from the organizational perspective, Kaushik Roy (2011) contests that the British won India simply by dint of their Westernized army. Rather, realizing that they had to adopt certain indigenous elements and adapt to the Indian conditions, the EIC's success stemmed from the creation of a hybrid military establishment. True, the British initiated a sort of revolution by abolishing the traditional Mughal *jagirs* and replacing them with cash payments and the regimental system of recruitment, which remains the norm today. Lacking enough British soldiers, the EIC enlisted Indians (called *sepoys*) for infantry units, drilling and equipping them in the Western fashion. Recognizing that regional operations also required large numbers of light cavalry, company officials also recruited Indian horsemen, called *sowars*.

These policies benefited from the absence of any Indian nationalism at that time. Rather, South Asian society was divided along regional, caste, class, and clan lines, and military service with a foreign power was nothing unnatural. EIC's army thus became the site for collaboration between certain indigenous communities and the British, and hence the lynchpin of Britain's colonial regime. Indians joined the British officered force because military service for them offered a sort of upward mobility. The British fused indigenous cultural practices with Westernized discipline and created martial identities like Gurkhas and Purbiyas for indigenous soldiers. The British officers also created a sort of hybrid institution, mixing Indian and Western elements, in order to establish loyalty with the *sepoys* and *sowars*. For instance, reflecting Western notions of leadership, British officers led from the front and displayed bravery to establish what John Keegan (1987) describes as the "heroic" mask of command. But they also paid respect to Indian culture by allowing the soldiers to follow their dietary customs, and often participated in Hindu religious festivals organized by their troops. The EIC also fused caste and clan ties along with religious ethos when constructing the Indian regiments. Especially in the field of logistics, the British indigenized their military establishment by using Indian porters, *banjaras* (mobile traders), and animal resources. The British thus initiated a military synthesis rather than a military revolution in India.

The EIC had to defeat three big indigenous powers: Mysore, Maratha Confederacy, and the *Khalsa* Kingdom of Punjab. Simultaneously, these indigenous powers were not passive but also Westernized their military establishments and centralized their state structures. Roy differs from the nationalist historian S. N. Sen's contention (1928) that the Marathas made a mistake in modernizing or Westernizing their army. They would have won, continues Sen, had they remained wedded to their traditional mode of guerrilla warfare. Sen is wrong because the Maratha *ganimi kava* (guerrilla warfare conducted by cavalry) failed during the First Anglo-Maratha War (1774–1783). In reaction, between 1780 and 1849, Mysore, Marathas, and the Sikhs utilized European mercenaries for raising Western-style drilled and disciplined infantry equipped with muskets and supported by field artillery. Sen notes that the Maratha Army's recruitment of non-Maratha soldiers from north India during the Second Anglo-Maratha War (1803–1805) weakened the cohesion of their force. Yet the EIC itself depended on soldiers drawn from north and south India. Nationalism or sub-nationalism was not a strong issue then. Moreover, French mercenaries helped make the Maratha field artillery superior compared to that of the EIC.

However, the weakness of the indigenous powers' modernized military establishments was that they failed to build a hierarchical professional officer corps. Unlike Petrine Russia, the Indian rulers never ordered their feudal aristocracy to train their sons in European institutions to create a Western-style officer corps (Hellie, 2005; Davies, 2007). Here, the EIC had a distinct advantage over the indigenous powers. Finally the Indian rulers did not disband their traditional military force structure. Whereas they and the EIC both used modernized as well as traditional military forces, the indigenous powers failed to maintain a balance between Indian and Western elements of warfare in creating a successful hybrid military establishment, unlike the British.

Historians who could be categorized as the maritime school have shifted the attention from armies to ships. They argue that the absence of a seagoing navy was one of the primary reasons for the collapse of the Indian powers. Hermann Kulke et al. (2010), for example, concentrate on the political economy of the Chola Empire, which formed in the tenth century. The Cholas developed a "blue water" navy and conquered Lakshadeep, Andaman, and the Nicobar Islands. In the eleventh century, they launched a maritime expedition against the Kingdom of Sri Vijaya in South-East Asia to control maritime trade in that region. The Cholas also conquered the northern part of Ceylon, today Sri Lanka. The Cholas' blue water navy was an exception in the Indian case. Most of the subcontinent's powers, facing real or potential enemies on their land frontiers, followed land strategies. The agricultural wealth of India's interior also weakened the South Asians' urge to indulge in extensive overseas trade and commerce, thus inhibiting the development of a dynamic mercantile community and pressure on indigenous governments to maintain large seagoing fleets. The Cholas themselves had to protect their territorial frontiers, and maintaining their fleet made them somewhat weaker on land. The Cholas suffered from imperial overstretch when they had to engage in war against the Chalukyas as well as against Orissa and Bengal, facilitating their decline by the twelfth century.

The Chola case had certain parallels with the Byzantine and the Chinese empires. The Byzantine Empire failed to maintain both a navy to control the Mediterranean and check the Venetians, and a land force strong enough to cope with myriad enemies, like the Avars, Bulgars, and the Persians (Decker, 2013). Similarly, given the threat posed by the mounted nomads to their northern frontier (especially along the Ordos loop), Chinese empires lacked the luxury of maintaining an ocean-going fleet. Zheng He's force in the fifteenth century was the exception, but thereafter the Mings disbanded it due to the rise of the Manchu threat along their north-eastern border (Levathes, 1994). Afterwards, the Manchus who succeeded them maintained only a "brown water" navy.

After the Cholas, no Indian power maintained a blue water navy. The Mughals had a riverine navy capable of conducting amphibious expeditions along the rivers and the coastline, and the south Indian rulers (Kurup, 1997) maintained a coastal navy for commerce raiding. Similarly the Marathas maintained a brown water navy, which political factionalism made ineffective. Only Tipu Sultan of Mysore (r. 1782–1799) had a grand plan of constructing a seagoing navy, but the Fourth Anglo-Mysore War (1799) prevented its completion.

Philip Macdougall (2014) argues that during its wars with indigenous powers, the EIC's supremacy in the sea enabled it to conduct land warfare successfully. Merchant vessels not only maintained the sea lines of communication with Britain but also carried on the company's lucrative maritime commerce, the profit from which enabled successful war making. Further, ships also enabled the EIC to communicate between its enclaves in Bombay, Madras, and Calcutta, and to concentrate troops as circumstances dictated.

The longevity of the Raj's rule stemmed partly from its Indian Army. Subaltern historiography, as represented by R. Mukherjee (1993) and T. Roy (1994), focuses on the "nameless and faceless" common men, looking at history from below to emphasize that soldiers were peasants in uniform, and to highlight conflict between the Indian soldiers and their British masters. Conversely social historians, like David Omissi (2007) and Tan Tai Yong (2005), following C. Bayly's collaboration thesis, note that the Raj separated its indigenous military collaborators from Indian society by using the martial race theory. The argument claims that the British deliberately divided Indian populations into martial and non-martial races. Recruiting the former, like Gurkhas, Punjabi Muslims, and Sikhs, they established a special collaborative relationship with these communities by offering them extra pay and cultural privileges. Only under the stress and strain of global warfare (1939–1945) did this symbiotic relationship disintegrate, paving the way for the dissolution of the British Empire in India.

Britain's rule in India remains a watermark in South Asian military history, as the regimental system continues to structure the independent armies of India and Pakistan. This remains true in spite of the influence of American military theories (e.g., air-land warfare, network-centric warfare) and the talk of revolution in military affairs in the present-day Indian Army. Despite some thinking about creating an Air Landing Division, Mechanized Infantry Division, and so forth, the British regimental system still reigns supreme. One chief of army staff of India in recent times concluded, "Our army inherited the traditions, culture and customs from the British Indian Army and therefore we share many values" (Singh, 2012: 247). But to what extent they shaped policies in the postcolonial period provides the crux of the change-continuity debate.

In 1947, the British Indian Army divided into the Indian and Pakistan armies. Numerous writers, including political scientists Stephen P. Cohen (1971, 1984), Sumit Ganguly (2002), and Steven I. Wilkinson (2015), organizational historians Pradeep P. Barua (2005) and Colonel Brian Cloughley (1999), and journalist Shuja Nawaz (2008), among others, emphasize continuity between the colonial era and the post-colonial period. They argue that both India's and Pakistan's armies maintained the regimental structure and recruitment doctrines of the British Indian Army, continuing to rely upon long-service volunteers. Both forces continue to recruit from the "martial races," like Punjabi Muslims and Pathans (for Pakistan), and Sikhs and Gurkhas (for India). Advocates for "continuity" also note that the Indian and Pakistani officer corps embody notions of secular professionalism, and the tactical doctrine, introduced by the British. They argue that both the Indian and Pakistan armies, in accordance with British training, can launch set-piece attacks and conduct linear attritional battles, but not engage in fast-paced manoeuvre warfare. Like that of the British, both armies' command structures emphasize top-down centralization, not the mission-oriented system required for overcoming the chaos inherent in manoeuvre warfare.

A weakness among this scholarship is that British perspectives emphasize civilian dominance of the military. However, these authors cannot reconcile this concept with the Pakistan Army's penchant for intrusion in the political sphere, along with the influence of Islamic ethos among its officer corps. Moreover, these authors fail to acknowledge that under the Raj, the top British commanders had control over the grand strategy and financial aspects of the regime.

In contrast to those who emphasize continuity, military officers writing about India's combat history after 1947, like the regimental historians, portray the glorious saga of Indian armed forces and emphasize differences with the colonial regime. Their narratives of operations during the three wars with Pakistan (1947–1948, 1965, 1971) and the border conflict with China (1962) stress, at times, sacrifice and suffering. Three representatives of this school are Lieutenant General S. L. Menezes (1993), Srinath Raghvan (2010), and Air Vice Marshal Arjun Subramaniam (2016), who argue that the Indian officer corps' professionalism is indigenous and self-generated due to the structure of democratic India. They note that India's first prime minister, Jawaharlal Nehru, undermined the power of the military officers with the aid of civilian bureaucrats. The presence of a strong urban middle class and a free media further facilitated civilian dominance over India's military. Advocates of discontinuity also consider the Pakistan Army the principal culprit for starting the conflicts in South Asia, claiming that it is not professional, given its frequent interventions in the civilian sphere. The Pakistani generals' repeated political manipulations in turn reflect the clan ethos of the Punjabi Muslims and Pathans who dominate the army, as well as the absence of a strong media and university-educated, urban middle class in the country. Recent work by the political scientist C. Christine Fair (2014) reiterates the view of the Pakistan Army as a predatory organization and product of rabid Islamism.

As of the time of this writing Pakistani tanks and gunships are fighting insurgents in Waziristan, and the *Tehrek-i-Taliban* in Punjab. Simultaneously, the Indian Army is countering insurgencies in north-east India and Kashmir. Notably, the Indian Army has now suffered more casualties in unconventional warfare than in the four conventional wars it has fought to this date. While Rajagopalan (2008) emphasizes continuity between the British and Indian counter-insurgency techniques, Gates and Roy (2014) harp on the innovations in the Indian doctrine with the passage of time.

However, scholarly attention generally neglects unconventional warfare to focus on South Asia's nuclear politics. One fifth of the world's population lives in the region, and both India and Pakistan are nuclear powers. While realists claim that nuclear weapons are necessary for security, Gandhian Liberals, such as Ramana and Rammanohar Reddy (2003), assert that security could best be achieved by both India and Pakistan renouncing their nuclear weapons. Neorealists, like Bharat Karnad (2002), argue that Pakistan has rejected a no–first use policy for these arms, and in the 1999 Kargil Crisis resorted to nuclear blackmail against India. They fear possible future scenarios in which Pakistani generals might launch a nuclear strike against India, or worse, that nuclear weapons might fall into the hands of the jihadis. Thus they argue that India must maintain a triad of nuclear weapons (deliverable by air, land, and sea) to achieve nuclear superiority over Pakistan. In contrast, Liberal realists, like Ashley J. Tellis (2001), argue that sustaining a credible nuclear deterrence or a recessed arsenal with a no–first use policy is adequate for Indian security in particular, and South Asian security in general.

The interrelationship between economy and warfare remains an understudied theme. About 33 per cent of the Raj's budget even in peacetime went to the British Indian Army or the colonial army. And at present, a large chunk of the scarce resources of India and Pakistan goes to their respective armed establishments. The army was the largest government employer in colonial India, and in postcolonial India the army remains the second largest government employer after the railways. Both the British Indian Army and the Indian and Pakistani armies recruit small peasants from the countryside. Again, both the colonial army and the post-colonial armies of India and Pakistan comprise long-term volunteers rather than short-service conscripts. Some of the questions which the historians should be asking are how pay and pension shape the socio-economic fabric of the rural regions. The rise of dominant peasant castes due to army recruitment in regions like Punjab and Uttar Pradesh and the emergence of identity politics (like the Gurkhas and the Sikhs demanding separate homelands in Darjeeling and Punjab) and so forth are some of the issues which historians and political scientists need to look into.

Bibliography

Abraham, Itty, ed. *South Asian Cultures of the Bomb: Atomic Publics and the State in India and Pakistan*. 2009. Reprint, New Delhi: Orient Blackswan, 2010.

Alam, M. *The Crisis of Empire in Mughal North India: Awadh and the Punjab, 1707–1748*. Delhi: Oxford University Press, 1986.

Athar, Ali. *Military Technology and Warfare in the Sultanate of Delhi*. New Delhi: Icon, 2006.

Barua, Pradeep P. *The State at War in South Asia*. Lincoln: University of Nebraska Press, 2005.

Bayly, C.A. *The New Cambridge History of India*, 2:1, *Indian Society and the Making of the British Empire*. Cambridge: Cambridge University Press, 1988.

Bhakari, S.K. *Indian Warfare: An Appraisal of Strategy and Tactics of War in Early Medieval Period*. New Delhi: Munshiram Manoharlal, 1981.

Chakravarti, P.C. *The Art of War in Ancient India*. 1941. Reprint, Delhi: Low Price, 1989.

Chattopadhyaya, Brajadulal. *The Making of Early Medieval India*. 1994. Reprint, New Delhi: Oxford University Press, 2005.

Cloughley, Brian. *A History of the Pakistan Army: Wars and Insurrections*. Karachi: Oxford University Press, 1999.

Cohen, Stephen, P. *The Indian Army: Its Contribution to the Development of a Nation*. 1971. Reprint, New Delhi: Oxford University Press, 1991.

———. *The Pakistan Army*. 1984. Reprint, Karachi: Oxford University Press, 1993.

Cooper, R.G.S. *The Anglo-Maratha Campaigns and the Contest for India: The Struggle for Control of the South Asian Military Economy*. Cambridge: Cambridge University Press, 2005.

Creasy, Edward. *Fifteen Decisive Battles of the World From Marathon to Waterloo*. 1851. Reprint, London: J.M. Dent & Sons, 1963.

Davies, Brian L. *Warfare, State and Society on the Black Sea Steppe, 1500–1700*. Oxon: Routledge, 2007.

Decker, Michael J. *The Byzantine Art of War*. Yardley: Westholme, 2013.

Digby, Simon. *War Horses and Elephants in the Delhi Sultanate*. Karachi: Oxford University Press, 1971.

Dikshitar, V.R.R. *War in Ancient India*. 1944. Reprint, Delhi: Motilal Banarasidas, 1987.

Drews, Robert. *The End of the Bronze Age: Changes in Warfare and the Catastrophe CA 1200 BC*. Princeton: Princeton University Press, 1993.

Eaton, Richard M., and Phillip B. Wagoner. *Power, Memory, Architecture: Contested Sites on India's Deccan Plateau, 1300–1600*. 2014. Reprint, New Delhi: Oxford University Press, 2015.

Fair, C. Christine. *Fighting to the End: The Pakistan Army's Way of War*. New York: Oxford University Press, 2014.

Ganguly, Sumit. *Conflict Unending: India-Pakistan Tensions Since 1947*. New Delhi: Oxford University Press, 2002.

Gates, Scott, and Kaushik Roy. *Unconventional Warfare in South Asia: Shadow Warriors and Counterinsurgency*. Surrey: Ashgate, 2014.

Gommans, J.J.L. *Mughal Warfare: Indian Frontiers and High Roads to Warfare, 1500–1700*. London: Routledge, 2002.

Gommans, J.J.L., and Dirk H.A. Kolff, eds. *Warfare and Weaponry in South Asia: 1000–1800*. New Delhi: Oxford University Press, 2001.

Habib, Irfan. *The Agrarian System of Mughal India: 1556–1707*. 1963. Reprint, New Delhi: Oxford University Press, 2010.

Habib, M. "The Urban Revolution in Northern India", In *Warfare and Weaponry in South Asia: 1000–1800*. Eds. Jos J.L. Gommans and Dirk H.A. Kolff. New Delhi: Oxford University Press, 2001.

Hanson, V.D. *The Western Way of War: Infantry Battle in Classical Greece*. 1994. Reprint, Berkeley: University of California Press, 2009.

Hasan, S. Nurul. *Religion, State, and Society in Medieval India: Collected Works of S. Nurul Hasan*. Ed. Satish Chandra. 2005. Reprint, New Delhi: Oxford University Press, 2009.

Hellie, Richard. "The Petrine Army: Continuity, Change, and Impact." In *Warfare in Europe 1650–1792*. Ed. Jeremy Black. Aldershot: Ashgate, 2005.

Irvine, William. *The Army of the Indian Moghuls: Its Organization and Administration*. 1896. Reprint, Delhi: Low Price, 1994.

Jackson, Peter. *The Delhi Sultanate: A Political and Military History*. Cambridge: Cambridge University Press, 1999.

Jha, D.N. *Ancient India in Historical Outline*. 1977. Reprint, Delhi: Manohar, 2003.

Jones, William. "The Fifth Anniversary Discourse, Delivered 21 February 1788: On the Tartars," *Asiatic Researches, vol. 2* (1788), pp. 18–34.

Karnad, Bharat. *Nuclear Weapons & Indian Security: The Realist Foundations of Strategy*. New Delhi: Palgrave Macmillan, 2002.

Keegan, John. *The Mask of Command*. 1987. Reprint, New York: Penguin, 1988.

Khan, Iqtidar Alam, *Gunpowder and Firearms: Warfare in Medieval India*. New Delhi: Oxford University Press, 2004.

Kulke, Hermann, K. Kesavapany, and Vijay Sakhuja, eds. *Nagapattinam to Suvarnadwipa: Reflections on the Chola Naval Expeditions to Southeast Asia*. New Delhi: Manohar, 2010.

Kurup, K.K.N., ed. *India's Naval Traditions: The Role of Kunhali Marakkars*. New Delhi: Northern Book Centre, 1997.

Levathes, Louise. *When China Ruled the Seas: The Treasure Fleet of the Dragon Throne (1405–1433)*. 1994. Reprint, New York: Oxford University Press, 1996.

Macdougall, Philip. *Naval Resistance to Britain's Growing Power in India 1660–1800: The Saffron Banner and the Tiger of Mysore*. Suffolk: Boysdell, 2014.

MacMunn, G.F. *The Armies of India*. 1911, Reprint, New Delhi: Heritage, 1991.

Majumdar, B.K. *The Military System in Ancient India*. Calcutta: Firma KLM, 1960.

Malleson, Colonel G.B. *The Decisive Battles of India From 1746 to 1849 Inclusive*. 1883. Reprint, Jaipur: Aavishkar, 1986.

M'Crindle, J.W. *The Invasion of India by Alexander the Great*. 1896. Reprint, New Delhi: Cosmo, 1983.

Menezes, S.L. *The Indian Army: From the Seventeenth to the Twenty-First Century*. New Delhi: Viking, 1993.

Mukherjee, Rudrangshu. "The Sepoy Mutinies Revisited." In *India's Colonial Encounter: Essays in Memory of Eric Stokes*. Eds. M. Hasan and N. Gupta. 1993. Reprint, New Delhi: Manohar, 2004, 193–204.

Nawaz, Shuja. *Crossed Swords: Pakistan, Its Army, and the Wars Within*. Karachi: Oxford University Press, 2008.

Omissi, David. *The Sepoy and the Raj: The Indian Army, 1860–1940*. Basingstoke: Macmillan, 1994.

Parker, Geoffrey. *The Military Revolution: Military Innovation and the Rise of the West, 1500–1800*. Cambridge: Cambridge University Press, 1988.

Peers, Douglas M. *Between Mars and Mammon: Colonial Armies and the Garrison State in Early Nineteenth-Century India*. London: I.B. Tauris, 1995.

Pinch, William, R. *Warrior Ascetics and Indian Empires*. Cambridge: Cambridge University Press, 2006.

Raghavan, Srinath. *War and Peace in Modern India: A Strategic History of the Nehru Years*. New Delhi: Permanent Black, 2010.

Rajagopalan, Rajesh. *Fighting Like a Guerrilla: The Indian Army and Counter Insurgency*. New Delhi: Routledge, 2008.

Ramana, M.V., and C. Rammanohar Reddy, eds. *Prisoners of the Nuclear Dream*. New Delhi: Orient Longman, 2003.

Richards, John F. *The New Cambridge History of India*, 1:5, *The Mughal Empire*. 1993. Reprint. New Delhi: Foundation Books, 2002.

Rosen, S.P. *Societies and Military Power: India and Its Armies*. New Delhi: Oxford University Press, 1996.

Roy, Kaushik. *War, Culture and Society in Early Modern South Asia, 1740–1849*. London: Routledge, 2011.

———. *Warfare in Pre-British India: 1500 BCE to 1740 CE*. London: Routledge, 2015.

Roy, T. *The Politics of a Popular Uprising: Bundelkhand in 1857*. Delhi: Oxford University Press, 1994.

Sandhu, G.S. *A Military History of Ancient India*. New Delhi: Vision, 2000.

Sarkar, Jadunath. *A Short History of Aurangzib: 1618–1707*. 1930. Reprint, New Delhi: Orient Longman, 1979.

———. *Military History of India*. 1960. Reprint, New Delhi: Orient Longmans, 1970.

Sarkar, J.N. *The Art of War in Medieval India*. New Delhi: Munshiram Manoharlal, 1984.

Sen, S.N. *The Military System of the Marathas*. 1928. Reprint, Calcutta: K.P. Bagchi, 1979.

Singh, General J.J. *A Soldier's General: An Autobiography*. New Delhi: HarperCollins, 2012.

Stein, Burton. *A History of India*. 1998. Reprint, New Delhi: Oxford University Press, 2004.

Streusand, Douglas E. *Islamic Gunpowder Empires: Ottomans, Safavids, and Mughals*. Boulder: Westview, 2011.

Subramaniam, Arjun. *India's Wars: A Military History, 1947–1971*. Noida: HarperCollins, 2016.

Tellis, Ashley, J. *India's Emerging Nuclear Posture: Between Recessed Deterrent and Ready Arsenal*. New Delhi: Oxford University Press, 2001.

Thapar, Romila. *The Penguin History of Early India: From the Origins to AD 1300*. 2002. Reprint, New Delhi: Penguin, 2003.

Thapliyal, U.P. *Warfare in Ancient India: Organizational and Operational Dimensions*. New Delhi: Manohar, 2010.

Trautmann, Thomas R., ed. *The Aryan Debate*. New Delhi: Oxford University Press, 2005.

———. *Elephants and Kings: An Environmental History*. New Delhi: Permanent Black, 2015.

Wilkinson, Steven I. *Army and Nation: The Military and Indian Democracy Since Independence*. New Delhi: Permanent Black, 2015.

Wink, Andre. Al-Hind: *The Making of the Indo-Islamic World*, vol. 2, *The Slave Kings and the Islamic Conquest 11th–13th Centuries*. 1997. Reprint, New Delhi: Oxford University Press, 2001.

Yong, Tan Tai. *The Garrison State: The Military, Government and Society in Colonial Punjab, 1849–1947*. New Delhi: SAGE, 2005.

Zaman, M.K. *Mughal Artillery*. New Delhi: Idarah-i-Adabiyat, 1983.

5 War and Society in the Middle East and North Africa

Eileen Ryan

Until recently, military history has been a missing piece in wider discussions of the history of the modern Middle East and North Africa, especially in languages easily accessible to most students of the subject in Western Europe and the Americas. As Thierry Gongora (1997) points out, the lack of scholarly attention to the subject is a glaring problem given the centrality of military power and military culture to the history of state formation in the region. Of the studies on the subject produced in recent years, the majority focus on some aspect of the over six-hundred-year history of the Ottoman Empire and the legacies of its collapse. Emerging around 1300 in northwest Anatolia, the Ottoman Empire expanded gradually in the first century of its existence. After about 1450, territorial acquisition increased in volume and speed and, at its greatest extent at the end of the seventeenth century, it stretched from the imperial center in Istanbul across North Africa to the coast of Algeria and reached into southeastern Europe, almost to Vienna (which the Ottoman military failed to capture in 1683). Among land-based empires of its kind, the Ottoman Empire was remarkable both for its strength and its longevity. In one form or another, it remained a major world power until 1923, when, following defeat in the First World War, its boundaries were reduced to the core of the empire in Anatolia and the Ottoman sultan was deposed in the subsequent formation of the Republic of Turkey.

During much of its existence, the Ottoman Empire set a standard for social and military progress that posed both a model and a potential threat to emerging European nations. For much of the twentieth century, however, historical scholarship of the Middle East and North Africa depicted it as inherently weak and following a path of inevitable decline. The task of identifying a date to mark the beginning of the downturn became something of an obsession for scholars of the region; it was easy enough to find evidence to support a decline narrative. Even Ottoman officials complained of decline as early as the seventeenth century, and Western European concerns of the inevitability of an eventual collapse of "the sick man of Europe" increased in the eighteenth and nineteenth centuries when a series of nationalist uprisings against Istanbul and a major Russian victory over the Ottoman army threatened to upset Europe's balance of power.

In the years following the First World War, a consensus emerged to identify the starting point during the seventeenth century, thus defining the last three hundred years of the Ottoman Empire as an era of decline. Historians H.A.R. Gibb and Harold Bowen established a standard explanation for the decline narrative in *Islamic Society and the West* (1950–1957). This work portrayed the demise of the Ottoman Empire in the First World War as an inevitable consequence of stagnant state structures and a failure to innovate at the same rate as Western European societies. Like Gibb and Bowen, most historians of the region posed the decline narrative in a comparative light, as an inability to adapt to a modern era. According to this theory, the vast size and wide variety of populations under Ottoman sovereignty, once seen as a sign of the empire's might, became a source of weakness in a modern era defined by centralized nation-states with uniform military structures.

Historians began to challenge the validity of the decline narrative in the 1970s, and save for few exceptions, scholarship of the military and social history of the modern Middle East and North Africa remains focused on repudiating the persistent perception of the inevitable decline of the Ottoman Empire. One of the most prominent early critiques came in 1975, when Oxford historian Roger Owen gave a talk at the annual meeting of the British Society for Middle East Studies. Owen targeted two assumptions underpinning the thesis of Gibb and Bowen in particular and the decline narrative more generally. First, he pointed out that they sought evidence of the weakness of the Ottoman Empire in signs of decentralization. During the era of Ottoman expansion, Istanbul often absorbed preexisting power structures into the imperial system, and in moments of financial or military strain, the sultan at times extended greater autonomy to the provinces. Owen argued against casting these tendencies as signs of weakness, and made the case instead for appreciating the flexibility of this approach to imperial expansion. Second, Owen discounted their comparative model for positing a strict separation between the Ottoman and European worlds until the eighteenth century, at which point, according to the decline narrative, the Ottoman Empire had already fallen too far behind the centralization and modernization of smaller European nation-states. Owen instead pointed to continuous contact across the Caucasus and the Mediterranean throughout the history of the Ottoman Empire as sources of mutual inspiration for changes in military, state, and social systems.

Most recent scholarship on the region's military history follows Owen's trend of contesting the decline narrative, focusing instead on finding evidence of flexibility in Ottoman state and military structures to explain the longevity of the empire. From a general perspective, historians Mesut Uyar and Edward J. Erickson (2009) reject any attempts to identify a point of decline in the sixteenth or seventeenth centuries as ahistorical, because doing so presumes an end to the Ottoman Empire that was far from a foregone conclusion at the time. Rhoads Murphey (1999) challenges one of the foundational assumptions of the decline narrative, that the Ottoman Empire was driven primarily by an atavistic sense of ideological struggle against the Christian world, a source of motivation that promoted risky behavior rather than strategic thinking. Focusing instead on motivations for Ottoman expansion, his study demonstrates that, rather than driven by ideological objectives of the sultan, it was more likely to result from sporadic conflicts in marginal areas. Localized conflicts on the frontiers might then become wider, large-scale conflicts, but only if local objectives fit with concurrent interests in the imperial center.

Murphey also overturns a common perception of the Ottoman Empire as technologically weak compared to European militaries. Localized conflicts, he demonstrates, worked particularly well against fragmented European powers like the Habsburgs, who often faced wars on multiple fronts, and he points to the persistent sense of Ottoman military superiority among Europeans well into the eighteenth century. While he does not refute the decline narrative entirely, Murphey argues that the loss of comparative advantage by the Ottoman military was due not to technological insufficiencies but rather to a shift in the power dynamics in Europe—one that was neither a foregone conclusion nor necessarily permanent. Having taken an innovative approach to military history in exploring processes of warfare and their broader social and psychological effects, Murphey inspired a historiographic trend of questioning assumptions that the Ottomans entered a modern military era any later or weaker than their European counterparts. This was a strain of scholarship taken up, for example, by Gábor Ágostan (2005) in his more recent examination of the development of a sophisticated armament industry in the Ottoman Empire in the sixteenth and seventeenth centuries. Challenging superficial assumptions that Islamic societies and governments, due to supposed Islamic conservatism or despotism, fell behind European counterparts, Ágostan examines armaments production in concrete detail to demonstrate comparable Ottoman technical skills and parity of Ottoman military technologies in the early modern world.

One specific aspect of the Ottoman army that developed in the era of imperial expansion, and that came to be seen as a source of structural weakness in the decline narrative, was the incorporation of mamluks. The practice of taking boys or young men as military slaves, or mamluks, was a widespread strategy for building armed forces in the region before Ottoman expansion, and it became a central component of the Empire's military structures in both Istanbul and the outer provinces. James Waterson (2007) provides an overview of Ottoman Mamluk policy beginning in the late fourteenth century with the recruitment of Christian boys taken prisoner during wars of expansion in the Caucuses. Some were directed into the Janissaries, an elite and highly unified force. Their training involved conversion to Islam and learning Turkish—thus contributing to the establishment of an Islamic and Turkish-speaking core of Ottoman culture—and they were provided with generous salaries and pensions. The result was a new class of elite subjects culled from a Christian minority and devoted to the Ottoman sultan. As time went on, the Janissaries gained political power, which they used to oppose military reforms that might threaten their position; they remained a dominant force in the military until Sultan Mahmud II managed to put down a Janissary uprising (with enormous amounts of bloodshed) and abolish the corps in 1826.

Due to their political activities, the Janissaries have received a great deal of blame for any perceived inabilities of the Ottoman military to keep up with the development of modern armies in Western Europe. However, Cemal Kafadar (2007) challenges characterization of the Janissaries as opposed to modernization and therefore damaging to the adaptability of the Ottoman state. Over time, Kafadar explains, the Janissaries expanded into a distinct social class that was not always strictly tied to the military. Merchants and tradesmen, Janissaries and a subsidiary realm of Janissary-affiliates developed a decidedly modern urban culture, especially in Istanbul. As such, Janissary protests or revolts—including the activities that led to the brutal repression of the Janissaries in 1826—could be seen as expressive of broader interests through political action, or even as a valuable check on autocratic tendencies, rather than the reactionary traditionalism of an elite military group.

Mamluks also played an important part in governance of the empire's largest province, Egypt. The Ottomans captured Egypt with their victory over the forces of the Mamluk Sultanate, a state governed by soldier slaves turned into a political class. Shai Har-El (1995) and C.Y. Muslu (2014) examine the relationship between the expanding empire and the Mamluk Sultanate as they alternated between open warfare and careful diplomacy, from the former's initial expansion until the Ottomans' conquest of Cairo in 1517. Their perspectives on the Ottoman-Mamluk competition are valuable both for understanding the rise of the Empire's military power and as a corrective to the common view of the Ottomans as obsessed with combating Christian powers out of a sense of ideological competition. Ruled by Sunni Muslims much like the Ottoman Empire, the Mamluk Sultanate was a primary concern for expansionist sultans in Istanbul, and their successful conquest of region, as the authors demonstrate, had ramifications for both Ottoman and Western European military history for a long time to come.

The conquest of Egypt in 1517 initiated a period of expansion throughout North Africa wherein more Mamluk elements were incorporated into Ottoman forces and administration. This practice has led many historians to see Ottoman rule in North Africa as inherently weak; by simply superimposing Ottoman claims to sovereignty over a preexisting Mamluk structure, the argument goes, Istanbul did little to integrate disparate provinces into its expanding empire. Jane Hathaway (1997) and M'hamed Oualdi (2016) have challenged this characterization. Examining sixteenth-century Egypt and eighteenth- and nineteenth-century Tunisia, respectively, they demonstrate the creative ways in which Ottoman sultans combined elements of preexisting Mamluk structures with innovative strategies to ensure Ottoman control in the North African provinces. Not only do their studies contribute greatly to our understanding of the relationship between the Ottoman imperial center and its outer regions, but also they open up the definition of who could be categorized as a Mamluk or part of a Mamluk structure. Much like the Janissaries in Istanbul, these were not a homogenous group of former slave soldiers, but rather disparate individuals linked through networks of blood and patronage to ruling households whose interconnectedness contributed to long-term stability even in eras of imperial decentralization.

Challenging the tendency to characterize the Ottoman Empire as a land-based empire, Egypt also served as a strategic base for the expansion of naval power in the Red Sea and the Mediterranean. By examining the use of naval resources for commercial and military activities, historians like Palmira Brummett (1994) and Svat Soucek (2015) have inserted the Ottomans fully into the competitive economic arena of the sixteenth-century Age of Discovery, with its resultant expansion of trade networks highlighting commonalities in the expansionism of Christian and Islamic powers in a broad Euro-Asian zone. Brummett pays particularly close attention to competing claims to universal principles and religious legitimacy among both Christian states and the Ottoman Empire. Doing so helps to dispel myths of Ottoman fanaticism in military competition by presenting religious claims as a shared discourse of naval rivalries.

In the historiography of Ottoman naval power, the activities of North African corsairs became a focal point of Western European criticism as representative of the lack of centralized control in Istanbul and signs of a supposed inability to modernize. According to the decline narrative, the French Revolution and the Napoleonic Wars established a new era of contact across the Mediterranean with a resultant inability of the Ottoman Empire to adequately respond to the pressures of modernization, manifested in part by a subsequent resurgence of North African piracy. Daniel Panzac (1999) instead characterizes this as a brief period of opportunity for North African merchants cut short by the European expansion following the resolution of the Napoleonic Wars. Linking trade to piracy, Panzac demonstrates how Europeans went from welcoming the involvement of corsairs in Mediterranean trade at the turn of the century to targeting piracy as justification for their own expansion in North Africa—such as France's occupation of Algeria in 1830. Panzac's study thus marks a departure from those who would define the corsairs as an inherent component of an Ottoman/North African system (or lack thereof) to portray it instead as the consequence of a failure to incorporate North African Muslim merchants into Southern European trade networks. Along similar lines, some of the chapters collected in the volume edited by Elizabeth Zachariadou (2002) are of interest for their emphasis on Ottoman regional authority in an era of naval competition, including chapters outlining Ottoman reactions to piracy and Mediterranean defenses.

Wherever individual historians stand on particular issues in the decline narrative, there is no question that the late eighteenth century began a period of remarkable upheaval for the Ottoman Empire, as it was for most of continental Europe. Virginia Aksan (2007) identifies the Russo-Ottoman War of 1768–1774 as a watershed moment for the Ottoman Empire. The loss of territory to the Russian Empire inaugurated a period of widespread concern over what contemporary observers called the "Eastern Question," referring to uncertainty over the future of the Ottoman Empire and the threat of Russian ascendancy. The military setback also sparked a range of military reforms, increased absolutism, and a Turkification of the Ottoman forces, all of which Aksan argues reached a pinnacle in the destruction of the Janissaries in 1826.

The increased absolutism of the Ottoman sultan, in particular, contributed to a rise in internal challenges to Ottoman central authority. These included new figures of local rule in the Ottoman Empire, the most famous of which would certainly be Mehmed (or Mohammed) Ali Pasha. Born in Albania to a mercantile family, Mehmed Ali rose through Ottoman military ranks and was considered widely responsible for reestablishing Ottoman rule in Egypt after Napoleon's brief occupation of 1798–1801. In the 1820s, Mehmed Ali introduced a system of conscription that incorporated thousands of Egyptian peasants into the region's military structures. As a result, historians have often posed Mehmed Ali as a force of opposition to the Ottoman sultan and as a founder of the Egyptian nation; by bringing peasants into the army, he fostered a regional identity independent of the Ottoman Empire. Khaled Fahmy (1997), however, has redefined Mehmet Ali's military reforms in Egypt as a quintessential component of Ottoman strategies. Rather than portraying Mehmed Ali as a proto-nationalist, Fahmy integrates him fully in an imperial history.

In a more recent study, John P. Dunn (2005) calls into question the utility of later attempts to reform the Egyptian military along Western European lines. He examines a program Mehmed Ali's grandson, the Khedive Ismail, introduced, inviting European mercenaries to advise on military restructuring in the late nineteenth century. Seen as a way of resurrecting his grandfather's reforms, these efforts were considered a catastrophic failure; facing widespread revolt and loss of international support, the Khedive Ismail was deposed in 1879. Observers at the time, and the dominant historical narrative since, have faulted the incomplete nature of the reforms, reflecting an inability to adapt fully to modern Western European military norms. Dunn, however, asserts the attempt to rely on European military advisors as a move that led to a crisis of leadership and a loss of political authority in Cairo.

The threats—or at least calls for reform—from within the Ottoman Empire in the nineteenth century accompanied an increase in external pressures. For the most part, those pressures came in the form of European possession of Ottoman territories in a series of wars of colonial occupation. Beginning with the French invasion of Algeria in 1830, Ottoman sovereignty in its North African territories faced periodic threats. Following the revolt that ousted the Khedive Ismail, the British declared Egypt a protectorate in 1882, one year after France expanded its holdings in Algeria into bordering Tunisia. The Ottomans lost their final North African provinces in contemporary Libya following a war with Italy in 1911–1912.

However, the nineteenth century was not a simple downward trajectory for the Ottoman Empire in terms of territorial control. As Candan Badem (2010) points out, in the Crimean War of 1853–1856 French and British soldiers fought alongside Ottoman troops to prevent Russian aggrandizement at Ottoman expense. Moreover, the conflict propelled Ottoman military modernization and integration into a European alliance system—though it also led to financial dependence on European support that would ultimately weaken the sultan's autonomy. The pressures of European aggression in North Africa, however, did contribute to shifts in Ottoman military strategies. An initial response to external pressure, as Selim Deringil (1998, 2003) has pointed out, was to mobilize nomadic forces in Ottoman provinces from Yemen to Libya. In an earlier study, Rachel Simon (1987) demonstrated the longevity of that approach in the Ottoman war with Italy over the Libyan provinces. When these strategies failed to produce results, the loss of territory contributed to a transformation of military identities as the Ottoman armed forces became more firmly Turkish in the years leading up to the First World War.

Given the preoccupation among historians with understanding either the factors leading to the demise of the Ottoman Empire or the conditions that contributed to its longevity, Ottoman involvement in the First World War has long held a great deal of interest for historians of the Middle East and North Africa. Widespread perceptions that many current social and political problems in the region originate from forces that emerged in the early twentieth century further contribute to this interest. Add to the mix the recent centennial of the beginning of the war, and the result has been a wealth of recent studies on the topic, making it easily the most active subfield in social and military histories of the Middle East and North Africa.

Focusing on the First World War in the Ottoman Empire and the wider region brings unexpected benefits for anyone interested in its broader history. Naci Yorulmaz (2014), for example, demonstrates the importance of Ottoman military reforms for a burgeoning German armaments industry in a codependent system that developed in the decades leading up to the First World War. The result was a threatening relationship to other European powers based on shared economic and expansionist interests in Berlin and Istanbul. Kristian Ulrichsen (2014) takes a comparative approach to examining how all of the various powers active in the region—not just the Ottomans—dealt with the strategic and tactical puzzle posed by the particular environmental and strategic concerns of the Middle East and North Africa as a campaign of war.

Sean McMeekin (2015) and Eugene Rogan (2015) emphasize how a focus on the Ottoman experience of the First World War challenges the temporal and geographical dimensions of the conflict, which is typically confined by a European perspective. Seen from Istanbul, the 1911 Italian occupation of the Ottoman Empire's provinces in contemporary Libya takes center stage as an important turning point in military history, presaging important tactical developments. One of the last of the colonial wars that typified European expansionism in the nineteenth century, the Italo-Ottoman struggle in Libya saw a number of innovations that would become common in the First World War, including the offensive employment of airplanes. Both McMeekin and Rogan emphasize that the Ottoman Empire, more than any other belligerent, pushed the First World War beyond the boundaries of Western Europe. To understand the complexity of conditions that led to the Ottoman Empire becoming involved in what was previously a European war, Aksakan (2008) examines the Ottoman decision to join the side of the Germans from a diplomatic point of view, while Erickson (2001) presents the Ottoman involvement in the war from a strictly military perspective.

Many recent studies on the First World War in the Middle East seek to answer questions about how the conflict transformed the relationship between Istanbul and the various provinces of the Ottoman Empire. Mehmet Beşikçi (2012), for example, examines Ottoman mobilization for clues. His account of volunteerism, conscription, and rising levels of desertion over the course of the First World War illuminates shifts in the relationships among various portions of the empire's population and the state itself, ultimately contributing to the emergence of a firmly Turkish nation-state. Much of his study expands on Erik Zürcher's earlier work (1999) on the longer history of military conscription in the Middle East. Despite its broad focus, Zürcher's study was crucial for identifying the importance of the First World War, in particular by examining how desertion and the experiences of ordinary soldiers linked the region's social and military history with long-lasting ramifications for the Ottoman state.

Since Zürcher's study, additional research into the everyday lives of Ottoman soldiers and civilians in the First World War has also connected the region's history with the broader conflict in Europe, especially regarding issues of experience and memory in the "total war" of the early twentieth century. Selim Tamari (2011), for example, looked to the diaries of three regular soldiers as a way of examining wider social and political changes the First World War brought about in the region known as "greater Syria" or Ottoman Syria—a region larger than contemporary Syria in the eastern Mediterranean. More recently, Leila Tarazi Fawaz (2014) has expanded on Tamari's investigation to provide a rare analysis of how it influenced the development of national, religious, and class identities in Ottoman Syria. Both studies offer compelling interventions in a realm of public memory that often overlooks the impact of the First World War in the region due to the overwhelming focus on more recent conflicts, such as Second World War or the Lebanese civil war.

Historians of the modern Turkish Republic that formed from the ashes of the Ottoman Empire consider the years leading up to the First World War, and the experiences of the war itself, as particularly crucial for understanding the historical processes that created the new nation. The early history of the Turkish Republic was dominated by one figure, Mustafa Kemal, also known as Atatürk (meaning "father of the Turks"). Mustafa Kemal was instrumental in the formation of the Turkish Republic from a military and diplomatic standpoint; an Ottoman officer during the First World War, he established a shadow parliament in the city of Ankara in 1920 to delegitimize the sultan in Istanbul. His parliament became the seed of the new Turkish Republic, and from his position, he finalized the international agreements securing Turkish sovereignty in the Treaty of Lausanne of 1923. Until his death in 1938, Atatürk embarked on an ambitious program of modernization that touched on social and political issues in the emerging Turkish nation-state. For example, he instituted a dramatic linguistic change by discarding the Ottoman practice of writing Turkish texts using Arabic characters in favor of a new alphabet using a modified version of Latin characters. He also promoted a general secularization of Turkish political and social culture through the education system, among other avenues. Most importantly for our consideration, Atatürk promoted a strong and modern military as essential for safeguarding the independence of Turkey.

Atatürk posed his modernization programs as an effort to distance the new Turkish state from what he painted as a failed Ottoman past. In positioning the military as an agent of modernization, he promoted an interpretation of late Ottoman history that would see the sultan's court and the military as diametrically opposed—the military as a force for change that was stymied by a reactionary sultan. Furthermore, by rejecting the use of the Arabic script and promoting secularization over the sultan's frequent claims to act in the position of the caliph, or spiritual leader of the Muslim world, Atatürk distanced the modern Turkish Republic from the Ottoman ties to its Arab provinces.

But there were also signs of continuity in Atatürk's modernization schemes that reflected ideals common among prewar Ottoman military commanders. Handan Akmese (2005) highlights these continuities by scrutinizing the development of a particular military culture among membership of the Committee of Unity and Progress (CUP). An organization that emerged in the late nineteenth century as a secret society, it later transitioned into a political movement calling for modernization in Istanbul when it combined with a movement referred to as the Young Turks. With prominent officers in leadership positions, the CUP called for liberalization and democratization within the Ottoman Empire. Moreover, Akmese traces how unique nineteenth-century military reforms shaped the subsequent program of this class of military elites with a particular ideology centered on modernization. Akmese thus attributes the popularity of Atatürk's modernization efforts and his emphasis on militarism to the groundwork laid by a decidedly Ottoman movement.

In an innovative attempt to counter this emphasis on militarism in Turkish nationalism, however, anthropologist Ayşe Gül Altinay (2004) calls into question the popularity of militarism in the early Turkish nation. Altinay argues that the Turkish state, both during and after the Atatürk regime, fostered a myth of the military-nation within Turkey's public culture through the writing of official history. Her study also offers a reading of the connection between gendered identities and this discourse of militarism through, for example, an examination of Turkey's first female combat pilot, Sabiha Gokcen. Altinay's work, based primarily on ethnographies, points to the ways in which the Turkish state developed systematic control over both male and female bodies in its definitions of the modern nation.

Outside of Turkey, transitions from Ottoman rule to postimperial nation-states in the decades after the First World War took different paths. Concern with the potential for regional instability after the dissolution of the empire led to the establishment of the mandate system, by which the League of Nations granted Great Britain a mandate to govern Iraq and Palestine, and France a mandate in Syria and Lebanon. Based on an understanding that these territories were not prepared for self-rule, the European powers were supposed to provide guidance and support on the path to sovereignty. The mandate system lasted longest in Palestine and Syria. Syria—divided into Lebanon and modern-day Syria—gained independence in 1945, and the territories of Palestine—partitioned into Jordan, Israel, and the West Bank—gained independence at various points between 1946 and 1948. Historians looking to understand the complex ethnic, religious, and national tensions of these countries tend to look to conditions established during the mandate period for how they influenced nation-state formation.

For example, historians have identified foundations in the mandate period for communal politics. This vein of scholarship attempts to answer an important question: in a region known for its diversity, how did religious, linguistic, and ethnic identities become so entrenched in the political landscape of the late twentieth century? In a now classic study of the French mandate in Syria, Philip Khoury (1987) tied the issue directly to class. He argued that in the nationalist movement that emerged during the French mandate period, elites were more interested in using communalism to maintain access to power rather than using unity to lead a political revolution against French rule. More recently, Benjamin White (2011) shows how the structures of the French mandate system promoted the idea of an Arab-Syrian national identity as a counterweight to French rule. White thus explains, although far from inherent divisions within Syrian society, the creation and reification of minority communities such as Druzes, Kurds, or Armenians as part of the emergence of a Syrian national movement.

Similarly, British policies contributed to sectarianism in the mandate territory of Palestine. Looking at Christian minority groups, Laura Robson (2011) and Noah Haiduc-Dale (2013) argue that early models of nationalism in the region used more inclusive language that became fractured, especially during the Arab Revolt of 1936–1939. Typically seen as a manifestation of long-standing Christian Muslim tensions, Haiduc-Dale examines divisions within each community during the revolt and highlights moments of unity across religious divisions. Robson traces the process by which a Palestinian intellectual nationalism excluded Christian participation after the early period of inclusiveness. But the most disruptive form of sectarian politics in the British mandate territory of Palestine emerged as a result of contradictory support for the establishment of a "national home" for European Jews in tension with promises to safeguard the sovereignty of all people of Palestine. The development of sectarian politics out of this tension has received enormous attention from scholars. In terms of understanding how religious divisions affected the subsequent territorial partition of the Palestinian territory, Gideon Biger (1995) wrote a key study in explaining geographic borders as a result of British strategic needs. More recently, James Renton (2007) looks back to the Barfour Declaration in 1917—the basis of British support for establishing a Jewish territory—as a piece of political propaganda the British employed during the First World War. Both authors contribute to an understanding of how the British mandate heightened or even created sectarian tensions.

Outside of scholarship on modern Turkey, there has been relatively little work done more specifically on the legacies of violence in the emergence of nation-states out of the former Ottoman Empire. In part, this is reflective of a lack of access to relevant sources, but it also stems from traditional biases toward the region's history that tend to prioritize the perspective of European colonial administrations and militaries in the era of nineteenth-century imperialism. The Eurocentric approach even dominates works analyzing the legacies of the anti-colonial struggles that contributed to an end to European colonial occupations during and after the Second World War. Notable exceptions include Matthew Connelly's (2002) treatment of diplomacy in the Algerian War of Independence of 1954 to 1962 and Alistair Horne's (1977) classic examination of the military activities of the Algerian National Liberation Front (FLN). Ali Ahmida (1994) and Lisa Anderson (1986) also contribute discussions of anti-colonial resistance in the history of nation building and state formation in Libya and Tunisia, but neither focuses their attention on the military, militarism, or violence as primary forces in the process.

Some recent scholarship indicates a potential increase in interest in the issue. The current era of political instability following the collapse of military regimes in the Middle East and North Africa after the wave of revolutions in 2010–2011 suggests the necessity of looking more closely at the local and national legacies of war, violence, and traditions of militarism in the process of state formation. The political scientist Daniel Neep (2012) has offered a theoretical intervention that could prove helpful in this effort. Examining the rise of Syrian nationalism during the mandate period in particular, Neep develops a model to think about the structures of colonial rule and the ways in which violence—which he defines variously as war, revolution, and resistance—shaped nation-state formation in the postcolonial era. Charles Tripp (2013) and Stephanie Cronin (2014) both hope to place the "Arab Spring" in a longer context of various types of resistance to state control in the modern Middle East and North Africa. Doing so offers a way to problematize Western stereotypes that date back to the Ottoman era of the peoples of the region as particularly susceptible to irrational violence or as victims of dictatorships. In contextualizing communalism and military culture, we can better understand the historical legacies of the rise and fall of Ottoman power and the emergence of postimperial nation-states.

Bibliography

Ágostan, Gábor. *Guns for the Sultan: Military Power and the Weapons Industry in the Ottoman Empire*. Cambridge: Cambridge University Press, 2005.

Ahmida, Ali Abullatif. *The Making of Modern Libya: State Formation, Colonization and Resistance, 1830–1932*. Albany: State University of New York Press, 1994.

Akmese, Handan Nezir. *The Birth of Modern Turkey: The Ottoman Military and the March to World War I*. London: I.B. Tauris, 2005.

Aksakal, Mostafa. *The Ottoman Road to War in 1914*. Cambridge: Cambridge University Press, 2008.

Aksan, Virginia H. *Ottoman Wars 1700–1800: An Empire Besieged*. Harlow: Pearson Longman, 2007.

Altinay, Ayşe Gül. *The Myth of the Military-Nation: Militarism, Gender, and Education in Turkey*. New York: Palgrave Macmillan, 2004.

Anderson, Lisa. *The State and Social Transformation in Tunisia and Libya, 1830–1980*. Princeton: Princeton University Press, 1986.

Badem, Candan. *The Ottoman Crimean War (1853–1856)*. Leiden: Brill, 2010.

Beşikçi, Mehmet. *The Ottoman Mobilization of Manpower in the First World War: Between Voluntarism and Resistance*. Leiden: Brill, 2012.

Biger, Gideon. *An Empire in the Holy Land: Historical Geography of the British Administration in Palestine, 1917–1929*. New York: St. Martin's Press, 1995.

Brummett, Palmira. *Ottoman Seapower and Levantine Diplomacy in the Age of Discovery*. Albany: State University of New York Press, 1994.

Connelly, Matthew. *A Diplomatic Revolution: Algeria's Fight for Independence and the Origins of the Post-Cold War Era*. New York: Oxford University Press, 2002.

Cronin, Stephanie. *Armies and State-Building in the Modern Middle East: Politics, Nationalism, and Military Reform*. New York: I.B. Tauris, 2014.

Deringil, Selim. "'They Live in a State of Nomadism and Savagery': The Late Ottoman Empire and the Post-Colonial Debate," *Comparative Studies in Society and History* 45 (April 2003): 311–342.

———. *The Well-Protected Domains: Ideology and Legitimation of Power in the Ottoman Empire, 1876–1909*. London: I.B. Tauris, 1998.

Dunn, John P. *Khedive Ismail's Army*. New York: Routledge, 2005.

Erickson, Edward J., ed. *Ordered to Die: A History of the Ottoman Army in the First World War*. Westport, CT: Greenwood Press, 2001.

Fahmy, Khaled. *All the Pasha's Men: Mehmed Ali, His Army, and the Making of Modern Egypt*. Cambridge: Cambridge University Press, 1997.

Fawaz, Leila Tarazi. *A Land of Aching Hearts: The Middle East in the Great War*. Cambridge: Harvard University Press, 2014.

Gibb, H.A.R., and Harold Bowen. *Islamic Society and the West*. London: Oxford University Press, 1950–1957.

Gongora, Thierry. "War Making and State Power in the Contemporary Middle East," *International Journal of Middle East Studies* 29 (August 1997): 323–340.

Haiduc-Dale, Noah. *Arab Christians in British Mandate Palestine: Communalism and Nationalism, 1917–1948*. Edinburgh: Edinburgh University Press, 2013.

Har-El, Shai. *Struggle for Domination in the Middle East: The Ottoman-Mamluk War, 1485–1491*. Leiden: E.J. Brill, 1995.

Hathaway, Jane. *The Politics of Households in Ottoman Egypt: The Rise of the Qazdağlıs*. Cambridge: Cambridge University Press, 1997.

Horne, Alistair. *The Savage War of Peace: Algeria 1954–1962*. New York: Viking Press, 1977.

Kafadar, Cemal. "Janissaries and other Riffraff of Ottoman Istanbul: Rebels Without a Cause?" *International Journal of Turkish Studies* 13 (October 2007): 113–134.

Khoury, Philip. *Syria and the French Mandate: The Politics of Arab Nationalism 1920–1945*. Princeton: Princeton University Press, 1987.

McMeekin, Sean. *The Ottoman Endgame: War, Revolution, and the Making of the Modern Middle East, 1908–1923*. New York: Penguin, 2015.

Murphey, Rhoads. *Ottoman Warfare, 1500–1700*. New Brunswick, NJ: Rutgers University Press, 1999.

Muslu, C. Y. *The Ottomans and Mamluks: Imperial Diplomacy and Warfare in the Islamic World*. London: I.B. Tauris 2014.

Neep, Daniel. *Occupying Syria Under the French Mandate: Insurgency, Space and State Formation*. Cambridge: Cambridge University Press, 2012.

Oualdi, M'hamed. "Mamluks in Ottoman Tunisia: A Category Connecting State and Social Forces," *International Journal of Middle East Studies* 48 (August 2016): 473–490.

Panzac, Daniel. *Les Corsaires barbaresques: La Fin d'une Épopée 1800–1820*. Paris: CNRS Éditions, 1999. Published in English as *Barbary Corsairs: The End of a Legend 1800–1820*. Leiden: Brill, 2005.

Renton, James. *The Zionist Masquerade: The Birth of the Anglo-Zionist Alliance, 1914–1918*. New York: Palgrave Macmillan, 2007.

Robson, Laura. *Colonialism and Christianity in Mandate Palestine*. Austin: University of Texas Press, 2011.

Rogan, Eugene. *The Fall of the Ottomans: The Great War in the Middle East, 1914–1920*. London: Allen Lane, 2015.

Simon, Rachel. *Libya Between Ottomanism and Nationalism: The Ottoman Involvement in Libya During the War With Italy (1911–1919)*. Berlin: Klaus Schwarz Verlag, 1987.

Soucek, Svat. *Ottoman Maritime Wars, 1416–1700*. Istanbul: Isis Press, 2015.

Tamari, Salim. *Year of the Locust: A Soldier's Diary and the Erasure of Palestine's Ottoman Past*. Berkeley: University of California Press, 2011.

Tripp, Charles. *The Power and the People: Paths of Resistance in the Middle East*. New York: Cambridge University Press, 2013.

Ulrichsen, Kristian Coates. *The First World War in the Middle East*. London: Hurst, 2014.

Uyar, Mesut, and Edward J. Erickson. *A Military History of the Ottomans: From Osman to Atatürk*. Santa Barbara, CA: Praeger, 2009.

Waterson, James. *The Knights of Islam: The Wars of the Mamluks*. London: Greenhouse Books, 2007.

White, Benjamin Thomas. *The Emergence of Minorities in the Middle East: The Politics of Community in French Mandate Syria*. Edinburgh: Edinburgh University Press, 2011.

Yorulmaz, Naci. *Arming the Sultan: German Arms Trade and Personal Diplomacy in the Ottoman Empire Before World War I*. London: I.B. Tauris, 2014.

Zachariadou, Elizabeth, ed. *The Kapudan Pasha: His Office and His Domain*. From *Halcyon Days in Crete IV*. Heraklion: Crete University Press, 2002.

Zürcher, Erik J. *Arming the State: Military Conscription in the Middle East and Central Asia, 1775–1925*. London: I.B. Tauris, 1999.

6   War and Society in Sub-Saharan Africa

Bruce Vandervort and Marilyn Zilli

If military history was until fairly recently viewed in some Western academic circles as the bastard child of the profession, its African subset was even more of an outcast. Writing on the eve of the new millennium, a military historian of precolonial Africa, John Thornton, observed that while many of the post–Vietnam War generation of Western historians shied away from the study of war because of "its suspect association with militarism," the African military historian had an even steeper hill to climb. African historians recounting the Continent's past, he wrote, did not include the military, "whose part in the colonial era is too darkly shadowed by their history as imperialist collaborators" (Thornton, 1999, 3). Four years later, John Lamphear, another military historian of precolonial Africa, wrote that African scholars had discouraged the study of Africa's military history because it could be used to evoke the image of "a dark continent of incessant conflict waged by savage warriors," precisely the sort of image that had been used to justify the Scramble by European powers to impose colonial rule on Africa 150 years earlier (Lamphear, 2003, 169). There were, of course, African military historians who swam against this tide of denial, but theirs were isolated voices (Ogot, 1972; Uzoigwe, 1975, 1977). More recently, Timothy Stapleton, author of the monumental three-volume *Military History of Africa*, laid some of the blame for the slow progress of African military history on the African military itself. African soldiers do not show a great deal of interest in military history, he wrote, in part because African military staff schools and academies are small and underdeveloped, but also because of "the secretive, politicized, and internal-security-oriented nature of African military culture" (Stapleton, 2013, v. 1: xi).

Recently, the future of African military history has brightened considerably. Heightened interest in the subject is reflected in the publication of Stapleton's multivolume history of African warfare and in the decision of the Dutch publishing company Brill to launch *Journal of African Military History* in 2017. The tenor of the "new African military history" is captured in its most emphatic and provocative form in the work of Richard J. Reid, especially in his latest book, *Warfare in African History* (2012). Reid wants "to give war its due place in the larger narrative of the continent's history . . . and to make sense of violence as a force both for construction and destruction" (x).

The Precolonial Period

Warfare was a permanent feature of life in precolonial Africa, and it affected all levels of society. The most common form of war in this period was the "raid," actions carried out by the autonomous villages that grouped most of the precolonial population and that gave Africans a reputation as "frontiersmen" (Clayton, 1999). Because no precolonial African society believed in private property, raids usually involved small groups of combatants whose aim was the seizure of resources. Acquiring resources was not intended to expand private wealth, as it was for Europeans, but, rather, to provide communities with new lands to till or graze or access to water. Even for the larger polities—the city-states of the Swahili coast of East Africa or the great kingdoms of the West African savannah—the aim of warfare was not to seize property but rather to add to the "assessable" population of the state. The seizure of people was the paramount goal. Since the African continent was underpopulated until the twentieth century, the aim of most military campaigns of these larger polities was to add to their productive or reproductive capacity (Reid, 2012).

This goal assured that the institution of slavery would be deeply entrenched in African society. The employment of slaves as soldiers was common to all civilizations at one time or another but was more widespread in Africa and persisted longer there than elsewhere (Thornton, 2006). The long life of the institution of slavery is said to have been due to its "reciprocal benefit" to slaveholders and slaves alike. Rulers gained fighting men whose loyalty was undiluted by ties to society, while the enslaved soldiers in turn could obtain power and wealth through military honor (Laband, 2016).

African societies were profoundly affected by the expansion of the Atlantic and Indian Ocean slave trades and by the parallel introduction of gunpowder. Guns made their initial impact on war making in West Africa in the 1500s; a century later they had become the mainstay of the expansion of regional kingdoms, such as Ashanti and Dahomey. Armies of musketeers helped make possible the tremendous surge in the slave trade in the eighteenth century.

Prior to the nineteenth century, very few African armies could be called "professional" in the European sense—that is, they were not standing armies. The larger armies of the wealthy savannah kingdoms, such as Mali, were similar to the levies of feudal Europe in that they were raised on orders from their prince by the "big men" of the provinces (Thornton, 1999). The wealth of these kingdoms made it possible to import horses from North Africa and to purchase the equipment necessary to create a mounted force. Thus the cavalry became the striking arm of these armies and an elite institution in African society (Elbl; Law; Fisher). However, even where horsemen served as the spearhead of an army, there was always a popular side to precolonial war making. Cavalries were never able to win victories without support from spear-wielding, missile-firing infantry drawn from the peasantry. In fact, in the forest zones of West and Central Africa and East Africa south of the marshes of southern Sudan, precolonial armies were composed almost entirely of infantry. Horses were almost unknown in these regions because of the presence of trypanosomiasis (sleeping sickness) spread by the tsetse fly, which was fatal to most domestic animals. Nature also dictated the tactics of the divergent forces: savannah armies relied heavily on the shock of charging cavalry, whereas the dense vegetation of the forest zones necessitated engagement by ambush and hand-to-hand combat (Ajayi and Smith).

Popular participation in warfare in Africa fell off considerably in the nineteenth century. The consolidation of political power in larger units—witness the indigenous empire-building in the Zulu kingdom in southern Africa; jihadi states in northern Nigeria and West Africa; the rejuvenated Solomonic empire in Ethiopia; the Mahdist theocracy in the Sudan—led to the decline of the citizen militias whose seasonal raids had been a distinguishing feature of African warfare over the previous centuries. The tendency of armies to become larger and more professional in some cases resulted from emulation of what became known of Western military practices. The army of Samori Ture, the Guinean empire-builder, was modeled on the French; arms production had been learned by artisans in French employ (Legassick). It was the seventy thousand trained soldiers of Ethiopia, the most "Westernized" of Africa's armies, who crushed the Italians at Adowa in 1896 (Vandervort, 1998). However, most of these new, larger military formations owed little or nothing to the influence of Western technology. The "foot cavalry" of Shaka's Zulu empire was an adaptation of the traditional practice of segmenting youth into age groups; these originally informal units were transformed into rigorously trained age-group regiments under direct command of the king (Gump, 1994; Laband, 1995). A similar transformation occurred among the Maasai and Turkana pastoralists of East Africa, who, in the nineteenth century, also used professionalized age-grade regiments to overawe neighboring peoples in the manner of the Zulus (Fukui and Turton). Both the Zulus and Maasai famously abjured firearms as weapons unworthy of warrior peoples and marched into the nineteenth century wielding the spears of their ances-tors. Recently, some historians have come to see the period we have just surveyed, the first three-quarters of the nineteenth century, as a "golden age" in the history of the Continent. This period saw "new forms of war, new military organizations, the expansion of state power and socio-economic change" (Reid, 2012, 110, 179).

The Colonial Period

Although the idea would have been considered outrageous in the years of the European Scramble for Africa, few historians today would argue with David Killingray's statement that "European empires in Africa were gained principally by African mercenary armies" (Killingray, 1989, 146). It is hard to see how it could have been otherwise. Although medical and other improvements had dramatically reduced deaths of European troops from tropical diseases, Africa continued to be "the white man's grave" in the minds of the European public in the nineteenth century (Curtin). The first and foremost constraint on the employment of European troops in Africa, however, was the continuously high level of demand for their services elsewhere. The tensions among continental European powers from 1870 to 1914 precluded sending more than small numbers of French and German troops to Africa during this period. Britain, preoccupied with protecting India, never sent more than a few hundred of her troops to serve in Africa (Vandervort, 1998).

This competing use dictated the raising of armies of African soldiers by the European powers, first of all to complete the Scramble for colonies and later to maintain order and, in the twentieth century, to function as a reserve army for service outside the Continent, including in Europe and Asia during the two world wars and in the wars of decolonization. But these were never large armies. Until late in the twentieth century, Africa was under-populated. Large armies were not required to carry out wars of conquest. The scale of the battles fought during the Scramble underscores this point. A typical example was the victory in 1895 of 812 Portuguese and allied African soldiers over some three thousand Tonga war-riors at the battle of Marrecuene in Mozambique (Vandervort, 1998).

But, most importantly, there are no examples of African states coalescing to form any-thing like a united front against the aggressors. This has been a sore point with some African military historians, who would much prefer to blame the relatively easy conquest of Africa by European powers on the great disparity in military technology (Headrick). "There is no doubt that the most crucial [European advantage] was military," wrote the Ghanaian historian A. Adu Boahen. "Above all, technologically and logistically, African armies were at a great disadvantage in comparison with the invaders" (Boahen, 1985, 56–57). This explana-tion for European success in the Scramble, which enjoyed some support in the early years after independence, has lost favor more recently, as evidence of the limitations of European military technology has grown and the decisive importance to European victory of African allies has become clearer.

The colonial powers not only exercised a monopoly of force, effectively "demilitarizing" the bulk of their subjects, but also recruited their African soldiers from narrow segments of the population. Though the French began to conscript subjects in West Africa on the eve of the First World War, most African colonial soldiers were volunteers recruited from among what were called the "martial races." Presumably these were peoples who were more warlike than the rest of society because they came from deep in the interior of the colony, far from the Westernized towns of the coast. More "primitive" and therefore supposedly more "martial," these "races" also were thought likely to be apolitical and therefore more loyal to their colonial masters (Kirk-Greene). Nigeria offers one example of how the "martial races" phenomenon played out. The British, who tended to view Muslim Africans as potentially more loyal and disciplined soldiers, recruited their Nigerian colonial army from among the isolated and conservative Hausa and Fulani peoples of the Muslim north, rather than from the Christian Igbo and Yoruba peoples of the south, who were deemed less "martial" because of their more "bourgeois" outlook (Stapleton, 2013, v. 2). That this pattern would have serious negative consequences for an independent Nigeria should come as no surprise.

Until recently, the African continent's interaction with the rest of the world has been largely coerced. First, there was the slave trade, and then the participation of thousands of African soldiers in the world wars fought by their colonial masters.

The French imposed conscription on their West African dependencies and by the end of the First World War had deployed around two hundred thousand African troops on the Western Front. When the Second World War broke out, there were seventy-five thousand West African troops on duty in France, some ten thousand of whom died in the German invasion of 1940 (Echenberg). Perhaps as many as three thousand of these soldiers, including wounded men, were shot out of hand by German troops (Scheck). Incidents such as this convinced many Africans that this conflict, unlike the 1914–1918 war, was in their interest. Large numbers of Africans volunteered for service in the armies of the Free French, and it is no exaggeration to state that until 1944, France was its empire, its armies being composed to a large extent of North and sub-Saharan Africans (Jennings).

Unlike their French allies, the British took exception to the idea of Africans fighting Europeans (Parsons, 1999a). World War I service was largely confined to the East African campaign to seize the colony of Tanganyika from Germany's colonial army, the *Schutztruppe* (Paice; Vandervort, 2009). In total, some 114,000 soldiers from the King's African Rifles and the Royal West African Frontier Force fought for Britain. In addition, large numbers of African civilians were recruited to serve as porters in the East African campaign. An estimated one hundred thousand of them died, almost entirely from overwork, disease, and malnutrition (Hodges). During the Second World War, still eschewing the notion of Africans fighting Europeans, the British diverted large numbers of their African troops to the war against the Japanese in Burma. In all, some eighty thousand West Africans and forty-six thousand East Africans saw service in Burma (Jackson; Killingray, 2010).

It may come as a surprise to learn that, despite the suffering undergone by Africans engaged in wars not of their making and not always in their interest, there was very little agitation for independence among veterans once the wars were over. Fears among Europeans that returning African soldiers, now schooled in Western ways of war, would revolt against their European rulers were nowhere realized.

The colonial army was one of the more troubled legacies passed on to the newly independent African states by the colonial powers. Recruited for the most part from the margins of the populations of the new nations, the soldiers expected their special status to be honored. They demanded that the perks they had enjoyed under the colonial regimes be continued. They made no secret of their intention to take matters into their own hands if their priorities were not met, a threat that, given their monopoly on force, they were able to make good on with disturbing frequency (Welch, 1987; Decalo, 1990, 1998).

Another presumed problematical legacy of the colonial period was the "demilitarization" of African society by reserving the "right to bear arms" exclusively for soldiers in the new colonial armies. For some African historians, forcing the bulk of the African male population to give up the gun and thus their traditional role as hunters and warriors had a profound social and psychological effect from which postcolonial Africa has yet to recover. Ali Mazrui, for example, has written that "Colonialism and Christianity not only demilitarized but demasculated [sic] Africa" (Mazrui, 1975, 196). While it is difficult to show the extent to which this erosion of the male warrior ethic may help explain the recent upsurge in armed conflict in Africa, it may nonetheless shed light on the widespread enthusiasm among militants in resistance movements in Africa in the 1960s and 1970s for Frantz Fanon's famous dictum that the Third World could be effectively "decolonized" only through what he called "counterviolence." As he put it in his much-read book, *The Wretched of the Earth*, "At the level of individuals, violence is a cleansing force. It frees the native from his inferiority complex and from his despair and inaction; it makes him fearless and restores his self-respect" (Fanon, 1996, 94).

Wars of Decolonization

The European imperial powers emerged seriously weakened from the Second World War to face colonial peoples emboldened by the promise of self-determination in the Atlantic Charter and by the knowledge that their aspirations for independence were supported by the new superpowers, the United States and the Soviet Union. By the mid-1960s most European colonies in Africa had become independent.

It is one of the ironies of recent African history that the weakest of all the European colonial powers, Portugal, was the last to leave the field. The Portuguese, for whom retention of their empire was indispensable to their place in the world, fought on against resistance movements in their colonies of Angola, Guinea-Bissau, and Mozambique from 1961 until 1974, when their soldiers, tired of the seemingly endless fighting, overthrew the neo-fascist regime in Lisbon and established a left-wing government that quickly granted independence to Portugal's African colonies (Cann; Pélissier; Venter).

Belgium was the least prepared of any of the colonial powers for the decolonization movement that swept Africa after 1945. Faced with the seeming impossibility of holding onto its colony in the Congo, Brussels hurriedly cut it free in 1960. The new Congo republic thus entered the world of self-governing nations in perhaps worse condition than any of its counterparts in Africa, being almost entirely bereft of a body of citizens who had either the education or political experience to govern. Almost immediately, the new state, today's Democratic Republic of Congo (DRC), became a battleground in the Cold War and subsequently has faced incessant challenges to its territorial integrity and national sovereignty (Nzongola-Ntalaja).

Some colonial peoples, however, had to fight for their freedom. These were, by and large, the African majority in colonies with large white settler populations: Kenya, Southern Rhodesia, and South Africa. Ending colonial rule in the first two cases came only after protracted warfare, led by British troops in tandem with local African supporters in the first instance and conducted in the second by white settler forces and allies in the African community and supported by South Africa. The African insurgents in Southern Rhodesia, today's Zimbabwe, enjoyed support from neighboring independent African states and, to some extent, from China and the USSR (Kriger; Moorcraft and McLaughlin; Ranger, 1985, 1996). South Africa occupies a situation apart, in that its white settler community, unlike that of Southern Rhodesia, had succeeded in achieving sovereignty and had imposed a system of extreme segregation called apartheid on the African majority of the country. Although the black opposition, the African National Congress, possessed an armed wing called the Spear of the Nation, which carried out some attacks within South Africa, it was in the end not violence but the severe dislocation of the domestic economy in the 1980s that brought down the regime and opened the way for majority rule under the leadership of Nelson Mandela. This monumental change occurred despite a bunker mentality among the white minority in the 1980s that effectively turned over government to the army and the intelligence services (Grundy; Sparks).

The African war of decolonization that has attracted the most attention from historians lately is the Mau Mau Rebellion in the British colony of Kenya in 1952–1960. The uprising was both a war of national liberation and a civil war. Mau Mau's support came largely from the poorer elements of the majority Kikuyu tribe: peasants, landless farm laborers, and urban slum dwellers (Maloba; Macharia and Kanyua). The guerrillas were opposed by the Kikuyu elite, led by Jomo Kenyatta, the future first president of independent Kenya. The backing of this elite was crucial to the British victory (Branch). British success in Kenya traditionally had been lauded as another triumph for the British Army's "hearts and minds" approach to counterinsurgency, similar to that achieved earlier in Malaya (Mumford). Lately, however, discovery of British government documents presumed to have been destroyed after the defeat of Mau Mau has exploded this cheerful view. That the British had used severe methods of repression, on a par with contemporary French practices in Algeria (Klose) and including large-scale resettlement of the Kikuyu population and concentration camps in which torture was routinely employed, was first laid bare in 2005 in Caroline Elkins' *Imperial Reckoning: The Untold Story of Britain's Gulag in Kenya*. More recent studies have confirmed and expanded upon its findings (Anderson; Bennett).

Warfare in Independent Africa

National independence did not always bring the peace that leaders of the struggle for freedom had envisaged. Although they have been few and far between, largely due to the 1963 declaration of the Organization of African Union guaranteeing the frontiers inherited from the colonial powers, interstate conflicts continued in Africa throughout the 1960s–1980s. The 1978–1979 invasion of Uganda by Tanzania to overthrow the dictator Idi Amin, who had attempted to annex Tanzanian territory, and the intervention of Rwanda in the neighboring DRC and Burundi as part of the Hutu/Tutsi conflict are examples.

Interstate warfare increased in the 1980s with the outbreak of a civil war in Angola that ended up as an international struggle with Cold War implications. The struggle for political power in Angola involved three domestic movements divided along ethnic/regional and ideological lines (Guimarães): the Marxist MPLA (People's Movement for the Liberation of Angola), opposed by two pro-Western nationalist factions, the FNLA (National Liberation Front of Angola), and the UNITA (National Union for the Total Independence of Angola). When this internal power struggle began to take on Cold War overtones, the MPLA secured material support from the Soviet Union and, ultimately, ground troops from Cuba. The FNLA, meanwhile, was supported internationally by Zaire (DRC) and the United States. Later, as the FNLA faded, the mantle of opposition to the MPLA passed to the UNITA, which would win material support from the United States and from South Africa, which sought to prevent Angola from being used as a sanctuary for opponents of South African rule in neighboring South West Africa (today Namibia). The international phase of the struggle ended in 1988, although the internal contest between the MPLA and the UNITA rumbled on until 2002 (Gleijeses, 2002, 2009, 2013).

Although secessionist movements bedeviled many of the new nations in Africa following independence, most notably the attempt by the province of Katanga, incited by foreign interests seeking to maintain their control of its mineral wealth, to break free of the DRC, the Biafran war of secession in Nigeria during 1967–1970 continues to have the most resonance today and thus has attracted the most attention from historians. The war claimed the lives of 90,000–120,000 soldiers and as many as three million civilians.

Britain, the former colonial power, must bear some responsibility for this civil war. For reasons largely of administrative convenience, the British had in the early 1900s effected a forced marriage among the three previously separate regions of its Nigerian colony, bringing together the two Christian, commercially oriented regions of the south, ultimately the center of Nigerian oil production, with the poorer Muslim, agrarian north. The British had also drawn their colonial army in Nigeria from among the presumably more "martial" population of the north. This was not an amalgamation that was destined to survive independence in 1960 without strife. The establishment by the Igbo population of the southeast of an independent state of Biafra in May 1967 prompted attack by the larger northern military. In large part because of the country's oil reserves, the civil war attracted international attention. The north received military support from Britain and the USSR, while Biafra received largely humanitarian aid from a number of predominantly Christian African countries and France. (French doctors who served in Biafra would go on to form the famous *Médecins sans Frontières* or Doctors Without Borders organization.) Although the weight of numbers and military hardware would bring on the collapse of Biafra in January 1970, memories of the war linger on in the country to complicate efforts to achieve national unity (Maier; Njokwu; Ekwe-Ekwe; Falola).

1980 to the Present: "Ragged Wars"

From the 1980s to the present, Africa has been subject to new types of conflicts. Most of these can be called low-intensity conflicts (LICs); they include coups, civil wars, wars of secession, and what have become known as "ragged wars." LICs are unconventional counterinsurgency struggles, featuring small cadres of soldiers, guerrillas, civilian militias, paramilitary forces, or private "security contractors," and are basically intrastate wars, though they have been known to spill across national frontiers, as in the recent case of Liberia and Sierra Leone (Reno, 1999, 2011). They usually occur when populations become alienated from the state that rules them or find themselves in conflict with fellow countrymen whom they have come to consider enemies because of their religion, ethnicity, or place of birth. Standing armies cease to function and give way to small cadres of fighters, instruments of the beleaguered ruler or some rebel leader. In response, communities organize their own militias to defend themselves and to fight opposing communities, often using "scorched earth" tactics. The classic analysis of the LICs that have dominated warfare in Africa in recent years is William Reno's *Warfare in Independent Africa*.

The states of Liberia, Sierra Leone, the DRC, Congo Brazzaville, Mali, Niger, Nigeria, and Mauritania have all experienced LICs. Indeed, the recent history of the DRC demonstrates that LICs in the different forms described here can plague a single country simultaneously. Gérard Prunier's aptly titled *Africa's World War: Congo, the Rwandan Genocide and the Making of a Continental Catastrophe* is the most comprehensive work on the conflicts in Central Africa. Prunier shows how the flight of Rwandan Hutu refugees into the DRC, Rwanda's invasion of that country in pursuit, and its subsequent efforts to topple the DRC's government created a power vacuum that drew Rwanda, Uganda, Angola, Zimbabwe, the Sudan, and other African countries into a chaotic war that continues to this day (Stearns).

Mali offers an excellent example of the more common intrastate LIC: ethnic, geographic, and religious in nature. Relations between the country's south, site of the capital, Bamako, and primarily Christian, and its desert north, inhabited by nomadic Muslim peoples or Tuaregs, have been tense for years. The north has long resented southern rule and its inhabitants have frequently rebelled, most recently in 2012. Despite a military response involving French troops and a UN-brokered peace, rebel activity can be expected to continue until the government in the south honors the terms of the "national pact" made in the 1990s, including plans to gradually demilitarize the north, to integrate rebels into special units of the army, and to reduce the severe economic gap between north and south (Perret; Galy; Harmon).

The LICs that have come to symbolize contemporary warfare in Africa are not, however, the more familiar battles between insurgents and conventional forces as in Mali, but, for example, the contests between warlords in Liberia and Sierra Leone that have only just ended. These are characterized by the wholesale "rearming" of the civilian population, the large-scale employment of child soldiers, and the targeting of women.

Africa today is awash in small arms to an even greater extent than it was in the days of the slave trade. These are mostly weapons like the AK-47 assault rifle cast off by the armies of the major military powers as they upgraded their arsenals. With the erosion of the state monopoly on armed force fostered by the colonial powers, small arms have once again begun finding their way into the hands of the population at large. Lest they find themselves unarmed in war zones occupied by rebel forces or government troops, or be themselves the targets of paramilitary operations, African civilians are availing themselves of easily attainable arms to defend themselves and their families. While the provenance of small arms in African conflicts is difficult to document, there are several studies of arms trafficking available to the researcher (Florquin and Berman; Krause).

Although some historians liken the employment of child soldiers in the current wave of low-intensity conflicts in Africa to the raising of age-grade regiments among pastoral peoples, such as the Zulus, in the nineteenth century, the present-day scale is considerably greater and the reasons for the phenomenon significantly different. The combination of overpopulation in parts of Africa and the mass unemployment it has fostered has produced a large body of children available for exploitation. They are cheap labor, and the weaponry available is relatively uncomplicated and easy to employ. Looting provides rewards for child soldiering; drugs and alcohol are available when motivation flags. The phenomenon of children in arms in Africa has received considerable attention from journalists and scholars (Honwana; Dallaire; Sommers; Boyden and de Berry). The specific topic of Joseph Kony's Lord's Resistance Army, largely composed of child soldiers and its run from its Uganda base, have been subjects (Cline; Eichstaedt; Allen and Vlasenroot).

African women have probably been more intimately associated with war making and revolutions, as both victims and protagonists, than their counterparts elsewhere in the world (Maloba). Because of their productive and reproductive capacities, they were prime targets for slave raiders in previous centuries. But they also went to war as caregivers alongside their husbands and formed combat units like the famed Amazons of Dahomey (Alpern; Edgerton). In the modern LICs, however, women have been especially targeted as victims, as rape has become a weapon of war to intimidate and overpower military foes and civilians alike. Though difficult to document, because women understandably are loath to discuss their victimization, a number of books have appeared on this distressing subject (Turshen and Twagiramariya; Coulter; Eze; Baaz and Stern). The subject of rape as a weapon of war has won increased scrutiny as a result of a ruling by the International Criminal Court in 2016 that rape in time of war is both a war crime and a crime against humanity and that commanders can be prosecuted for the crime of rape committed by soldiers under their orders. This ruling was the outcome of a case brought against the vice president of the Central African Republic.

The Future

The topics of African military history that merit further study seem to us to be subsumed in the primary question of change vs. continuity in the relationship between war and society. Is Richard

Reid correct in arguing that warfare in Africa since independence is essentially of a piece with the history of warfare in Africa since earliest times—that is, that there is nothing new in the LICs and other forms of post-independence warfare? If the present, however, does not simply reflect a reassertion of long-term historical dynamics, what factors have created the "new" warfare? Development of these propositions must include study of at least the following areas:

African way of war: Was/is there a distinctive African way of war? What impact has the rise of nationalism had on the goals of warfare in Africa? Are Africans now fighting like everyone else to acquire property or to gain or defend territory, or are other goals now being pursued?

Men and women: There is a great need for the "gendering" of African military history. Showing continuities over time in the role of women in war and changes that have taken place in those roles in the context of the new ways of war since independence would be very useful. In addition, serious attention should be given to Ali Mazrui's suggestion that reactions to the erosion of masculine roles in African society during the colonial period are one of the roots of the upsurge in violence across the Continent since independence. The victimization of women and the use of sexual abuse as an instrument of war have increased over time and appear to be inversely proportional to the existence of female fighting forces. Is this actually the case and, if so, why? Does increased victimization reflect a resumption by African males of their paternalistic role as warriors before colonial rule or simple depravity? Why has the respect inherent in matrilinealism not protected women?

Children: The use of child soldiers has increased over time. Why? Is this use evidence of a change in the nature and goals of warfare since independence that inures children to violence? Which makes violence attractive to them? Or is it the result simply of massive youth unemployment or the breakdown of traditional African society caused by continual warfare that leaves children without families and open to exploitation?

Culture: As the destruction of the ancient shrines and manuscripts of Timbuktu has demonstrated, war in Africa today often includes attacks on an enemy's culture. Have these attacks always been a weapon of war? If not, why are such attacks being made now? How have the goals of war changed to mandate such destruction? Has cultural production played any role in determining or changing the nature of war?

Environment: Traditionally environment has had a determinative impact on warfare in Africa. Is that still the case today? To what extent has the destruction of the environment (water, forests, arable land) been used as an instrument of war? Is its importance greater today than in the past?

Religion: Is the pervasive role of religion in African warfare today a continuation of or a departure from the past? Have "Lord's Resistance"–type armies always existed? If not, what changes in the nature and goals of war explain their existence today? To what extent do armed Islamist factions like Boko Haram represent movements of regional or ethnic dissent as well as religious fundamentalism?

Bibliography

Ade Ajayi, J. F., and R. S. Smith. *Yoruba Warfare in the Nineteenth Century*. Cambridge: Cambridge University Press, 1964.

Allen, Tim, and Koen Vlassenroot. *The Lord's Resistance Army: Myth and Realities*. London: Zed Books, 2010.

Alpern, Stanley B. *Amazons of Black Sparta: The Women Warriors of Dahomey*. New York: New York University Press, 1998.

Anderson, David. *Histories of the Hanged: The Dirty War in Kenya and the End of Empire.* New York: W.W. Norton, 2005.

Baaz, Maria Erickson, and Maria Stern. *Sexual Violence as a Weapon of War? Perceptions, Prescriptions, Problems in the Congo and Beyond.* New York: Zed Books, 2013.

Beachey, R.W. *The Slave Trade of Eastern Africa.* New York: Barnes & Noble, 1976.

Bennett, Huw. *Fighting the Mau Mau: The British Army and Counter-Insurgency in the Kenya Emergency.* Cambridge: Cambridge University Press, 2012.

Boahen, A. Adu. *African Perspectives on Colonialism.* Baltimore: Johns Hopkins University Press, 1985.

Boyden, Jo, and Joanna de Berry, eds. *Children and Youth on the Front Line: Ethnography, Armed Conflict and Displacement.* New York: Berghahn Books, 2004.

Branch, Daniel. *Defeating Mau Mau, Creating Kenya: Counterinsurgency, Civil War, and Decolonization.* Cambridge: Cambridge University Press, 2009.

Cann, John. *Counterinsurgency in Africa: The Portuguese Way of War, 1961–1974.* Westport, CT: Greenwood Press, 1997.

Clayton, Anthony. *Frontiersmen: Warfare in Africa Since 1950.* London: UCL Press, 1999.

Cline, Lawrence. *The Lord's Resistance Army.* Santa Barbara, CA: Praeger, 2013.

Coulter, Chris. *Bush Wives and Girl Soldiers: Women's Lives Through War and Peace in Sierra Leone.* Ithaca, NY: Cornell University Press, 2009.

Curtin, Philip D. *Death by Migration: Europe's Encounter With the Tropical World in the Nineteenth Century.* Cambridge: Cambridge University Press, 1990.

Dallaire, Roméo. *They Fight Like Soldiers, They Die Like Children.* New York: Walker, 2011.

Decalo, Samuel. *Civil-Military Relations in Africa.* Gainesville: Florida Academic Press, 1998.

———. *Coups and Army Rule in Africa: Motivations & Constraints.* 2nd ed. New Haven: Yale University Press, 1990.

Echenberg, Myron. *Colonial Conscripts: The Tirailleurs Sénégalais in French West Africa, 1857–1960.* Portsmouth, NH: Heinemann, 1991.

Edgerton, Robert B. *Warrior Women: The Amazons of Dahomey and the Nature of War.* Boulder, CO: Westview Press, 2000.

Eichstaedt, Peter. *First Kill Your Family: Child Soldiers of Uganda and the Lord's Resistance Army.* Chicago: Lawrence Hill Books, 2013.

Ekwe-Ekwe, Herbert. *The Biafra War: Nigeria and the Aftermath.* Lewiston, NY: Edwin Mellen, 1990.

Elbl, Ivan. "The Horse in Fifteenth Century Senegambia," *International Journal of African Historical Studies* 24, no. 1 (1991): 85–110.

Elkins, Caroline. *Imperial Reckoning: The Untold Story of Britain's Gulag in Kenya.* New York: Henry Holt, 2005.

Eze, Pauline Aweto. *Wartime, Rape, African Values at Crossroads.* Iperu-Remo, Nigeria: Ambassador, 2012.

Falola, Toyin. *The Military Factor in Nigeria, 1966–1985.* Lewiston, NY: Edwin Mellen Press, 1994.

Fanon, Frantz. *The Wretched of the Earth.* New York: Grove Press, 1996 [1961].

Fisher, Humphrey. "'He swalloweth the ground with fierceness and rage': The Horse in the Central Sudan," *Journal of African History* 13, no. 3 (1972) and 14, no. 3 (1973).

Florquin, Nicolas, and Eric G. Berman, eds. *Armed and Aimless: Armed Groups, Guns and Human Security in the ECOWAS Region.* Geneva: Small Arms Survey, Graduate Institute of International and Development Studies, 2005.

Fukui, Katsuyoshi, and David Turton, eds. *Warfare Among East African Herders: Papers Presented at the First International Symposium, Osaka, September, 1977.* Osaka, Japan: National Museum of Ethnology, 1979.

Galy, Michel André. *La guerre au Mali: comprendre la crise au Sahel et au Sahara: enjeux et zones d'ombre.* Paris: Découverte, 2013.

Gleijeses, Piero. *Conflicting Missions: Havana, Washington and Africa, 1959–1976.* Chapel Hill: University of North Carolina Press, 2002.

———. *The Cuban Drumbeat: Castro's World View: Cuban Foreign Policy in a Hostile World.* New York: Seagull, 2009.

———. *Visions of Freedom: Havana, Washington, Pretoria and the Struggle for Southern Africa, 1976–1991.* Chapel Hill: University of North Carolina Press, 2013.

Grundy, Kenneth W. *The Militarization of South African Politics.* Bloomington: Indiana University Press, 1986.

Guimarães, Fernando Andresen. *The Origins of the Angolan Civil War: Foreign and Domestic Political Conflict.* New York: St. Martin's Press, 1998.

Gump, James O. *The Dust Rose Like Smoke: The Subjugation of the Zulu and the Sioux.* Lincoln: University of Nebraska Press, 1994.

Harmon, Stephen A. *Terror and Insurgency in the Sahel-Sahara Region: Corruption, Contraband, Jihad, and the Mali War of 2012–2013.* Burlington, VT: Ashgate, 2014.

Headrick, Daniel R. *The Tools of Empire: Technology and European Imperialism in the Nineteenth Century.* New York: Oxford University Press, 1981.

Hodges, Geoffrey. *Kariakor: The Carrier Corps: The Story of the Military Labour Forces in the Conquest of German East Africa.* Nairobi: University of Nairobi Press, 1999.

Honwana, Alcinda Manuel. *Child Soldiers in Africa.* Philadelphia: University of Pennsylvania Press, 2006.

Jackson, Ashley. *The British Empire and the Second World War.* London: Hambledon, 2006.

Jennings, Eric T. *Free French Africa in World War II.* Cambridge: Cambridge University Press, 2015.

Killingray, David. "Colonial Warfare in West Africa." In *Imperialism and War.* Ed. J. A. De Moor and H. L. Wesseling. Leiden: Brill, 1989.

———. *Fighting for Britain: African Soldiers in the Second World War.* Rochester, NY: James Currey, 2010.

Kirk-Greene, A.H.M. "Damnosa Hereditas: Ethnic Ranking and Martial Race Imperatives in Africa," *Ethnic and Racial Studies* 3 (1980): 393–411.

Klose, Fabian. *Human Rights in the Shadow of Colonial Violence: The Wars of Independence in Kenya and Algeria.* Trans. Dona Geyer. Philadelphia: University of Pennsylvania Press, 2013.

Krause, Keith, ed. *Small Arms Survey: 2015.* Geneva: Graduate Institute of International and Development Studies, 2016

Kriger, Norma. *Zimbabwe's Guerrilla War: Peasant Voices.* Cambridge: Cambridge University Press, 1992.

Laband, John. *Rope of Sand: The Rise and Fall of the Zulu Kingdom in the Nineteenth Century.* Johannesburg: Jonathan Ball, 1995.

———. "The Slave Soldiers of Africa," *Journal of Military History* 81, no. 1 (2017): 9–38.

Lamphear, John, "Sub-Saharan African Warfare." In *War in the Modern World Since 1815.* Ed. Jeremy Black. London: Routledge, 2003.

Law, Robin. "Horses, Firearms and Political Power in Pre-Colonial West Africa," *Past and Present* 72 (1976): 112–132.

Legassick, Martin. "Firearms, Horses and Samorian Army Organization," *Journal of African History* 71, no. 1 (1966): 95–115.

Macharia, Kinuthia, and Muigai Kanyua. *The Social Context of the Mau Mau Movement in Kenya (1952–1960).* Lanham, MD: University Press of America, 2006.

Maier, Karl. *This House Has Fallen: Nigeria in Crisis.* Boulder, CO: Westview Press, 2002.

Maloba, Wunyabari O. *Africa Women in Revolution.* Trenton, NJ: Africa World Press, 2007.

———. *Mau Mau and Kenya: An Analysis of a Peasant Revolt.* Bloomington: Indiana University Press, 1993.

Mazrui, Ali A. "The Resurrection of the Warrior Tradition in African Political Culture," *The Journal of Modern African Studies* 13, no. 1 (1975): 67–84.

Moorcraft, Paul L., and Peter McLaughlin. *Chimurenga! The War in Rhodesia, 1965–1980.* Marshalltown, Zimbabwe: Sygma/Collins, 1982.

Mumford, Andrew. *The Counter-Insurgency Myth: The British Experience of Irregular Warfare.* New York: Routledge, 2012.

Njokwu, H. M. *Tragedy Without Heroes: Nigerian-Biafran War.* Enugu, Nigeria: Fourth Dimension, 1987.

Nzongola-Ntalaja, George. *The Congo, From Leopold to Kabila: A People's History.* London: Zed Books, 2002.

Ogot, Bethwell A., ed. *War and Society in Africa.* London: Frank Cass, 1972.

Paice, Edward. *Tip and Run: The Untold Tragedy of the Great War in Africa.* London: Weidenfeld & Nicolson, 2007.

Parsons, Timothy H. *The African Rank-and-File: Social Implications of Colonial Military Service in the King's African Rifles, 1902–1964.* Portsmouth, NH: Heinemann, 1999a.

———. "'All *Askaris* Are Family Men': Sex, Domesticity and Discipline in the King's African Rifles, 1902–1964." In *Guardians of Empire: The Armed Forces of the Colonial Powers c. 1700–1964.* Eds. David Killingray and David Omissi. Manchester: Manchester University Press, 1999b.

Pélissier, René. *Le naufrage des caravelles: études sur la fin de l'empire portugaise, 1961–1975.* Orgéval, France: Pélissier, 1979.

Perret, Thierry. *Mali: une crise au Sahel.* Paris: Karthala, 2014.

Prunier, Gérard. *Africa's World War: Congo, the Rwandan Genocide and the Making of a Continental Catastrophe.* New York: Oxford University Press, 2008.

Ranger, Terence O. *Peasant Consciousness and Guerrilla War in Zimbabwe: A Comparative Study.* Berkeley: University of California Pres, 1985.

———. *Society in Zimbabwe's Liberation War.* Portsmouth, NH: Heinemann, 1996.

Reid, Richard J. *Warfare in African History.* New York: Cambridge University Press, 2012.

Reno, William. *Warfare in Independent Africa.* Cambridge: Cambridge University Press, 2011.

———. *Warlord Politics and African States.* Boulder, CO: Lynne Rienner, 1999.

Scheck, Raffael. *Hitler's African Victims: The German Army Massacres of Black French Soldiers in 1940.* Cambridge: Cambridge University Press, 2006.

Smith, Robert S. *Warfare and Diplomacy in Pre-Colonial West Africa.* Madison: University of Wisconsin Press, 1989.

Sommers, Marc. *The Outcast Majority: War, Development and Youth in Africa.* Athens: University of Georgia Press, 2015.

Sparks, Allister. *Tomorrow Is Another Country: The Inside Story of South Africa's Road to Change.* New York: Hill & Wang, 1995 [1994].

Stapleton, Tim. *A Military History of Africa*, 3 vols. Santa Barbara, CA: Praeger, 2013.

Stearns, Jason. *Dancing in the Glory of Monsters: The Collapse of the Congo and the Great War of Africa.* New York: Public Affairs, 2012 [2011].

Thornton, John K. "Armed Slaves and Political Authority in Africa in the Era of the Slave Trade, 1450–1800." In *Arming Slaves From Classical Times to the Modern Age.* Eds. Christopher Leslie Brown and Philip D. Morgan. New Haven, CT: Yale University Press, 2006, 79–94.

———. *Warfare in Atlantic Africa, 1500–1800.* London: University College London Press, 1999.

Turshen, Meredith, and Clotilde Twagiramariya, eds. *What Women Do in Wartime: Gender and Conflict in Africa.* New York: Zed Books, 1998.

Uzoigwe, G. N. "Pre-Colonial Military Studies in Africa," *Journal of Modern African Studies* 13, no. 3 (1975): 469–481.

———. "The Warrior and the State in Precolonial Africa," *Journal of African and Asian Studies* 12, nos. 1–4 (1977): 20–47.

Vandervort, Bruce. "New Light on the East African Theater of the Great War: A Review of English-language Sources." In *Soldiers and Settlers in Africa, 1850–1918.* Ed. Stephen M. Miller. Leiden: Brill, 2009.

———. *Wars of Imperial Conquest in Africa, 1830–1914.* Bloomington: Indiana University Press, 1998.

Venter, Al J. *Portugal's Guerrilla Wars in Africa: Lisbon's Three Wars in Angola, Mozambique and Portuguese Guinea 1961–1974.* Solihull: Helion, 2013.

Welch, Claude E., Jr. *No Farewell to Arms? Military Disengagement From Politics in Africa and Latin America.* Boulder, CO: Westview Press, 1987 [1986].

———, ed. *Soldier and State in Africa: A Comparative Analysis of Military Intervention and Political Change.* Evanston, IL: Northwestern University Press, 1970.

7   War and the Ancient Mediterranean

Matthew Trundle

War was a central aspect of antiquity. The close ideological connection between war, society, politics, and wealth meant that military service was intertwined with status and power. The period from 1000 BCE–500 CE presents a cohesive image of war and society in Greek and Roman contexts. War was conducted primarily by infantrymen, often heavily armored and drawn from a class of society that held land and had the resources to furnish their own equipment and to spend time in campaigning away from their property. The historiography of military developments, primarily in Greece and Rome from the so-called Dark Age (1000 BCE) to the end of the Roman Empire in the fifth century of our era, is rich. The Mediterranean Basin and the western parts of Asia (from the Indus Valley westward) were highly interconnected through trade, intellectual exchange, and of course military encounters. Several major empires (notably those of the Assyrians, Persians, Macedonians, and Romans) also did much to unite these regions. Interconnection through technology, such as iron weapons, dominated the age as the principal metal of edged weapons.

Some authors have attempted to draw broad and cohesive pictures of warfare in antiquity. Thus, J. E. Lendon's *Soldiers and Ghosts* (2005) attempts to show how the common cultural-historical roots of Greek and Roman soldiers drew their inspiration from Homer's poems, and ancient myths and legends cast a spell over all subsequent Greek and Roman warfare as Classical Greeks, Macedonians, and Roman legionaries drew inspiration from the past. Both Greek and Roman warfare have good textbook-style overviews that serve as both introductions and gateways to further study as well as discussions of useful ongoing debates. Hans van Wees' (2004) *Greek Warfare: Myths and Realities* does this with aplomb for the Archaic and Classical Greek world, while Louis Rawlings' *Ancient Greek Warfare* (2007) also presents a good introduction to a range of issues, including types of soldiers, battlefield experiences, and logistical support. Roman warfare is similarly well served in this regard, most recently by Jeremy Armstrong's *War and Society in Early Rome: From Warlords to Generals* (2016b). But the Republic and Empire together have also enjoyed good general analyses in recent years (see Keppie, 1998; Goldsworthy, 2000; Southern, 2007; Erdkamp, 2007; Rawlings, 1999), while several works also focus on early Roman military developments (Smith, 2006; Rich, 2007; Armstrong, 2016a), the Republic (Sekunda, 1996; Sage, 2008; Brice, 2014), and the Empire (Jones, 1964; Campbell, 1984; Ferrill, 1986; Le Bohec, 2000).

Edited collections of essays on classical warfare have also suggested common links and connections for Greece and Rome, like that of Rich and Shipley (1993), or Garrett Fagan and Matthew Trundle's collection (2010), exploring the legacies of research on chariots and cavalry, economies, and generalship, while most recently Jeremy Armstrong (2016b) pairs Greek and Roman topics, especially on the economic aspects of warfare. David Pritchard (2010) examines the Athenian empire's military legacy, and the central question of the aggressive nature of the Athenian democracy looms large.

Changing academic interests have shifted away from examining military technology, equipment, and engagements in isolation to more broad and incorporative studies that embrace the relationship of war and society. Philologists have given way to more social historians and, in recent years, economic historians. With this in mind, the five major works of W. K. Pritchett published during 1971–1990 under the titles of *The Greek State at War* retain great significance. These volumes explore every aspect of Greek warfare, from a philological and terminological perspective of every reference in texts to specifically isolated phenomena, like military pay, arms and armor, rituals, and hierarchies. It remains an excellent starting point for any study of specific aspects of Greek warfare.

Most significantly for the debates concerning Archaic and Classical Greek warfare are Victor Davis Hanson's books emanating from his doctoral thesis, "Warfare and Agriculture in Ancient Greece," (1980). Hanson built on work done to connect war and society in socioeconomic contexts, like that of French scholars Yvon Garlan (1975), Jean-Pierre Vernant (1974), and Pierre Ducrey (1986). The connection between citizenship and hoplites has been well argued (see Salmon, 1977; Ridley, 1979). Hanson (1983) has argued that the way the Greeks fought wars in the polis period during c. 750–500 BCE was a product of their socioeconomic environment. He showed that Greek farming practices in the Archaic and Classical Ages shaped how farmers, who owned their land and worked it themselves with one or two assistant slaves, defended their territories as a community of heavily armored infantrymen. These farmers could in turn afford heavy armor, which made possible their participation in warfare. War became tailored to socioeconomic conditions. Farmers could not spend much time away from their land, and so developed a form of warfare that settled disputes quickly and efficiently: the pitched battle. Called "hoplite warfare," such battles required limited training and skill. Two groups of heavily armored infantrymen met on a chosen plain, fought a short, sharp encounter in which the losers withdrew—throwing away heavy equipment as they fled—and the victors won a limited but symbolically decisive victory. Neither side lost devastating numbers of men or materials, and both could then return to their farms. Wars appear therefore as *agônal*, almost symbolic affairs, rather than fights to the death or interminable struggles that required long periods of time away from homes and hearths. Wars were swift encounters that settled arguments over marginal borderland. Hoplites as farmers, landholders, and citizens symbiotically emerged as an ideal in the world of the Archaic Greek city (polis). To Hanson, therefore, farming and fighting justified status and cemented the Greek ideal of amateur-citizen-soldier-farmers. Wars in turn were limited, seasonal, and controlled.

Hanson's work has spawned much debate that still rages today. Despite challengers, Adam Schwartz's *Reinstating the Hoplite* (2010) recently supported Hanson's basic arguments, especially with regard to the largely static, defensive, untrained, and immobile nature of the phalanx. Such a formation feeds the idea that hoplites needed little specialization to function independently and relied only on the morale and collective cohesion of the group. Christopher Matthew (2012) has also recently discussed the mechanics of hoplite warfare from a practical perspective. Matthew confirms the defensive qualities of the hoplite and the shield's essential nature for his role on the battlefield, but also the hoplites' superiority against more lightly armored soldiers, like archers and slingers.

Although images of aristocrats dominated warfare, Latacz (1977) was the first to promote the idea that Homer's poems illustrate the significant place of non-elite soldiers on the battlefield. That is to suggest that lines or phalanxes of men took as much a part in battle as their heroic and elite chieftain-aristocrats. Homer's chariots are used to transport heroes to and from the battlefield, though once they appear in battle when Patroclus uses his spear to haul an opponent "like a fish" from his car. Cavalry are almost entirely absent from early Greek descriptions of fighting, though they do appear in Archaic Greek art. As noted earlier,

Lendon (2005) argued that the spell of Homer's heroic poetry and the mythic traditions of the Romans held later generations of Greeks and Romans in thrall to their legendary military past, at one moment hampering their development and at others aiding technological change and improvement.

The emergence of hoplite warfare remains a vexed debate. Sometime after 700 BCE Greek cities adopted a new method of warfare, probably first in the northeast Peloponnese and then spreading elsewhere (Lorimer, 1947; Salmon, 1977; Snodgrass, 1965). Most significant was the introduction of a new type of shield called an *aspis*, often misnamed by modern writers as the *hoplon* (Lazenby and Whitehead, 1996; see also Lazenby, 1977; Holladay, 1982). This shield differed from earlier types in that it had a central armband (a *porpax*) and handgrip inside its rim (an *antilabe*). It covered a user's body from neck to knee, but also protected much of the side of the man to his left. The shield itself could be used as a weapon to push over the enemy, with a front of men presenting a solid wall of shields. Each man in this new phalanx theoretically relied on the others around him to stand their ground for mutual protection.

Scholars have long debated both the introduction of this type of hoplite warfare and the way in which it actually operated in practice. Some believe that it was entirely new, replacing individual heroic combatants fighting with different types of weapons (but only as soloists) with mass-band fighters almost overnight (Snodgrass, 1965). Others believe that mass-band fighting had always existed, but that the introduction of identical arms and armor replaced other types of shields, and other types of soldiers in the phalanx (Latacz, 1977; Salmon, 1977). In this second model it took some time for the lighter infantry to be removed from the phalanx battle line. Scholars who favor this idea still argue about how long it took, with some suggesting that only after the Persian Wars were light infantry finally removed from the phalanx proper (van Wees, 2004; Krentz, 1985). Those who believe this latter point also disagree over whether the change from heterogeneous troop-types to identically armored hoplites happened slowly or suddenly.

The evidence for the transformation from Homeric-style fighting to that of the hoplite phalanx makes it difficult to know for sure what occurred from the eighth to the fifth century BCE. Homer's poems present phalanxes of men armed differently, with archers and other light infantry standing beside more heavily armored fellow-fighters. The poems of Tyrtaeus present light-armed men—called naked ones (*gumnetes*) due to their lack of armor—hiding behind the shields of more heavily armored soldiers and throwing missiles at their enemies. It is clear that by the second quarter of the fifth century at the latest the various different types of troops found in Greek poleis' armies had become separated into distinct units. Hoplites, archers, slingers, and javelin-throwers each now operated in their own space on the battlefield (van Wees, 2004; Trundle, 2011).

The heavily armored hoplite became the paradigm of Greek polis identity. Hoplites came, theoretically, from a specific status-group within Greek communities. Due to the expensive nature of the equipment, shield, helmet, body armor, spear, and sword, only those wealthy enough to afford them could fight as hoplites. The similarity of their equipment reinforced their social-political and economic equality (Hanson, 1983; Runciman, 1998). Some, like Hans van Wees (2004), have challenged the rigidity of this relationship, but in many states where land ownership and civic status were closely connected, hoplites came from the citizen body alone. Hoplites were the farming-landholders (*autourgoi*). At Athens, they came from the class of men called the yoke-men (*zeugitês*), who could afford a pair of oxen to plough their land. Scholars have suggested that the *zeugitai* were also "yoked" together in hoplite service on the battlefield. The paragon of such hoplite equality and identity was the Spartan elite citizens, the *Spartiatai*, also referred to as the *homoioi*: the equals or peers. Their similarity once again reinforced their social and political identity, as well as their military equality as *hoplitai* and vice versa (Cartlegde, 1977; Hanson, 1983; Schwartz, 2009). The Spartan army became the most revered in the later Archaic and early Classical period (Lazenby, 1985; Cartledge, 1987).

As previously noted, hoplite warfare, with its emphasis on single pitched battles, had an *agônal* and ritualistic quality (Runciman, 1998). Some work challenges the notion of ritualized warfare as simply idealism (Krentz, 2002; van Wees, 2004; Trundle, 2010). In a similar vein, hoplites clearly were not the only type of soldier in these emerging Greek poleis and in Archaic Greek warfare. Our evidence, limited though it is, does suggest that wars were more limited in scale than those of later Greek history, but probably due to the ability of states to prosecute major wars rather than their intention (Pritchard, 2010). Scholars have questioned whether there ever was a ritualized "good old days" of ritualized battles fought between hoplite armies alone for marginal land (van Wees, 2004; Trundle, 2010). It is clear that light troops, archers, slingers, and stone throwers had always played some role on Greek battlefields. Our first detailed literary descriptions of battles only really appear in the mid-fifth century BCE, at a time when battles had also clearly become far more complex affairs involving all manner of different troop-types, rather than just hoplite amateurs.

As hoplite warfare became dominant among Greek communities in the later sixth century BCE, larger and more organized navies became increasingly significant. Early Greek warships were designed for raiding coastlines and for piracy, but smaller raiding vessels were replaced by the trireme in the sixth century BCE. Triremes required a great amount of centralized resources to build and maintain. They were far more expensive than their predecessors, and also required a great deal of skill in their maneuvers. According to Thucydides, several states of western Asia Minor, like Miletus, alongside the Phoenicians, the Egyptians, and later Corinth on the Greek mainland, possessed the first trireme navies. Tyrants like Polycrates also possessed triremes in some number. Athens developed its trireme navy later than these states. In 483 BCE the Athenians discovered a rich vein of silver in Laurium, which they invested in building a fleet of new triremes. By the time of the Great Persian War, the Athenians possessed around two hundred triremes, more than all the other states of mainland Greece combined. It was these ships that laid the foundation for the naval victory over the Persian-led Phoenician and Ionian fleets at the great naval Battle of Salamis in 480 BCE.

Athenian triremes look to us like mini-Athenian communities (Morrison, 2000). Strauss (1996) argued that the trireme contributed to the democratic feelings of ordinary Athenians as they mixed with rich ship-owners and fellow oarsmen. Wallinga (1993) and others have challenged this idealized notion. The ship's commander came from the highest socioeconomic group, the men called the *pentekosiomedimnoi*, or five hundred measure men, and had often contributed in part to the building or fitting out of the ship. On the deck a group of specialists (*hyperesiai*) oversaw the rigging, steering, and general maintenance of the ship, while heavy armored infantry, called passengers (*epibatai*) by our sources, protected the ship from boarders (van Wees, 2004). In addition to these *epibatai*, the Athenians also established an archer corps drawn from the poorer citizens of Athens (Trundle, 2011). Archers were especially useful on ships for disrupting enemy oarsmen, clearing enemy decks, and getting men already capsized at sea to surrender. Beneath the deck, the three banks of oarsmen might well have been divided socially. The top bank of rowers (*thranitai*) pulled a longer oar and received higher pay; perhaps these came from Athenian poor citizens or *thetes*. The middle-bank oarsmen (perhaps called *zygioi*) and the bottom-bank, the least pleasant place to row, were called the men in the hold (*thalamioi*). Athenian *thetes* rowed in these ships, though oarsmen also came from the resident foreigners living in Athens (*metics*) and other foreigners (*xenoi*) from within the Athenian Empire, and even slaves, whose pay went to their masters. We have some evidence to show that slaves and masters rowed on the same ship. Thus, contra Strauss,

Athenian triremes were cross sections of Athenian society from top to bottom rather than floating ships of equals (Morrison, 2000).

Triremes maneuvered to ram their opponents, or perhaps more likely to shear off one side of an enemy ship's oars, thus crippling the vessel. Over time maneuver-and-ram gave way to grapple-and-board tactics whereby one ship's crew would try to board an enemy ship and take it by force. By the end of the fifth century BCE triremes were being replaced by fours and fives, bigger vessels still with yet more oarsmen, crew, and marines. The Romans conquered the Mediterranean with fives, sometimes armed with the raven (*corvus*), which was a boarding plank with a grappling hook that acted as both a pinning mechanism and gangplank along which enemy ships could be attacked (Pitassi, 2012).

Despite their significance, historians have often overlooked the importance of navies. Thus, Chester Starr's (1989) significant contribution to the subject noted that navies played a secondary role in the conquests of great empires, like the Persian, the Roman, and finally those that resulted in the fall of Rome. Even a naval power like Carthage is best remembered for Hannibal's territorial victories in Italy, rather than any naval prominence. Yet, naval power enabled Athens to control the Aegean, Carthage and Rome in turn to control the western Mediterranean, and ultimately the Romans to dominate all of the Mediterranean. Our ancient sources played down naval power's importance, due principally to the social status of heavy infantrymen. Recent work by Louis Rawlings (2010) explored the Carthaginian navy's significance.

Many (Gabrielsen, 1994; Trundle, 2004, 2010) now accept that Athenian navies, and naval conflict more generally, led the way in professionalizing and centralizing warfare in the eastern Mediterranean in the fifth century BCE. Oarsmen, who came from lower-status groups within Athenian society, required payment for their services. The evidence of payments and much concerning early trireme navies is heavily Athenocentric, but it would be interesting to know more about how the Phoenicians (and others) who clearly played a prominent role in early trireme warfare paid for and maintained their fleets. It was not long before payments became common among the infantry, even among citizen-hoplites. At the same time increasing numbers of specialist troops appeared in Greek contexts. As we have already noted Greek states were often reluctant to utilize the services of slaves in military activity, for obvious reasons, but they also often preferred to hire in specialists rather than arm and train the poor with weapons. Thus, mercenaries often provided much-needed specialist support to heavy armed hoplites in the form of more lightly armed troops carrying lighter wicker or smaller wooden shields and javelins, archers, and slingers (Parke, 1933; Griffith, 1935; Trundle, 2004).

Larger numbers of outsiders serving for pay appeared in Greek warfare at the end of the fifth century BCE, fueled by the role of coin money in the eastern Mediterranean. Traditionally, historians explained the rise of professionals in terms of supply, the result of the series of Peloponnesian Wars in the later fifth century (Parke, 1933). More recently, Trundle (2004) argued that mercenary service, if it can be called such, was both demand-driven and connected closely to traditional networks of aristocratic, ritualized friendship called by the Greeks *xenia*. As wars proliferated and the western Persian Empire became less stable in the latter fifth and through the fourth century BCE, Greek hoplites served with the Persian king and governors, and Egyptian pharaohs who lacked heavy infantry—but well-connected elites—facilitated this service. At the same time, light troops came into Greece to augment polis-hoplite armies. The regions of the northern Aegean, principally Thrace, became synonymous with specialist light infantrymen known as *peltasts* on account of the wicker shield they carried, called a *pelta* (Best, 1976). They threw javelins and could outrun and outmaneuver hoplites easily. These types of troops first appear in the later sixth century, but became increasingly important at the end of the fifth and throughout the fourth century BCE. Greek states regularly hired these men to supplement their heavy infantrymen on the battlefield. Crucial for pitched battles, hoplites remained central to any army, but were now just one arm of increasingly complex military systems on the Greek mainland in the more interconnected world of the Classical Age (Cartledge, 1987; Schwartz, 2009).

Major wars in the later fifth and early fourth centuries BCE transformed the Greek world and Greek warfare. The Second or Great Peloponnesian War (431–404 BCE) saw Athens and its respective "allies" clash with Sparta. In the end Sparta, supported by Persian resources, defeated Athens, which spawned a series of wars for the hegemony of the Greek world. For a while Thebes emerged as the most powerful Greek state, defeating Sparta's revered hoplites in two major set-piece battles, Leuctra (371 BCE) and Mantinea (362 BCE). These victories illustrated the increasingly important role of tactics and generalship on the battlefield. Concentrating their best troops on one wing, while refusing the other, the Thebans could win a battle by targeting a single point in the enemy line without exposing weaker or less well-trained men to battle at all. Once broken in a single place, it was then easy to roll up the rest of the enemy line by turning against the rear and the flank (Buckler, 1980; Cartledge, 1987).

It is perhaps misleading to see Archaic and Classical Greek warfare as *agônal* or even "sporting." The Greeks seem to have had few team sports—and none at the Olympic Games, which focused almost exclusively on individual achievement. Nigel Crowther (1995, 133) stated, "In a way the Greeks did have a competitive 'team game' and that was war." Like Runciman's and Hanson's views of war as ritualized, it makes for dangerous generalizations to see war as anything other than it was. As stated in the work of Christopher Matthew, synthesizing practical reenactment using reconstructed arms and armor with ancient source materials shows what might be achieved in understanding the nature of ancient fighting techniques. Similarly, Jason Crowley's *The Psychology of the Athenian Hoplite* (2012), using techniques of modern academic psychology, explored the intricate relationships that connected hoplites with their comrades and their communities.

The Kingdom of Macedonia emerged as a major military force in the mid-fourth century BCE (Launey, 1987; Bosworth, 1988; Chaniotis, 2005; Bugh, 2006). The Macedonian kings, notably Philip II (359–336 BCE) and his son Alexander III (336–323 BCE), could utilize enormous mineral resources to pay soldiers who themselves had a national sense of identity. The Macedonian army they forged overcame first the Greek cities to the south (356–338 BCE) and then the whole of the Persian Empire (334–325 BCE). At the heart of Macedonian military power was the cavalry, made up of aristocrats styled as companions (*hetairoi*), and the phalanx or foot-companions (*pezhetairoi*). This phalanx contained pikemen organized into squares of 256 (*syntagma* or *speira*), 16 deep, carrying a 5–6-meter spear (*sarissa*) and a small round buckler-type shield that enabled both hands to grasp the weapon. The cavalrymen—armed with thrusting spears rather than javelins, longer slashing swords, breastplates, and helmets, but probably not shields—held the right wing. An infantry unit called the shield-bearers (*hypaspistai*), perhaps armed like traditional hoplites, linked the phalanx and the cavalry. The battlefield tactics of the Macedonians mirrored that of the Thebans: refusing one wing from the enemy, cavalry and shield-bearers led the other and drove home the initial attack. Ideally, this wing would easily punch through the enemy line at a crucial point (e.g., Alexander targeted the Persian king) and then roll it up from one side and its vulnerable rear (Bosworth, 1988). Historians like Donald Engels (1978) have explored the significant role of logistical support behind the Macedonian military machine. Engels showed how the Macedonians were able to convey their food and equipment with less need for cumbersome baggage trains. This enabled Alexander's forces to move quickly over rough terrain as they traversed Asia.

The successors of Alexander who ruled over the separate kingdoms of Macedonia, Egypt, Pergamum (in northwestern Asia Minor), and the Seleucid Kingdom of western Asia refined further Macedonian tactics. Elephants, which Alexander had much admired in India, now augmented the battle line (Chaniotis, 2005; Launey, 1987). War had by the third century become more specialized, sophisticated, and professional than ever (Bugh, 2006; Griffith, 1935). The larger geographic stage rendered the old Greek hoplite-style warfare redundant. Alexander the Great's legacy to his successors, and subsequently to the Romans, was to make wars bigger, still more professional and efficient, and destructive. At Gaugamela in 331 BCE, the Persians could have lost as many as one hundred thousand men, similarly the Romans at Cannae fighting Hannibal's genius Romans perhaps could have lost as many as sixty thousand. In turn the Romans won great victories at little cost to themselves over the Carthaginians, Macedonian, and Seleucid kings. The secret was the linking of heavy infantry, lighter armed troops, and both heavy and light cavalry with increasingly sophisticated tactics. Chaniotis (2005) examined war and society in the Hellenistic world, but much more could be done to explore this period and the regions of Alexander's conquests.

The Macedonians also appeared adept with siege warfare. Winter (1971) explored the topic of Greek fortifications in detail. Other studies have examined catapult and other aspects of Greek siege technology (Kern, 1999; Campbell, 2003; de Souza, 2009). Most scholars accept that the Greeks developed siege technology from contact with the eastern powers. The Assyrians mastered siege technology in the first half of the first millennium BCE. Their great bas-reliefs testify to their engines, rams, and sappers demolishing city walls. Such siege technology spread among the peoples of the Near East as kings learned from one another. The Persians inherited Assyrian siegecraft skills, though the Greeks were relatively slow to develop such technology, which may explain why walled cities appear on the Greek mainland only in the later sixth century. In contrast, the Greeks of western Asia Minor, probably in response to siege technology of their eastern neighbors, built walls earlier. While the Spartans and Athenians had some siege machines and towers during the Great Peloponnesian War, they usually constructed lines of circumvallation and waited to starve a city into surrender. This of course took time and exposed the besiegers to a great deal of danger themselves, as the Athenians discovered to their detriment and total annihilation against Syracuse in 413 BCE.

The first Greek known to have developed sophisticated siege engines was Dionysius I, the Tyrant of Syracuse during 405–367 BCE. Almost certainly stimulated by proximity to the Phoenician-Carthaginians, Diodorus records an elaborate series of sieges, involving impressive machines and countermeasures against these, in the wars between Syracuse and the Carthaginians in the early years of the fourth century BCE. Philip and Alexander adopted siege technology with aplomb, and Alexander in particular successfully took great cities like Tyre and Gaza by storm, and several daunting mountaintop cities in the east, using impressive siege technology. As with other aspects of military science, siege technology enabled attackers to defeat their opponents more quickly than through starvation and so facilitated the great and relatively quick conquest of the east by Alexander the Great (English, 2009).

Into the world of Alexander's successor kingdoms came the Romans. The Roman army developed in an Italian context, influenced by Greeks to the south, Etruscans to the north, the peoples of central Italy, and, as increasingly appreciated by modern scholarship, the Celts of the far north. The Romans thus combined the qualities of heavy infantry, carrying large rectangular shields (the *scutum*) combined with missile weapons and short stabbing swords. Despite the size of the shield (though it probably weighed less than the Greek *aspis*), Roman formations seem more flexible and fluid than those of Greek and Etruscan hoplites. Historians, both ancient and modern, have often remarked on the flexibility and mobility of Roman armies, and in recent years the paradigm of the "cloud" to describe bodies of men, fluidly expanding and contracting in a motion-filled battlefield environment, has become common (e.g., Quesada Sanz, 2006). In this environment horsemen might well have played a significant role in the fighting. By the time of the so-called expulsion of the Etruscan kings, traditionally in 509 BCE, Rome was already a major city commanding the important crossing point on the river Tiber. Nevertheless, it was not until 270 BCE that Rome finally emerged as the dominant center of almost all of Italy.

In 264 BCE the first of Rome's great transmarine wars began. Twenty-three years later it had conquered almost all of Sicily (Goldsworthy, 2001; Hoyos, 2011; Lazenby, 1996). In the process, Rome had become a naval power. The subsequent Second Punic War (218–202 BCE) saw Hannibal the Carthaginian cross the Alps into Italy and defeat the Romans in several major battles on home soil before being recalled and defeated on the field of Zama (Lazenby, 1998; Daly, 2002). The third (149–146 BCE) and final war saw Carthage besieged and destroyed. Rome in the meantime had been drawn into Spain, and into wars against the successor kings of Alexander in Macedonia (215–207 BCE, 200–196 BCE, and 171–167 BCE) and the Seleucids of Syria (192–188 BCE). By the middle of the second century BCE Rome was the indisputable master of the Mediterranean world. As in the Greek world social and military status appears symbiotic in this period. Aristocratic senators provided the commanders of the army as annually elected consuls and praetors who commanded troops through *Imperium*, while the backbone of the army came in theory from citizen-farmers.

Both the Roman cavalry and navy have received less attention than they ought outside of large compendia volumes. The cavalry came from a class called the knights, but as with the Greek states, the cavalry made up the weakest part of the military forces, at least by the third century BCE. Indeed, after the Roman cavalry was driven from the field at Cannae in 216 BCE the Romans abandoned a citizen cavalry force entirely, and the title of horseman became a symbolic and aristocratic title (*eques*) only (Connolly, 1988). The Roman navy was rowed by men of the lowest census class, sometimes called the *proletarii* or men whose heads counted only in the census (*capite censi*). The navy had enabled the conquest of Sicily and the subsequent movement of troops into the eastern Mediterranean (most recently Pitassi, 2012).

The Roman Republican army was organized into legions of 4,000–5,000 men. The infantry of the early Roman army was evidently differentiated by wealth and military equipment, with the core of the army, the "first class," supposedly featuring hoplite-style equipment, with the rest of the army (the 2nd through 5th classes) utilizing incrementally less equipment and acting in a supporting role. By the middle Republic and the rise of the manipular legion, this timocratic system of military differentiation had become a tactical one, with Rome's army featuring heavily armed *hastati* (literally spearmen) and *principes* (the first men) armed with the large oblong shield (*scutum*), two javelins each called a *pilum* (which had a long, thin iron spike attached to a wooden shaft), and a short thrusting sword (*gladius*). By the late Republic, the legionary would throw each of his two *pila* at a charge and follow up with his short sword, which was particularly effective against the ponderous heavily armed pikemen fielded by the Successor Kingdoms of Alexander's empire. Finally, a group of spearmen, the third-line men (*triarii*), acted as both a last line of attack and defense, armed as they were with a long thrusting spear. Light troops (*velites* or *rorarii*) supported the maniples of heavy infantry and often began battles with skirmishes.

As Rome's Mediterranean empire grew, so did the need for more soldiers. Over time poorer men entered the army, which became increasingly professional and less connected to the landed citizen-class. Marius was supposedly the first to formally enroll the poor into the legions at the end of the second century BCE, but it is likely that in practice many men had served in the previous century who did not meet the census requirement. Traditionally, scholars saw a neat fit between the changing status of the men who made up the legions and the decline of the Roman Republic. Thus, scholars like Scullard (1980) saw the professionalization of the army connected closely to the rise of independent generals able to challenge the senate. Recent works on the changing relationship between landholding and soldiering have challenged traditional beliefs, such as Brunt's examination of the status and background of the men in the legions during the middle Republic (1971). Rosenstein (2004) challenged Brunt's figures and argued that as a result of Roman success in the Mediterranean, small-holders who contributed much to the army had more children, thus contributing to a population expansion. The resulting land shortage led increasing numbers of poorer (but not entirely destitute) men to seek livelihoods either in the urban center or in service in the legions.

In time, professional soldiers started to see their generals rather than the state as potentially their best interest for retirement and pensions. In addition, the Italian allies who supplemented Roman forces grew increasingly disgruntled at what they saw as the burdens of empire without the benefits. Conflict between Rome and the Italians (the Social War of 90 BCE), and then between the generals and the state, followed. These in turn laid the conditions for a series of civil wars that rocked the Republic and ultimately culminated with Octavian's eventual victory over Antony and Cleopatra at Actium in 31 BCE. Following their deaths Octavian was free to reorganize the state and install a new order.

Octavian took the name Imperator Caesar Augustus in 27 BCE. He reformed the Republic and became the first Roman emperor. (Traditionally, the name Imperator was bestowed upon a victorious general after battle.) He reorganized the army, reducing it to just twenty-eight legions of just under five thousand men each. Thereafter the number of legions fluctuated from twenty-five to thirty through the life of the empire, employing about 150,000 men (notably 15 percent of the empire's total male population) as professional soldiers. All legionaries swore an oath to the emperor and he paid them personally, even those stationed in provinces still controlled by members of the senate. A further 150,000 non-Roman citizens formed the auxiliaries drawn from within the empire, who provided skilled light infantry, archers, slingers, and cavalrymen. On their retirement these men received Roman citizenship. Under Augustus the empire doubled in size. Little further expansion followed, and Rome's borders increasingly came under pressure as peoples from beyond the borders sought to enter the empire in greater numbers. Roman emperors commanded and paid all the legions and claimed the right to triumphs and to glory from war as the ultimate commander-in-chief. Roman history shows that the last non-imperial family member to hold a triumph in the city was Balbus in 19 BCE. Subsequently only the emperor or his nominees could triumph, and increasingly only they would lead armies in the field—ensuring that only these men could indeed receive a triumph or the great martial glory.

Edward Luttwak in his *The Grand Strategy of the Roman Empire* (1976) argued that the Romans employed a well-planned policy whereby client kingdoms protected the frontiers and absorbed invasions before they actually entered the empire proper. Because Luttwak was not a professional classicist, his ideas concerning Roman strategy caused much controversy. Thus scholarship has challenged his views, delineating an approach to the frontiers that was far more ad hoc and regional, and also reflected the whims of imperial leadership. For example, in response to Luttwak, Benjamin Isaac (1990) argues for a more haphazard policy through examination of Palestine and the east, though Roman history more broadly seems to support such randomness. In a two-part article, Everett Wheeler (1993) published a good synthesis of these debates. Stephen Dyson (1985) suggested that, from early times, the frontier was essentially wherever the army could get to quickly. Walls, like that of Hadrian, were as much conduits and roads as barriers to access, and appeared only in specific parts of the Roman Empire.

By the third century CE Rome found itself challenged by both internal revolts and external invasion. From this period of turmoil the later Roman army emerged (recently see Lee, 2007). In the east a new and aggressive Persian or Sassanid Empire replaced the earlier Parthian Kingdom, threatening the eastern Roman world, while new northern tribes began pressuring the Rhine frontier. Under these challenges the Roman Empire appears to have collapsed during the so-called third-century crisis, a fifty-year span in which separate states emerged within the imperial sphere. These included the Roman Empire of the Gauls in what is now France, and a separate eastern empire based on the caravan city of Palmyra. Thanks to energetic emperors like Aurelian and then Diocletian, the Gallic states and Palmyra were brought back within the imperial fold. But the empire, and the Roman army, had been transformed for good. The old legions on the frontiers now became secondary to the main focus of military strategy. The emperors formed elite units ("companions," or *comitatenses*) in reactive mobile armies of infantry and cavalry that accompanied them to threatened points within the empire. Meanwhile legionnaires on the frontier (*limitanei*) became second-class soldiers defending local spaces on the frontier. They were lower paid than the troops who accompanied the emperors, who provided the shock troops of combat.

In addition, and little by little, the empire began to increasingly rely on non-Roman troops to defend parts of the empire. There is a vigorous debate about what it meant to be Roman in this period. Emperors allowed tribes to settle in regions on condition they defend that frontier area. These peoples were called allies (*foederati*) but there was an increasingly blurred line regarding their status as to whether they were Rome's subjects or allies who were independent of the imperial system. Arther Ferrill (1986) argued that modern scholars have discounted the military causes for the downfall of the empire in favor of focusing on internal problems, like taxation and an overbearing civil service. Similarly Christianity was often cited as a cause of Rome's collapse, encouraging as it did investment in the afterlife and disengagement from the here and now. This idea went back to Edward Gibbon's monumental *The Decline and Fall of the Roman Empire* (1994), which argued essentially for the triumph of both barbarism and religion. A.H.M. Jones (1964) observed the importance of external pressures on late imperial structures. Whatever the truth of this debate, the empire of the fourth century CE had changed; Rome was sacked twice in the fifth century CE. As the empire's strength and focus became increasingly eastern in outlook and location, its western regions disintegrated into a series of non-Roman kingdoms. The old Roman Empire and the Roman armies were no more.

Bibliography

Armstrong, J. *Circum Mare: Themes in Ancient Warfare*. Leiden: Brill, 2016a.

———. *War and Society in Early Rome: From Warlords to Generals*. Cambridge: Cambridge University Press, 2016b.

Best, J. *The Thracian Peltast and His Influence on Greek Warfare*. Groningen: Wolters-Noordoff, 1976.

Bosworth, A. B. *Conquest and Empire. The Reign of Alexander the Great*. Cambridge: Cambridge University Press, 1988.

Brice, L. *Warfare in the Roman Republic From the Etruscan Wars to the Battle of Actium*. Santa Barbara, CA: ABC-Clio, 2014.

Brunt, P. A. *Italian Manpower*. Oxford: Oxford University Press, 1971.

Buckler, J. *The Theban Hegemony*. Cambridge: Harvard University Press, 1980.

Bugh, G. R. "Hellenistic Military Developments." In *The Cambridge Companion to the Hellenistic World*. Ed. G. R. Bugh. Cambridge: Cambridge University Press, 2006.

Campbell, D. *Greek and Roman Artillery 399 BC–AD 363*. Oxford: Oxford University Press, 2003.

Campbell, J. B. *The Emperor and Roman Army, 31 BC–AD 235*. Oxford: Oxford University Press, 1984.

Cartledge, P. *Agesilaos and the Crisis of Sparta*. London: Johns Hopkins University Press, 1987.

———. "Hoplites and Heroes: Sparta's Contribution to the Technique of Ancient Warfare," *Journal of Hellenic Studies* 97 (1977): 11–23.

Chaniotis, A. *War in the Hellenistic World*. Oxford: Oxford University Press, 2005.

Connolly, P. *Tiberius Claudius Maximus: The Cavalryman*. Oxford: Oxford University Press, 1988.

Crowley, J. *The Psychology of the Athenian Hoplite*. Cambridge: Cambridge University Press, 2012.

Crowther, N. "Team Sports in Ancient Greece: Some Observations," *The International Journal of the History of Sport* 12 (April 1995): 127–136.

Daly, G. *Cannae: The Experience of Battle in the Second Punic War*. London: Routledge, 2002.

De Souza, P. "Greek Warfare and Fortification." In *The Oxford Handbook of Engineering and Technology in the Classical World*. Ed. J. P. Oleson. Oxford: Oxford University Press, 2009.

Ducrey, P. *Warfare in Ancient Greece*. New York: Schocken, 1986.

Dyson, S. *The Creation of the Roman Frontier*. Princeton: Princeton University Press, 1985.

Engels, D. *Alexander the Great and the Logistics of the Macedonian Army*. Berkeley: University of California Press, 1978.

English, S. *The Sieges of Alexander the Great*. Bradford: Pen and Sword, 2009.

Erdkamp, P. *A Companion to the Roman Army*. Malden, MA: Blackwell, 2007.

Fagan, G., and M. Trundle, ed. *New Perspectives on Ancient Warfare*. Leiden: Brill, 2010.

Ferrill, A. *The Fall of the Roman Empire: The Military Explanation*. London: Thames and Hudson, 1986.

Gabrielsen, V. *Financing the Athenian Fleet*. Baltimore: Johns Hopkins University Press, 1994.

Garlan, Y. *War in the Ancient World: A Social History*. London: W. W. Norton, 1975.

Gibbon, E. *The History of the Decline and Fall of the Roman Empire in Six Volumes 1776–1788*. London: Allen Lane, 1994.

Goldsworthy, A. K. *The Punic Wars*. London: Cassell, 2001.

———. *Roman Warfare*. London: Cassell, 2000.

Greenhalgh, P.A.L. *Early Greek Warfare*. Cambridge: Cambridge University Press, 1973.

Griffith, G. T. *The Mercenaries of the Hellenistic World*. Cambridge: Cambridge University Press, 1935.

Hanson, V. D. "Warfare and Agriculture in Ancient Greece." PhD diss., Stanford, 1980.

———. *Warfare and Agriculture in Classical Greece*. Pisa: Giardini, 1983.

———. *The Western Way of War: Infantry Battle in Classical Greece*. Berkeley: University of Los Angeles Press, 1989.

Holladay, A. J. "Hoplites and Heresies," *Journal of Hellenic Studies* 102 (1982): 94–103.

Hoyos, D. A. *Companion to the Punic Wars*. Malden, MA: Blackwell, 2011.

Hunt, P. *Slaves Warfare and Ideology in the Greek Historians*. Cambridge: Cambridge University Press, 1998.

Isaac, B. *The Limits of Empire: The Roman Army in the East*. Oxford: Oxford University Press, 1990.

Jones, A.H.M. *The Later Roman Empire 284–602: A Social, Economic and Administrative Survey*. Norman: University of Oklahoma Press, 1964.

Keppie, L.J.F. *The Making of the Roman Army From Republic to Empire*. Norman: Oklahoma University Press, 1998.

Kern, P. B. *Ancient Siege Warfare*. Bloomington: Indiana University Press, 1999.

Krentz, P. "Fighting by the Rules: The Invention of the Hoplite Agôn," *Hesperia* 71 (January–March 2002): 23–39.

———. "The Nature of Hoplite Battle," *Classical Antiquity* 4 (April 1985): 50–61.

Latacz, J. *Kampfparänese, Kampfdarstellung und Kampfwirklichkeit*. Munich: Zetemata 66, 1977.

Launey, M. *Recherches sur les armées Hellénistiques* (réimpression avec addenda et mise à jour, en postface, par Y. Garlan, P. Gauthier, & C. Orrieux), 2 vols. Paris: De Boccard, 1987.

Lazenby, J. *The First Punic War: Military History*. Stanford: Stanford University Press, 1996.

———. *Hannibal's War: Military History of the Second Punic War*. Norman: Oklahoma University Press, 1998.

———. "Hoplites and Heroes: Sparta's Contribution to the Technique of Ancient Warfare," *Journal of Hellenic Studies* 97 (1977): 11–27.

———. *The Spartan Army*. Warminster: Aris and Phillips, 1985.

Lazenby, J., and David Whitehead. "The Myth of the Hoplite's Hoplon," *Classical Quarterly* 46 (September 1996): 27–33.

Le Bohec, Y. *The Imperial Roman Army*. London: Routledge, 2000.

Lee, A. D. *War in Late Antiquity*. Oxford: Oxford University Press, 2007.

Lendon, J. E. *Soldiers and Ghosts. A History of Battle in Classical Antiquity*. New Haven: Yale University Press, 2005.

Lorimer, H. L. "The Hoplite Phalanx With Special Reference to the Poems of Archilochus and Tyrtaeus," *Annual of the British School of Athens* 42 (1947): 76–138.

Luttwak, E. N. *The Grand Strategy of the Roman Empire*. Baltimore: Johns Hopkins University Press, 1976.

Matthew, C. *A Storm of Spears: Understanding the Greek Hoplite at War*. Barnsley: Pen and Sword, 2012.

Morrison, J. S., J. F. Coates, and B. Rankov. *The Athenian Trireme*. 2nd ed. Cambridge: Cambridge University Press, 2000.

Parke, H. W. *Greek Mercenary Soldiers From the Earliest Times to the Battle of Ipsus*. Oxford: Oxford University Press, 1933.

Pitassi, M. *The Roman Navy: Ships, Men and Warfare 380 BC–AD 475*. Barnsley: Seaforth, 2012.

Pritchard, D., ed. *War, Culture and Democracy in Classical Athens*. Cambridge: Cambridge University Press, 2010.

Pritchett, W. K. *The Greek State at War, Volumes I–V*. Berkeley: University of California Press, 1971–1990.

Quesada Sanz, F. "Not So Different: Individual Fighting Techniques and Small Unit Tactics of Roman and Iberian Armies Within the Framework of Warfare in the Hellenistic Age," *Pallas* 70 (2006): 245–263.

Rawlings, L. "Alternative Agonies. Hoplite Martial and Combat Experiences Beyond the Phalanx." In *War and Violence in Ancient Greece*. Ed. H. van Wees. Swansea: The Classical Press of Wales, 1999.

———. *The Ancient Greeks at War*. Manchester: Manchester University Press, 2007.

———. "The Carthaginian Navy: Questions and Assumptions." In *New Perspectives on Ancient Warfare*. Eds. G. Fagan and M. Trundle. Leiden: Brill, 2010.

Rich, J. "Warfare and the Army in Early Rome." In *A Companion to the Roman Army*. Ed. P. Erdkamp. Oxford: Oxford University Press, 2007.

Rich, J., and G. Shipley, eds. *War and Society in the Greek World*. London: Routledge, 1993.

Ridley, R. T. "The Hoplite as Citizen," *Antiquité Classique* 48 (1979): 508–548.

Rosenstein, N. *Rome at War: Farms, Families and Death in the Middle Republic*. Chapel Hill: University of North Carolina Press, 2004.

Runciman, W. G. "Greek Hoplites, Warrior Culture, and Indirect Bias," *The Journal of the Royal Anthropological Institute* 4 (December 1998): 731–751.

Sage, M. *The Republican Roman Army: A Sourcebook*. London: Routledge, 2008.

Salmon, J. "Political Hoplites?" *Journal of Hellenic Studies* 97 (1977): 87–122.

Schwartz, A. *Reinstating the Hoplite: Arms, Armour and Phalanx Fighting in Archaic and Classical Greece*. Stuttgart: Franz Steiner Verlag, 2010.

Scullard, H. H. *A History of the Roman World 753–146*. London: Methuen, 1980.

Sekunda, N. *The Republican Roman Army 200–104 BC*. London: Osprey, 1996.

Smith, C. E. *The Roman Clan. The Gens From Ancient Ideology to Modern Anthropology*. Cambridge: Cambridge University Press, 2006.

Snodgrass, A. M. "The Hoplite Reform and History," *Journal of Hellenistic Studies* 85 (1965): 110–122.

Southern, P. *The Roman Army: A Social and Institutional History*. Oxford: Oxford University Press, 2007.

Starr, C. *The Influence of Sea Power on Ancient History*. Oxford: Oxford University Press, 1989.

Strauss, B. S. "The Athenian Trireme, School of Democracy." In *Dēmokratia: A Conversation on Democracies, Ancient and Modern*. Eds. J. Ober and C. W. Hedrick. Princeton: Princeton University Press, 1996, 313–326.

Trundle, M. *Greek Mercenaries: From the Late Archaic Period to Alexander*. London: Routledge, 2004.

———. "Light Troops in Classical Athens." In *War, Democracy and Culture*. Ed. D. Pritchard. Cambridge: Cambridge University Press, 2011.

———. "Money and the Transformation of Greek Warfare." In *New Perspectives on Ancient Warfare*. Eds. G. Fagan and M. Trundle. Leiden: Brill, 2010.

Vernant, J.-P. *Mythe et société en Grèce ancienne*. Paris: François Maspero, 1974.

Wallinga, H. *Ships and Sea Power Before the Great Persian War*. Leiden: Brill, 1993.

Wees, H. van. *Greek Warfare: Myths and Realities*. London: Duckworth, 2004.

Wheeler, E. "Methodological Limits and the Mirage of Roman Strategy, Part I," *The Journal of Military History* 57 (January 1993a): 7–42.

———. "Methodological Limits and the Mirage of Roman Strategy, Part II," *The Journal of Military History* 57 (April 1993b): 215–240.

Winter, F. E. *Greek Fortifications*. London: Routledge, 1971.

8 War and Society in Early Modern Europe

Aaron Graham

The relationship between warfare and society in early modern Europe between about 1450 and 1815 was complex and diverse, and marked by fluid interchange and exchange. This was because warfare was not a specialised and isolated aspect of wider society, but an undifferentiated aspect of it. The technical practice of warfare was not a self-contained process but grew and developed, as this chapter will show, in dialogue with technological innovation, social and cultural disciplining and education, economic entrepreneurship, and even political experimentation. In so doing it helped to shape these forces, and to influence how people in early modern Europe conceived of themselves in social, cultural, and political terms. The result was a process of unchecked and transformative change, especially in paradigmatic countries such as Britain, France, and the Dutch Republic. Historiographical attention, once fixed mainly on the "military revolution," has now broadened to encompass a wide range of interactions between warfare and society in early modern Europe, which flowed in both directions throughout this period.

While the links between warfare and society have been the staple of historians since the early modern period itself, perhaps the first attempt to elevate it to the level of a broader argument—it might not be excessive to call it a doctrine—was the concept of the "military revolution." Michael Roberts argued in 1954 that the effect of various tactical and technological shifts between 1560 and 1660, particularly the rise of the musket and the focus on manoeuvring thin lines of troops to concentrate firepower and secure decisive victories, had been to make European armies larger and more effective (Roberts, 1995). The efforts of military pioneers such as Gustavus Adolphus in Sweden and Prince Maurice in the Dutch Republic were crowned in the Thirty Years War (1618–1648), a conflict which engulfed almost the entire continent and saw large-scale military mobilisation that persisted and even increased after 1660. "Purely military developments, of a strictly technical kind, did exert a lasting influence upon society at large," he later noted, "[and] they were the agents and auxiliaries of constitutional and social change, and they bore a main share of responsibility for the coming of that new world which was to be so very unlike the old" (Roberts, 1995: 13). That new world was seen broadly in political terms. The growing size of armies both required and enabled absolutist states such as France or Sweden to mobilise larger shares of their national resources, using increasing bureaucratic structures to assess and collect taxes with the aid of military force at the expense of representative institutions, resulting in the enlightened absolutism of the eighteenth century. Roberts therefore fused two key elements of German sociological thought from the early twentieth century, best represented by Otto Hintze and Max Weber, that military force was the main element behind the development of the European nation state, and that bureaucratic structures were the most efficient way to organise it.

The attraction of the "military revolution" as a model was that it brought together the most important developments in early modern warfare and society into a plausible synthesis, and although the triggers that Roberts had identified rapidly came under attack, the model was so seductive that it was accepted wholesale. Geoffrey Parker's study in 1972 of the Spanish Army of Flanders between 1567 and 1659 examined an army that seemed to be treading the same path toward professionalism that Roberts identified; "I looked for evidence that would support his model of a backward, benighted, ineffectual force," Parker has recalled, "but I failed to find it . . . it was most puzzling" (Parker, 1972, and 1988: 155). He subsequently highlighted the importance of gunpowder, artillery, and the new *trace italienne* or artillery fortress in northern Italy and the Low Countries between about 1460 and 1560 (Parker, 1988). Arranged with mathematical precision to deflect incoming artillery fire, the bastions, ravelins, hornworks, crownworks, scarps, and counterscarps could destroy enemy attacks, but were also ruinously expensive to construct and besiege, and Parker argued in 1988 that the financial pressures on states instead arose from the cost of constructing and maintaining these fortresses and their artillery trains. This in turn facilitated a "gunpowder revolution" as successful states such as Portugal, Spain, England, France, and the Dutch Republic turned their attention outwards to empire in the seventeenth and eighteenth centuries, and used their disciplined firepower and state structures to crush the military challenges they faced in Asia and the Americas.

With the underlying thesis of the military revolution still accepted but its causal mechanisms seemingly up for grabs, the field developed into a welter of competing arguments that bring forward specific technological and tactical triggers to explain various events. Jeremy Black argued in 1991 that the real military revolution took place after 1660 as army sizes ballooned and the pike and matchlock musket were superseded by the bayonet and flintlock musket, pushing back the political changes to the century between 1660 and 1760, though his more recent work has, as noted below, also highlighted superior state structures for the control of resources and information (Black, 1991). Other studies, by contrast, pushed back the changes to the "infantry revolution" of the fourteenth century, which undermined the assumptions of the feudal system and triggered four destabilising centuries of "punctuated equilibrium evolution" as successive change catalysed further developments (Rogers, 1995c, 77). Although Parker examined the spread of tactics and technologies from a European perspective, other work inverted this and asked how regimes outside Europe such as the Ottoman and Mughal empires adopted gunpowder weaponry and tactics to defend their own territories (Ralston, 1990; Murphey, 1999). Translating the paradigm of military revolution into an extra-European idiom, Weston Cook argued that the early and considered adoption of gunpowder weapons in Morocco in the sixteenth century was instrumental in allowing local rulers not only to resist Portuguese and Ottoman incursions but also to consolidate their own internal political power (Cook, 1994).

With one eye on the rise of parliamentary states, such as England and the Dutch Republic, some historians, such as Jan Glete and Nicholas Rodger, have identified a parallel "naval revolution" between about 1500 and 1700, which saw the rise of specialised sailing warships and galleys armed with iron or copper cannon, and the "line of battle" tactics necessary to deploy them (Glete, 2000; Rodger, 2011). Naval warfare required larger and more long-term planning and investment, both in fixed capital, such as ships, stores, and dockyards, and in human capital, such as sailors, shipwrights, and administrators, which favoured parliamentary states rather than the absolutist regimes of the Continent, which had much shorter attention spans. These gradually superseded the tactics and technologies of galley warfare, which had dominated naval conflict in the mediaeval and early modern Mediterranean, but which had generated similarly large industrial complexes in places such as Venice, whose *Arsenale* remained a model for other states (Guilmartin, 1974; Mallett and Hale, 1984). Other work has also focused on the importance of diplomacy in creating coalitions for warfare that could overcome the advantages of military manpower and resources possessed by one side, as in the

Thirty Years' War, or the efforts to maintain a balance of power in Europe in the eighteenth century (Wilson, 2009; Scott, 2006)

War and society therefore seemed to be inextricably linked in early modern Europe, the various technological and tactical pressures of warfare ratcheting up the fiscal demands that sovereigns made of their subjects, and contributing to the rise of the absolutist and imperial states of the seventeenth and eighteenth centuries. State structures acted "as social power containers of advanced competencies that would otherwise not exist," enabling states to "sell" protection to their subjects in return for resources (Glete, 2002, 214). Renewed interest in state formation from sociologists and political scientists in the 1980s and 1990s gave the military revolution much wider coverage, incorporating it into wider arguments about the development of state structures but in some ways inverting its causality, arguing that although these states all faced comparable military pressures, the specific form that state formation took could be linked to other, external, factors. Brian Downing (1992) argued that absolutist regimes, such as France or Brandenburg-Prussia, tended to develop bureaucratic state structures, while constitutional or parliamentary regimes, such as England or the Dutch Republic, relied on personal or "patrimonial" rather than institutional power to control warfare, while Thomas Ertman (1997) disaggregated these elements entirely and noted the existence of early modern patrimonial absolutisms, such as Spain, or bureaucratic parliamentary regimes, such as Britain. Charles Tilly (1975) conceded that "war made the state, and the state made war" (42). Nevertheless, he argued that regimes could be "capital-intensive" or "coercion-intensive," or even "capital-coercive," such as Britain and France, which shaped in turn how they responded to the challenges of changing technologies and tactics (Tilly, 1990).

By the 1970s and the 1980s the military revolution had therefore hardened into a contested but still ultimately accepted consensus, and most of the synoptic or synthetic works produced in this period broadly agreed that the early modern period saw a military revolution as armed forces became more professional and state structures more bureaucratic (Howard, 1976; Corvisier, 1979; Strachan, 1983; Hale, 1985). In pushing deeper into this approach, however, problems began to emerge. For example, John Brewer's study (1989) of Britain's "sinews of power" was written partly in reaction to Tilly,

> to produce a social history or sociology of politics that was not reductive, that refused to treat political institutions and ideologies as mere reflections of social forces. . . [and] to dispel the complacent liberal platitude . . . about a light or non-existent British state when contrasted with the draconian oppressiveness of absolutist regimes.
>
> (Brewer, 2016, 29)

It therefore noted that the legitimacy offered by Parliament made excise taxes acceptable, but argued that it was the bureaucratic nature of the excise system that made it efficient, which, according to Brewer, "more closely approximated to Max Weber's idea of bureaucracy than any other government agency in eighteenth-century Europe" (1989, 56). The renewed interest this generated in the rise of the "fiscal-military state" in the British Isles and Europe, however, broke apart the assumptions of the military revolution, and the axiomatic links between bureaucracy, warfare, and absolutism (Storrs, 2009).

Work on the three paradigmatic states of Britain, France, and the Dutch Republic each exposed challenges. Though acknowledging areas of bureaucratic growth, studies of the British Isles have stressed the cooperative and consensual nature of British politics and state formation, the importance of Parliament as a political actor, and the major shortfalls in bureaucratic administration (Graham and Walsh, 2016). In France, it is now clear that neither absolutist kings nor revolutionary regimes ever succeeded in imposing fully bureaucratic administration onto the French army. Rounds of concessions were required in order to maintain the support of key interest groups within the army, the navy, the revenue services, the financial sector, and the central and local administration, and Louis XIV was often best served when he could devolve power to trusted ministerial allies, such as Jean-Baptiste Colbert and Michel Le Tellier, who would then use their own patronage networks to carry out the royal will (Rowlands, 2002). Revolutionary regimes were forced to similar expedients (Brown, 1995). The process consolidated, rather than replaced, very similar types of state building carried out by Richelieu on behalf of Louis XIII in the 1620s and 1630s (Parrott, 2001; James, 2004). Indeed, far from being a forcing-house of political and administrative change, the stresses and strains of warfare sometimes came close to undermining the Louis-quatorzian regime entirely (Dee, 2009; Rowlands, 2012). The Dutch Republic faced perhaps the greatest strains between 1572 and 1672, but fought off French and Spanish invasions despite having highly decentralised and extremely unbureaucratic political structures compared to its closest counterparts in Spain, France, and England ('t Hart, 1993; Gunn et al. 2007).

Rather than spurring political and social change, the pressures of warfare in the early modern period therefore did much to entrench existing patterns of social and political behaviour, especially the widespread use of contractors. The military entrepreneur, who hired out companies or regiments of hired guns to princes, had been common in the late mediaeval period, but the military revolution seemed to presage his decline as standing national armies replaced unreliable mercenary forces (Redlich, 1964). In fact, although larger and autonomous military enterprisers such as Albrecht von Wallenstein had fallen from favour by 1648—Wallenstein was assassinated by his employer, Holy Roman Emperor Ferdinand II in 1634—early modern armies still relied heavily on intermediaries to recruit troops for them (Parrott, 2012). The French army under Louis XIV in effect turned its own officers into loyal military enterprisers, while the British Army hired troops in wartime by treaty from princes in Germany, or used the clan connections of its Scottish officers to raise regiments in the Highlands after 1745 (Parrott, 2001; Dziennik, 2016; Wilson, 1995). Revised focus on the logistical pressures of war, by Martin van Creveld (1997), Geoffrey Parker (1972), and others, has also shown how states were forced to depend on military contractors as well as their own resources to meet the urgent need for food, fodder, and other essentials. Even large and complex naval installations in Britain depended on contractors to supply raw materials and manufactured goods (Knight and Wilcox, 2010; Bannerman, 2008; Conway, 2006), while the military power of the Dutch Republic in this period similarly reflected the capacity of the advanced economy of the Netherlands to supply cheaply the materials that Dutch naval and military forces required, creating in effect a public-private partnership (Brandon, 2015).

Most recently, attention has also shifted (back) to the financial sector, where it is clear that the growing size and scale of military mobilisation in early modern Europe both stimulated and, to some degree, depended on the wider expansion in the European economy that occurred in this period. On the one hand, historians have agreed that the fiscal demands of warfare prompted not only increased taxes but also, in Britain, France, and the Dutch Republic, new and highly sophisticated instruments of public credit. In England in the 1650s and 1690s, for instance, increased military expenditure forced English regimes to move beyond existing financial structures and culminated in the creation of the Bank of England in 1694, which anchored the burgeoning English financial sector and helped to support growth elsewhere in the economy (Dickson, 1967; Coffman, 2013). On the other hand more recent work has also stressed the underlying strength of the financial sector in England and the Dutch Republic, which drew on existing systems of commercial and agricultural credit. Military finance could strengthen and expand these networks, but also had to conform itself to them, and work within their limits. In Britain this could mean condoning informal practices that were often indistinguishable from outright corruption (Graham, 2015). In France, the pressure of warfare was enough to overwhelm these systems entirely between 1709 and 1711, bringing the country to the brink of collapse as financial networks broke under the strain of remitting funds to French armies during a period of harvest failure (Rowlands, 2015). The shape of the state could therefore determine the level of military mobilisation, but it is now clear that these cannot be reduced to the broad categories offered by Ertman or Tilly.

Other links in the chain of military revolution have also been examined again, and their complexity revealed, such as the claim that the self-reinforcing process of technological innovation was sufficient to drive both tactical change—in the form of larger, professional armies—and the political and social effects that followed. It is clear that many early modern states recognised the importance of technological innovation, and deliberately put structures in place to cultivate it. In Britain, for example, the Royal Observatory at Greenwich was founded in 1675 with support from the Admiralty and the Ordnance Office to allow research into astronomical data for maritime navigation, and the Board of Longitude existed between 1714 and 1828 to encourage the creation of a reliable means for calculating longitude at sea, and these efforts were mirrored by other European countries (Dunn and Higgitt, 2016). The importance of artillery encouraged new techniques for gunfounding and metalworking, and sponsored chemical research that was intended to improve the qualities of gunpowder and its key ingredient, saltpetre. In France, for example, the state set up the *régie des poudres et salpêtres* in 1775 under the direction of Antoine Laurent Lavoisier, one of the leading figures in chemical research in Europe, who was able to develop new techniques to support a domestic saltpetre industry (Buchanan, 2006). These monumental shifts drew on the profound cultural and intellectual developments of the scientific revolution and Enlightenment, which provided the intellectual and practical context within which successful scientific advances could take place.

The same was true at lower levels. Recognising that the major inventions of military technology could not have taken place without a series of much smaller incremental innovations, which depended on having an educated and experienced workforce capable of translating scientific discoveries into practical results, technological change is now increasingly seen to have taken place within a wider culture favourable to this process. For example, although there was nothing directly comparable to the *régie des poudres* in eighteenth-century Britain, the Ordnance Office appointed key figures such as Sir William Congreve and Richard Blomfeld to major posts where they could make incremental improvements to the production of gunpowder and armaments (Cole, 2012). Although fortress construction and engineering in France arguably reached its peak under Sébastien Le Prestre de Vauban in the late seventeenth century, who instituted a new professional culture of engineering, Jamel Ostwald has shown that his advice on siegecraft and artillery was widely ignored at the time in favour of frontal attacks that reflected contemporary values of bravery and honour (Ostwald, 2007). This only gradually changed under the wider cultural influence of the Enlightenment, which made Vauban's rational approach both legitimate and fashionable, and gave engineers the leverage needed to effect rational reforms of their service (Alder, 1997; Langins, 2004). This failed, though, because the new culture had not percolated down into the lower levels of the administration, where artisans and workmen rejected untried experiments in favour of established techniques and working practices. Technological innovation therefore arose not just from military necessity but also from complex social factors, suggesting in turn that military change was the result rather than the outcome of change.

The same approach has also challenged the role of tactics, which were not merely the adjunct of technological changes but also had their own independent logic. Roberts suggested that the early modern period saw a shift toward linear infantry formations that maximised firepower, which only professional Dutch and Swedish armies could perform effectively, but in practice many of these tactics were optimised for very specific battlefield situations, and were already being widely practised across Europe as officers and soldiers moved from one army to another (Parrott, 1995). In regions such as Eastern Europe, characterised by wide open plains that favoured cavalry forces and highly mobile warfare, they were not at all helpful. Robert Frost has argued that the decline of the Polish-Lithuanian Commonwealth in this period in fact reflected underlying political problems, which frustrated the creation of a large standing army, rather than their failure to adopt new military tactics that were in any case inappropriate for the challenges they faced from the Swedish armies of Gustavus Adolphus (Frost, 2000). On the other hand, the supposed superiority of light infantry tactics in North America has also been assessed again, with more recent work laying stress on the superior execution of established tactics by professional soldiers combined with more efficient logistical infrastructure (Chet, 2003; Spring, 2010), while even the concept of the decisive battle has been questioned (Ostwald, 2000; Harari, 2007). Tactical education and the opportunity to practise discipline and drill were therefore crucial, and states such as Britain that could find the time to support constant exercises found themselves at a distinct advantage on the battlefield (Houlding, 1981). Erik Lund has also stressed how important mundane logistical skills were to troops, who were often expected to live off the land on campaign (Lund, 1999).

The practice of warfare in the early modern period was therefore deeply affected not only by the capacities of states but also, as will have become clear, by the wider social and cultural environment of early modern Europe. This has emerged largely in reaction or response to the concept of the military revolution itself, which since 1956 has provided a central analytical concept, now increasingly threadbare, around which work could define itself. At first influenced only by wider historiographical changes, such as the rise of social history in the 1950s and 1960s, the cultural turn of the 1970s and 1980s, and the intellectual turn of the 1990s, scholarship has increasingly moved away from the terms of debate set by the military revolution to consider the history of warfare in early modern Europe in relation to other historiographies. The result has been the rise of "new military history," which has focused on exploring the different types of complex interactions that could occur between military forces and the wider society, including civil and intellectual culture, identity, gender, medicine, race, and the body (Chambers, 1991; Citino, 2007). This has lent undoubted and undeniable empirical depth and analytical sophistication to the history of warfare, but has also demolished the central analytical model of the military revolution without yet offering an overarching model of warfare and society that can draw these elements together.

For example, recent work has shown soldiers in this period drew heavily on wider cultural values to guide and legitimate their actions. Ideas such as duty, honour, and loyalty could help regulate behaviour between social and cultural equals, and when these norms were set aside—as with the English in Ireland—it was due to wider cultural views of the Gaelic Irish as uncivilised savages living outside the values of European society (Donagan, 2008; Lee, 2011). In the English Civil War itself, the treatment of foreign soldiers was influenced by how they were perceived in English culture (Stoyle, 2005). The preoccupation of many early modern military officers with honour, courage, and reputation, demonstrated most explicitly by the widespread practice of duelling, was not alien to European culture but a greatly heightened expression of some of its core values (Peltonen, 2003; Billacois, 1990). It came under threat in the late seventeenth century, though, challenged by new codes of military conduct that stressed duty, loyalty, and sobriety that reflected wider cultural shifts in European society toward renewed understandings of civility, politeness, and duty. Cultural representations in this period shifted to accommodate these new values (Paret, 1997; Smith, 2011). On the other hand, both Lynn and Parrott have shown that the culture of honour, bravery, and manliness in the French army was one of the factors that encouraged French nobles to enlist and even to support their own units financially, and Hanlon has shown that Italian nobles began to withdraw from military participation in the eighteenth century as military culture in the peninsula shifted from the chivalric to the professional (Lynn, 1997; Parrott, 2001; Hanlon, 1998).

The growing interest in the interaction between warfare, culture, and society, prompted at least in part by the cultural turn of the 1980s, has also focused attention on the means of transmission in early modern Europe, and the links between warfare and the growing public sphere. In seventeenth-century England, for example, warfare created an immense appetite for news which was met by a burgeoning supply of newssheets and pamphlets that drew on the existing infrastructure but also helped to support it (Randall, 2008). This was matched by a growing level of intellectual sophistication as writers refined how they presented their news in response to audiences who were becoming more nuanced and discerning in their consumption. The highly militarised culture of the period has also been shown to have created a demand for literature such as drill manuals and military treatises (Lawrence, 2009; Gruber, 2010). This in turn made it easier to disseminate standardised military knowledge of discipline and drill, facilitating many of the tactical changes noted earlier. Although the history of military thought itself has always been of interest, recent work has begun to highlight its very close connections to wider intellectual phenomena, such as the Enlightenment, which not only promoted the technological developments that have already been identified but also suggested that the strategy and tactics of warfare itself were open to logical and rational description (Quimby, 1957; Gat, 1989 and 2001). In their edited translation of Carl von Clausewitz's 1832 work *Vom Kriege* (*On War*), military historians Michael Howard and Peter Paret have argued that intellectual process culminated in a bold effort to systematise and rationalise the practice of warfare based on his experiences in the Napoleonic Wars (Clausewitz, 1984).

The militia or citizen army was a particularly important focus for changing cultural expectations in this period. Both Britain and Ireland saw revived interest in the militia from the 1750s as a volunteer force that served as a focus for national identity, culminating in the Volunteers of the 1780s and Yeomanry or Fencibles of the 1790s, which were volunteer regiments motivated by a profound sense of patriotism and national identity (Cookson, 1997; McCormack, 2015). The transition was eased because many of these national cultures were, as noted earlier, already laden with key martial concepts, such as honour, duty, and loyalty, that could then be appropriated by these forces to justify and legitimate their actions. Indeed, military victory over the Catholic powers—especially France—was a key component in the formation of a British identity in the eighteenth century, and militia service allowed vast numbers of Britons to associate themselves with this powerful new identity (Colley, 1992). The same was true in France, where the conscript army or the *levée en masse* created after the French Revolution in 1792 was seen as a vehicle for the revolutionary and nationalistic fervour it had unlocked, embodying its qualities of liberty, equality, and fraternity (Lynn, 1984; Bell, 2007). These forces were increasingly embodied in artwork and other cultural representations (de Bruyn and Regan, 2014; Russell, 1995). Yet in some ways this process was little more than a continuation of existing trends, since ideology—in its religious form—has now been restored as one of the most important influences on the conduct and character of early modern warfare. Drawing on wider historiographical reassessments of early modern religion from the 1970s, which have stressed the power (and malleability) of religious belief as a motivating force, the religious dimensions to seminal early modern conflicts from the mediaeval period until well into the eighteenth century have been restored (Holt, 1993; Onnekink, 2009; Housely, 2002).

Recent scholarship has also complicated the picture of war and society in other areas. The early modern world saw changing views of gender and masculinity, especially under the influence of civility and politeness in the early eighteenth century, but also something of a backlash in mid-century as contemporaries worried about the declining status of manliness. It is now clear that the militia in Britain was increasingly seen as an instrument of both national and masculine revival: the reforms proposed in Britain in the 1750s were

> not just a question of national defence but of national regeneration: the means of reinvigorating the polity, of reviving public spirit, and—at the root of it all—of restoring a gender order that some commentators alleged was on the verge of collapse.
>
> (McCormack, 2015, 14)

The *levée en masse* generated similar expectations, drawing clear and heavily gendered links between military service, patriotism, and citizenship and reinforcing in some respects established gender norms that consigned women to separate and very subordinate spheres. On the other hand, recent work has emphasised not only the divergent masculinities that could emerge from this process but also the opportunities that warfare offered for women on the battlefield and the home front to assert their own priorities and principles (Lynn, 2008; Hagemann et al. 2010). Many served alongside men as wives, nurses, and camp followers, helping to support the complex logistical infrastructure that kept armies in the field, and some even dressed as men and served directly.

These examples point to the fluid overlap between warfare, society, and culture, in which the conduct of war interacted with ongoing social trends in complex and often unpredictable ways that historians are still in the process of teasing out. Changing views of medicine and the body found their expression in new practices of military medicine, as states worked to use their human resources more efficiently and reduce losses from disease and disability. During the Seven Years' War (1756–1763), for instance, the British Army and Navy drew on new medical theories and practice to try to address the problem of disease, adapting their plans of campaign and their logistical infrastructure to follow best practice (Charters, 2014). This fed back into the wider medical profession, offering opportunities for physicians to study tropical diseases in closer detail. Closer historiographical interest in the topic of race has also focused attention on the links with warfare in the early modern world. On the one hand, race itself underwent a process of definition and redefinition by the eighteenth century, led in part by a growing awareness of slavery. Black slaves imported from Africa to the West Indies were seen as savage and violent, but their military victories in the Haitian Revolution between 1791 and 1804 also suggested that they were capable of military discipline and political citizenship (Brown and Morgan, 2006). Twelve regiments of black slaves were raised by Britain in the West Indies on this basis in 1795, mirroring the many thousands of sepoys recruited in South Asia (Buckley, 1979).

Some of these movements are, as yet, unfinished. As noted earlier, the gendered and racial dimensions of early modern warfare have not yet been fully explored, while much work remains to be done on topics such as identity and memory (Harari, 2008). A small but nuanced literature on warfare and the environment has emerged, mainly in response to studies of naval mobilisation, timber, and ecological policies in France, Spain, and Venice, pointing to another fruitful direction for research (Bamford, 1956; Appuhn, 2009; Wing, 2015). Parker's recent work on the "global crisis" of the seventeenth century has approached this theme from the other direction, and even suggested a model for how all of these elements might be tied together (Parker, 2013). Symptomatic of a renewed interest in transnational or global history as well as environmental history, other work has also begun to offer an important corrective to histories of warfare focused on Europe, not least because the tactics and technology of the military revolution rapidly diffused to other regimes with equivalent levels of social and economic development (Ralston, 1990). This has directed renewed interest in the political structures that may have allowed European regimes, such as the East India Company in South Asia, to manage warfare more effectively (Black, 2011, 2012). The East India Company, for example, used its commercial and financial assets better than the Mughals while adapting itself to local conditions by employing local provisions contractors and tapping into military labour markets for troops (Cooper, 2003; Roy, 2011). Even European warfare itself has not been fully studied, with many historians ignoring the impact of war and society in eastern and south-eastern Europe, or even in the minor German and Italian states, despite encouraging moves in that direction (Davies, 2012; Storrs, 2009).

The intersection and interaction of warfare were therefore complex and fluid in this period, complicating the clear and sharply delineated categories of analysis and interplay that were offered by the "military revolution" in its various iterations. The result has been an explosion, in the most positive sense of the word, in the nuance and sophistication of scholarship, whose breadth and depth have increased massively in the last few decades. These scholars have breached the walls and bastions of the military revolution and even the overrunning of its historiographical citadel, but without offering a replacement model that can draw together these separate elements and suggest how and why they interacted with each other, and if there was a conceptual unity to them. This remains the last and most challenging goal for historians of warfare and society in early modern Europe.

Bibliography

Alder, Ken. *Engineering the Revolution: Arms and Enlightenment in France, 1763–1815*. Princeton: Princeton University Press, 1997.

Appuhn, Karl. *A Forest on the Sea: Environmental Expertise in Renaissance Venice*. Baltimore: The Johns Hopkins University Press, 2009.

Bamford, Paul. *Forests and French Sea Power, 1660–1789*. Toronto: University of Toronto Press, 1956.

Bannerman, Gordon. *Merchants and the Military in Eighteenth-Century Britain: British Army Contracts and Domestic Supply, 1739–1763*. London: Pickering & Chatto; 2008.

Bell, David. *The First Total War: Napoleon's Europe and the Birth of Modern Warfare*. London: Bloomsbury, 2007.

Billacois, Francois. *The Duel: Its Rise and Fall in Early Modern France*. Trans. Trista Selous. New Haven: Yale University Press, 1990.

Black, Jeremy. *Beyond the Military Revolution: War in the Seventeenth Century*. Basingstoke: Palgrave Macmillan, 2011.

———. *A Military Revolution?: Military Change and European Society, 1550–1800*. Basingstoke: Palgrave Macmillan, 1991.

———. *War in the Eighteenth Century World*. Basingstoke: Palgrave Macmillan, 2012.

Brandon, Pepijn. *War, Capital and the Dutch State (1588–1795)*. Leiden: Brill, 2015.

Brewer, John. "Revisiting *The Sinews of Power*." In *The British Fiscal-Military States, 1660–c. 1783*. Eds. Aaron Graham and Patrick Walsh. London: Routledge, 2016.

———. *The Sinews of Power: War, Money and the English State, 1688–1783*. London: Unwin Hyman, 1989.

Brown, Christopher Leslie, and Philip D. Morgan, eds. *Arming Slaves: From Classical Times to the Modern Age*. New Haven: Yale University Press, 2006.

Brown, Howard. *War, Revolution and the Bureaucratic State: Politics and Army Administration in France, 1791–1799*. Oxford: Oxford University Press, 1995.

Buchanan, Brenda, ed. *Gunpowder, Explosives and the State: A Technological History*. Aldershot: Ashgate, 2006.

Buckley, Norman. *Slaves in Red Coats: The British West India Regiments, 1795–1815*. New Haven: Yale University Press, 1979.

Chambers, John. "The New Military History: Myth and Reality," *Journal of Military History* 55 (1991): 395–406.

Charters, Erica. *Disease, War and the Imperial State: The Welfare of the British Armed Forces in the Seven Years War*. Chicago: University of Chicago Press, 2014.

Chet, Guy, *Conquering the American Wilderness: The Triumph of European Warfare in the Colonial Northeast*. Amherst: University of Massachusetts Press, 2003.

Citino, Robert. "Military Histories Old and New: A Reintroduction," *American Historical Review* 112 (2007): 1070–1090.

Clausewitz, Carl von. *On War*. Ed. and trans. Michael Howard and Peter Paret, Princeton, NJ: Princeton University Press, 1984.

Coffman, D'Maris. *Excise Taxation and the Origins of Public Debt*. Basingstoke: Palgrave Macmillan, 2013.

Cole, Gareth. *Arming the Royal Navy, 1793–1815: The Office of Ordnance and the State*. London: Pickering & Chatto, 2012.

Colley, Linda. *Britons: Forging the Nation, 1707–1837*. New Haven: Yale University Press, 1992.

Conway, Stephen. *War, State and Society in Mid-Eighteenth Century Britain and Ireland*. Oxford: Oxford University Press, 2006.

Cook, Weston. *The Hundred Years War for Morocco: Gunpowder and the Military Revolution in the Early Modern Muslim World*. Boulder, CO: Westview, 1994.

Cookson, J. E. *The British Armed Nation, 1793–1815*. Oxford: Oxford University Press, 1997.

Cooper, Randolf G. S. *The Anglo-Maratha Campaigns and the Contest for India: The Struggle for Control of the South Asian Military Economy*. Cambridge: Cambridge University Press, 2003.

Corvisier, André. *Armies and Societies in Europe, 1494–1789*. Bloomington: Indiana University Press, 1979.

Creveld, Martin van. *Supplying War: Logistics From Wallenstein to Patton*. Cambridge: Cambridge University Press, 1997.

Davies, Brian, ed. *Warfare in Eastern Europe, 1500–1800*. Leiden: Brill, 2012.

De Bruyn, Frans, and Shaun Regan. *The Culture of the Seven Years War: Empire, Identity and the Arts in the Eighteenth-Century Atlantic World*. Toronto: University of Toronto Press, 2014.

Dee, Darryl. *Expansion and Crisis in Louis XIV's France: Franche-Comté and Absolute Monarchy, 1674–1715*. Rochester, NY: University of Rochester Press, 2009.

Dickson, P.G.M. *The Financial Revolution in England: A Study of the Development of Public Credit, 1688–1756*. London: Palgrave Macmillan, 1967.

Donagan, Barbara. *War in England, 1642–1649*. Oxford: Oxford University Press, 2008.

Downing, Brian. *The Military Revolution and Political Change: Origins of Democracy and Autocracy in Early Modern Europe*. Princeton: Princeton University Press, 1992.

Dunn, Richard, and Rebekak Higgitt, eds. *Navigational Enterprises in Europe and Its Empire, 1730–1850*. Basingstoke: Palgrave Macmillan, 2016.

Dziennik, Matthew. *The Fatal Land: War, Empire and the Highland Soldier in British America*. New Haven: Yale University Press, 2016.

Ertman, Thomas. *Birth of Leviathan: Building States and Regimes in Medieval and Early Modern Europe*. Cambridge: Cambridge University Press, 1997.

Frost, Robert. *The Northern Wars: War, State and Society in North Eastern Europe, 1558–1721*. Harlow: Longman, 2000.

Gat, Azar. *A History of Military Thought: From the Enlightenment to the Cold War*. Oxford: Oxford University Press, 2001.

———. *The Origins of Military Thought From the Enlightenment to Clausewitz*. Oxford: Clarendon Press, 1989.

Glete, Jan. *War and the State in Early Modern Europe: Spain, the Dutch Republic and Sweden as Fiscal-Military States, 1500–1650*. London: Routledge, 2002.

———. *Warfare at Sea, 1500–1650*. London: Routledge, 2000.

Graham, Aaron. *Corruption, Party and Government in Britain, 1702–1713*. Oxford: Oxford University Press, 2015.

Graham, Aaron, and Patrick Walsh, Patrick, eds. *The British Fiscal-Military States, 1660–c. 1783*. London: Routledge, 2016.

Grenier, John. "Recent Trends in the Historiography on Warfare in the Colonial Period (1607–1765)," *History Compass* 8 (2010): 358–367.

Gruber, Ira. *Books and the British Army in the Age of the American Revolution*. Chapel Hill: University of North Carolina Press, 2010.

Guilmartin, John. *Gunpowder and Galleys: Changing Technology and Mediterranean Warfare at Sea in the Sixteenth Century*. Cambridge: Cambridge University Press, 1974.

Gunn, Steven, David Grummitt, and Hans Cools. *War, State and Society in England and the Netherlands, 1477–1559*. Oxford: Oxford University Press, 2007.

Hagemann, Karen, Gisela Mettele, and Jane Rendall, eds. *Gender, War and Politics; Transatlantic Perspectives, 1775–1830*. Basingstoke: Palgrave Macmillan, 2010.

Hale, J. R. *War and Society in Renaissance Europe, 1450–1620.* London: Fontana Press, 1985.

Hanlon, Gregory. *The Twilight of a Military Tradition: Italian Aristocrats and European Conflicts, 1560–1800.* London: University College London Press, 1998.

Harari, Yuval Noah. "The Concept of 'Decisive Battles' in World History," *Journal of World History* 18 (2007): 251–266.

———. *The Ultimate Experience: Battlefield Revelations and the Making of Modern War Culture, 1450–2000.* Basingstoke: Palgrave Macmillan, 2008.

Holt, Mack. "Review Article: Putting the Religion Back into the Wars of Religion," *French Historical Review* 18 (1993): 524–551.

Houlding, J. A. *Fit for Service: The Training of the British Army, 1715–1795.* Oxford: Clarendon Press, 1981.

Housley, Norman. *Religious Warfare in Europe, 1400–1536.* Oxford: Oxford University Press, 2002.

Howard, Michael. *War in European History.* Oxford: Oxford University Press, 1976.

James, Alan. *The Navy and Government in Early Modern France, 1572–1661.* Woodbridge: Boydell & Brewer, 2004.

Knight, Roger and Martin Wilcox *Sustaining the Fleet, 1793–1815. War, the British Navy and the Contractor State.* Woodbridge: Boydell Press, 2010.

Langins, Janis. *Conserving the Enlightenment: French Military Engineering From Vauban to the Revolution.* Cambridge: MIT Press, 2004.

Lawrence, David. *The Complete Soldier: Military Books and Military Culture in Early Stuart England, 1603–1645.* Leiden: Brill, 2009.

Lee, Wayne E. *Barbarians and Brothers: Anglo-American Warfare, 1500–1865.* Oxford: Oxford University Press, 2011.

Lund, Erik. *War for the Every Day: Generals, Knowledge and Warfare in Early Modern Europe, 1680–1740.* Westport, CT: Greenwood Press, 1999.

Lynn, John. *The Bayonets of the Republic: Motivation and Tactics in the Army of Revolutionary France, 1791–1794.* Urbana: University of Illinois Press, 1984.

———. *The Giant of the Grand Siècle: The French Army, 1610–1775.* Cambridge: Cambridge University Press, 1997.

———. *Women, Armies and Warfare in Early Modern Europe.* Cambridge: Cambridge University Press, 2008.

Mallett, M. E., and J. R. Hale. *The Military Organisation of a Renaissance State: Venice, 1400–1617.* Cambridge: Cambridge University Press, 1984.

McCormack, Matthew. *Embodying the Militia in Georgian England.* Oxford: Oxford University Press, 2015.

Murphey, Rhoads. *Ottoman Warfare, 1500–1700.* London: University College London Press, 1999.

Onnekink, David, eds. *War and Religion After Westphalia, 1648–1713.* Farnham: Ashgate, 2009.

Ostwald, Jamel. "The 'Decisive' Battle of Ramilies, 1706: Prerequisites for Decisiveness in Early Modern Warfare," *Journal of Military History* 64 (2000): 649–678.

———. *Vauban Under Siege: Engineering Efficiency and Martial Vigour in the War of the Spanish Succession.* Leiden: Brill, 2007.

Paret, Pete. *Imagined Battles: Reflections of War in European Art.* Chapel Hill: University of North Carolina Press, 1997.

Parker, Geoffrey. *The Army of Flanders and the Spanish Road, 1567–1659: The Logistics of Spanish Victory and Defeat in the Low Countries' Wars.* Cambridge: Cambridge University Press, 1972.

———. *Global Crisis: War, Climate Change and Catastrophe in the Seventeenth Century.* New Haven: Yale University Press, 2013.

———. *The Military Revolution: Military Innovation and the Rise of the West, 1500–1800.* Cambridge: Cambridge University Press, 1988.

Parrott, David. *The Business of War: Military Enterprise and Military Revolution in Early Modern Europe.* Cambridge: Cambridge University Press, 2012.

———. *Richelieu's Army: War, Government and Society in France, 1624–1642.* Cambridge: Cambridge University Press, 2001.

———. "Strategy and Tactics in the Thirty Years War: The 'Military Revolution.'" In *The Military Revolution Debate: Readings on the Military Transformation of Early Modern Europe, History and Warfare.* Ed. Clifford Rogers. Boulder, CO: Westview Press, 1995.

Peltonen, Marrku. *The Duel in Early Modern England: Civility, Politeness and Honour.* Cambridge: Cambridge University Press, 2003.

Quimby, Robert. *The Background of Napoleonic Warfare: The Theory of Military Tactics in Eighteenth-Century France*. New York: Columbia University Press, 1957.

Ralston, David. *Importing the European Army: The Introduction of European Military Techniques and Institutions into the Extra-European World, 1600–1914*. Chicago: University of Chicago Press, 1990.

Randall, David. *Credibility in Elizabeth and Early Stuart Military News*. London: Pickering & Chatto, 2008.

Redlich, Fritz. *The German Military Enterpriser and His Workforce: A Study in European Economic and Social History*. Wiesbaden: F. Steiner, 1964.

Roberts, Michael, "The Military Revolution, 1560–1660." In *The Military Revolution Debate: Readings on the Military Transformation of Early Modern Europe, History and Warfare*. Ed. Clifford Rogers. Boulder, CO: Westview Press, 1995.

Rodger, N.A.M. "From the 'Military Revolution' to the 'Fiscal-Naval State.'" *Journal for Maritime Research* 13 (2011): 119–128.

Rogers, Clifford, ed. *The Military Revolution Debate: Readings on the Military Transformation of Early Modern Europe, History and Warfare*. Boulder, CO: Westview Press, 1995a.

———. "The Military Revolution in History and Historiography." In *The Military Revolution Debate: Readings on the Military Transformation of Early Modern Europe, History and Warfare*. Ed. Clifford Rogers. Boulder, CO: Westview Press, 1995b.

———. "The Military Revolutions of the Hundred Years War." In *The Military Revolution Debate: Readings on the Military Transformation of Early Modern Europe, History and Warfare*. Ed. Clifford Rogers. Boulder, CO: Westview Press, 1995c.

Rowlands, Guy. *Dangerous and Dishonest Men: The International Bankers of Louis XIV's France*. Basingstoke: Palgrave Macmillan, 2015.

———. *The Dynastic State and the Army Under Louis XIV: Royal Service and Private Interest, 1661–1701*. Cambridge: Cambridge University Press, 2002.

———. *The Financial Decline of a Great Power: War, Influence and Money in Louis XIV's France*. Oxford: Oxford University Press, 2012.

Roy, Kaushik. *War, Culture and Society in Early Modern South Asia, 1740–1849*. London: Routledge, 2011.

Russell, Gillian. *The Theatres of War: Performance, Politics and Society, 1793–1815*. Oxford: Oxford University Press, 1995.

Scott, Hamish. *The Birth of a Great Power System, 1740–1815*. Harlow: Pearson, 2006.

Smith, Hannah, "Politics, Patriotism and Gender: The Standing Army Debate on the English Stage, Circa 1689–1720," *Journal of British Studies* 50 (2011): 48–75.

Spring, Matthew. *With Zeal and Bayonets Only: The British Army on Campaign in North America, 1775–1783*. Norman, OH: University of Oklahoma Press, 2010.

Storrs, Christopher, ed. *The Fiscal-Military State in Eighteenth-Century Europe: Essays in Honour of P.G.M. Dickson*. Farnham: Ashgate, 2009.

Stoyle, Mark. *Soldiers and Strangers: An Ethnic History of the English Civil War*. New Haven: Yale University Press, 2005.

Strachan, Hew. *European Armies and the Conduct of War*. London: Allen & Unwin, 1983.

't Hart, Marjolein. *The Making of a Bourgeois State: War, Politics and Finance in the Dutch Revolt*. Manchester: Manchester University Press, 1993.

Tilly, Charles. *Coercion, Capital and European States, A.D. 990–1992*. Oxford: Basil Blackwell, 1990.

———. "Reflections on the History of European State-Making." In *The Formation of National States in Western Europe*. Ed. Charles Tilly. Princeton: Princeton University Press, 1975.

Van Creveld, Martin. *Supplying War: Logistics From Wallenstein to Patton*. Cambridge: Cambridge University Press, 1977.

Wilson, Peter. *Europe's Tragedy: A History of the Thirty Years War*. London: Penguin, 2009.

———. *War, State and Society in Württemberg, 1677–1793*. Cambridge: Cambridge University Press, 1995.

Wing, John. *Roots of Empire: Forests and State Power in Early Modern Spain, c. 1500–1750*. Leiden: Brill, 2015.

9 War and Society in Modern Europe

Adam R. Seipp

This essay considers the relationship between war and social change in Europe since the Wars of the French Revolution. At the broadest level, the study of war and society asserts that we can learn a great deal about human societies by studying the ways that they prepare for and wage war. At the same time, we can better understand war by studying the societies that engage in it. This converse relationship distinguishes war and society from the broader field of social history from which it emerged. As this essay will show, the field has been further enriched by scholarly interest in the cultural history of war, a phenomenon largely dating from the 1990s. These interrelated fields of inquiry have allowed scholars to move beyond state-centered conceptions of war and to understand the interrelationship between armed conflict and social change.

During the period under consideration, the increasing scale, scope, and deadliness of war produced divergent and seemingly contradictory outcomes. First, the expansion of violence during this period more thoroughly enmeshed European societies within the state, an institution that Max Weber famously defined as "an association that claims a monopoly of the legitimate use of violence" (Weber, 2009, 334). At the same time, the rapidly increased frightfulness of war helped to catalyze efforts to find secular alternatives to war. This essay will try to consider both in tandem.

Writing a history of war and society in Europe since 1789 presents a number of interpretive challenges. On one hand, there are a number of common developments that make it superficially easy to consider the Continent as a whole. The long-running political conflict over the French Revolution and its legacies, industrialization, the rise of national states and simultaneous decline of empires, the creation of mass armies, and the twentieth-century antagonism between communism and anti-communism all provide convenient frameworks for a continental approach to the topic.

However, beneath the surface the view is considerably muddier. Leaving aside the teleological views of traditional national historiographies, there is little about any of these processes that was inevitable or that followed a straight trajectory. A history of Europe that focuses on war and warfare should necessarily be one that takes into account regional and subregional variations, dynamic interdependencies, and wide variation between national experiences. If we take "modernity" as the defining characteristic of the European order after the collapse of the ancien régime, then we must accept that Europeans employed a range of strategies to confront the challenges of modern life (Jarausch, 2015). The same tensions that produced the dynamism, innovation, and hope of a continent that found itself at the center of world events also produced terror, destruction, and death on an unprecedented scale.

The dramatic collapse of the French monarchy and the beginning of revolutionary upheaval in Western Europe are a relatively clear beginning point for an exploration of this topic. Scholarship on the "military revolution" has, if anything, made the importance of this transitionary period even more distinct. Driven by a mutually reinforcing combination of organizational, bureaucratic, technological, and economic changes, European rulers and states had, by the end of the eighteenth century, substantially transformed the nature of warfare. European militaries grew larger, more complex, and more thoroughly integrated into the machinery of state, which was itself expanding. There is some debate among historians about whether 1789 was the culmination of centuries of warfare-driven development (Parker, 1996) or if the late eighteenth century saw European states finally reach a level of organizational capacity needed to implement the kind of centralization that some had been working toward for years (Parrott, 2012).

The revolutionary moment in France produced, largely out of necessity, a series of dramatic military reforms. The revolutionary armies, while still drawing heavily from the methods and personnel of their royal predecessors, also reflected a series of innovations. The *levée en masse* of 1793 ostensibly created one of the world's first mass armies, made up of conscripts and supported by a mobilized "nation at arms" (Cobb, 1987; Lynn, 1996) Despite widespread resistance to conscription, troop strength swelled to more than eight hundred thousand within a year, dwarfing even the large armed forces mobilized during the wars of the mid-eighteenth century. This was also an army suffused with revolutionary ideology and arrayed against foreign and domestic enemies. While the relationship between ideological zeal and battlefield performance is debatable, there is little doubt that this new army was successful against its external enemies and capable of horrific brutality against internal enemies like the insurgents in the Vendée (Secher, 2003).

During the decades that followed, the seesaw of coalitions and alliances that both supported and attacked revolutionary and Napoleonic France saw warfare spread throughout Europe and across the globe-spanning empire. Battles raged from the Atlantic coast of Portugal to South Africa and from Russia to New Orleans. Something like four million people died in the revolutionary wars. Familiar political features like the Holy Roman Empire vanished, and many Europeans found themselves living under new rulers and new systems of government, even if only for a time.

David Bell (2008) has called this global struggle "the first total war." The term "total war" is notorious for its analytical ambiguity, and typically refers to the spatial and ideological expansion of warfare in the modern world. States grew more capable of mobilizing their societies, ideologically committed to conflicts of national survival, able to use modern weaponry to inflict substantially more damage on enemy militaries and civilians, and more willing to sustain and inflict massive casualties to achieve victory (Boemeke et al., 2006). Total war is probably best considered as an idea perpetually in being, an ideal type of warfare that might express the desire of states and rulers, but not necessarily the lived reality of those who experience it.

The postwar settlement that followed Napoleon's defeat crafted an uneasy stability that proved remarkably durable. After a period of endemic conflict since at least the reign of Louis XIV, Europe enjoyed a reprieve from general war until the middle of the century. However, the social transformations of this period were profound and closely tied to the development of states and their war-making capability.

One of the most powerful intellectual forces to emerge from the revolutionary period was the idea of nationalism. The notion of a large "imagined community" of belonging expressed in political terms was not completely novel in the early nineteenth century, but the foment of those years allowed a generation of intellectuals across Europe to begin seeing a future of vernacular states that expressed an atavistic and primordial cultural identity in political terms (Anderson, 1991). New forms of patriotism, whether in Britain (Colley, 1992), North Germany (Aaslestad, 2005), or post-partition Poland (Davies, 2005), created novel loci for mass mobilization. Nationalism began as an intellectual and cultural project, but soon began to morph into a political program that, particularly in Central and Eastern Europe, challenged existing political and social structures.

At the same time, European societies began to encounter what would, with hindsight, be called the Industrial Revolution. As Sidney Pollard (1981) argued in his classic study, industrialization was highly uneven and asynchronous. It began in Britain and northwest Europe, then developed in some regions further east and south. Industrialization was not state-driven in its early stages, but rather reflected a new liberal sensibility. This did not last, and by mid-century the state, and with it the military, became deeply enmeshed in industrial developments.

The social and economic transformation of the early nineteenth century, however uneven, produced growing tensions in Europe's growing cities and towns. Restoration monarchies struggled to reassert their legitimacy after the revolutionary tumult of the century's early years. A series of mostly failed uprisings shook the Continent's apparent stability, from the Iberian Peninsula to the Russian Empire, culminating in the upheavals of 1848, when conservative monarchies used troops against their own populations. These regimes supported each other militarily, using the new industrial technology of the railroad to move troops quickly. But the uprisings also highlighted the dynamism of European society, the fragility of the restoration regimes, and the persistent power of notions of liberty and sovereignty (Sperber, 2005). From London, Karl Marx and Friedrich Engels marked that epochal year with this call: "Let the ruling classes tremble at a Communistic revolution. The proletarians have nothing to lose but their chains. They have a world to win" (Marx and Engels, 1908, 63).

The events of 1848 also proved to be an impetus for a political transformation of Europe. The post-1815 peace fell apart as France's Napoleon III helped to engineer a war against Russia on the Crimean Peninsula, at least in part to consolidate the legitimacy of his regime at home (Figes, 2012). In the two decades after 1848, shifting coalitions of liberal, conservative, and radical forces engaged in a series of national state-building projects. By century's end, the map of Europe looked quite different as states like Italy, Germany, the hybrid Austria-Hungary, and a series of Balkan polities emerged from a series of political compromises and armed struggles. The most dramatic example of unification by force was in Prussia-Germany. From a position of relative weakness compared with its neighbors, successive Prussian rulers used industrial capacity, technological sophistication, and the strategic deployment of its well-trained army to win a series of victories over the Danes, Austrians, and French in quick succession (Showalter, 1975).

One of the key features of this period was the institution of the mass army. Many of the continental states, from the most powerful to minor players, embraced the lessons of the French Revolutionary model and employed some sort of conscription. While the terms of service varied and often engendered considerable debate within European states, it gave rulers considerable flexibility in training and deploying large armed forces. By making military service into a relatively common feature in the lives of young (and not-so-young) men through a system of active and reserve duty, conscription fostered what historians have termed "societal militarism" or "military culture," in which military ideas, practices, and behaviors became instantiated in the civilian sphere. This militarization of civilian life took place in states that are readily identified as highly militarized, like Prussia (Frevert, 2004), but also in places not commonly thought of in those terms, such as Austria (Cole, 2014).

Mass service had other effects as well. Even some of the most established national states, like France, contained an extraordinary degree of internal diversity. Regional and local identities, dialects, and customs merged into national military service. Europe's armies, in their function as a "school of the nation," began to chip away at these differences, instilling at the same time a sense of national purpose. As Eugen Weber (1976) famously put it, the army helped to turn "peasants into Frenchmen." In the multiethnic imperial polities, issues like choosing a language of command turned armies into microcosms of the empires themselves. In his classic study, István Deák (1990) describes the Hapsburg officer corps as a small but dedicated elite, who by necessity tried to overcome the endemic divisions in their polyglot army, and with it the empire itself, until that empire collapsed under the weight of defeat and its own contradictions.

During this period, the ability of European states to project power overseas expanded immeasurably. A complex web of factors, including technological improvements in communication, transportation, weaponry, and medicine, along with the need for raw materials and concomitant search for markets for finished goods, gave Europeans both the desire and the ability to expand their foothold in Africa and Asia (Headrick, 1981). Added to these factors was the rise of Social Darwinism, and "scientific racism" contributed as well by fostering the idea of biological competition for space and resources. In addition, powerful interest groups, particularly in the new national states, began to argue that colonies were a sine qua non for would-be powers. The resulting "colonial whirl," as Bismarck termed it, brought Europeans into the African and Asian interior as never before. Historians have long debated the relationship between colonial violence and the wars fought in Europe during the twentieth century. Isabel Hull (2005) persuasively argues that, in the German case, the violence of the wars in Southwest Africa reflected, but did not generate, the "institutional extremism" of the Wilhelmine army.

Many Europeans, elite and non-elite alike, grew concerned about the increasing militarization of the Continent and the potential for new states to emerge as bellicose and internally repressive. The period after 1871 also saw the first international efforts to establish minority protection and to search for ways to protect civilians in times of war (Fink, 2004). The emerging socialist movement in Europe consistently campaigned against the race to arms, while liberals like the Russian industrialist Ivor Bloch argued that war was no longer in the interest of European states and their highly integrated economies (Cooper, 1991).

Despite these countervailing voices, the combustible mixture of national chauvinism, militarism, and geopolitical rivalry contributed to an increasing instability in European power politics at the end of the nineteenth century. The causes of the First World War have been analyzed since the war's outbreak in 1914, but sophisticated historiography has struck a balance between seeing the war as the result of the decisions taken by a small number of political and military elites and of vast sociocultural forces welling up across a dynamic and fast-changing continent (Clark, 2013). The armies that went to war in 1914 were the products of Europe's uneven industrialization and the apotheosis of the massive investment in social militarization undertaken by Europe's rulers (Herrmann, 1996).

Current historiography on the First World War has been energized over twenty years by a methodological turn toward cultural history and by the heightened international interest associated with the centennial. Anglophone scholarship on the war long suffered under a relatively narrow focus on the Western Front. Thankfully, this has shifted dramatically toward a more global understanding of the conflict (Xu, 2011) and one that is much more engaged with the multiple fronts across which combatants fought (Watson, 2014).

The literature on the First World War, perhaps more than any other conflict in European history, has focused on the experience of individual soldiers and small units. Increased literacy meant more correspondence between front and home front, while a wave of memoirs and novels a decade after the war provides a rich trove of sources, at least from the Western Front. The result has been a rich and nuanced historiography that examines the day-to-day realities of combat (Leed, 1981), the psychological challenges of risking and inflicting death (Bourke, 2000), and the life of prisoners of war (Feltman, 2015). Alexander Watson (2008) draws extensively on this literature to ask, essentially, why the British Army was able to remain in the field while the German Army ultimately cracked in the spring of 1918.

Just as important, historians of the First World War have become keenly interested in the sociocultural history of the war. This begins as recognition of what two French historians termed "the great paradox" of the war: that it was enormously popular in combatant states, and only in retrospect became a symbol of pointless slaughter (Audoin-Rouzeau and Becker, 2003). Historians are now much more comfortable using the methods of cultural history to understand the self-conception of soldiers and civilians (Ziemann, 2007). There has also been a surge of urban histories of the war, studies that attempt to break down the artificial division between front and home front and that try to understand the relationship between domestic consent, everyday life, and the ability of states and societies to endure the strains of war (Healy, 2007). In his history of Freiburg im Breisgau at war, Roger Chickering (2007) ambitiously took on the challenge of writing "total history" to understand a "total war."

Finally, the literature on the First World War has been enhanced by an interest in the human upheaval of the conflict, particularly for civilians. This new focus on violence against civilian populations (Horne and Kramer, 2001) and on displacement (Gatrell, 2005) has championed two important perspectives. First, it has re-centered our understanding of the war away from the trenches. Second, it reminds us of its long-term impact and helps to re-periodize the conflict beyond the limits of the years 1914–1918. The First World War did much to shape the rest of the century to come.

The end of the First World War did little to resolve the underlying issues that had created European political instability. In fact, the creation of new states along notionally ethnic lines, the inability of peacemakers to agree on how best to sanction Germany, the presence of a revolutionary state in Russia, the withdrawal of the United States into isolation, and the dissatisfaction of nearly all of the participant powers with the conflict's results likely made Europe a more dangerous place (Boemeke et al. 2006). The French military *supremo* Ferdinand Foch famously and presciently remarked that the postwar settlement "is not peace. It is an armistice for twenty years" (quoted in Sharp, 2011, 3). The demands of wartime mobilization left enduring legacies as postwar victorious and defeated states consistently failed to live up to the promises that they had made to their populations during wartime. This "crisis of reciprocity" (Seipp, 2009) weakened the legitimacy of postwar governments across the Continent. As George Mosse (1991) and others have shown, the conflict's legacies suffused interwar culture. The vast and unprecedented personal and familial losses of the war, along with efforts by states to commemorate and give meaning to those losses, fueled a "myth of war experience" that glorified death and helped to fuel radical political and social visions on the Right and Left. In this environment, moderates stood little chance of achieving their own visions of stability.

After 1919, almost all European states between the Iberian Peninsula and the border of Russia had some form of representative government. Very few managed to survive the next twenty years. Democracy in interwar Europe was fragile and ultimately failed to meet the challenges of economic uncertainty and interstate tension. The most dynamic states of the interwar period were the dictatorships of the Left and Right, with the communist Soviet Union at one extreme and the fascist states on the other. While these experiments were very different in ideology, they shared a number of key features. Critically, both embraced a constant state of mobilization, in which those who were accepted as members of the community were brought into institutions of the state and the party that mirrored and amplified the conditions of wartime. This rejection of demobilization was not unique to communists and fascists, but do suggest that the common features of these deeply antagonistic states deserve as much study as do their differences (Gerwarth and Horne, 2012).

Much like the era before the First World War, prominent voices in Europe attempted to ameliorate interstate tensions between the wars. The League of Nations, while ultimately ineffective, emerged as a forum for resolving disputes on the Continent and in the imperial realm (Steiner, 2007). Parties of the center-left found themselves squeezed between ultra-nationalists and the extreme left, which now had an ideological home in Moscow. Most ordinary Europeans appear to have been reticent about the possibility of another war, not least because of fears of aerial bombardment and the demographic losses of 1914–1918 (Weber, 1996). As the civil war in Spain amply demonstrated, the mobilizing fascist states were far more committed to the path of rearmament and war than the Western democracies or the Soviet Union were to stopping them (Preston, 2007).

The Second World War, the largest conflict in human history, is difficult to assess briefly. This essay will consider four related but distinct themes from the war that were particularly important from the perspective of social change. First, for ordinary Europeans, the normative experience of the Second World War was not combat, but rather occupation. In some cases, this occupation lasted for years, or saw one occupier defeated and replaced by another. In practice, this meant a series of moral and political choices that were, all too often, little better than zero-sum (Sweets, 1986). Debates about who had been a collaborator and who had been a resister came to dominate politics in post-occupation societies. These distinctions, which were often quite messy in practice, had deadly consequences in the score-settling that followed (Deák, 2015). Timothy Snyder (2012) has identified a vast swath of Eastern Europe that he calls the "Bloodlands," where the Soviets and Germans occupied the same territory at different times. This double occupation, he argues, helped to radicalize already bitterly divided local cleavages and created conditions for ethnic cleansing and genocide.

Second, the war created solidarities in wartime society that had not previously existed. While every wartime state had to contend with dissenters and internal opponents, it is clear that the conflict brought vast swaths of society into the national project. In some places, like Britain, this resulted in labor peace and a perhaps unprecedented sense of social leveling (Field, 2014). In Nazi Germany, the war brought the regime as close as it ever came to realizing its dream of a race-based *Volksgemeinschaft*, or "national community," albeit a community based on expropriation, legal estrangement, and murder (Kershaw, 2008). For decades, a substantial number of Germans continued to believe that the *Wehrmacht*, the chief military instrument of the Third Reich, fought a relatively "clean" war; today, after numerous books and public exhibitions, such a view is clearly untenable (Wette, 2007; Megargee, 2007). The Soviet regime initially stumbled in its response to the war before adopting a more pragmatic approach that fused older Russian-nationalist ideas with the dynamism of revolutionary socialism (Merridale, 2006). In many cases, this solidarity did not survive the war. By 1945, civil wars broke out across the Continent in places like Greece, Italy, and contentious border zones of East Central Europe (Statiev, 2010).

Third, the years of 1939–1945 saw the terrible radicalization of war against civilians, both in rhetoric and in practice. This is a difficult area about which to write, because it risks creating moral equivalences where they do not belong. The Germans and their allies developed, implemented, and sustained a vast and lethal campaign of murder, primarily in Eastern Europe, directed at "racial enemies." While the 5.7 million Jews murdered in the Holocaust were certainly not the only group of victims, it is equally unquestionable that no other group was targeted for wholesale global extinction (Bauer, 2002). The German campaign also included the killing of social elites and the starvation of vast numbers of Eastern Europeans (Kay et al., 2014).

At the same time, new technology and the imperatives of a desperate war freed air warfare planners from the constraints under which they had previously operated (Biddle, 2004). Both sides engaged in the "strategic" bombing of enemy cities, but the Allies were much better at it. By late 1944, the Allies had air supremacy over Western Europe. The result was the comprehensive destruction of Germany's urban areas and the deaths of about five hundred thousand civilians. Historians still debate the efficacy of this campaign, which did not really begin to inhibit German industrial production until 1944 and did little to quash civilian morale (Overy, 2014).

Finally, the war's effects lingered in popular memory for a very long time, arguably until the present moment. The enormous sacrifices of wartime legitimized the Soviet regime (Weiner, 2002), and became ensconced in memory as a "People's War" in Britain (Rose, 2004). It also exposed political and cultural fault lines in places like France and Italy (Ventresca, 2004), and served as a site of national trauma, mourning, and questioning in a divided postwar Germany (Herf, 1997).

The extraordinary destruction of the Second World War further complicates how we conceptualize a distinct "postwar" era. The physical, economic, and moral devastation of the Continent, along with the unresolved conflicts that were then subsumed by the Cold War, has led scholars to increasingly argue for a very long period of transition (Judt, 2006). A number of patterns emerge that follow the broader trajectory of this period.

First, the war engendered a profound transformation of everyday life for millions of Europeans in its wake. Some of this was tragic, and stemmed from the wartime policies of combatant states. Most of Europe's Jews were dead or living outside of the Continent, a fate shared by millions of others from groups targeted by Nazi racial policy. More than ten million Europeans became displaced persons (DPs) at the end of the war, the vast majority of whom came from areas of Eastern Europe then dominated by the Soviet Union (Holian, 2015). At the same time, more than twelve million ethnic Germans and other minorities associated with the Axis fled or were driven out of homes in the East (Thum, 2011). These diaspora communities ended up in defeated Germany, but also spread throughout the world (Seipp, 2013).

James Sheehan (2008) argues that postwar Europe became a continent of "civilian states," which turned away from the militaristic legacies of the nineteenth and early twentieth centuries. Superpower hegemony and the threat of nuclear war no doubt contributed to this development. Reconstruction in Eastern Europe took place under Soviet military and economic domination (Applebaum, 2013). Soviet leadership ensured that the rebuilding of the region directly supported the reconstruction of the devastated Soviet Union. The rebuilding of armed forces for external defense and internal security took place for both pragmatic and ideological reasons. The German Democratic Republic rearmed, in effect if not in name, as early as 1949, seven years before the Federal Republic of Germany, in the form of the Garrisoned People's Police (*Kasernierte Volkspolizei*) (Lockenour, 2001; Corum, 2011). Yet apparent pacification of a previously bellicose continent owed much to endogenous developments as well.

In Western Europe, one response to the dual challenge of preventing the recurrence of war and of containing communism was to work toward economic integration. More than six decades later, given the long and winding road toward European integration, it is striking to read the text of the 1951 Paris Treaty establishing the European Coal and Steel Community and its concern with simply preventing replication of conditions that might lead to another general war (Dinan, 2014). The signing states declared that they were "resolved to substitute for age-old rivalries the merging of their essential interests, to create, by establishing an economic community, the basis for a broader and deeper community of peoples long divided by bloody conflict" (Treaty of Paris).

At the same time, Western European states variously sought to create new economic systems that maintained market-oriented capitalism but protected citizens from the instability of the boom and bust cycle, which many held responsible for the instability of the interwar years. These systems varied from country to country, but the legacy of ideas like the social market in Germany and the Beveridge Report in Britain have proven durable. There is no question that these policy responses stemmed from the mobilization of the war years and reflected an attempt to find a new arrangement between government and the governed in the wake of war (Behrend, 2006).

Europe's overseas empire went into retreat after 1945. Literature on the wars of decolonization highlights two themes that are critical to this essay. First, that while these conflicts between colonial and anti-colonial forces were relatively small, they were often extraordinarily violent. France found itself embroiled in, among other places, Indochina and Algeria, which proved ultimately resistant to military solutions and absorbed an increasing share of national resources. Moreover, decolonization exposed a Republic created from the ashes of wartime collaboration to criticism that it was employing methods not dissimilar to those practiced by the Germans and their French allies during the occupation (Drake, 2002). Recent scholarship has also exposed the extent to which British colonial authorities were willing to engage in atrocities in places like Kenya during the Mau Mau insurgency (Elkins, 2005; Maloba, 1988).

At the same time, other research has emphasized how profoundly the violence of the anti-colonial wars affected political and social life in Europe. The new French Republic's struggle in Algeria energized opponents in metropolitan France, leading to a state of near civil war in 1958 and engendering an often brutal response from the state (Horne, 2006). Dutch politics, also bitterly divided over the legacies of occupation and collaboration during the Second World War, came close to collapse over the ultimately failed effort to retain the Dutch East Indies, now Indonesia (Foray, 2012). In 1974, Portugal, since the 1930s ruled by a radical authoritarian movement that styled itself the *Estado Novo* (New State), saw a largely peaceful military coup and transition to democracy following growing frustration with colonial wars in Mozambique and Angola (Maxwell, 1995).

Not only did these colonial conflicts create political unrest in metropolitan states, but also they helped to spur an enormous influx of migrants and refugees from colonial and postcolonial areas. While this varied tremendously by country, the sometimes painful debates about how to legally manage migration and integration of European- and non-European-descended newcomers from the former empires helped to transform citizenship laws and, more profoundly, ideas about national belonging (Shepard, 2008). Today, in the wake of escalating fears of terrorism and concerns about the integration of migrants (particularly those from the Islamic world), these debates continue to resonate (Scott, 2010).

The student movement of the late 1960s and early 1970s grew in no small part out of the tensions that have been discussed in this essay. While the targets of student outrage varied by place, activists saw important solidarities across Europe, in North America, and around the world (Davis et al. 2012). Most considered themselves to be anti-imperialist and were particularly aggrieved at America's war in Indochina (where they had followed the defeated French). Especially in West Germany, students and young people sharply criticized their parents, whom they held responsible for the crimes of national socialism. As the wave of student radicalism crested, a small but violent minority turned to revolutionary agitation and, in some cases, violence. The physical and psychological violence unleashed by the Red Army Faction in Germany and the Red Brigades in Italy, along with other organizations in Europe and North America, added a violent coda to a period of tumultuous change (Varon, 2004).

The politics of Europe's Cold War division played an important part in shaping the experiences of people on the Continent. The "golden age" of postwar recovery in the West was accompanied by real advances in living standards in the East, at least until the global recession and oil shocks of the 1970s. It is difficult to assess how much of that growth was directly related to issues of war and defense, but it is clear that defense spending was relatively high and that the need to provision large military machines along the Iron Curtain had tangible, if uneven, economic benefits. The impact of Warsaw Pact militarization remains understudied, while historians have begun to present a nuanced picture of the role of the NATO alliance in everyday life in West Germany (Naimark, 1997; Maulucci and Junker, 2015).

Over all of this hung the specter of nuclear warfare. The threat of atomic war, particularly the arrival of tactical nuclear weapons in the European theater beginning in the 1950s, raised alarm among wide swaths of the European population. Conventional descriptions of "limited" war, dating back to Clausewitz, appeared outdated in light of the potential for massive destruction. Even a brief nuclear exchange posed an existential challenge to those living in the places where it was likely to happen. This tension between superpower nuclear strategy and local responses reached its apogee in the early 1980s, when millions of Europeans, west and east, protested against the deployment of "Euromissiles" (Nuti et al., 2015). Almost three decades after the end of the Cold War, historians and political scientists continue to debate the importance of popular protest and the grass-roots claims to human rights in bringing about the end of the conflict. While some scholars rightly point to the failures of communism's "uncivil societies" (Kotkin, 2010), it is clear that the bravery of millions of people in Eastern Europe ultimately created a crisis of legitimacy that these regimes could not successfully placate or repress (Stokes, 2011).

The decades since the end of the Cold War have shown the continued salience of questions of war and peace in European societies. In a sense, the end of the Cold War marked both a divergence from and a return to Europe's troubled history. Historians have taken on the task of explaining how Europeans might grow together through novel political and economic arrangements dating from the period after 1945. At the same time, sometimes agonizing debates over Europe's role in the world revolve around the Continent's history of division, conflict, and exploitation. Initial euphoria over the sudden unwinding of four decades of communist domination of East Central Europe was tempered by anxieties about what a post–Cold War order might look like. The sudden reemergence of Germany as a substantial continental power engendered considerable anxiety in European states whose twentieth-century history was partly shaped by struggle with German attempts at hegemony. The continued commitment of the United States to act as an offshore balancer helped to alleviate at least some of those concerns (Sarotte, 2014).

One of the first crises faced by the new Europe was the violent dissolution of Yugoslavia. The response of NATO and the UN proved muddled and hesitant, with results like the tragic loss of civilian life in Operation Storm and the massacre at Srebrenica in 1995 (Glenny, 2012). European states and societies have in some ways accelerated the retreat from the militarism of the post-1945 period. Military budgets have declined, remaining conscription programs have been abandoned, and defense establishments have shrunk. Shaped by their history, Europeans hold a range of often contrasting views about issues like military intervention, democracy promotion, and international responsibility for enforcing human rights norms. These cleavages became evident in the early 2000s when the United States and Britain sought partners for the invasions of Afghanistan and Iraq.

In 2016, troops from a united Germany entered Poland for the first time since the Second World War as part of exercises designed to show NATO resolve against an increasingly bellicose Russia. Public opinion in Europe regarding issues of defense remains widely varied, but there is little durable support for increasing expenditures on defense. Europeans remain conscious of the lessons of their history, both of its tremendous material and political progress and of the legacies of murder and destruction. Those dual inheritances will likely continue to shape European responses to the challenges of war and peace for some time to come.

Bibliography

Aaslestad, Katherine B. *Place and Politics: Local Identity, Civic Culture, and German Nationalism in North Germany*. Leiden: Brill, 2005.

Anderson, Benedict. *Imagined Communities: Reflections on the Origin and Spread of Nationalism*. New York: Verso, 1991.

Applebaum, Anne. *Iron Curtain: The Crushing of Eastern Europe, 1944–1956*. New York: Anchor, 2013.

Audoin-Rouzeau, Stéphane, and Annette Becker. *14–18: Understanding the Great War*. New York: Hill and Wang, 2003.

Bauer, Yehuda. *Rethinking the Holocaust*. New Haven: Yale University Press, 2002.

Behrend, Ivan T. *An Economic History of Twentieth-Century Europe: Economic Regimes From Laissez-Faire to Globalization*. New York: Cambridge University Press, 2006.

Bell, David A. *The First Total War: Napoleon's Europe and the Birth of Warfare as We Know It*. New York: Mariner, 2008.

Biddle, Tami Davis. *Rhetoric and Reality in Air Warfare: The Evolution of British and American Ideas About Strategic Bombing, 1914–1945*. Princeton: Princeton University Press, 2004.

Boemeke, Manfred F., Roger Chickering, and Stig Förster, eds. *Anticipating Total War: The German and American Experiences, 1871–1914*. New York: Cambridge University Press, 2006.

Boemeke, Manfred F., Gerald Feldman, and Elisabeth Glaser, eds. *The Treaty of Versailles: A Reassessment After 75 Years*. New York: Cambridge University Press, 2006.

Bourke, Joanna. *An Intimate History of Killing: Face-to-Face Killing in Twentieth-Century Warfare*. New York: Basic, 2000.

Chickering, Roger. *The Great War and Urban Life in Germany: Freiburg, 1914–1918*. Cambridge: Cambridge University Press, 2007.

Clarke, Christopher. *The Sleepwalkers: How Europe Went to War in 1914*. New York: Harper Collins, 2013.

Cobb, Richard. *The People's Armies*. New Haven: Yale University Press, 1987.

Cole, Laurence. *Military Culture and Popular Patriotism in Late Imperial Austria*. Oxford: Oxford University Press, 2014.

Colley, Linda. *Britons: Forging the Nation, 1707–1837*. New Haven: Yale University Press, 1992.

Cooper, Sandi E. *Patriotic Pacifism: Waging War on War in Europe, 1815–1914*. New York: Oxford University Press, 1991.

Corum, James S., ed. *Rearming Germany*. Leiden: Brill, 2011.

Davies, Norman. *God's Playground: A History of Poland*. New York: Oxford University Press, 2005.

Davis, Belinda, et al., eds. *Changing the World, Changing Oneself: Political Protest and Collective Identities in West Germany and the U.S. in the 1960s and 1970s*. New York: Berghahn, 2012.

Deák, István. *Beyond Nationalism: A Social and Political History of the Habsburg Officer Corps, 1848–1918*. Oxford: Oxford University Press, 1990.

———. *Europe on Trial: The Story of Collaboration, Resistance, and Retribution during World War II*. Boulder, CO: Westview, 2015.

Dinan, Desmond. *Europe Recast: A History of European Union*. 2nd ed. Boulder: Lynne Reinner, 2014.

Drake, David. *Intellectuals and Politics in Post-War France*. New York: Palgrave, 2002.

Elkins, Caroline. *Imperial Reckoning: The Untold Story of Britain's Gulag in Kenya*. New York: Henry Holt, 2005.

Feltman, Brian. *The Stigma of Surrender: German Prisoners, British Captors, and Manhood in the Great War and Beyond*. Chapel Hill: University of North Carolina Press, 2015.

Field, Geoffrey. *Blood, Sweat, and Toil: Remaking the British Working Class, 1939–1945*. Oxford: Oxford University Press, 2014.

Figes, Orlando. *The Crimean War: A History*. New York: Picador, 2012.

Fink, Carole. *Defending the Rights of Others: The Great Powers, the Jews, and International Minority Protection, 1878–1938*. New York: Cambridge University Press, 2004.

Foray, Jennifer. *Visions of Empire in the Nazi-Occupied Netherlands*. New York: Cambridge University Press, 2012.

Frevert, Ute. *A Nation in Barracks: Modern Germany, Military Conscription and Civil Society*. London: Bloomsbury, 2004.

Gatrell, Peter. *A Whole Empire Walking: Refugees in Russia During World War I*. Bloomington: Indiana University Press, 2005.

Gerwarth, Robert, and John Horne. *War in Peace: Paramilitary Violence in Europe After the Great War*. Oxford: Oxford University Press, 2012.

Glenny, Misha. *The Balkans: Nationalism, War, and the Great Powers, 1804–2011*. New York: Penguin, 2012.

Headrick, Daniel R. *The Tools of Empire: Technology and European Imperialism in the Nineteenth Century*. New York: Oxford University Press, 1981.

Healy, Maureen. *Vienna and the Fall of the Habsburg Empire: Total War and Everyday Life in World War I*. New York: Cambridge University Press, 2007.

Herf, Jeffrey. *Divided Memory: The Nazi Past in the Two Germanys*. Cambridge: Harvard University Press, 1997.

Herrmann, David G. *The Arming of Europe and the Making of the First World War*. Princeton: Princeton University Press, 1996.

Holian, Anna. *Between National Socialism and Soviet Communism: Displaced Persons in Postwar Germany*. Ann Arbor: University of Michigan Press, 2015.

Horne, Alistair. *A Savage War of Peace: Algeria, 1954–1962*. New York: New York Review of Books, 2006.

Horne, John, and Alan Kramer. *German Atrocities, 1914: A History of Denial*. New Haven: Yale University Press, 2001.

Hull, Isabel V. *Absolute Destruction: Military Culture and the Practices of War in Imperial Germany*. Ithaca: Cornell University Press, 2005.

Jarausch, Konrad. *Out of Ashes: A New History of Europe in the Twentieth Century*. Princeton: Princeton University Press, 2015.

Judt, Tony. *Postwar: A History of Europe Since 1945*. New York: Penguin, 2006.

Kay, Alex, Jeff Rutherford, and David Stahel, eds. *Nazi Policy on the Eastern Front, 1941*. Rochester: University of Rochester Press, 2014.

Kershaw, Ian. *Hitler, the Germans, and the Final Solution*. New Haven: Yale University Press, 2008.

Kotkin, Stephen. *Uncivil Society: 1989 and the Implosion of the Communist Establishment*. New York: Modern Library, 2010.

Leed, Eric J. *No Man's Land: Combat and Identity in World War I*. New York: Cambridge University Press, 1981.

Lockenour, Jay. *Soldiers as Citizens: Former Wehrmacht Officers in the Federal Republic of Germany, 1945–1955*. Lincoln: University of Nebraska Press, 2001.

Lynn, John. *The Bayonets of the Republic: Motivation and Tactics in the Army of Revolutionary France, 1791–1794*. Boulder, CO: Westview Press, 1996.

Maloba, Wunyabari O. *Mau Mau and Kenya: Analysis of a Peasant Revolt*. Bloomington: Indiana University Press, 1988.

Marx, Karl, and Friedrich Engels. *The Communist Manifesto*. Chicago, Charles Kerr, 1908.

Maulucci, Thomas, and Detlef Junker, eds. *GIs in Germany: The Social, Economic, Cultural, and Political History of the American Military Presence*. New York: Cambridge University Press, 2015.

Maxwell, Kenneth. *The Making of Portuguese Democracy*. New York: Cambridge University Press, 1995.

Megargee, Geoffrey P. *War of Annihilation: Combat and Genocide on the Eastern Front, 1941*. New York: Rowman and Littlefield, 2007.

Merridale, Catherine. *Ivan's War: Life and Death in the Red Army, 1939–1945*. New York: Picador, 2006.

Mosse, George L. *Fallen Soldiers: Reshaping the Memory of the World Wars*. New York: Oxford University Press, 1991.

Naimark, Norman M. *The Russians in Germany: A History of the Soviet Zone of Occupation, 1945–1949*. Cambridge: Belknap, 1997.

Nuti, Leopoldo, Frédéric Bozo, Marie-Pierre Rey, and Bernd Rother, eds. *The Euromissile Crisis and the End of the Cold War*. Washington: Woodrow Wilson Center Press, 2015.

Overy, Richard. *The Bombers and the Bombed: Allied Air War Over Europe 1940–1945*. New York: Penguin, 2014.

Parker, Geoffrey. *The Military Revolution: Military Innovation and the Rise of the West, 1500–1800*. New York: Cambridge University Press, 1996.

Parrott, David. *The Business of War: Military Enterprise and Military Revolution in Early Modern Europe*. New York: Cambridge University Press, 2012.

Pollard, Sidney. *Peaceful Conquest: The Industrialization of Europe, 1760–1970*. Oxford: Oxford University Press, 1981.

Preston, Paul. *The Spanish Civil War: Reaction, Revolution, and Revenge*. New York: Norton, 2007.

Rose, Sonya. *Which People's War?: National Identity and Citizenship in Wartime Britain 1939–1945*. New York: Oxford University Press, 2004.

Sarotte, Mary. *The Collapse: The Accidental Opening of the Berlin Wall*. New York: Basic, 2014.

Scott, Joan Wallach. *The Politics of the Veil*. Princeton: Princeton University Press, 2010.

Secher, Reynald. *Vendée: A French Genocide*. South Bend: University of Notre Dame Press, 2003.

Seipp, Adam R. *The Ordeal of Peace: Demobilization and the Urban Experience in Britain and Germany, 1917–1921*. Farnham: Ashgate, 2009.

———. *Strangers in the Wild Place: Refugees, Americans, and a German Town, 1945–1952*. Bloomington: Indiana University Press, 2013.

Sharp, Alan. *Consequences of Peace: The Versailles Settlement and Legacy, 1919–2010*. London: Haus, 2010.

Sheehan, James M. *Where Have All the Soldiers Gone?: The Transformation of Modern Europe*. New York: Mariner, 2008.

Shepard, Todd. *The Invention of Decolonization: The Algerian War and the Remaking of France*. Ithaca: Cornell University Press, 2008.

Showalter, Dennis. *Railroads and Rifles: Soldiers, Technology, and the Unification of Germany*. Hamden: Archon, 1975.

Snyder, Timothy. *Bloodlands: Europe Between Hitler and Stalin*. New York: Basic, 2012.

Sperber, Johnathan. *The European Revolutions, 1848–1851*. New York: Cambridge University Press, 2005.

Statiev, Alexander. *The Soviet Counterinsurgency in the Western Borderlands*. New York: Cambridge University Press, 2010.

Steiner, Zara. *The Lights That Failed: European International History 1919–1933*. Oxford: Oxford University Press, 2007.

Stokes, Gale. *The Walls Came Tumbling Down: Collapse and Rebirth in Eastern Europe*. New York: Oxford University Press, 2011.

Sweets, John. *Choices in Vichy France: The French Under Nazi Occupation*. New York: Oxford University Press, 1986.

Thum, Gregor. *Uprooted: How Breslau Became Wroclaw During the Century of Expulsions*. Princeton: Princeton University Press, 2011.

Treaty of Paris, 1951. Found at: www.consilium.europa.eu/uedocs/cmsUpload/Treaty%20constituting%20the%20European%20Coal%20and%20Steel%20Community.pdf

Varon, Jeremy. *Bringing the War Home: The Weather Underground, the Red Army Faction, and Revolutionary Violence in the Sixties and Seventies*. Berkeley: University of California Press, 2004.

Ventresca, Robert. *From Fascism to Democracy: Culture and Politics in the Italian Election of 1948*. Toronto: University of Toronto Press, 2004.

Watson, Alexander. *Enduring the Great War: Combat, Morale and Collapse in the German and British Armies, 1914–1918*. New York: Cambridge University Press, 2008.

———. *Ring of Steel: Germany and Austria-Hungary in World War I*. New York: Basic, 2014.

Weber, Eugen. *The Hollow Years: France in the 1930s*. New York: Norton, 1996.

———. *Peasants Into Frenchmen: The Modernization of Rural France, 1870–1914*. Stanford: Stanford University Press, 1976.

Weber, Max. *From Max Weber: Essays in Sociology*. New York: Routledge, 2009.

Weiner, Amir. *Making Sense of War: The Second World War and the Fate of the Bolshevik Revolution*. Princeton: Princeton University Press, 2002.

Wette, Wolfram. *The Wehrmacht: History, Myth, and Reality*. Trans. Deborah Lucas Scheider. Cambridge: Harvard University Press, 2007.

Xu, Guoqi. *Strangers on the Western Front: Chinese Workers in the Great War*. Cambridge: Harvard University Press, 2011.

Ziemann, Benjamin. *War Experiences in Rural Germany: 1914–1923*. London: Bloomsbury, 2007.

10 War and Society in England, Ireland, Scotland, and Wales

Margaret Sankey

The military history of Britain in the modern world, especially as addressed by the "new" scholarship on war and society, is a study of four interlocking themes that stretch from the outbreak of the English Civil War in the mid-seventeenth century to the current deployments of British forces in the Middle East. First, this period of British history has marked the rise of a relatively small, underpopulated island to become a powerful state and then, following the Second World War, struggle to define itself as a member of the EU, in a special relationship with the United States and as a former hegemon. Second, during its height, the British were also *the* imperial power, with a reach into all corners of the globe and into the lives of hundreds of millions of subject people. Many of them performed significant military roles in war production or as "martial race" soldiers and sailors. Others' resistance against Britain justified the maintenance and expansion of vast armies and navies answering to both the state and corporate masters. Third, the sheer vastness of the empire, from its roots in the unification of the British Isles themselves, and the diversity of populations meant that military service, conquest, and traditions have been baked in as a cultural identity and assimilation tool since the Hundred Years' War. Fourth, the means by which this dominance came about—undergirded by signature institutions in Britain, such as the Bank of England's constitutional limits on the military's political influence, military investments that stoked the Industrial Revolution, or strategic campaign decisions that redrew the world's maps—shape a legacy of Anglophone legal, military, cultural, and economic structures around the globe.

It has often been useful to cast the English Civil War (1642–1651) as the last religious war and the first political revolution. Certainly, the bulk of historiography has been an effort to push the causes of the war into tensions between the Catholic tendencies of Charles I and the Puritan orientation of Parliament, or the struggle of an emerging bourgeoisie to seize political power from an absolutist monarchy. An emerging trend in the scholarship, however, considers how the eventual explosion of hostilities played out not just on the battlefields of England but also in Scotland, Ireland, and Wales. Lois Schwoerer (2016) sets the stage of military knowledge and access of firearms in the Stuart polity, as improvements in drill manuals, manufacturing, and chemistry trickled across the channel via consumer purchasing, service in mercenary companies, or a print culture centered in Protestant Holland. Mark Kishlansky (1983) demonstrated how Parliament, with deep ties to this innovative military culture, began the conflict with significant advantages over the Stuart monarchy.

Peter Gaunt (2014) gives primacy to the organization, training, and battlefield tactics of each side, along with the way in which this reflected Royalist or Parliamentarian resources, political outlook, and strategic goals. Taking a different tack, Charles Carlton offers gripping descriptions of personal experiences of soldiers in brutal early modern warfare, their motivations and the aftereffects of their time under arms. Elsewhere, the kaleidoscope vision of the Civil War as a confluence of issues in England, Scotland, Ireland, and Wales is Trevor Royle's thesis (2004), which explains the Covenanters, the Bishops' War, Cromwell's campaigns in

Ireland, and the Levellers and Diggers as manifestations of regional issues bundled into a great conflagration. Mark Fissel's dissection of Charles I's campaigns in Scotland (1999) lays bare the weaknesses not just of Stuart capability in the field but also of the king's strategic and political misunderstanding of his northern kingdom. New scholarship on the turbulent mid-seventeenth century has likewise blended military history and other fields as evinced by Laura A. M. Stewart's examination of politics and conflict in the urban center of Edinburgh, Scotland (2006).

Whatever the complaints against the overreach of Charles I had been, Protector Cromwell and the period of the Commonwealth saw strict and large-scale reorganizations of British financial and political power. Jeremy Black (1998) argued that these vigorous domestic reforms significantly increased the efficacy of the state's military power, rather than the military empowering the extraction of resources. Mark Fissel (1991) and James Scott Wheeler (2002) further explain that these efforts included more efficient collection of taxes, professionalization of the armed forces both on land and at sea, and the increasing involvement of the military in governance in the form of martial law in Scotland and Ireland. The success of this reform-minded Commonwealth regime extended to naval conflicts. Beginning in the mid-seventeenth century, as explored by J.R. Jones (1996), the wars against the Dutch at sea led to both the acquisition of Caribbean outposts and further Restoration expansion into continental North America, as well as to the institution of the Navigation Acts and to preserve naval and economic supremacy. The latter pair eventually became a major source of friction between Britain and her colonies.

Despite the many changes wrought by the Civil War, the unfinished political and religious settlement continued to play itself out in 1688. James II violated the unwritten expectations of Parliament that had been made at the Restoration in 1660 by stacking military offices with Catholic favorites and threatening to undermine Church of England primacy. Britain's elite feared encroachment by Louis XIV's France on Britain's colonial possessions and absolutist domestic government by James II himself. As studied in Steven Saunders Webb's four volumes (1979, 1984, 1995, and 2013), the politically active and savvy military leaders in Britain and the North American colonies took increasingly larger roles in shaping decisions in favor of an officer class hailing from powerful, land-owning, aristocratic, Protestant backgrounds.

This rising group decisively intervened in 1688 to assure the deposition of James II and his replacement with William III and Mary II. Once ensconced with Parliament's approval, these new monarchs delivered on their backers' expectation of continued evolution for British political, financial, and military institutions that favored Parliament and the oligarchy against a potentially absolutist king. Within the Bill of Rights of 1689 that established subsequent constitutional monarchy, Joyce Malcolm finds the roots of the deeply held Anglo-American connection to guns as a corrective to overreach of government tyranny. Another connection can be seen in how Dutch mercantilism and financial innovation influenced William III of Orange. John Brewer's *The Sinews of Power* (1989), for example, situates the advantages of establishing a national Bank of England, extensive naval yards, and colonial supply networks as foundations to the rise of Britain as a world power.

The next generation of warfare followed a cyclic political pattern. During King William's War, Queen Anne's War, the Great Northern War, the War of the Spanish Succession, the War of the Polish Succession, and finally the War of the Austrian Succession, the British formed coalitions with the Dutch, Savoyards, Austrians, or Prussians to contain the expansion of France. Although Paul Langford's survey (2010) of the period refers to the British as a "Polite and Commercial People," that mentality was made possible by a fearsome commitment to military investment and involvement with the Continent. These early-eighteenth-century conflicts saw years-long campaigns because of the relative stalemate in technology and numbers made possible by coalitions. The resulting mixture of siege and maneuver warfare relied on the ability of the belligerent states to keep paying and supplying their soldiers.

Britain's mastery of maneuver warfare reached its apogee in the Battle of Blenheim in 1704. Richard Holmes (2008) recounts how John Churchill, Duke of Marlborough, and Austrian counterpart Eugene of Savoy defeated a Franco-Bavarian enemy coalition. Despite this decisive victory, Marlborough fell victim to the politics of the late Stuart court and fell from power as Britain emerged from a fourteen-year war having accomplished little except *status quo ante bellum*. This pattern repeated several more times in Europe, with drawn-out conflicts of great cost, little British direct advantage, and peace treaties that benefited continental allies rather than Great Britain. As pointed out by Karl Schweitzer (1998), the War of the Austrian Succession in the 1740s identified Prussia as a rising power to which the monarchy of Great Britain had marital and martial ties, if not yet strategic interests in common.

Although the English Civil War did bring Puritan settlers in New England back to the home country to enlist in the New Model Army, and the Anglo-Dutch engagements spread into the Atlantic and Caribbean, British Wars after 1689 would become inevitably global in scope as colonial assets became more strategic and vulnerable targets. Recent military historiography has grown more cognizant of the roles played by settler colonists and the indigenous people as both allies and antagonists. For example, Haefeli and Sweeney (2003) has given equal attention to the motivations of all participants in the Deerfield raid of 1704.

For merchants in Britain, the most pressing of these overlapping conflicts was hostility between Britain and Spain. One of the few advantages Britain had gained from the Peace of Utrecht in 1714 was the *asiento* allowing British merchants to trade with Spanish America, opening lucrative slave markets and undermining Spanish mercantile control of her empire. Spanish attempts to restrict British expansions of the treaty's terms incited widespread anger among the British public. After decades of festering because of the reluctance of long-serving prime minister Robert Walpole to fight, war erupted when the Spanish customs inspectors confronted a British captain who literally lost his ear in the ensuing struggle. This event in turn led to public outcry against the traditional Spanish Catholic enemy. Philip Woodfine's work (1998) is an analysis of not just the increasing presence of colonies and their profits in the minds of the British public but also the degree to which print propaganda and popular enthusiasm could push the country into war.

As one side effect of the military stalemate between France and Britain, the French tended to induce political and religious unrest within Great Britain through their support of deposed Stuart king James II and his exiled descendants. Louis XIV sent James II back to Ireland in 1690 with a French fleet and soldiers only to be defeated by the British at the Battle of the Boyne. Nevertheless, this did prove that the French could cause great disruption, distraction, and diversion of resources. Bruce Lenman (1980) offers a broad survey of these insurgencies, while Daniel Szechi (2006) and Stuart Reid (1996) examine specific examples of the insurgencies in subsequent studies. In each case, a relatively minor investment of French support drew conservative Catholics in Scotland, Ireland, and the north of England to fight for James Francis Stuart ("The Old Pretender") and his son Charles Edward Stuart ("Bonnie Prince Charlie") against the Hanoverian monarchy. These clashes, the last battles to take place on British soil, fanned continued religious hostility to Catholics, prompted the disarming and dismantling of the Highland clan system and led to calls for better defense preparation of county militias in the form of the Norfolk Drill handbook (1759).

Margaret Sankey's book on Jacobitism (2005) and her collaborative work with Daniel Szechi (2001) outline the whole-of-government approach adopted by the Hanoverian government against the rebels. Although able to pull more troops from Europe to crush Jacobitism militarily, George I and his ministers were keen to avoid accusations of tyranny, absolutism, or the grisly brutality of James II's suppression of the Monmouth Rebellion in 1685. Instead, after executing a handful of key leaders, they chose a strategy of complex arrangements for leniency with loyalist relatives of rebels, and invested money in nation-building projects like road networks and schooling for the heirs of clan chiefs while deporting 632 of the rebels to Britain's North American colonies as an alternative to mass executions.

As chronicled by Fred Anderson, during the 1740s, increased populations and movement of colonists from the coastline into the interior of North America put pressure on the borderlands between French and British-claimed territory. The French, who had sent far fewer settlers to the Continent but had more successfully established military and diplomatic links with native peoples, began a more aggressive policy of fortifying the back country of the Ohio River valley. This prompted the governor of Virginia to send a surveying party, led by a very young George Washington, to investigate. The resulting incident, recounted by David Clary (2011), rippled out to London and Paris and into open warfare, and made a celebrity of the young officer. The British responded by sending an expedition under Major General Edward Braddock that met disaster at the Battle of the Monongahela in 1755. This action also proved to be the formative military experience for George Washington and other leaders of the American Revolution, as explained by David Preston (2015). Subsequent French attacks, aided by their native allies, on the frontier British forts threw settlers into a panic and swept away British authority.

In Europe, other crucial changes were afoot. Austria believed that her long-standing alliance with Britain had cost too dearly in British demands to make the Treaty of Aix-la-Chappelle happen (Britain did not receive these lands), and looked elsewhere for a diplomatic partner. In 1756, this resulted in a drastic revision of alliances through a pact between Austria, Russia, and France. This posed a grave danger to Britain, and left an opening into which Prussia could step as a potential replacement. William Pitt the Elder, the British prime minister, took this opportunity to propose a new, high-risk strategy: Britain bankrolled Prussian forces in continental Europe, while the British Army and Navy concentrated on breaking French colonial power, a move studied by Richard Middleton (1985) and Daniel Baugh (2011). In their edited anthology, Mark Danley and Patrick Speelman (2012) fully display the breadth of the Seven Years' War that extended from the Plains of Abraham at Quebec to the trading forts of Bengal in the Indian Ocean. Pitt was also keen to use the war to reintegrate Scottish and Irishmen to British advantage as soldiers, allowing regiments to keep their bagpipes, kilts, and other domestically banned items of cultural heritage as regimental uniforms to promote morale. This approach paid off in 1759 as the sons of disgraced Jacobites, like Simon Fraser, broke French lines with a Highland charge at the battle for Quebec.

Britain's eventual 1763 victory, however, set into motion more problems than it solved. Friction with the many native peoples formerly allied to the French, in the form of Pontiac's Rebellion, led the British government to restrict white settlement to the edge of the Appalachians, enraging land speculators and reminding settlers of royal authority, as Colin Calloway (2006) recounts. Although the British treasury had measures in place to handle financing huge and long wars, the war debt needed to be addressed through increased taxation of the colonies. New demands for money and the presence of British regulars in American cities rubbed raw sensibilities and expectations on both sides, although for trade and security reasons Andrew O'Shaughnessy (2000) studied, some of them reacted more contentiously than others. Stephen Brumwell (2002) includes many such confrontations over quartering, religion, public order, and deference, all of which fueled the widening divisions between the thirteen American colonies and Britain.

A particularly British perspective on the American Revolution has emerged from "new" military history, highlighting the many dissenters within the British government, including Col. Isaac Barre and John Wilkes, many drawing on military experience to inform their political outlook. Nick Bunker's book (2014) is a masterful study of Britain as a flawed military-imperial system unable to respond to the legitimate complaints of colonists in the wake of 1763, partly because the greater size and scope of the empire made North America a low priority until it was on fire, and partly because of an unwillingness in London to consider how second- and third-order effects of decisions like the British East India Company bailout, or the Quebec Acts, would be provocative in New England.

For a view of the war from the perspective of a British regular soldier, Stanley Weintraub (2005) wrote in the shadow of American involvement in Iraq and Afghanistan (as well as Vietnam), but the comparison is not unwarranted. Kevin Linch and Matthew McCormick (2014) provide further context for a redcoat's experience—from enlistment through discipline, sickness, and personal relationships to death. At the top of the political pyramid, the "deciders" are represented in Andrew O'Shaughnessy's *The Men Who Lost America* (2013), which also follows the rest of the careers of men like General Charles Cornwallis and Admiral George Rodney, both of whom had other roles to play in the larger empire. One of the most hated aspects of British military power was its employment of German mercenaries (usually labeled generically as "Hessians"), facilitated through Britain's possession of the Kingdom of Hanover, an expensive but significant way to avoid overburdening domestic recruiting and the manpower repercussions of a long and unpopular war, as seen by Brady Crytzer (2015).

Before the American Revolution, the deeply entwined relationship of American shipbuilding and sailing expertise, London merchants, and transatlantic trade contributed to Britain's maritime dominance, but this power came at the price of Americans feeling exploited by demands for manpower and the Navigation Acts. Christopher Magra (2016) takes on the resentment of impressment, while Nathan Perl-Rosenthal (2015) attempts to explain how men who had previously thought of themselves as inhabiting an Atlantic seaborne world chose to be Americans.

A decade after the American Revolution, the British Empire was once again fully engaged in war, this time against Revolutionary France (and after Toulon, with its emerging leader Napoleon Bonaparte). That this sustained effort, involving land, sea, and economic resilience, could happen at all was the result of a century of Britain's mastery of material production, financial institutions, and imperial trade networks. Roger Knight (2013) studies the production infrastructure available to Britain's military. Kirkpatrick Sale (1995), however, serves as a counterbalance by describing the Luddite movement as a significant cry from the common people who made it possible. In 1798, inspired by the ideals of the French Revolution, Irish rebels took up arms, expecting French assistance in throwing off British rule. Guy Beiner (2009) unpacks not just what actually happened as the movement failed but also how the Irish remembered these events and built on them in subsequent rebellions. For a panoramic view of the effects this generation of warfare had on British society, Jenny Uglow (2014) follows a cross section of families to reconstruct a quarter century of home front transformation, while Linda Colley (1992) sees the Napoleonic Wars as the culmination of Britain's efforts to cohere Ireland, Scotland, and Wales by "othering" France. In Britain, the Napoleonic Wars also include the War of 1812, seen not as the "Second American Revolution" (so dubbed by some recent American historians) but as a sidelight complicated by Britain's 1809 wartime decision to outlaw the slave trade as a form of economic warfare against France. This confluence of military and abolitionist policy is the centerpiece of a recent book by Alan Taylor (2013).

Because of Napoleon's control of continental Europe, the British Navy was particularly crucial in maintaining control of the trade lanes and colonial resources and communications, a military feat made possible by the institutional structures scrutinized by N.A.M. Rodger (1986), the training and tactics studied by James Davey (2015), and the skilled manpower recruited—press-ganged or otherwise acquired—examined in *Hornblower's Historical*

Shipmates by Heather Noel-Smith and Lorna Campbell (2016). The navy was also largely responsible for the care of prisoners, often in prison hulks or in prisoner-of-war camps, which both antagonized and offered opportunities for local populations, as seen in Paul Chamberlain (2008). Andrew Lambert (2005) demonstrates that in national memory, however, these gritty realities of shipboard life and death were overshadowed by the glamorous career and memorialized afterlife of Admiral Horatio Nelson, whose 1805 victory at Trafalgar assured his place as a heroic martyr for Britain's survival.

Britain's primary land commitment was to the "Spanish Ulcer." Its forces operated from fortified Portugal to bleed Napoleon's brother Joseph and detract from Napoleon's preferred strategy. This vicious conflict spawned not only the term "guerilla" but also the brutal visual representation by Goya and, from the perspective of Charles Esdaile (2003), offers the roots of Latin American independence wars and even the Spanish Civil War of 1936–1939. In a recent work Allan Forrest (2015) synthesizes his own scholarship, the historiographical view of the final military showdown with Napoleon at Waterloo, and its afterlife of commemoration and reverberation. The man most responsible for the peninsular campaign, Arthur Wellesley, the Duke of Wellington, took on a nationalistic stature like Nelson's, although a two-volume biography of Wellington by Rory Muir (2013a, 2013b) explores a process of heroic memory-making that was complicated by Wellington's survival and subsequent forty years as a Tory politician.

Between the Congress of Vienna and the Crimean War, Britain's isolationist politics removed her from continental conflict for nearly a generation. During the early nineteenth century, the army and navy engaged in Britain's vast empire (discussed ahead), but also played a vital role in the expansion of science and technology. In the eighteenth century, it had been royal funding that spurred inventors to find an accurate way to measure longitude, which Dava Sobel examines in a book of the same name (1995). Naval explorations like Captain Cook's also sought botanical solutions like breadfruit to the challenges of provisioning colonies while conducting mapping and exploration missions. Richard Drayton (2000) studies the British government's use of the military to conduct science missions, particularly collecting for Kew Gardens, including the voyage of the HMS Beagle, enabling Darwin's study of the Galapagos. The cartography and oceanographic expertise, like Captain Blight's astounding dead reckoning to the Australian coast, can be found in the research of David MacKay (1985).

By the mid-nineteenth century, the plight of Turkey at the hands of an expansionist Russia and the potential for an upset of the European "balance of power" should Russia seize the straits combined to draw Britain into the Crimean War. Andrew Lambert explains the humanitarian and diplomatic reasoning behind Britain's engagement in the Crimean War, but most people remember the war for its semi-senile generals, the senseless charge of the Light Brigade, and the grotesque neglect of medical care that produced both Florence Nightingale's image as the "Lady With the Lamp" and her hard-nosed, statistically based demands for reform of hygiene and nursing. Stefanie Markovitz (2009) examines Britain's failure to modernize training, institutional leadership, and planning, and Amanda Foreman (2010) deals with the effect of the American Civil War on Britain, from the incendiary speeches of Frederick Douglass to the disruption of the northern textile industry by the Union blockade.

The second half of the nineteenth century saw the acceleration of a concept begun with the co-option of Welsh, and then Scottish and Irish, soldiers—that these non-English men were "martial races," unruly and rebellious, but whose natural tendency to instability and violence could be bent to the empire's advantage with discipline and enlistment in Britain's military. Over time, this came to encompass the Marathas, Pathans, Gurkhas, and Sikhs in a propaganda-boosted view of race that meshed with prevailing popular ideas of Social Darwinism and a self-serving classification of these troops as childlike and in need of British mastery and supervision. Heather Street-Salter (2010) analyzes the construction and promulgation of this idea, and its legacy in the postcolonial armies of India, Pakistan, and Afghanistan. In the upper ranks, this period saw a parallel ideal forming to explain defeats in far-flung colonial outposts, particularly the heroic self-sacrifice and martyrdom of General Charles Gordon in the Sudan, Sir John Franklin in the Arctic, or Dr. David Livingston's failure to find the correct source of the Nile. This rationalization and its absorption into the British character can be seen in books by Stephanie Barczewski (2016) and Melvin Smith (2008) on citations for Britain's highest military decoration as a benchmark for public expectations of heroism.

These "martial races," along with working-class men from Britain's industrial cities, fought the imperial conflicts that kept the Union Jack flying in areas so far flung that the sun never "set" on the British empire. War and society scholars have advanced studies of these wars by seeking out and including the records and perspective of the indigenous side, beginning with Donald Morris' (1986) consideration of Zulu society and military institutions with the same gravity as those of the British. Others, like William Dalrymple's view of the 1857 Mutiny (2006) and Song-Chuan Chen's radical reassessment of the Opium War (2017), draw extensively from non–British sources. Thomas Pakenham (1979) wrangled with the conduct of the Boer War, offering a clear-eyed view of British strategic use of concentration camps, and the imperial tensions that produced draft riots, the Breaker Morant case, and fears that poor health and city living would yield terrible conscripts, as well as the founding of the Boy Scouts to try to avoid this fate.

War and society scholars have flourished in the shade of traditional World War I operational and unit histories, offering fresh insights to the experience of the Lost Generation. David Silbey (2005) attempted to explain why men who had believed "working men have no country" flocked to volunteer in the military. Nina Edwards (2015) rendered a sensory history of clothing used in World War I, from wet wool uniforms full of trench lice to the glee of women wearing short skirts and no corsets. The terrible price paid in lives and health emerges from Ben Shephard's study of mental hospital records (2001) to chronicle the tortured history of military psychiatry and rehabilitation; and Virginia Nicholson (2007) scrutinizes the far-reaching demographic effects of the millions of men lost and the lives of the women who would never marry them. Adam Hoschild (2011) used the Pankhurst family to illustrate the tensions between pacifist and military branches of one family amid the war's overwhelming losses and tragedy.

Paul Fussell took on the seismic changes WWI wrought on British society in *The Great War and Modern Memory* (1975), tracking its aftershocks through materials as varied as poetry, social class distinctions, marital advice books, and local memorials. A year later, John Keegan published his *The Face of Battle* (1976). One section examines the horrific experiences of soldiers during the Battle of the Somme, and two earlier sections focus on the Battles of Agincourt and of Waterloo. Keegan helps put the reader on the ground during combat. Between them, these two works formed a vanguard of war and society scholarship by critical examination of World War I's mythology, portrayal of the effect of the war's weapons and tactics on the mental and physical experience of soldiers, and willingness to depart from traditional military history's focus on generalship.

Britain's role at the Treaty of Versailles and her self-serving dual assumption of mandate supervision and discouragement of colonial independence in troop-supplying India are the focus of David Fromkin's analysis (2009), zeroing in on the ways in which the partitioning of the former Ottoman Empire by Britons like Gertrude Bell festered into modern conflicts. Tom Buchanan (1997) takes on one facet of Britain's response to the rising tide of fascism in Europe—turning a blind eye to volunteers (including Churchill's nephew Esmond Romilly) going off armed on Pyrenees holidays, while Keith Robbins' *Appeasement* (1997) looks at another rather more hands-on.

In 1937, the Mass Observation Project began collecting anthropological information from everyday people in Britain. Continuing through the war years (when the government used some of this data to craft propaganda and measure morale), it offers war and society scholars a view into the lived experience of World War II. Among the many works from this perspective are Mark Roodhouse's study about the evasions of rationing (2013), and Julie Summers' book (2013) about the virtues of the Women's Institute canning and victory gardening their way to survival. Children's voices come through in the works by Stephen Wade (2011) and John Welshman (2010) about the thousands of urban evacuees, whose time as foster children was hardly as magical as in *The Lion, the Witch and the Wardrobe*, by C.S. Lewis. Richard Toye (2013) used diaries and the Mass Observation Project to measure the effect of Winston Churchill's speeches and legendary persona on the wartime British public, finding that the day-to-day reality for people under fire was not as they later remembered from the safety of the 1950s.

At the University of Exeter, an ongoing project to measure the effect of the Combined Bomber Offensive (CBO) via aerial photographs, insurance records, and GIS mapping allowed Richard Overy to produce *The Bombers and the Bombed* (2013), which wrestled with the impact of civilian targeting on both the population and decision makers. Martin Francis (2008) ties RAF pilot culture to the traditions of the British armed forces, including the pride in sangfroid and chivalric masculinity. But the pressure to meet public expectations created a Victorian duality requiring these men to be both agents of ruthless destruction and civilized lunch companions in the same afternoon, producing stress that exacted extraordinary physical and emotional costs.

Sinclair McKay (2012) reveals the only recently declassified activities of the mathematicians, code clerks, and early computer engineers, like Alan Turing, who masterminded the war's greatest deception—the breaking and use of encrypted Nazi communication. The lives of the men and women who spent five years in war and then attempted to shift into civilian life is the focus of Alan Allport (2009), who examines the oft-repeated ejection of children, many of whom had never met their fathers, from their mother's bed and into a cold cot. He also explores the difficulties in reasserting traditional gender roles once women had been so widely employed in uniform and in factory work.

Britain's hard look at the postwar world comes into focus in the research of Peter Speiser, in which a restive, conscript occupation force in the constant shadow of the Americans attempted to keep the peace and cultivate West Germany as a potential ally, one inebriated and horny twenty-year-old at a time. Unlike the aftermath of WWI, the late 1940s pressed Britain with the inevitable decline and dissolution of much of the empire. John Darwin offers a large-scale study, especially of the military's role in planning and executing withdrawal. In the mold of studies about the acquisition of these lands that use indigenous sources, Shouchou Yao's *Malayan Emergency* (2016) and Huw Bennett's *Fighting the Mau Mau* (2013) look at these conflicts extensively from both sides, using documents and interviews that deconstruct the mythology of low-casualty British counterinsurgency operations and instead reveal widespread coercion and mass detentions. Keith Kyle (2011) explains Britain's reaction to the Suez Crisis as an overestimation of generations of colonial British expertise on the Middle East and misreading of the special relationship with the United States, while Roger Hardy expands the analysis of the fading British and French empires to the rest of the region.

One of the greatest military challenges to Britain in the postwar period was its contentious relationship with both the sovereign Republic of Ireland and Northern Ireland as part of Great Britain. Long a back door for enemy exploitation, Ireland's own armed independence movements were a continuing threat in the nineteenth century. The mass emigration of Irish people during the potato famine proved to be a boon to American military recruitment in the Mexican-American and Civil Wars, where numerous Fenians gained military training and access to arms. The pressures of WWI, including German plans to exploit Irish rebellion, sparked an armed uprising in 1916, laid in parallel to the Somme as depicted by Richard Grayson and Fearghal McGarry (2016), part of a wave of centennial studies of this seminal event. Britain sent many demobbed WWI veterans to Ireland in 1919 as the hated and feared militarized police "Black and Tans" to subdue the Irish Republican Army, examined by D. M. Leeson (2011). Even after the independence of the Republic, and the WWII employment of thousands of Irish nationals in wartime industry, Ireland's desire for full unification continued to be an opportunity for enemies, as Carolle Carter revealed (1977) in a study of Nazi contact with neutral Ireland. The 1960s and 1970s, in contexts with the Cold War, global civil rights, and decolonization movements, saw a resurgent IRA, which Tim Pat Coogan's magisterial history puts in its military, economic, and political context (2002). This experience of the British military in Irish counterinsurgency, A.R.B. Linderman (2016) argues, has been a deep well for expertise, doctrine, and lessons learned.

Britain's military resurgence came in 1982, with the Falklands War against Argentina. This ten-week conflict, which became deeply entwined with Prime Minister Margaret Thatcher's image as "The Iron Lady" and Britain's status as a partner to the United States (and Ronald Reagan), spawned a wave of nationalism that included praise for the participation of Prince Andrew and revived interest in the historical and contemporary power of the Royal Navy. Daniel Gibran's history (1998) subtitles it "Britain vs. the Past" and stresses the national and cultural interests of reviving pride in a seafaring and powerful nation having little to do with the possession of small islands in the south Atlantic. With a generation's distance, Sarah Maltby (2016) considers this patriotic resurgence, as well as the backlash by non-Thatcherite, punk, and anti-militarist groups who opposed the popular embrace of war and Thatcher's Cold War relationship with Ronald Reagan's America.

Subsequent conflicts have involved Britain as a member of NATO (North Atlantic Treaty Organization) or a coalition partner. In 1999, Prime Minister Tony Blair pushed hard for ground intervention in Kosovo, and although an air campaign ultimately became NATO's primary tool for coercing the Serbs, British troops entered the country as part of the peacekeeping force. Dag Henriksen's analysis of the campaign (2007) emphasizes the crisis of identity NATO suffered at the end of the Cold War, and the extent to which Kosovo was a test of its repurposing as a European security force with Britain playing a significant role. Blair was quick to pledge British support to the United States after the terrorist attack on September 11, 2011, and the British military invaded Afghanistan and Iraq as part of the U.S.-led "Coalition of the Willing." The length of the war, the special relationship with the United States, waning public support, and the value to British national strategy of participating have come in for scathing criticism of the leadership by Christopher Elliot (2015), and the seeming failure of counterinsurgency doctrine learned in decolonization by Frank Ledwidge (2011). Charles Miller's tracking of public opinion (2010) also offers significant insights, tying these military actions to anti-NATO and Brexit feelings.

The future of war and society scholarship about Great Britain lies in continued interpretation of those four themes with a twenty-first-century care to seek out and give voice to British subjects whose lives have been difficult to reconstruct, or whose place in history has been overshadowed by elites. Digitization of primary documents and tools like geospatial information systems make it possible to piece together scattered references, like the Old Bailey Online, which houses searchable eighteenth-century court archives and crowdsources the work of joining these legal records to army enlistment rolls, parish birth and death registers, and emigration lists. Similarly, searchable archives of nineteenth-century newspapers reveal letters home from Britain's soldiers meant to inform their local communities about the conquest and sustainment of the empire. These, combined with conscientious efforts to seek out sources in the archives of independent parts of the former empire, like India, Sri Lanka, and South Africa, offer a more nuanced picture of conflict and community.

From a twenty-first-century vantage point, scholars must reckon with Britain's meteoric rise in the context of its subsequent decline as a world power, and its responsibility in the lands it formerly ruled. The consolidation or elimination of Highland Regiments raises questions about Britain's use of its military as an assimilator of people into "British" identity, especially in a time of Brexit ambivalence about the nation's ability to manage a multicultural population. Finally, the rich records of Britain's institutions—the tools of power like the Bank of England, the East India Company, the Norfolk Naval Shipyards, and the National Coal Board—are available for researchers to use as they examine the human cost of both the rise to power and a current place in a globalized contemporary world that Great Britain pioneered but no longer leads.

Bibliography

Allport, Alan. *Demobbed: Coming Home After the Second World War*. New Haven: Yale University Press, 2009.

Anderson, Fred. *Crucible of War: The Seven Years' War and the Fate of Empire in British North America, 1754–1766*. New York: Vintage Books, 2001.

Bannerman, Gordon. *Merchants and the Military in Eighteenth-Century Britain: British Army Contracts and Domestic Supply, 1739–1763*. London: Pickering and Chatto, 2008.

Barczewski, Stephanie L. *Heroic Failure and the British*. New Haven: Yale University Press, 2016.

Baugh, Daniel A. *The Global Seven Years War, 1754–1763: Britain and France in a Great Power Contest*. Harlow, England: Longman, 2011.

Beiner, Guy. *Remembering the Year of the French: Irish Folk History and Social Memory*. Madison: University of Wisconsin Press, 2009.

Bennett, Huw C. *Fighting the Mau Mau: The British Army and Counter-Insurgency in the Kenya Emergency*. Cambridge: Cambridge University Press, 2013.

Black, Jeremy. *A Military Revolution?: Military Change and European Society 1550–1800*. Basingstoke: Palgrave Macmillan, 1998.

"Bombing, States and Peoples in Western Europe, 1940–1945." Centre for the Study of War, State and Society—University of Exeter. http://humanities.exeter.ac.uk/history/research/centres/warstateandsociety/projects/bombing/.

Brewer, John. *The Sinews of Power: War, Money, and the English State, 1688–1783*. New York: Knopf, 1989.

Brumwell, Stephen. *Redcoats: The British Soldier and War in the Americas, 1755–1763*. Cambridge: Cambridge University Press, 2002.

Buchanan, Tom. *Britain and the Spanish Civil War*. Cambridge: Cambridge University Press, 1997.

Bunker, Nick. *An Empire on the Edge: How Britain Came to Fight America*. New York: Alfred A. Knopf, 2014.

Calloway, Colin G. *The Scratch of a Pen: 1763 and the Transformation of North America*. Oxford: Oxford University Press, 2006.

Carlton, Charles. *Going to the Wars: The Experience of the British Civil Wars, 1638–1651*. London: Routledge, 1992.

Carter, Carolle J. *The Shamrock and the Swastika: German Espionage in Ireland in World War II*. Palo Alto, CA: Pacific Books, 1977.

Chamberlain, Paul. *Hell Upon Water: Napoleonic Prisoners of War in Britain, 1793–1815*. Stroud, UK: Spellmount, 2008.

Chen, Song-Chuan. *Merchants of War and Peace: British Knowledge of China in the Making of the Opium War*. Hong Kong: Hong Kong University Press, 2017.

Clary, David A. *George Washington's First War: His Early Military Adventures*. New York: Simon and Schuster, 2011.

Colley, Linda. *Britons: Forging the Nation, 1707–1837*. New Haven: Yale University Press, 1992.

Coogan, Tim Pat. *The IRA*. New York: Palgrave for St. Martin's Press, 2002.

Crytzer, Brady. *Hessians: Mercenaries, Rebels, and the War for British America*. Yardley, PA: Westholme, 2015.

Dalrymple, William. *The Last Mughal: The Fall of a Dynasty, Delhi, 1857*. London: Bloomsbury, 2006.

Danley, Mark, and Patrick Speelman, eds. *The Seven Years' War: Global Views*. Leiden: Brill, 2012.

Darwin, John. *Britain and Decolonization: The Retreat From Empire in the Post-War World*. New York: St. Martin's Press, 1988.

Davey, James. *In Nelson's Wake: The Navy and the Napoleonic Wars*. New Haven: Yale University Press, 2015.

Drayton, Richard Harry. *Nature's Government: Science, Imperial Britain, and the "Improvement" of the World*. New Haven: Yale University Press, 2000.

Edwards, Nina. *Dressed for War: Uniforms, Civilian Clothing and Trappings, 1914–1918*. London: I.B. Tauris, 2015.

Elliott, Christopher L. *High Command: British Military Leadership in the Iraq and Afghanistan Wars*. Oxford: Oxford University Press, 2015.

Esdaile, Charles J. *The Peninsular War: A New History*. New York: Palgrave Macmillan, 2003.

Fissel, Mark Charles. *The Bishops' Wars: Charles I's Campaigns Against Scotland, 1638–1640*. Cambridge: Cambridge University Press, 1999.

———, ed. *War and Government in Britain: 1598–1650*. Manchester, UK: Manchester University Press, 1991.

Foreman, Amanda. *A World on Fire: Britain's Crucial Role in the American Civil War*. New York: Random House, 2010.

Forrest, Alan. *Waterloo: Great Battles Series*. Oxford: Oxford University Press, 2015.

Francis, Martin. *The Flyer: British Culture and the Royal Air Force, 1939–1945*. Oxford: Oxford University Press, 2008.

Fromkin, David. *A Peace to End All Peace: The Fall of the Ottoman Empire and the Creation of the Modern Middle East*. New York: Holt, 2009.

Fussell, Paul. *The Great War and Modern Memory*. New York: Oxford University Press, 1975.

Gaunt, Peter. *English Civil War: A Military History*. London: I.B. Tauris, 2014.

Gibran, Daniel K. *The Falklands War: Britain Versus the Past in the South Atlantic*. Jefferson, NC: McFarland, 1998.

Grayson, Richard S., and Fearghal McGarry. *Remembering 1916: The Easter Rising, the Somme and the Politics of Memory in Ireland*. Cambridge: Cambridge University Press, 2016.

Haefeli, Evan, and Kevin Sweeney. *Captors and Captives: The 1704 French and Indian Raid on Deerfield*. Amherst: University of Massachusetts Press, 2003.

Hardy, Roger. *The Poisoned Well: Empire and Its Legacy in the Middle East*. New York: Oxford University Press, 2017.

Henriksen, Dag. *NATO's Gamble: Combining Diplomacy and Airpower in the Kosovo Crisis, 1998–1999*. Annapolis: Naval Institute Press, 2007.

Hochschild, Adam. *To End All Wars: A Story of Loyalty and Rebellion, 1914–1918*. Boston, MA: Houghton Mifflin Harcourt, 2011.

Holmes, Richard. *Marlborough: England's Fragile Genius*. London: Harper Press, 2008.

Jones, J. R. *The Anglo-Dutch Wars of the Seventeenth Century*. London: Longman, 1996.

Keegan, John. *The Face of Battle: A Study of Agincourt, Waterloo, and the Somme*. New York: Viking Press, 1976.

Kishlansky, Mark. *The Rise of the New Model Army*. Cambridge: Cambridge University Press, 1983.

Knight, Roger J. B. *Britain Against Napoleon: The Organization of Victory, 1793–1815*. London: Allen Lane, 2013.

Kyle, Keith. *Suez: Britain's End of Empire in the Middle East*. London: I.B. Tauris, 2011.

Lambert, Andrew. *The Crimean War: British Grand Strategy, 1853–1856*. Manchester: Manchester University Press, 1990.

———. *Nelson: Britannia's God of War*. London: Faber and Faber, 2005.

Langford, Paul. *A Polite and Commercial People: England, 1727–1783*. Oxford: Oxford University Press, 2010.

Ledwidge, Frank. *Losing Small Wars: British Military Failure in Iraq and Afghanistan*. New Haven: Yale University Press, 2011.

Leeson, David. *The Black and Tans: British Police and Auxiliaries in the Irish War of Independence, 1920–1921*. Oxford: Oxford University Press, 2011.

Lenman, Bruce. *The Jacobite Risings in Britain, 1689–1746*. London: Eyre Methuen, 1980.

Linch, Kevin, and Matthew McCormack, eds. *Britain's Soldiers: Rethinking War and Society, 1715–1815*. Liverpool: Liverpool University Press, 2014.

Linderman, A.R.B. *Rediscovering Irregular Warfare: Colin Gubbins and the Origins of Britain's Special Operations Executive*. Norman: University of Oklahoma Press, 2016.

Lodge, Richard. *Great Britain and Prussia in the Eighteenth Century*. New York: Octagon Books, 1972.

Mackay, David. *In the Wake of Cook: Exploration, Science, and Empire, 1780–1801*. New York: St. Martin's Press, 1985.

Magra, Christopher. *Poseidon's Curse: British Naval Impressment and Atlantic Origins of the American Revolution*. Cambridge: Cambridge University Press, 2016.

Malcolm, Joyce Lee. *To Keep and Bear Arms: The Origins of an Anglo-American Right*. Cambridge: Harvard University Press, 1994.

Maltby, Sarah. *Remembering the Falklands War: Media, Memory and Identity*. London: Palgrave Macmillan, 2016.

Markovits, Stefanie. *The Crimean War in the British Imagination*. Cambridge: Cambridge University Press, 2009.

McKay, Sinclair. *The Secret Lives of Codebreakers: The Men and Women Who Cracked the Enigma Code at Bletchley Park*. New York: Penguin Group, 2012.

Middleton, Richard. *The Bells of Victory: The Pitt-Newcastle Ministry and the Conduct of the Seven Years' War, 1757–1762*. Cambridge: Cambridge University Press, 1985.

Miller, Charles A. *Endgame for the West in Afghanistan?* Carlisle, PA: Strategic Studies Institute, U.S. Army War College, 2010.

Morris, Donald R. *The Washing of the Spears: A History of the Rise of the Zulu Nation Under Shaka and Its Fall in the Zulu War of 1879*. New York: Simon and Schuster, 1986.

Muir, Rory. *Wellington: The Path to Victory 1769–1814*. New Haven: Yale University Press, 2013a.

———. *Wellington: Waterloo and the Fortunes of Peace 1814–1852*. New Haven: Yale University Press, 2013b.

Nicholson, Virginia. *Singled Out: How Two Million Women Survived Without Men After the First World War*. London: Viking, 2007.

Noel-Smith, Heather, and Lorna M. Campbell. *Hornblower's Historical Shipmates: The Young Gentlemen of Pellew's Indefatigable*. Woodbridge, UK: The Boydell Press, 2016.

O'Shaughnessy, Andrew Jackson. *An Empire Divided: The American Revolution and the British Caribbean*. Philadelphia: University of Pennsylvania Press, 2000.

———. *The Men Who Lost America: British Leadership, the American Revolution, and the Fate of the Empire*. New Haven: Yale University Press, 2013.

Overy, R. J. *The Bombers and the Bombed: Allied Air War Over Europe 1940–1945*. New York: Viking, 2013.

Pakenham, Thomas. *The Boer War*. New York: Random House, 1979.

Perl-Rosenthal, Nathan. *Citizen Sailors: Becoming American in the Age of Revolution*. Cambridge, MA: Belknap Press of Harvard University Press, 2015.

Preston, David L. *Braddock's Defeat: The Battle of the Monongahela and the Road to Revolution*. Oxford: Oxford University Press, 2015.

"The Proceedings of the Old Bailey." *Old Bailey Online—The Proceedings of the Old Bailey, 1674–1913—Central Criminal Court*. www.oldbaileyonline.org/.

Reid, Stuart. *1745: A Military History of the Last Jacobite Rising*. New York: Serpedon, 1996.

Robbins, Keith. *Appeasement*. Oxford: Blackwell, 1997.

Rodger, N.A.M. *The Wooden World: An Anatomy of the Georgian Navy*. London: Collins, 1986.

Roodhouse, Mark. *Black Market Britain: 1939–1955*. Oxford: Oxford University Press, 2013.

Royle, Trevor. *The British Civil War: The Wars of the Three Kingdoms, 1638–1660*. New York: Palgrave Macmillan, 2004.

Sale, Kirkpatrick. *Rebels Against the Future: The Luddites and Their War on the Industrial Revolution*. Reading, MA: Addison-Wesley, 1995.

Sankey, Margaret. *Jacobite Prisoners of the 1715 Rebellion: Preventing and Punishing Insurrection in Early Hanoverian Britain*. Aldershot: Ashgate, 2005.

Sankey, Margaret, and Daniel Szechi. "Elite Culture and the Decline of Scottish Jacobitism 1716–1745," *Past and Present* 173, no. 1 (2001): 90–128.

Schweitzer, Karl. *England, Prussia, and the Seven Years War: Studies in Alliance Policies and Diplomacy*. Lewiston, NY: Mellen, 1998.

Schwoerer, Lois G. *Gun Culture in Early Modern England*. Charlottesville: University of Virginia Press, 2016.

Shephard, Ben. *A War of Nerves: Soldiers and Psychiatrists in the Twentieth Century*. Cambridge: Harvard University Press, 2001.

Silbey, David. *The British Working Class and Enthusiasm for War, 1914–1916*. London: Frank Cass, 2005.

Smith, Melvin Charles. *Awarded for Valour: A History of the Victoria Cross and the Evolution of British Heroism.* Houndmills, UK: Palgrave Macmillan, 2008.

Sobel, Dava. *Longitude: The True Story of a Lone Genius Who Solved the Greatest Scientific Problem of His Time.* New York: Walker, 1995.

Speiser, Peter. *The British Army of the Rhine: Turning Nazi Enemies Into Cold War Partners.* Urbana: University of Illinois Press, 2016.

Stewart, Laura A. M. *Urban Politics and the British Civil Wars.* Leiden: Brill, 2006.

Streets-Salter, Heather. *Martial Races: The Military, Race and Masculinity in British Imperial Culture, 1857–1914.* Manchester: Manchester University Press, 2010.

Summers, Julie. *Jambusters: The Story of the Women's Institute in the Second World War.* London: Simon and Schuster, 2013.

Szechi, Daniel. *1715: The Great Jacobite Rebellion.* New Haven: Yale University Press, 2006.

Taylor, Alan. *The Internal Enemy: Slavery and War in Virginia, 1772–1832.* New York: W.W. Norton, 2013.

Toye, Richard. *The Roar of the Lion: The Untold Story of Churchill's World War II Speeches.* Oxford: Oxford University Press, 2013.

Uglow, Jennifer S. *In These Times: Living in Britain Through Napoleon's Wars, 1793–1815.* New York: Farrar, Straus and Giroux, 2014.

Wade, Stephen. *Air-Raid Shelters of World War II: Family Stories of Survival in the Blitz.* Barnsley, UK: Remember When, 2011.

Webb, Stephen Saunders. *1676, The End of American Independence.* New York: Knopf, 1984.

———. *The Governors-General: The English Army and the Definition of the Empire, 1569–1681.* Chapel Hill: University of North Carolina Press, 1979.

———. *Lord Churchill's Coup: The Anglo-American Empire and the Glorious Revolution Reconsidered.* New York: Knopf, 1995.

———. *Marlborough's America.* New Haven: Yale University Press, 2013.

Weintraub, Stanley. *Iron Tears: America's Battle for Freedom, Britain's Quagmire, 1775–1783.* New York: Free Press, 2005.

Welshman, John. *Churchill's Children: The Evacuee Experience in Wartime Britain.* Oxford: Oxford University Press, 2010.

Wheeler, James Scott. *The Irish and British Wars, 1637–1654: Triumph, Tragedy, and Failure.* London: Routledge, 2002.

Willis, Sam. *The Struggle for Sea Power: A Naval History of the American Revolution.* New York: W.W. Norton, 2016.

Woodfine, Philip. *Britannia's Glories: The Walpole Ministry and the 1739 War With Spain.* Woodbridge, UK: Royal Historical Society/Boydell Press, 1998.

Yao, Souchou. *The Malayan Emergency: A Small, Distant War.* Copenhagen: NIAS Press, 2016.

11 War and Society in Latin America

Ellen D. Tillman

One of the greatest difficulties facing any scholar of Latin American war and society—or perhaps broader Latin American history—is how to understand the centuries-long drift of tendencies toward military dictatorships or militarized state control, which permeates Latin America in many periods and across regions. Indeed, military control has been at the core of Latin American state building, internal relations, and often identities, and this despite a relative lack of involvement in major foreign wars. While struggling with Latin America's repetitive economic and political crises, scholars have sought the origin(s) of militarization in any given society, or across Latin America more broadly. Have the many iterations come from the same historical source, a continuation of colonial institutions and ideologies? Or have they been more a product of coincidences of geography and various external influences and impositions, such as foreign loans and interventions, over time?

Militarization, at the heart of state building, diplomacy, and regional and national security, greatly complicates our understanding of Latin American war and society. Throughout Latin American history, and particularly the twentieth century, English-language scholarship has tended to focus on civil-military relations. The topics of militarism and US-Latin American relations, and the growth of US interest in the study of Latin America after World War II, came together to virtually guarantee an emphasis on the role of military institutions and dictators in increasingly restrictive civil societies. From historical and anthropological studies to news items to Hollywood films, the image of Central and South American and Caribbean dictatorships evolved, through the Cold War, to a depiction that portrays ever-larger and more prominent militaries, corrupt and more interested in centralized, internal control than concerned about external wars. The portrayal of Latin American societies as inherently militaristic—and often as thus inferior to the United States—became commonplace.

Academic research in the past three decades has provided revelations about US and especially CIA interventionism, the complexities of Latin American race relations, and class and gender development (to name some of the more prominent lines of inquiry), but students' assumptions about Latin American government, war, and militaries continue to be dominated by ideas of militarization and dictatorship. While oversimplified and problematic, they are not entirely unfounded; these tendencies are real and long-standing in Latin American societies. The complex origins of militarism and the lack of primary records or earlier works on certain groups in society have contributed to a scholarly emphasis on individuals such as charismatic military and political leaders—often one and the same. The national and foreign tendency was long to focus inquiries on the roles and biographies of the "great heroes" of independence or the caudillos and dictators of other periods and militarized elements of political and societal structures. The distinct trajectory and culture of militarism in society and government through Latin American history, from colonial oversight through twentieth-century Cold War dictatorships, has understandably perpetuated a focus on the role of the military in government and politics, which often eclipses other considerations of war in

Latin America. In fact, it has traditionally been difficult *not* to overemphasize military and war when referencing the political, due to the amalgamation of the two in so many aspects of Latin American national and regional histories.

The growing interest in social histories of the military has brought a prolific and valuable shift in considerations of Latin American history—from the reexamination of militaristic traditions to explanations for the continued poor standard of living and often-militarized conditions of life and politics throughout Latin America. For many, as with earlier studies emphasizing the role of the military in political policy, this has meant increased attention to civil-military relations—including the relation of the military to daily civilian lives—from the colonial period through the twentieth century. The expansion of historical approaches and research subjects has helped to demonstrate that the causes of militarism so deeply affecting Latin American society and diplomacy stem from a varied and complex interaction of consistent and highly intrusive external influences and the persistence and particularities of other internal structures stretching back into the colonial and even precolonial periods.

In addition to the problem of overcoming, or dealing with the origins of, militarism and the preconceptions of militarism, the study of war and society in Latin America is greatly complicated by the extent of geography, chronology, and perspectives to be covered—not to mention the difficulties in defining Latin America. In its broadest definitions, "Latin America" encompasses South and Central America, the Caribbean, and Mexico, over forty countries and dependencies, with diverse histories, cultures, government structures, and geographical characteristics. Languages also vary, from Spanish and Portuguese to French, Haitian Creole, and English, to Quechua, Guaraní, Aymara, Nahuatl, and many other indigenous languages and dialects. The diversity of identities and ethnicities in each country or region has been accentuated in many places by a traditional lack of integration of many groups into wider national and state projects. Chronologically, the field is also daunting. From precolonial cultures through centuries of colonial experience and what in many locations now encompasses two centuries of national histories, historians have sought to identify patterns and trends that define "Latin America." This chapter will, of necessity, focus mostly on Spanish America and Brazil.

Because of the distinctive nature of Latin American societies and militaries, many consistencies and influences from colonial times have had an especially powerful sway in decisions made in the modern period. A relative paucity of early written sources precludes extensive documentary research on precolonial war and society, but some fruitful examinations into precolonial societies, especially in the past three decades, have helped us to understand those societies' development and how and why earlier influences so drastically affect it. Newer works allow for some possible comparisons between European military systems and those of early Mesoamerica and South America. While precolonial societies often lacked militaries that could dominate large areas by force, some empires managed to extend their power far through tradition, kinship, compromise, and co-optation of populations—as by use of financial or status-based rewards. The Aztecs, for example, seem to have used war to develop much of their empire and institutions, then maintained influence through other means, such as control of commodities (Hassig, 1992). Such control systems established patterns and traditions that were to persist into the colonial period and be readily recognizable to European nobles. Those charged with controlling societies and vast distances in the colonial era depended on adaptation and co-optation of tradition and kinship—essential compromises that made the Spanish colonial system work in so many places for so long.

The centrality of compromise and negotiation, and the lack of consistent external threats, as from competing European empires, made possible the coincidence of a Spanish imperial system alongside autonomous and semiautonomous indigenous cultures and communities. While representatives of the latter negotiated Spanish imperial institutions in order to represent—or, at times, take advantage of—their constituents, the lack of mainstream integration meant a distant and less intrusive Spanish system that allowed for continued indigenous community autonomy and tradition. In many parts of Spanish and Portuguese America, this tenuous but effective system was at the center of imperial core-periphery communications and interactions for nearly three centuries. In effect, this system of compromise and indirect control meant that precolonial models persisted and became ever more institutionalized, in the long term, because the cultures of native societies continued almost independently in so many colonial regions (Thomson, 2002).

Many Latin American structures were formed in the context of colonial governments unconcerned with major invasions. Further, both conquest and internal control were facilitated by such wide-ranging factors as disease—which removed immense portions of population in short periods—and access to global trade and advanced weaponry. The relative ease of running such colonies changed with the European imperial wars of the mid-1700s, which forced change in Spanish and Portuguese expectations and understandings—as epitomized, for example, in the British 1762 occupation of Havana and Spanish responses. This late colonial period has broadly come to be associated with, and named after, the Spanish Bourbon Reforms (coinciding roughly with the Pombaline Reforms in Portuguese America). As a highly documented period of major transition that led into some of the region's greatest wars—the wars of independence—the decades-long period of reforms has come to dominate much of the period's historiography. To the mid-eighteenth century, the Spanish Empire's military and security concerns had centered largely around state and local control of populations and labor, often through negotiation, which has provided a fruitful, fascinating, and growing body of work (Jacobsen and Aljovín de Losada, 2005; Ganson, 2003) about the legal origins of many state structures in Latin America. During roughly the final half century of the Spanish colonial period, the Bourbon and other related reforms attempted to consolidate greater control in order to shore up the empire against external as well as internal security threats due to other colonial powers' incursions into the sphere of the Spanish Empire. These reforms created new, stronger military forces, including the creation of militia systems, as well as other new state centralized institutions, from which would stem a variety of uprisings and internal wars (Kuethe and Marchena, 2005).

Late colonial Latin American society, while often deeply resisting central reform, was also in some ways transformed during these decades. Through the military especially, the reforms disrupted the balance of power and the often tenuous balance of cultures and semiautonomous communities throughout the Latin American colonial system. In the end, they pleased few and helped—alongside external factors of change, such as Enlightenment thought and US and Haitian independence—to upset the colonial balance. The Bourbon Reforms attempted, and eventually failed, to intensify Spanish imperial control. In the end, taken together, the reforms acted as one of the most influential turning points in Latin American identity, state structure, and war and society. While exposing the weaknesses and imbalances in the imperial institutions, they sped independence even while exacerbating vital rifts in society. Inspired by both external and internal military and security concerns, many of the late eighteenth-century reforms were gradual, building up over time as solutions (or suggestions of solutions) leading toward more cohesive state control. Yet, they also increased awareness and societal division, led to experimentation with new military structures, and pushed many colonists in the direction of independence—or at least a rejection of tightening imperial control.

Recent reevaluations of the Bourbon Reforms and late colonial Spanish military especially address civil-military relations and seek the origins of militaristic systems. The late colonial past shaped early national militaries through the connections of the military to state and local economies, and the formation of citizen militias—though all, of course, was complicated by regional variation (Kuethe and Marchena, 2005). These works correlate with other studies that demonstrate a clear connection of institutions between the colonial past and the early national period—carried on through institutional continuity and possibly even inspiring an enduring pessimism about Latin American nations' futures (Adelman, 1999). The Enlightenment influence, too, was preserved in powerful ways through its ideological prominence in the independence wars; it encouraged both some level of elitism (fears of lower classes and disorder) and the search for formulaic answers to society's problems. Simón Bolívar himself—one of the "fathers of Latin American independence"—was an Enlightenment thinker who emphasized more representative systems, but also explicitly drew out the differences he perceived between Spanish colonial societies and the United States. He and others feared race war with the ending of slavery and the arming of underprivileged populations, and believed their societies necessitated a greater state control than did other Enlightenment governments (Quintero Montiel, 2002; Thibaud, 2003).

Importantly, the reforms of the late colonial period demographically pulled in populations that had been consistently or increasingly excluded. The shift to inclusion was to be sped up by the wars of independence. The sometimes desperate need for local defense and control meant the creation of militias that often included people seen as having suspect loyalty, such as the creoles—Spanish people born in the Americas rather than the *peninsulares* born in the Iberian Peninsula. Controversies over the wisdom of recruiting, arming, and training these potentially disloyal populations into such units increased division even while providing arms, perspective, and military training to many potential revolutionaries. Recent works have begun to dig more deeply into the motivations for, and implications of, such particular military organizations, and demonstrate how thoroughly they were able to upset the hierarchical and exclusivist order of colonial society even as independence loomed (Marchena Fernández, 1992). Yet many factors forced the hand of the Spanish authorities who were *peninsulares* and administrators. In addition to the very serious question of manpower, *peninsulares* were more likely—for example—to suffer from local disease than were locals.

The controversy over how best to extend military control without overly empowering people of potentially suspect loyalty led to various creative adaptations and institutions, many of which had long-term effects on society and security organization that carried over into the formation of national institutions and identities. Most importantly, perhaps, were such creations as the colonial militia and the *fuero militar*, which allowed the creoles and even *mestizos* (those of mixed indigenous and European heritage) to serve militarily outside of the major central forces; these institutions also allowed such groups to gain and retain arms and military training. To those who served in the colonial military, and particularly during the last decades before independence, came certain important benefits. Such organizations as the *fuero militar*, which often exempted those who served from other lay and clerical laws and taxes, extended into the national period for military service. The experience allowed by these late colonial adaptations also aided many in later years in the independence wars against Spain. Colonial militaries benefited certain classes and created regional class imbalances that upset tenuous traditional power relations (McAlister, 1957; Archer, 1977). The regional variants, economic factors, and peninsular-creole divisions and negotiations have colored the study of the context and causes of the independence wars as they were influenced by the late colonial period. These studies deepen and complicate the picture of Latin American military development leading into the independence period, especially along the frontiers of newly founded countries (Lynch, 1973).

The Napoleonic Wars in Europe sparked the final outbreak of the wars of independence, whose currents were building through the final decades of the colonial period. As the rearrangements and expansion of the Bourbon Reforms worsened some conditions in the Americas, the Spanish were unable to cope with Napoleonic challenges to their power throughout the empire. Rebellion and warfare varied across Latin America, including civil war, insurgency, and counterinsurgency. But whatever its regional form, conflict affected people of all classes and drastically changed society (Kuethe and Marchena, 2005). How "revolutionary" it was has been up for serious debate; as with studies of the United States, recent work has shown increasingly that most of the change was, at least by intent, much less revolutionary and more gradual than long imagined. An overview of the independence wars, spanning roughly 1810–1825, shows a wide geographical overlap of themes—such as Enlightenment thought—throughout the region of the former Spanish Empire in the Americas (Lynch, 1973; Langley, 1997). Building on earlier histories of independence, many of which were memoirs of participants and often politically biased, historians have since the 1980s increasingly shown a greater ideological role than formerly believed for Spanish liberalism and creole conservative fears, as well as extensive participation by women, the poor, the enslaved and formerly enslaved, and ethnic minorities.

The independence wars and subsequent societal and political divisions incited a century of debates about religion, liberalism vs. conservatism, civil wars, revolutions, repressions, and struggles over the meaning of independence and why the wars had been fought. Contested ideas of nation, race, identity, and each country's place in the world, as well as the meaning of the very independence movements, continue to be deeply colored and sometimes shadowed by the independence movements. Scholarship in recent decades has interrogated the development of those identities and disputed meanings, and begun to argue for the place of Latin American independence wars in a more global, or at least Atlantic, context. New work has also begun to interpret the daily experiences of the wars, including those of long-underrepresented groups, such as indigenous communities, blacks, women, and the lower classes. Not only were the wars extremely influential in the daily lives of all of these groups, but also these groups shaped the wars in ways that have been ignored—fighting actively and integrally for both sides, and forging their own powerful national identities in the process. While many such groups were again politically and often economically marginalized in the early national period, the heritage and legacies of their military service continued to shape their identities and their understanding of their communities' place within the now independent nation-state (Santoni, 2008; Foote and Harder Horst, 2010).

Late colonial reforms and other attempts to change or consolidate central state control over the colonial system and its people—particularly the most autonomous (and excluded) of them—led to quite a few formative uprisings, including the independence movements and attempts to consolidate new states. The reforms and changes also led to major indigenous revolts before, during, and after independence, most especially powerful because of the long isolation and autonomy of indigenous communities. Centralized control from the same state that had long compromised was unacceptable to peoples who had enjoyed long-term autonomy that had left so much of their indigenous culture intact.

Much can be learned about Latin American state and identity formation from the growing number of investigations into why indigenous populations rebelled during the colonial period and after, the form taken by those uprisings, and their consequences. When central state incursions into indigenous communities increased, major insurrections sought either to reestablish the status quo or—in some more drastic instances—to eliminate the Spanish and their institutions in the region. Studies of these uprisings allow the retrieval of multiple, difficult-to-access perspectives on colonial military history, but also demonstrate that these were formative events in modern Latin American development. Native insurrections against central state rule clearly show a strong tendency to combine indigenous and Spanish traditions, most particularly those related to aspects of Catholicism as the native populations in any given region had adopted it, even while seeking a "return" to pre-Spanish culture. An evolving adaptation of culture shaped the insurrections and the often-widespread battles they inspired, though the resulting suppressions further damaged indigenous autonomy in many cases—enabling the state to consolidate control further (Gosner, 1992; Robins, 2005; Kuethe and Marchena, 2005).

The extent and types of devastation, and the mobility of revolting groups, meant that violence thoroughly affected large swaths of society from the seventeenth century into the early nineteenth. First, both uprisings and colonial/state responses affected everyday lives through sieges, massacres, rapes, cultural humiliation, disease, and the economic effects of ebb-and-flow struggles for power. The colonial authorities' responses (which often reflected fear) ranged from local programs and change, in which colonial powers worked to curtail native autonomy, to broader centralization reforms. While efforts to reinstate authority or order often included the predictable increase of centralized military and state control, utilizing both force and political rearranging, many were more innocuous—as through co-optation by godparent relationships and the use of symbolism to equate state power with indigenous communities' cultures and needs (Van Young, 2012). Indigenous movements and the uprisings they sometimes supported therefore resulted in increased conformity to state directives and also provided the excuse for a more military and widespread system of state control.

Wars in the Latin American national period, roughly the nineteenth and twentieth centuries, have varied broadly and been characterized by the features developed earlier, as well as increasingly by global power politics. Based on societal norms and practices developed largely in the colonial and independence periods, wars have ranged from major external wars and civil wars to frequent border conflicts, involvement in global wars, and invasions and military occupations. Insurgencies and counterinsurgencies, growing larger in the twentieth century, sometimes led to extensive guerrilla warfare. Certain patterns predominated in all or most of these sorts of conflicts, many of them clear products of the particular Latin American colonial past: a tendency toward state repression and control of media; a proliferation of foreign control, alliances, and economic involvement; race and class divisions; and the maintenance (at best) of a precarious social balance.

Brazil's transition to independence from Portugal differed significantly from the various paths of Spanish American countries, being gradual and relatively bloodless (1822). In contrast to many other locations, however, it was thereafter heavily characterized by its dependence on slavery and its massive, largely indefensible borders. In many ways, these issues defined the course of early Brazil as a nation. As in the United States, nineteenth-century Brazilian warfare challenged the traditional slave system, which had ended or largely ended with independence throughout much of the rest of Latin America. In the Brazilian case, interference in the Uruguayan Civil War (1830s–1851) and involvement in the War of the Triple Alliance (1864–1870) showed the complications of recruitment and the ongoing slave system. As in the US Civil War (1861–1865), the state had to adjust policy to allow for recruitment and reorganization sufficient to conduct the war—changes that continued to influence society, and especially race and class relations, long after the fighting had officially ended (Beattie, 2001; Izecksohn, 2014). Thereafter, much of Brazilian war and military experience revolved around trying to address recruitment needs amid tense race relations, as well as trying to balance the traditional Brazilian military ethos with civil-military race relations. While much nineteenth- and twentieth-century recruitment occurred through force—the first Brazilian draft was conducted during World War I—the state also used patronage, notions of honor and gender, and ideals of social reform to entice enlistments (Beattie, 2001; McCann, 2004; Love, 2012).

Throughout the former colonies of Spanish America, the greatest issues after independence revolved around colonial borders, replacing or preserving colonial institutions, and the tendency toward militarization—a product of both heritage and perceived necessity. For Latin America at large, the nineteenth century followed a general trajectory of independence wars, to conflicts about the borders of new states, to the forms those states would take. The latter were most frequently defined by the liberal-conservative debate, leadership through the charismatic military rule of *caudillos*, and a gradual professionalization of military forces that—alongside loans contracted to fight independence or to build new states—often opened societies to the heavy influence of foreign powers. While studies long emphasized the US influence on Latin American military professionalization and warfare, the European influence was stronger in some places and just as strong in others (Lieuwen, 1967; Nunn, 1983). European interference with the heritage of the colonial system combined with independence wars usually kept warfare in the region somewhat limited, and internal repression was at least as formative for the military as were external threats (Centeno, 2002). Those charismatic and paternalistic military leaders, the *caudillos*, for example, were a product of direct Spanish heritage, internal colonial structures, the independence wars, geography, and regional compromise that included the consistent threat of minor border wars (Lynch, 1992).

The confluence of European-style military professionalism and the old *caudillo*-style politics culminated in such major late-nineteenth-century wars as the War of the Triple Alliance (1864–1870) and the War of the Pacific (1879–1883). These conflicts were central to the consolidation of states, their power, and their permanent institutions; the War of the Pacific, for example, significantly shifted borders and centralized Peruvian and Chilean military power (Farcau, 2000; Sater, 2007). Also formative (and a major, recent historiographical focus of this influential period) are the ways that such conflicts and state centralization forced Latin American nations to seek coping mechanisms for the wide gaps between classes and racial or ethnic groups. New works show conclusively that these decisive years for national identity opened ever-growing junctures for the inclusion of communities and groups that had long remained excluded. Black and indigenous groups, for example, capitalized on states' and parties' need for backing, especially as global markets transformed societies and their needs. The military, as one of the most organized and essential elements of the state, was one of the most important avenues for challenging identity and definitions of the nation—especially when wars allowed the underrepresented to contest their status (Foote and Harder Horst, 2010). While early national wars, the influence of military leaders, and patronage often impeded respect for the political process, voices and bodies loudly—and often effectively—challenged such patterns during this period of internationalization (Lynch, 1992; Loveman, 1999).

The same global trends and integration forced change within long-standing precolonial, colonial, and early national institutions and traditions. Most overtly, major powers' imperial competition and naval expansion at the turn of the twentieth century directly influenced Latin American nations. From US industrialization seeking a ready source for raw materials and markets to hemispheric security questions shifting US military interests farther and farther south, nations throughout the Americas scrambled to adjust and find their places in a changing world order. Even when they sought neutrality, these nations were increasingly drawn into rapid industrial development as well as global military conflicts, such as the world wars. The more rapid the change and the more pervasive the external influences, of course, the more conflict engendered at home: communities were forced to integrate or fight back as more workers and soldiers were needed to support the centralizing nation-state, and international integration through ready loans further drew countries into global markets. These turn-of-the-century transformations—including large influxes of European immigrants welcomed in for industrial labor—exacerbated long-standing racial and class tensions (Love, 2012).

Global military and economic power shifts also spelled revolutionary change for areas that had remained more rooted in the past. This included many deep interior indigenous communities, but also the remainder of the Spanish American Empire in lands such as Cuba, which was to become one of the central players in the twentieth-century Americas. Though many Cubans traveled in the late nineteenth century to recruit international intellectuals and print media to the cause of independence, much of the island's population preferred the colonial status quo. Recent debate has enlivened discussion about the contemporary popularity of Cuba's independence movement, powerfully shifting our understanding of decades-long wars and the early formation of the Cuban nation. In addition to regional and racial exclusion, Cuba formed under the shadow of competing imperial powers. Most important of the latter was the United States, then in its most blatant period of imperial expansionism. The old colonial systems and Spanish institutions, already tied by long-term trade with the United States, met head-on with the external influences of the imperially competitive epoch and the demands of international politics. Intersections of race and identity became conspicuous in the formation of modern Cuba as it entered the twentieth-century global economy (Helg, 1995; Ferrer, 1999; Tone, 2006).

Global industrialization drastically shifted broader Latin American national identities, often through war. The Mexican Revolution provides one of the clearest examples of the disruption and multiple intersecting elements of tradition and change. The disorder of the revolution (1910–1920), permanently changing Mexican society and government, derives partly from the complexity of a disparate range of influences. Intellectual agitation helped spark the overthrow of Porfirio Díaz that began the revolution. But it could not have occurred without the resistance of formerly excluded and relatively autonomous communities, enraged by the government's use of traditional hierarchies and abuse of traditional patron-client relationships to forcibly integrate them into the international market economy. Responses to the Díaz regime and the revolution were also shaped by a rapid, industrial transformation of the landscape, a reorientation of the Mexican economy and countryside toward an aggressively expansionist United States, large foreign loans to encourage capitalist development, and a growing global demand for oil. Mexican revolutionaries were also heavily influenced by international ideologies such as Marxism, also being shaped during these years in reaction to global industrialization. In the end, both despite and because of these many ideologies, the revolution upheld at least as much of traditional society as it overthrew, but did allow institutional change that used the revolution's legacy for legitimacy (Knight, 1986; Centeno, 2002; Joseph and Buchenau, 2013).

The United States exercised influence well beyond Mexico in the early decades of the century, though especially in Central America and the Caribbean. Whether by economic controls, short-term military interventions, or military occupations, deployed US forces sought and at times fully expected to impose a model set of institutions and form of society on forcibly controlled nations. The actual results were of course complicated by resources, institutional realities, resistance, and negotiation, as well as unforeseen nationalist and proto-nationalist resistance, encompassing both international protests and prolonged guerrilla warfare. All of this together made interventions economically and diplomatically costly for the United States, which by the 1920s sought to withdraw and find other modes of influence. In the long term, it also led to amalgamations of local traditions and US-sponsored societal norms in those regions (Gobat, 2005; McPherson, 2015; Tillman, 2016). These interventions, followed by aid for US-friendly dictatorships, also brought to the fore such long-standing controversies and national questions as race relations, class inequalities, and gender inequalities, all thrown into clear relief by the clashing of ideologies and societal ideals during occupation (Suárez Findlay, 2000; Renda, 2001).

As US policy change, Latin American resistance, and global economic trends forced the end of occupations, the tenuous balance and U.S. financial support gave way to a period of U.S.-allied dictatorships under the "Good Neighbor Policy." Like the earlier occupations and later Cold War, this period encouraged centralized state repression. While the United States played a great role in the stabilization of, and the violent repression under, these dictatorships,

Latin American traditions and institutions contributed at least equally to making such forms of government acceptable and functional. A heritage of centralized and militarized states, charismatic *caudillo* leadership, repression, and exclusion all played into the new formation of right-wing and capitalist investment–friendly governments even before World War II. For Latin American countries, therefore, many aspects and patterns of the Cold War began much earlier in the twentieth century, while alliances between these dictatorships and US involvement meant a relative lack of foreign wars (Grandin and Joseph, 2010).

The turbulent years during and following World War II tested Latin American loyalties, institutions, and alliances. Following the war—often framed as a global effort against totalitarian governments—and with growing tensions between the global superpowers during the Cold War, U.S. government and academic funding of Latin American studies rose, only to increase exponentially again following the successful 1959 Cuban Revolution. While extensive studies improved understanding of the persistence of Latin American traditions, as well as U.S. influence, investigations from without sometimes overemphasized those elements to the exclusion of others. Importantly, such studies demonstrated conclusive connections between nineteenth-century dictatorship, and even colonial central state control, and the authoritarian systems that so defined the twentieth century. They also demonstrated in depth the complex nature of civil-military relations—such as how much of social development was conducted and protected by the military, and how closely tied such states' militaries had become to U.S. policy and the U.S. economy (Lieuwen, 1967; Johnson, 1964).

Responding in large part to poverty and unrest relating to U.S. interference and economic domination, and the aggressive military policies coinciding with early Cold War concerns, insurgent and intellectual opposition groups rose throughout Latin America. Most importantly for long-term development, of course, was the Cuban Revolution. Already an example for those who looked to recent precedent and sought to push back against U.S. influence and dictatorships, Fidel Castro's revolution exhibited staying power despite the general power imbalance in the Americas. Through the failed U.S.-backed Bay of Pigs invasion and the Cuban Missile Crisis, Cuba inspired revolutionaries throughout the Americas to question U.S. hegemony and take up the mantle of Cuban-style guerrilla warfare. Continuing the Latin American tendency of responding and adapting to external influences while maintaining powerful aspects of a colonial and repeatedly revolutionary past, insurgents in the guerrilla warfare of the 1960s, 1970s, and 1980s—regardless of whether Soviet-backed—employed a variety of Spanish, eighteenth- and nineteenth-century guerrilla techniques alongside the modern teachings of Mao and the Soviets. The lived experience of people through these events and upheavals, and consistent foreign interference, was destabilizing enough to perpetuate civil unrest and challenge centralized government control, while ideological debates continued to rage across decades (Grandin and Joseph, 2010).

Latin American states' Cold War policies deftly combined local traditions, and global politics and economics, through ever more repressive dictatorships that fed expanding resistance movements. The latter also combined traditions and global trends. Again, external economic influences on states—including direct U.S. financial support for the most repressive of dictatorships, because of their perceived stability—led to fewer explicitly foreign wars. Centeno has argued that in the long term, this repeated trend has actually hindered the consolidation of central states and national identities (Centeno, 2002). During the second half of the twentieth century, it certainly shut down many popular movements, made populations more ambivalent about the role of the military in society, and challenged optimism and nationalisms. Most notably, the Cold War years in Latin America were characterized by the infamous "Dirty Wars" of the Southern Cone, and especially Argentina, which emphasized and intensified the armed forces' centrality within state institutions. While the global developments of the 1960s (including the Cuban Revolution) inspired utopian student and worker movements, right-wing reaction fed increasingly brutal right-wing centralized states—reflecting centuries-old and to an extent even pre-Conquest traditions (Nunn, 1992; Loveman, 1999; Dávila, 2013).

The late Cold War and early post–Cold War years saw a turn toward historical investigations of U.S. interventionism in Latin America, demonstrating how the United States colluded expansively with repressive dictatorships to shut down guerrilla wars, uprisings, and popular movements, all with long-term implications for Latin American states and U.S.-Latin American relations (LaFeber, 1983; Smith, 1996; Grandin, 2006). The coincident increased funding for such studies, and social turns in the study of history, have also increased awareness and research of such understudied subjects as colonial military histories, race and gender studies, and the social composition of militaries. This rapidly growing historiography addresses long-neglected studies of race and gender during war, revolution, and times of conflict, especially for the Caribbean and Colombia, but much remains to be done. Newer such studies have explored the development of national consciousness, have reached into transnational elements, and have deeply examined the extent to which race and identity issues have affected the response to tensions of national growth, industrialization, world wars, and the Cold War, as well as to U.S. involvement (Joseph and LeGrand, 1998).

Post–Cold War histories have also begun to examine war and society for specific countries, contextualizing historical analysis of the period after 1980—especially historicizing the Cold War and pre–Cold War origins, dynamics, and consequences of current state systems (Stern, 1998). To uncover long-neglected voices, and thereby examine social elements and changes, scholars have begun more widely to deal with complex questions of identity through a creative reading of precolonial and colonial sources and oral histories.

A general lack of scholarship on inter–Latin American wars of the twentieth century, such as the Chaco War (1932–1935), leaves a great deal to be done in the field, although recently a renewed emphasis on wars and their relation to broader society promises a greater understanding of Latin American systems as well as a meaningful contribution to the wider historiography of war and society (Centeno, 2002; Scheina, 2003). A strong set of written and philosophical traditions that examine Latin America's place in broader global trends and ideologies, and the diversity and breadth of its experience, allows historians and social scientists to contextualize the region's conflicts. They also show its role in international diplomacy—especially through the extensive (and too often ignored) cultural and social as well as diplomatic and economic ties between the United States and its "southern neighbors." These trends have worked to tie the Americas together, examining the international, transnational, and borderlands in ways that establish the influence of the foreign on the national and the influence of Latin American national conflicts on wider global movements and ideologies (Joseph and LeGrand, 1998; Joseph and Buchenau, 2013).

Post–Cold War international and regional attention to human rights issues led to incremental growth in funding to uncover some of the most buried social realities of Latin American warfare and militarism since the 1970s and especially the 1980s. But the growing role of the region's states and militaries in global peacekeeping should also be accompanied by increased study of the topic. Such roles have led to new missions for the armed forces, increased deployments abroad, and new conceptions of the Latin American military, all of which will be important and formative in the years to come (Loveman, 1999). Collaboration in the "War on Drugs," likewise, changes the roles of states and their institutions even as the international drug trade itself forces changes in the security approaches of these nations. Finally, whereas civil-military relations have long formed one of the focal points for scholarship from within Latin America and from abroad, the focus of most such studies has been on political structures, individuals, political science, state-level diplomacy, and the Cold War influence. The growing transnational approach points scholars toward the need for understanding one of the most neglected intersections in historiography: that of warfare and labor. One historian has provided a notable example of how this might be approached, showing the ways that impressment and conscription form a vital part of broader Brazilian labor and race relations (Beattie, 2001). Generally, from territorial conflicts to immigration issues, Latin American labor history has been a central concern, but seldom has the study of labor, or even the expanding studies of race, ethnicity, class, and gender, been connected to the military and warfare.

Bibliography

Adelman, Jeremy. *Colonial Legacies: The Problem of Persistence in Latin American History*. New York: Routledge, 1999.

Archer, Christon. *The Army in Bourbon Mexico, 1760–1810*. Albuquerque: University of New Mexico Press, 1977.

Beattie, Peter. *The Tribute of Blood: Army, Honor, Race, and Nation in Brazil, 1864–1945*. Reprint Edition. Durham: Duke University Press, 2001.

Centeno, Miguel Angel. *Blood and Debt: War and the Nation-State in Latin America*. Pennsylvania Park: The Pennsylvania State University Press, 2002.

Dávila, Jerry. *Dictatorship in South America*. West Sussex, UK: Wiley-Blackwell, 2013.

Farcau, Bruce W. *The Ten Cents War: Chile, Peru, and Bolivia in the War of the Pacific, 1879–1884*. Westport: Greenwood, 2000.

Ferrer, Ada. *Insurgent Cuba: Race, Nation, and Revolution, 1868–1898*. Chapel Hill: University of North Carolina Press, 1999.

Foote, Nicola, and René Harder Horst. *Military Struggle and Identity Formation in Latin America: Race, Nation, and Community During the Liberal Period*. Gainesville: University Press of Florida, 2010.

Ganson, Barbara. *The Guaraní Under Spanish Rule in the Río de la Plata*. Stanford: Stanford University Press, 2003.

Gobat, Michel. *Confronting the American Dream: Nicaragua Under U.S. Imperial Rule*. Durham, NC: Duke University Press, 2005.

Gosner, Kevin. *Soldiers of the Virgin: The Moral Economy of a Colonial Maya Rebellion*. Tucson: University of Arizona Press, 1992.

Grandin, Greg. *Empire's Workshop: Latin America, the United States, and the Rise of the New Imperialism*. New York: Metropolitan Books, 2006.

Grandin, Greg, and Gilbert M. Joseph, eds. *A Century of Revolution: Insurgent and Counterinsurgent Violence During Latin America's Long Cold War*. Durham: Duke University Press, 2010.

Hassig, Ross. *War and Society in Ancient Mesoamerica*. Berkeley: University of California Press, 1992.

Helg, Aline. *Our Rightful Share: The Afro-Cuban Struggle for Equality, 1886–1912*. Chapel Hill: University of North Carolina Press, 1995.

Holden, Robert H. *Armies Without Nations: Public Violence and State Formation in Central America, 1821–1960*. New York: Oxford University Press, 2006.

Izecksohn, Vitor. *Slavery and War in the Americas: Race, Citizenship, and State Building in the United States and Brazil, 1861–1870*. Charlottesville: University of Virginia Press, 2014.

Jacobsen, Nils P., and Cristóbal Aljovín de Losada, eds. *Political Culture in the Andes, 1750–1950*. Durham: Duke University Press, 2005.

Johnson, John J. *The Military and Society in Latin America*. Stanford: Stanford University Press, 1964.

Joseph, Gilbert, and Jürgen Buchenau. *Mexico's Once and Future Revolution: Social Upheaval and the Challenge of Rule Since the Late Nineteenth Century*. Durham, NC: Duke University Press, 2013.

Joseph, Gilbert, and Catherine C. LeGrand, eds. *Close Encounters of Empire: Writing the Cultural History of U.S.-Latin American Relations*. Durham, NC: Duke University Press, 1998.

Knight, Alan. *The Mexican Revolution*. 2 vols. Cambridge: Cambridge University Press, 1986.

Kuethe, Allan J., and Juan F. Marchena, eds. *Soldados del rey: El ejército borbónico en América en vísperas de la independencia*. Castelló de la Plana, Spain: Publicaciones de la Universidad Jaume, 2005.

LaFeber, Walter. *Inevitable Revolutions: The United States in Central America*. New York: W.W. Norton, 1983; second edition, revised and expanded, 1993.

Langley, Lester. *The Americas in the Age of Revolution, 1750–1850.* New Haven: Yale University Press, 1997.

Lieuwen, Edwin. *Arms and Politics in Latin America.* New York: Frederick A. Praeger, 1967.

Love, Joseph. *The Revolt of the Whip.* Stanford: Stanford University Press, 2012.

Loveman, Brian. *For la Patria: Politics and the Armed Forces in Latin America.* Wilmington, DE: Scholarly Resources, 1999.

Lynch, John. *Caudillos in Spanish America, 1800–1850.* Oxford: Clarendon Press, 1992.

———. *The Spanish American Revolutions, 1808–1826.* London: Wiedenfield & Nicolson, 1973.

Marchena Fernández, Juan. *Ejército y milicias en el mundo colonial Americano.* Madrid: Editorial MAPFRE, 1992.

McAlister, Lyle. *The Fuero Militar in New Spain, 1764–1800.* Gainesville: University of Florida Press, 1957.

McCann, Frank D. *Soldiers of the Pátria: A History of the Brazilian Army, 1889–1937.* Stanford: Stanford University Press, 2004.

McPherson, Alan. *The Invaded: How Latin Americans and Their Allies Fought and Ended U.S. Occupations.* New York: Oxford University Press, 2015.

Nunn, Frederick. *The Time of the Generals: Latin American Professional Militarism in World Perspective.* Lincoln: University of Nebraska Press, 1992.

———. *Yesterday's Soldiers: European Military Professionalism in South America, 1890–1940.* Lincoln: University of Nebraska Press, 1983.

Quintero Montiel, Inés Mercedes. *La conjura de los Mantuanos: Ultimo acto de fidelidad a la monarquía española (Caracas, 1808).* Caracas: Universidad Católica Andrés Bello, 2002.

Renda, Mary. *Taking Haiti: Military Occupation and the Culture of U.S. Imperialism, 1915–1940.* Chapel Hill: University of North Carolina Press, 2001.

Robins, Nicholas A. *Native Insurgencies and the Genocidal Impulse in Latin America.* Bloomington: Indiana University Press, 2005.

Rodríguez, Linda Alexander. *Rank and Privilege: The Military and Society in Latin America.* Wilmington, DE: Scholarly Resources, 1994.

Santoni, Pedro. *Daily Lives of Civilians in Wartime Latin America: From the Wars of Independence to the Central American Civil Wars.* Westport: Greenwood Press, 2008.

Sater, William F. *Andean Tragedy: Fighting the War of the Pacific, 1879–1884.* Lincoln: University of Nebraska Press, 2007.

Scheina, Robert L. *Latin America's Wars: The Age of the Professional Soldier, 1900–2001.* 2 vols. Washington, DC: Brassey's, 2003.

Smith, Peter. *Talons of the Eagle: Dynamics of U.S.-Latin American Relations.* New York: Oxford University Press, 1996.

Stern, Steve J. *Shining and Other Paths: War and Society in Peru, 1980–1995.* Durham: Duke University Press, 1998.

Suárez Findlay, Eileen J. *Imposing Decency: The Politics of Sexuality and Race in Puerto Rico, 1870–1920.* Durham: Duke University Press, 2000.

Thibaud, Clément. *Repúblicas en armas: los ejércitos bolivarianos en la Guerra de Independencia de Colombia y Venezuela.* Bogotá: Instituto Francés de Estudios Andinos, 2003.

Thomson, Sinclair. *We Alone Will Rule: Native Andean Politics in the Age of Insurgency.* Madison: University of Wisconsin Press, 2002.

Tillman, Ellen D. *Dollar Diplomacy by Force: Nation-Building and Resistance in the Dominican Republic.* Chapel Hill: University of North Carolina Press, 2016.

Tone, John Lawrence. *War and Genocide in Cuba, 1895–1898.* Chapel Hill: University of North Carolina Press, 2006.

Van Young, Eric. *Writing Mexican History.* Stanford: Stanford University Press, 2012.

Wasserman, Mark. *Everyday Life and Politics in Nineteenth-Century Mexico: Men, Women, and War.* Albuquerque: University of New Mexico Press, 2000.

12 War and Society in North America to 1877

Matthew S. Muehlbauer

North America, from the seventeenth to the late nineteenth centuries, offers intriguing contexts in which to study the relationship between war and society. Historians of the continent in this period employ the same frameworks as those considering other times and places, such as race, class, gender, and local/regional studies. They also examine topics more germane to war per se, such as combatant and noncombatant experiences. But their work also explores special circumstances in early North America. For example, until the end of the nineteenth century some parts of the continent contained borderlands, regions in which (1) peoples of different and distinct cultures—including Native Americans—interacted and (2) political authority was unsettled or contested. In these areas war was a central concern for all groups, and research has examined how intercultural experiences altered perspectives and conceptions of conflict, as well social norms and identities. Another distinct feature of early North America was the nature of political authority. The limited extent of imperial, later national, authority raises intriguing questions about the nature of military service and sacrifice among the broader population, and how particular American and U.S. military institutions evolved up to the late nineteenth century.

For this era, the North American war and society literature is vast, with works on topics such as the American Civil War themselves too numerous for a short essay to adequately consider. As such, the discussion can only touch lightly on most subjects, and unfortunately limited space precludes addressing many worthy historiographical themes and publications (e.g., naval institutions). The chapter's overall emphasis will be on the United States and the regions it ultimately encompassed, as such dominates the field, though some books on other areas will be noted. While it will note research and methodologies that have been prominent over the past few decades, the essay will also highlight new and developing approaches.

One area of fruitful research has been Native American peoples, particularly regarding the issue of captivity. Early in the period being considered, captives were a crucial benefit of indigenous warfare, and often an outright objective, but interactions with European traders and settlers challenged and changed native captivity practices. Daniel Richter (1992) has explored the "mourning war" traditionally practiced by the Five Nations of the Iroquois in light of seventeenth-century developments such as epidemics, the spread of firearms, and more frequent fighting to control trade. Evan Haefeli and Kevin Sweeney's work on the 1704 Deerfield raid (2003) demonstrates the importance of captive taking in the American northeast. Allan Gallay (2002) has examined how English traders cultivated a market for Indian slaves in the American southeast of the late seventeenth and early eighteenth centuries, facilitating more frequent warfare among the region's native peoples. Christina Snyder (2010) has investigated how captivity practices of southeastern Indians changed from the seventeenth to the nineteenth centuries in a manner that gave more emphasis to race. For the southwestern borderlands, James Brooks' research (2002) illustrates how women and children captives fostered intercultural networks and enabled communication between native and European-descended societies with male power structures.

Treatments of captive taking are a subset of work over the past generation that has given greater attention to native agency, and how Indian peoples interacted with colonists and settlers in early American history against the backdrop of war. One of the most prominent of these is Richard White's examination of the Great Lakes region (1991), which argues that Indian groups and the Canadian French constructed a "middle ground," a means of respectful social interaction to which both contributed and could understand and derive meaning from, prior to British victory in the Seven Years' War. The experiences of peoples in the American southeast such as the Creek (Saunt, 1999) have also been studied. With respect to New England, a recent biography (Fisher and Silverman, 2014) scrutinizes the life of the Niantic sachem Ninigret, whereas for the southwest Andrew Knaut (1995) has surveyed the seventeenth-century Pueblo, noting how interactions with Spanish colonizers led to the Pueblo revolt of 1680.

Many of these works grapple with the evolution of distinct racial identities in early America. Gregory Dowd (1992) has considered efforts to create a pan-Indian identity against the background of wars and other events from the mid-eighteenth century to 1815. Conversely, Jill Lepore (1998) has examined how accounts of King Philip's War were written to bolster colonists' sense of English identity by depicting Indian foes as "savage" and "heathen." More recently, Peter Silver (2008) has examined how native raids in the Pennsylvania backcountry during the French and Indian War sparked a whole series of political texts that catalyzed notions of race and nationalism in early America.

Another intriguing interaction in borderlands regions is how peoples altered their approaches to warfare to better fight one another. Patrick Malone (1993) argues that New England Indians quickly recognized the value of firearms for their "skulking" form of warfare, but that warriors adopted a more lethal and destructive approach to warfare against English colonists in King Philip's War, based on the devastation settlers had wrought previously in the Pequot War. John Grenier (2005) claims that the "first American way of war" involved settlers using irregular tactics and techniques, as well as inflicting great violence against Indian populations as a means of subjugation, throughout the period 1607 to 1814. Examining the period 1689–1748, Steven Eames (2011) argues that the warfare practiced by northern New England's provincial soldiers incorporated both Western and Native American approaches to tactics and operations. Eames also considers the relationship between colonial New England society and its military system, noting how men regarded military service as a civic duty but struggled to rectify it with obligations to family and community. Kyle Zelner (2009) notes that earlier, in King Philip's War, Essex County officials relied more on impressing poor and marginalized men for service, challenging the belief that all contributed equally to the common defense.

A number of works consider how various regions coped with the challenge of the French and Indian War, as the Seven Years' War was known in North America. Fred Anderson (1984) demonstrates that in Massachusetts, contracts guaranteed conditions under which militiamen would volunteer for war service. James Titus (1991) examines how the war revealed the limits of deference in Virginia's highly hierarchical society, with authorities employing bounties to spur volunteer enlistments when impressment failed to provide adequate military manpower. Matthew C. Ward (2003) focuses specifically on the peoples in the Pennsylvania and Virginia backcountry from the start of hostilities in 1754 through to the conclusion of Pontiac's War in 1765. Stephen Brumwell (2002) has done the same for British soldiers, examining their garrison, campaigning, combat, and captivity experiences.

With respect to military customs and conduct, Ian Steele (1990) has demonstrated that the 1757 massacre outside Fort William Henry stemmed from a discrepancy between European norms by which victors extended courtesies to defeated opponents, and native culture in which warriors needed goods and captives to attest to their military prowess. More recently, Christian Ayne Crouch (2014) has examined the evolution of distinct military cultures in both metropolitan France and colonial New France before the French and Indian War, and how they clashed following the arrival of large numbers of regular troops in Canada during the conflict.

Fred Anderson's expansive overview of the war (2000) also considers how, in its aftermath, the British government pursued policies that sparked resistance and revolt in its North American colonies. T. H. Breen (2010) scrutinizes the years 1774–1776, wherein he claims large segments of the colonial population became radicalized and supported an insurgency against British rule. Michael McConnell (2004) takes a different perspective of the period leading up to the revolution, examining the challenges faced by British garrisons tasked with preserving order in the continent's frontier zones.

Regarding the War of the American Revolution, much scholarly attention has addressed the Continental Army. In 1982 James Kirby Martin and Mark Edward Lender (2015, third edition) challenged the popular emphasis on citizen soldiers, asserting that over time Continental troops came more from poor and dependent segments of society. Charles Neimeyer (1997) has extended this critique, noting large numbers of Irish and German immigrants and African American men in the ranks, and the importance of financial incentives for their enlistments. Beyond its social composition, E. Wayne Carp (1984) has analyzed how distrust of standing forces inherent in republican ideology led members of the Continental Congress to question and impede efforts of the men charged with supplying the Continental Army. John Resch (1999) traces how popular attitudes toward Continental Army veterans changed in the decades following the war.

Examining records for a particular New England town, Resch also questions the perspective of Martin and Lender and of Neimeyer, noting that men from all social classes bore arms, but that most did so in the militia rather than with the Continentals. The question of who served in militia units and their usage still requires more research, and few book-length treatments have addressed them since Robert Gross' social history of the minutemen who responded to the British advance upon Lexington and Concord (1976). Mark Kwasny (1996) is an exception, having examined militia units in northern areas, including George Washington's reliance on and complicated relationship with them.

American Loyalists have also been a popular area of study. Some recent works explore loyalism in places such as New York City (Chopra, 2011). Others have reflected the growing interest in world and Atlantic history, tracing the impact of Loyalist refugees and ideas on other parts of Britain's empire (Jasanoff, 2011). As for regular forces that fought for the Crown, Sylvia Frey (1981) offers an analysis of British troops similar to that provided by Martin and Lender for Continental units, while Rodney Atwood (1980) provided a rigorous study of the background and service of soldiers from Hessen-Kassel. More recently, Daniel Krebs (2013) has examined the relationships German prisoners of war developed with the communities in which they spent their captivity, which were often mutually beneficial: captives could provide needed service and labor, and take advantage of opportunities for better treatment and even to acquire their freedom.

Numerous works examine how various distinct regions experienced the Revolutionary War. Michael McDonnell (2007) has examined how the efforts to raise military manpower generated class and racial tensions in Virginia, undermining deference to social elites. Wayne Lee (2001) scrutinizes the cultural norms that gave meaning to and restrained public violence in North Carolina before and during the war, and how they broke down. Wayne Bodle (2002) has analyzed the experiences of, and interactions between, Continental Army soldiers and civilians in southeastern Pennsylvania during the Valley Forge winter of 1777–1778, while Judith Van Buskirk (2002) has illuminated how people in and around New York City continued to interact with each other despite the British occupation of the city—and the presence of Continental Army or Patriot forces just beyond—for the majority of the war.

With respect to gender frameworks, much scholarship has considered how the revolution offered enhanced status and opportunities for women (Norton, 1980; Berkin, 2005). Holly Mayer (1996) has delved more deeply into the experiences of women who accompanied armies on campaign. As for considerations of masculinity, John Ruddiman (2014) has dissected the expectations young recruits had of army service and how it would bolster their efforts to become independent men—as well as the impact of such notions on their military capabilities and performance.

With regards to race, many works have situated African American revolutionary experiences within broader temporal or geographic frameworks. Sylvia Frey (1991) has examined slaves' impact on both the course of the war in the American South and socioeconomic developments in the region in the years thereafter, while Douglas Egerton (2009) has examined more broadly how slavery expanded and became entrenched after 1783—and black efforts to resist that development. Other research includes that of Cassandra Pybus (2006), who traced the diaspora of slaves who left North America when the war ended among areas of the British Empire, and Judith L. Van Buskirk's (2017) examination of black Continental Army soldiers during the conflict and their subsequent treatment as veterans.

Few studies in recent decades have examined American Indian peoples during the revolution per se, though many works noted earlier address the period within a broader time frame (White, 1991; Dowd, 1992; Saunt, 1999). Colin Calloway (1995) offers the widest scope, contrasting the experiences of eight native communities in different areas of the continent, whereas Karim Tiro (2011) has examined the wartime and postwar experiences of the Oneida people, one of the two Iroquois nations to ally with the Patriot cause. Other studies emphasize complex interactions between Indians and other racial groups within a particular region. Woody Holton (1999) has examined how tensions among natives, slaves, and indebted whites in the years leading up to the war prompted Virginia's elites to support independence, while Jim Piecuch (2008) has noted that the British inability to understand the interests and tensions among their erstwhile Indian allies, as well as the agency of slaves, led the Crown to rely primarily on white Loyalists in the American South.

Having secured its independence with the Peace of Paris in 1783, the United States subsequently faced the prospect of developing permanent military institutions. Various research has assessed the political context that shaped the latter in the Republic's earliest years. Examining the years 1783 to 1802, Richard Kohn (1975) describes how Federalists created a constitution that allowed for national armed forces, but that it was the wars against native peoples in Northwest Territory that overcame the suspicions of Democratic Republicans and enabled the permanent establishment of the U.S. Army—primarily as a frontier constabulary for the first decades of its existence. Conversely Lawrence Delbert Cress (1982), assessing different strains of English Whig ideology, explains the emphasis of Anti-federalists and Democratic Republicans on state militias for the young country's defense needs, and their efforts to limit federal control over those organizations. Theodore Crackel (1987) reconsiders Thomas Jefferson's policy toward the U.S. Army, arguing that the 1802 Military Peace Establishment Act, along with the creation of the U.S. Military Academy and the Army Corps of Engineers, served to depoliticize the army after years of Federalist control, making it a reliable force for enforcing national policies in U.S. territories.

In the early republic's borderlands, William Bergmann (2012) demonstrates how both military force against native peoples and civilian agencies and policies served to integrate the Ohio River valley region into the expanding U.S. national market, and compares these developments to concomitant ones in British Upper Canada. Alan Taylor (2006) contrasts the experiences of Mohawk chief Joseph Brant and missionary Samuel Kirkland to offer a similar account of the northern borderlands encompassing Upper Canada and what is today western New York state.

The War of 1812 was the first major test of the policy whereby the United States maintained a small standing army in peacetime to be supplemented by large numbers of militiamen in the case of a major war. It did not fare well, at least with respect to President James Madison's goal of invading Canada. J.C.A. Stagg (1983) details the latter and how the country's administrative and military institutions proved too weak to achieve it. Edward Skeen (1999) has scrutinized the performance of militia units, noting their failures reflected the absence of a system that could provide militiamen with a minimum, satisfactory degree of combat proficiency.

Whereas traditional interpretations emphasize the futility of the U.S. effort against Great Britain, more recent scholarship has considered the advancement of national interests in borderlands regions at the expense of native peoples. These include a biography of William Henry Harrison's command of army and militia forces in the Old Northwest (Skaggs, 2014), and a treatment of the mobilization and campaigns of Tennessee field forces—primarily for the First Creek War of 1813–1814 (Kanon, 2014). Alan Taylor (2010) again focuses on the northern borderlands to argue that, in many aspects, the War of 1812 was a civil war among the former Patriots and Loyalists, Irish immigrants, and native peoples of the region.

Many works focus on the institutional development of the early U.S. Army. Here Edward Coffman's examination of peacetime soldiers, officers, and their dependents from 1784 until the end of the nineteenth century remains a standard (1986). Other scholars have scrutinized the evolution of the officer corps. William Skelton (1992) argued that it was only after 1815 that conditions allowed for officership to become a viable, professional career. Two volumes by Samuel Watson (2012, 2013) delve deeper into this question, assessing how officers developed a more professional ethos as they strove to enforce national policies in borderlands during the nineteenth century's first decades. The army's broader frontier experiences and interactions with local communities have been also been revisited by Michael Tate (1999).

In older areas of white settlement, state militia organizations became all but defunct in the antebellum period. But some authors have examined their continuing importance in borderlands regions. Mary Ellen Rowe (2003) traces militia systems as they were transplanted from Kentucky to Missouri, and then to Washington territory, highlighting their contributions to local order and defense in each area. Henry Laver (2007) has focused more particularly on Kentucky during the early republic, and how militia service encompassed broader civic and political involvement, in addition to reinforcing male gender roles. However, even in areas where regular militia training had become dormant, early U.S. political culture remained important for stimulating enlistments in wartime. Rick Herrera (2015) has dissected the "military ethos of Republicanism" by which citizenship remained intimately tied to military service for generations of young men between the American Revolution and the Civil War.

While traditional accounts of U.S. military history usually skip from the War of the 1812 to the Mexican War, in recent decades scholars have explored some of the intervening borderlands conflicts. Patrick Jung's account of the Black Hawk War (2007) ties its origins to the pan-Indian movements of the late eighteenth and early nineteenth centuries described by Gregory Dowd (1992), whereas John Hall (2009) examines the motivations of native groups that assisted the U.S. government during the conflict. Similarly, John Ellisor (2010) offers a novel approach to the Second Creek War, describing the multiethnic and racially diverse borderlands society that was developing before conflict erupted in 1836, and how many whites and blacks sought to assist Creeks resisting encroachment and expulsion from their lands.

Other scholarship emphasizes the experiences of races and nationalities beyond white U.S. citizens, but over a broader temporal scope. Pekka Hämäläinen (2008) presents the history of late eighteenth- and nineteenth-century North America from the perspective of the Comanche empire. Sean McEnroe (2012) ties the evolution of independent Mexico back to the alliance between Spanish and Tlaxcalan peoples that expanded Spain's American frontiers beginning in the sixteenth century, whereas Mark Wasserman (2000) focuses on the nineteenth century, examining how common Mexicans struggled to maintain political autonomy in light of political crises and wars, as well as the latter's impact on gender relations.

These perspectives have also shaped recent treatments of the Mexican-American War. David Clary (2009) goes back to Mexico's Wars of Independence, assessing how clashes between it and the United States set the context for the war of 1846–1848, and then addresses that conflict from the standpoint of both belligerents. Brian Delay (2008), taking a similarly expansive temporal scope, focuses on Indian raiding in the southwest borderlands and how it shaped the U.S.-Mexico conflict. The experience of U.S. soldiers in the Mexican-American War has remained a topic of interest, though Paul Foos (2002) arguably goes the furthest in illuminating the racism and class pressures that influenced enlistments and shaped soldiers' conduct. More recently, Tyler Johnson (2012) has examined the service of immigrant soldiers and how such was used to both promote and undermine anti-immigrant views of the era.

The years between the Mexican-American War and the American Civil War remain a fairly neglected period. It has received some attention in literature addressing the U.S. Army and is the explicit focus of Durwood Ball (2001), who addresses the myriad duties and challenges frontier units faced in this period. A new anthology (Lahti, 2017) explores the experiences of individual soldiers of varying ethic and racial backgrounds, and serving with different forces in the southwestern borderlands (U.S. Army, Mexican Army, territorial units, etc.) from 1848 to 1886. Robert May (2002) scrutinizes the private military adventurers known as filibusters, who plagued Latin America in the decade after the Mexican-American War. Assessing their popularity among expansionist-minded elements of the public, he also notes the federal government's inability to check them, how they contributed to the sectional crisis that led to the Civil War, and the problems they generated for U.S. foreign policy and the country's standing in the world.

War and society scholarship on the American Civil War is extensive, and can only be touched upon in a short review. The task is even more complicated because individual books often, both explicitly and implicitly, simultaneously address multiple themes within the field. Studies of women and the war, for example, often grapple with class as well. In the Confederacy, for example, Drew Gilpin Faust (1996) has focused on the upper class and efforts to preserve their status, whereas others have illuminated the circumstances of poor white and black women (Schwalm, 1997; Edwards, 2000). Similarly, works have scrutinized how organizing volunteer efforts in the North raised issues for female middle-class gender roles (Attie, 1998), and the ways in which working-class and African American women grappled with and challenged wartime federal government policies (Giesberg, 2009).

Wartime black experiences have become a prominent area of study since the original 1965 publication of James McPherson's sweeping overview (2003). In recent years, the literature has been supplemented by work such as David Williams' emphasis on slaves' agency in securing their own freedom, or self-emancipation (2014), and Jim Downs' study (2012) on how illness ravaged African Americans during and after the Civil War. But an especially prominent topic has been the experiences of black soldiers, with many monographs examining troops from particular areas (Spurgeon, 2014; Quinn, 2015). Other notable works include Joseph Glatthaar's on relationships between white officers and African American troops in the regiments of the U.S. Colored Troops, or USCT (Glatthaar, 1990), and Barbara Gannon's study (2011) of how white and black enlisted soldiers supported one another during and after the war.

Scholarship on whites of distinct ethnic backgrounds mirrors that on African Americans, in that many focus on soldiers. These include overviews of immigrant troops and their experiences (Öfele, 2008) as well as considerations of Irish volunteers in the Union (Bruce, 2006) and the performance of German officers in the USCT (Öfele, 2004). Other works address, more broadly, ethnic populations in particular areas, such as the factors that led German immigrants in St. Louis to support abolition before and during the war (Anderson, 2016) and the ambiguities of Irish support for the Confederacy in the South (Gleeson, 2013).

Numerous works consider the wider military service of white soldiers beyond particular ethnic populations. James McPherson (1997), for example, has considered troops' motivations for service. Gerald Linderman (1987) discusses how the realities of combat destroyed soldiers' expectations of demonstrating their courage, producing disillusionment—but not so much that veterans sought to challenge romantic views of battle that became popular after the war. More recent research has examined guerrilla warfare (Sutherland, 2009). Some scholars have considered the issue of the Union draft (Geary, 1991), while others have probed factors behind voluntary enlistments, tying them to community and family values in the North (Mitchell, 1993) and South (Sheehan-Dean, 2007). Mark Weitz (2005) offers a critical assessment of desertion and its impact on the Confederacy.

Regarding civilian experiences in the Civil War, among the most novel and impressive books is Drew Gilpin Faust's work (2008) on how the great number of fatalities challenged prevailing cultural perspectives on death. Another problem faced in both Union and Confederacy was dissent. Whereas Jennifer Weber (2006) has reassessed the threat northern Copperheads posed to the Union war effort, other studies have focused specifically on how local circumstances fostered defiance and protest in regions such as the Pennsylvania Appalachians (Sandow, 2009). A number of works have similarly addressed the challenges Unionism posed to Confederate authorities in various regions (Storey, 2004). The issue of southern dissent touches on older scholarship that argues defeat stemmed from a lack of Confederate nationalism (Escott, 1978). More recently, William Blair (1998) has focused on Virginia to illustrate the factors that undermined support for continuing the war, while Gary Gallagher (1997) argues that southern hopes for victory came to reside in Robert E. Lee and the Army of Northern Virginia, whose surrender hence ended hostilities.

Civilian experiences differed significantly with respect to potential for interacting with enemy troops, in that numerous southerners had to face the prospect of real or potential Union invasion, whereas Confederate incursions north were few and short-lived. Mark Grimsley (1995) has examined how broader Union policy changed during the war, going from lenient to harsher, particularly with respect to property seizures and destruction, but abiding within certain limits. Grimsley's arguments include Sherman's campaigns, which other scholars have examined with an explicit focus on the experiences of civilian and soldiers (Campbell, 2003), including the impact on women (Frank, 2015). Kentucky and Tennessee have also been studied for the social impact of military operations (Cooling, 1997, 2011). The latter state also included wartime occupation by Union forces, another topic that has grown more prominent. Stephen Ash (1995) has provided a broader overview of the occupation efforts in the entire South, while others have examined particular areas more closely (Browning, 2011).

Much research remains within temporal boundaries that separate the Civil War from the periods preceding and following. But authors have been challenging these divisions, particularly in respect to regional studies. Stephen Ash (2006), for example, scrutinized the war's impact on Middle Tennessee, tying those experiences to issues the area faced in the first years of Reconstruction. Edward Ayers' focus is earlier (2003), comparing how two communities in Pennsylvania and Virginia reacted to events that initiated the conflict, and the challenges each faced after it began. Victoria Bynum (2010) addresses developments in southern dissenting regions during Reconstruction and afterwards, while Aaron Astor (2012) has looked at wartime and postwar developments in the border states of Missouri and Kentucky.

Once regarded as something of an afterthought to the Civil War, scholars have increasingly been emphasizing the importance of Reconstruction as a period in its own right, particularly for understanding how white racial supremacy became entrenched in the South after 1865. Initial studies addressed the U.S. Army's role in Reconstruction policies, with Mark Bradley's recent examination of North Carolina (2009) giving more attention to soldier–civilian interactions. But newer research has focused on the violence conservative white southerners used to suppress and intimidate African Americans and white Republicans in the absence of army units, or after they had been withdrawn. George C. Rable (1984) provided an early treatment of such developments in the South, whereas Mark Grimsley (2012) has more recently approached both Reconstruction and the civil rights movement (which he calls the Second Reconstruction) as insurgencies. Various studies have examined how Reconstruction unfolded in specific states, such as South Carolina (Zuczek, 1996) and Texas (Moneyhon, 2004), whereas treatments of Tennessee have examined how the short-lived State Guard attempted to cope with racial violence (Severance, 2005) and the 1866 Memphis riot (Ash, 2013).

The last of the North American Indian wars has received significant attention in works for general readers, but could use more attention from scholars. More research is needed to situate these wars against the backdrop of interactions between native peoples, settlers of varying ethnicities and races, and federal authorities amid the turmoil of the Civil War and its aftermath. That said, these relationships should receive more scholarly focus in general, for it is the role of the U.S. government that distinguishes borderlands history after 1783 from what preceded it. In particular, more work is required on the relationships between territorial, state, and local governments, and federal authorities, with respect to native peoples and prospects for regional conflict (and peace). For the colonial and revolutionary eras, additional research is needed on the role, function, and composition of militia units and organizations, especially outside of New England, with work on southern areas accounting for the impact of slavery. While the Civil War does not lack for scholarly attention, the great number of studies on particular regions, or emphasizing particular methodologies, has created a need for more works of synthesis to integrate these varying approaches. That said, the growing focus on Reconstruction and its relationship to preceding and subsequent eras must continue, for it is the events of that period that shaped subsequent prevalent understandings of the Civil War and the factors that produced it.

Bibliography

Anderson, Fred. *Crucible of War: The Seven Years' War and the Fate of Empire in British North America, 1754–1766*. New York: Knopf, 2000.

———. *A People's Army: Massachusetts Soldiers and Society in the Seven Year's War*. Chapel Hill: University of North Carolina Press, 1984.

Anderson, Kristen Layne. *Abolitionizing Missouri: German Immigrants and Racial Ideology in Nineteenth-Century America*. Baton Rouge: Louisiana State University Press, 2016.

Ash, Stephen V. *A Massacre in Memphis: The Race Riot That Shook the Nation One Year After the Civil War*. New York: Hill & Wang, 2013.

———. *Middle Tennessee Society Transformed, 1860–1870: War and Peace in the Upper South*. New ed. Knoxville: University of Tennessee Press, 2006.

———. *When the Yankees Came: Conflict and Chaos in the Occupied South, 1861–1865*. Chapel Hill: University of North Carolina Press, 1995.

Astor, Aaron. *Rebels on the Border: Civil War, Emancipation, and the Reconstruction of Kentucky and Missouri*. Baton Rouge: Louisiana State University Press, 2012.

Attie, Jeanie. *Patriotic Toil: Northern Women and the American Civil War*. Ithaca: Cornell University Press, 1998.

Atwood, Rodney. *Hessians: Mercenaries From Hessen-Kassel in the American Revolution*. New York: Cambridge University Press, 1980.

Ayers, Edward L. *In the Presence of Mine Enemies: War in the Heart of America, 1859–1863*. New York: W.W. Norton, 2003.

Ball, Durwood. *Army Regulars on the Western Frontier, 1848–1861*. Norman: University of Oklahoma Press, 2001.

Bergmann, William G. *The American National State and the Early West*. New York: Cambridge, 2012.

Berkin, Carol. *Revolutionary Mothers: Women in the Struggle for America's Independence*. New York: Knopf, 2005.

Blair, William. *Virginia's Private War: Feeding Body and Soul in the Confederacy, 1861–1865*. New York: Oxford, 1998.

Bodle, Wayne. *The Valley Forge Winter: Civilians and Soldiers in War*. University Park: Pennsylvania State University Press, 2002.

Bradley, Mark L. *Bluecoats and Tar Heels: Soldiers and Civilians in Reconstruction North Carolina*. Lexington: University Press of Kentucky, 2009.

Breen, T. H. *American Insurgents, American Patriots: The Revolution of the People*. New York: Hill & Wang, 2010.

Brooks, James F. *Captives & Cousins: Slavery, Kinship, and Community in the Southwest Borderlands*. Chapel Hill: University of North Carolina Press, 2002.

Browning, Judkin. *Shifting Loyalties: The Union Occupation of Eastern North Carolina*. Chapel Hill: University of North Carolina Press, 2011.

Bruce, Susannah Ural. *The Harp and the Eagle: Irish-American Volunteers and the Union Army, 1861–1865*. New York: New York University Press, 2006.

Brumwell, Stephen. *Redcoats: The British Soldier and War in the Americas, 1755–1763*. New York: Cambridge University Press, 2002.

Bynum, Victoria E. *The Long Shadow of the Civil War: Southern Dissent and Its Legacies*. Chapel Hill: University of North Carolina Press, 2010.

Calloway, Colin G. *The American Revolution in Indian Country: Crisis and Diversity in Native American Communities*. New York: Cambridge, 1995.

Campbell, Jaqueline Glass. *When Sherman Marched North From the Sea: Resistance on the Confederate Home Front*. Chapel Hill: University of North Carolina Press, 2003.

Carp, E. Wayne. *To Starve the Army at Pleasure: Continental Army Administration and American Political Culture, 1775–1783*. Chapel Hill: University of North Carolina Press, 1984.

Chopra, Ruma. *Unnatural Rebellion: Loyalists in New York City During the Revolution*. Charlottesville: University of Virginia Press, 2011.

Clary, David A. *Eagles and Empire: The United States, Mexico, and the Struggle for a Continent*. New York: Bantam, 2009.

Coffman, Edward M. *The Old Army: A Portrait of the American Army in Peacetime, 1784–1898*. New York: Oxford University Press, 1986.

Cooling, Benjamin Franklin. *Fort Donelson's Legacy: War and Society in Kentucky and Tennessee, 1862–1863*. Knoxville: University of Tennessee Press, 1997.

———. *To the Battles of Franklin and Nashville and Beyond: Stabilization and Reconstruction in Tennessee and Kentucky, 1864–1866*. Knoxville: University of Tennessee Press, 2011.

Crackel, Theodore J. *Mr. Jefferson's Army: Political and Social Reform of the Military Establishment, 1801–1809*. New York: New York University Press, 1987.

Cress, Lawrence Delbert. *Citizens in Arms: The Army and Militia in American Society to the War of 1812*. Chapel Hill: University of North Carolina Press, 1982.

Crouch, Christian Ayne. *Nobility Lost: French and Canadian Martial Cultures, Indians, and the End of New France*. Ithaca: Cornell University Press, 2014.

Delay, Brian. *War of a Thousand Deserts: Indian Raids and the U. S.-Mexican War*. New Haven: Yale University Press, 2008.

Dowd, Gregory Evans. *A Spirited Resistance: The North American Indian Struggle for Unity, 1745–1815*. Baltimore: Johns Hopkins University Press, 1992.

Downs, Jim. *Sick From Freedom: African-American Illness and Suffering During the Civil War and Reconstruction*. New York: Oxford, 2012.

Eames, Steven C. *Rustic Warriors: Warfare and the Provincial Soldier on the New England Frontier, 1689–1648.* New York: New York University Press, 2011.

Edwards, Laura F. *Scarlett Doesn't Live Here Anymore: Southern Women in the Civil War Era.* Chicago: University of Illinois Press, 2000.

Egerton, Douglas R. *Death or Liberty: African Americans and Revolutionary America.* New York: Oxford, 2009.

Ellisor, John T. *The Second Creek War: Interethnic Conflict and Collusion on a Collapsing Frontier.* Lincoln: University of Nebraska Press, 2010.

Escott, Paul D. *After Secession: Jefferson Davis and the Failure of Confederate Nationalism.* Baton Rouge: Louisiana State University Press, 1978.

Faust, Drew Gilpin. *Mothers of Invention: Women of the Slaveholding South in the American Civil War.* Chapel Hill: University of North Carolina Press, 1996.

———. *This Republic of Suffering: Death and the American Civil War.* New York: Knopf, 2008.

Fisher, Julie A., and David J. Silverman. *Ninigret, Sachem of the Niantics and Narragansetts: Diplomacy, War, and the Balance of Power in Seventeenth-Century New England and Indian Country.* Ithaca: Cornell University Press, 2014.

Foos, Paul. *A Short, Offhand, Killing Affair: Soldiers and Social Conflict During the Mexican-American War.* Chapel Hill: University of North Carolina Press, 2002.

Frank, Lisa Tendrich. *The Civilian War: Confederate Women and Union Soldiers During Sherman's March.* Baton Rouge: Louisiana State University Press, 2015.

Frey, Sylvia R. *The British Soldier in America: A Social History of Military Life in the Revolutionary Period.* Austin: University of Texas Press, 1981.

———. *Water From the Rock: Black Resistance in a Revolutionary Age.* Princeton: Princeton University Press, 1991.

Gallagher, Gary W. *The Confederate War.* Cambridge: Harvard University Press, 1997.

Gallay, Alan. *The Indian Slave Trade: The Rise of English Empire in the American South, 1670–1717.* New Haven: Yale University Press, 2002.

Gannon, Barbara A. *The Won Cause: Black and White Comradeship in the Grand Army of the Republic.* Chapel Hill: University of North Carolina Press, 2011.

Geary, James W. *We Need Men: The Union Draft in the Civil War.* DeKalb: Northern Illinois University Press, 1991.

Giesberg, Judith. *Army at Home: Women and the Civil War on the Northern Home Front.* Chapel Hill: University of North Carolina Press, 2009.

Glatthaar, Joseph T. *Forged in Battle: The Civil War Alliance of Black Soldiers and White Officers.* Baton Rouge: Louisiana State University Press, 1990.

———. *The Green and the Gray: The Irish in the Confederate States of America.* Chapel Hill: University of North Carolina Press, 2013.

Grenier, John. *The First Way of War: American War Making on the Frontier.* New York: Cambridge, 2005.

Grimsley, Mark. *The Hard Hand of War: Union Military Policy Toward Southern Civilians, 1861–1865.* New York: Cambridge University Press, 1995.

———. "Wars for the American South: The First and Second Reconstructions Considered as Insurgencies," *Civil War History* 58 (2012): 6–36.

Gross, Robert A. *The Minutemen and Their World.* New York: Hill & Wang, 1976.

Haefeli, Evan, and Kevin Sweeney. *Captors and Captives: The 1704 French and Indian Raid on Deerfield.* Amherst: University of Massachusetts Press, 2003.

Hall, John W. *Uncommon Defense: Indian Allies in the Black Hawk War.* Cambridge: Harvard University Press, 2009.

Hämäläinen, Pekka. *The Comache Empire.* New Haven: Yale University Press, 2008.

Herrera, Ricardo A. *For Liberty and the Republic: The American Citizen as Soldier, 1775–1861.* New York: New York University Press, 2015.

Holton, Woody. *Forced Founders: Indians, Debtors, Slaves, and the Making of the American Revolution in Virginia.* Chapel Hill: University of North Carolina Press, 1999.

Jasanoff, Maya. *Liberty's Exiles: American Loyalists in the Revolutionary World.* New York: Knopf, 2011.

Johnson, Tyler V. *Devotion to the Adopted Country: U.S. Immigrant Volunteers in the Mexican War.* Columbia: University of Missouri Press, 2012.

Jung, Patrick J. *The Black Hawk War of 1832*. Norman: University of Oklahoma Press, 2007.

Kanon, Tom. *Tennesseans at War, 1812–1815: Andrew Jackson, the Creek War, and the Battle of New Orleans*. Tuscaloosa: University of Alabama Press, 2014.

Knaut, Andrew L. *The Pueblo Revolt of 1680: Conquest and Resistance in Seventeenth-Century New Mexico*. Norman: University of Oklahoma Press, 1995.

Kohn, Richard H. *Eagle and Sword: The Federalists and the Creation of the Military Establishment in America, 1783–1802*. New York: The Free Press, 1975.

Krebs, Daniel. *A Generous and Merciful Enemy: Life for German Prisoners of War During the American Revolution*. Norman: University of Oklahoma Press, 2013.

Kwasny, Mark V. *Washington's Partisan War, 1775–1783*. Kent, OH: Kent State University Press, 1996.

Lahti, Janne, ed. *Soldiers in the Southwest Borderlands, 1848–1886*. Norman: University of Oklahoma Press, 2017.

Laver, Henry S. *Citizens More Than Soldiers: The Kentucky Militia and Society in the Early Republic*. Lincoln: University of Nebraska Press, 2007.

Lee, Wayne E. *Crowds and Soldiers in Revolutionary North Carolina: The Culture of Violence in Riot and War*. Gainesville: University of Florida Press, 2001.

Lepore, Jill. *The Name of War: King Philip's War and the Origins of American Identity*. New York: Knopf, 1998.

Linderman, Gerald. *Embattled Courage: The Experience of Combat in the American Civil War*. New York: The Free Press, 1987.

Malone, Patrick M. *The Skulking Way of War: Technology and Tactics Among the New England Indians*. Baltimore: Johns Hopkins University Press, 1993.

Martin, James Kirby, and Mark Edward Lender. *A Respectable Army: The Military Origins of the Republic, 1763–1789*. 3rd ed. Malden, MA: Wiley & Sons, 2015.

May, Robert. *Manifest Destiny's Underworld: Filibustering in Antebellum America*. Chapel Hill: University of North Carolina Press, 2002.

Mayer, Holly A. *Belonging to the Army: Camp Followers and Community During the American Revolution*. Columbia: University of South Carolina Press, 1996.

McConnell, Michael N. *Army and Empire: British Soldiers on the American Frontier, 1758–1775*. Lincoln: University of Nebraska Press, 2004.

McDonnell, Michael A. *The Politics of War: Race, Class, and Conflict in Revolutionary Virginia*. Chapel Hill: University of North Carolina Press, 2007.

McEnroe, Sean F. *From Colony to Nationhood in Mexico: Laying the Foundations, 1560–1840*. New York: Cambridge University Press, 2012.

McPherson, James M. *For Cause and Comrades: Why Men Fought in the Civil War*. New York: Oxford, 1997.

———. *The Negro's Civil War: How American Blacks Felt and Acted During the War for the Union*. New York: Pantheon Books, 1965; New York: Knopf, 2003.

Mitchell, Reid. *The Vacant Chair: The Northern Soldier Leaves Home*. New York: Oxford, 1993.

Moneyhon, Carl H. *Texas After the Civil War: The Struggle of Reconstruction*. College Station: Texas A&M University Press, 2004.

Neimeyer, Charles Patrick. *America Goes to War: A Social History of the Continental Army*. New York: New York University Press, 1997.

Norton, Mary Beth. *Liberty's Daughters: The Revolutionary Experience of American Women, 1750–1800*. Boston: Little, Brown, 1980.

Öfele, Martin W. *German-Speaking Officers in the US Colored Troops, 1863–1867*. Gainesville: University of Florida Press, 2004.

———. *True Sons of the Republic: European Immigrants in the Union Army*. Westport, CT: Praeger, 2008.

Piecuch, Jim. *Three Peoples, One King: Loyalists, Indians, and Slaves in the American Revolutionary South, 1775–1782*. Columbia: University of South Carolina, 2008.

Pybus, Cassandra. *Epic Journeys of Freedom: Runaway Slaves of the American Revolution and Their Global Quest for Liberty*. Boston: Beacon, 2006.

Quinn, Edythe Ann. *Freedom Journey: Black Civil War Soldiers and the Hills Community, Westchester County, New York*. Albany: State University of New York Press, 2015.

Rable, George C. *But There Was No Peace: The Role of Violence in the Politics of Reconstruction*. Athens: University of Georgia Press, 1984.

Resch, John. *Suffering Soldiers: Revolutionary War Veterans, Moral Sentiment, and Political Culture in the Early Republic.* Amherst: University of Massachusetts Press, 1999.

Richter, Daniel K. *The Ordeal of the Longhouse: The Peoples of the Iroquois League in the Era of European Colonization.* Chapel Hill: University of North Carolina Press, 1992.

Rowe, Mary Ellen. *Bulwark of the Republic: The American Militia in Antebellum West.* Westport, CT: Praeger, 2003.

Ruddiman, John A. *Becoming Men of Some Consequence: Youth and Military Service in the Revolutionary War.* Charlottesville: The University of Virginia Press, 2014.

Sandow, Robert M. *Deserter Country: Civil War Opposition in the Pennsylvania Appalachians.* New York: Fordham, 2009.

Saunt, Claudio. *A New Order of Things: Property, Power, and the Transformation of the Creek Indians, 1733–1816.* Cambridge: Cambridge University Press, 1999.

Schwalm, Leslie Ann. *A Hard Fight for We: Women's Transition From Slavery to Freedom in South Carolina.* Chicago: University of Illinois Press, 1997.

Severance, Ben H. *Tennessee's Radical Army: The State Guard and Its Role in Reconstruction, 1867–1869.* Knoxville: University of Tennessee Press, 2005.

Sheehan-Dean, Aaron. *Why Confederates Fought: Family and Nation in Civil War Virginia.* Chapel Hill: University of North Carolina Press, 2007.

Silver, Peter. *Our Savage Neighbors: How Indian War Transformed Early America.* New York: W.W. Norton, 2008.

Skaggs, David Curtis. *William Henry Harrison and the Conquest of the Ohio Country: Frontier Fighting in the War of 1812.* Baltimore: Johns Hopkins University Press, 2014.

Skeen, C. Edward. *Citizen Soldiers in the War of 1812.* Lexington: University of Kentucky Press, 1999.

Skelton, William B. *An American Profession of Arms: The Army Officer Corps, 1784–1861.* Lawrence: University Press of Kansas, 1992.

Snyder, Christina. *Slavery in Indian Country: The Changing Face of Captivity in Early America.* Cambridge: Harvard University Press, 2010.

Spurgeon, Ian Michael. *Soldiers in the Army of Freedom: The 1st Kansas Colored, the Civil War's First African American Combat Unit.* Norman: University of Oklahoma Press, 2014.

Stagg, J.C.A. *Mr. Madison's War: Politics, Diplomacy, and Warfare in the Early American Republic, 1783–1830.* Princeton: Princeton University Press, 1983.

Steele, Ian K. *Betrayals: Fort William Henry and the Massacre.* New York: Oxford, 1990.

Storey, Margaret M. *Loyalty and Loss: Alabama's Unionists in the Civil War and Reconstruction.* Baton Rouge: Louisiana State University Press, 2004.

Sutherland, Daniel E. *A Savage Conflict: The Decisive Role of Guerrillas in the American Civil War.* Chapel Hill: University of North Carolina Press, 2009.

Tate, Michael, L. *The Frontier Army in the Settlement of the West.* Norman: University of Oklahoma Press, 1999.

Taylor, Alan. *The Civil War of 1812: American Citizens, British Subjects, Irish Rebels, & Indian Allies.* New York: Knopf, 2010.

———. *The Divided Ground: Indians, Settlers and the Northern Borderland of the American Revolution.* New York: Knopf, 2006.

Tiro, Karim. *The People of the Standing Stone: The Oneida Nation From the Revolution Through the Era of Removal.* Amherst: University of Massachusetts Press, 2011.

Titus, James. *The Old Dominion at War: Society, Politics and Warfare in Late Colonial Virginia.* Columbia: University of South Carolina Press, 1991.

Van Buskirk, Judith L. *Generous Enemies: Patriots and Loyalists in Revolutionary New York.* Philadelphia: University of Pennsylvania Press, 2002.

———. *Standing in Their Own Light: African American Patriots in the American Revolution.* Norman: University of Oklahoma, 2017.

Ward, Matthew C. *Breaking the Backcountry: The Seven Years' War in Virginia and Pennsylvania 1754–1765.* Pittsburgh: University of Pittsburg Press, 2003.

Wasserman, Mark. *Everyday Life and Politics in Nineteenth Century Mexico: Men, Women, and War.* Albuquerque: University of New Mexico Press, 2000.

Watson, Samuel J. *Jackson's Sword: The Army Officer Corps on the American Frontier, 1810–1821.* Lawrence: University of Kansas Press, 2012.

————. *Peacekeepers and Conquerors: The Army Officer Corps on the American Frontier, 1821–1846*. Lawrence: University of Kansas Press, 2013.

Weber, Jennifer L. *Copperheads: The Rise and Fall of Lincoln's Opponents in the North*. New York: Oxford, 2006.

Weitz, Mark A. *More Damning Than Slaughter: Desertion in the Confederate Army*. Lincoln: University of Nebraska Press, 2005.

White, Richard. *The Middle Ground: Indians, Empires, and Republics in the Great Lakes Region, 1650–1815*. Cambridge: Cambridge University Press, 1991.

Williams, David. *I Freed Myself: African American Self-Emancipation in the Civil War Era*. New York: Cambridge, 2014.

Zelner, Kyle F. *A Rabble in Arms: Massachusetts Towns and Militiamen During King Philip's War*. New York: New York University Press, 2009.

Zuczek, Richard. *State of Rebellion: Reconstruction in South Carolina*. Columbia: University of South Carolina Press, 1996.

13 War and Society in North America Since 1877

Antulio J. Echevarria II

It is something of a cliché to say warfare influences society as much as society influences warfare. The motives that drive war are socially constructed, and the norms and practices by which war is waged are socially derived. The conduct of war is as much a social experience as it is a personal and political one; it is at once shared, singular, and exploitable. The history of war and society helps readers understand these three dimensions, and not only these. But perhaps the most important contribution such history can make, as Michael Howard (1975) has demonstrated, is that the extent to which we appreciate how much war and society transform each other is contingent on what we know about each. And what we know about each will change, so says Howard, as we gain distance from the defining historical events of a period, such as world wars and radical cultural movements.

The paragraphs that follow discuss what has changed regarding our understanding of the citizen soldier concept so germane to North American military history; the recurring themes of militarism, antimilitarism, and imperialism in America's modern wars; some new perspectives on race, class, and gender relations in wartime; and our appreciation of what the American way of war is or is not.

Since the end of the Vietnam conflict, historians and social scientists have taken a much closer look at the divisions that existed in North American societies, particularly in the United States. The political fracturing caused by the direction (or misdirection) of the Vietnam War, the rise of the antiwar movement, the spread of "counterculture" values, and the pervasive racial violence that characterized the struggle over civil rights in the 1960s all revealed an American society that was more deeply divided, and one that apparently remains so, than previous scholarly works had acknowledged. The "melting pot" effect that was supposed to have fused the many classes, races, religious, and ethnic groups that populated the United States at the dawn of the modern era into a single American identity never did so. Instead, the picture of American society that post-Vietnam-era research has uncovered is one that defies the homogeneity and wistful appeal of a Norman Rockwell painting. Similar national myths were dispelled by Desmond Morton (1985) for the Canadian military and its civil society; Canadians were apparently no more peace-loving than their American counterparts, though each continues to see themselves as such.

It is not that racial, ethnic, religious, and class divisions went unnoticed in North America before the Vietnam War. Gerald Linderman (1974) demonstrates divisions over class and race were abundantly evident during the Spanish-American War. We can find similar divisions in Canada as well; Constance Backhouse (1997), for instance, reveals how the Canadian legal system has long been "color-coded" against racial minorities. Such divisions were surely captured, if unsystematically, in popular historical literature. One example of such literature is Studs Terkel's "*The Good War*" (1984), an intentionally ironic title for a brilliantly arranged oral history of the Second World War, which revealed many of the racial and ethnic divisions that existed in the America of the 1940s. To highlight one account: Princeton historian Arno

Mayer vividly described the anti-Jewish sentiment he endured at the hands of his comrades in the U.S. military intelligence corps, many of whom also openly despised the social policies of President Roosevelt (469–470). Such divisions were always present in U.S. society, but the Vietnam War and the racial discord of the 1960s brought them to American televisions, cinemas, and playhouses—for the world to see.

Citizen Soldiers: Conscripts and Volunteers

The concept of the citizen soldier, which has long implied a social contract between the government and its populace, rarely applied to all U.S. citizens; or, perhaps more accurately, it was unevenly applied across racial and gender lines at least until the early decades of the twenty-first century. To understand America's malleable and contestable concept of the citizen soldier, one must appreciate the various systems of conscription America has employed from colonial times to the present. Readers could do much worse than beginning with John Chambers (1987), who examines the origins and mechanisms of the draft as an institution in the United States. Stephen Ambrose (1997) popularized the American citizen soldier concept as it played out during the Second World War; his *Citizen Soldiers* is for the U.S. military what Tom Brokaw's *Greatest Generation* (1998) is for American society during the era of its "good war." Both were designed to appeal to the nostalgia market, and both succeeded in doing so, perhaps too well. Peter Kindsvatter (2003) provides a more grounded portrayal of the U.S. combat soldier from the First World War through the Vietnam conflict; he takes apart many of the lofty generalizations about why American citizen soldiers served, and he concludes that for most the reason was a complex combination of "comrades, cause, country and self" (285). For the Canadian army, Daniel Byers (2016) offers an anti-nostalgic account of those Canadian citizen soldiers who volunteered and those who were conscripted to serve in the Second World War. Byers' study reflects the Canadian theme of "two nations" (one English-speaking and one French-speaking) under one flag, and describes how that theme influenced Canadian defense policies, and when it failed to do so.

Michael Foley (2003) explores the several ways in which the draft was resisted during the Vietnam War. Foley aptly distinguishes between draft-dodgers, who left the United States to escape conscription, and draft resisters, who remained and actively demonstrated against the war, often accepting incarceration as a means of raising the profile of the antiwar movement. In contrast, Beth Bailey (2009) covers how the U.S. Army transitioned from conscript force to an all-volunteer one in the wake of Vietnam; it is the story not only of recovery and restoration but also of redefinition. The question remains, however, of whether America's understanding of citizen soldiers was redefined during this period, or whether it returned to its original meanings, or some combination of the two. Penny Lewis (2013) shows us the antiwar movement was not restricted to upper- and middle-class white society; blacks, Chicanos, and other members of the working class also participated, though their "voices" might not have been as well organized and thus not heard as loudly or as often as those of college students and professors with better access to publishing outlets.

Although the American working class also held strong antiwar sentiments, statistics show some 80 percent of U.S. service members came from working-class families. Christian Appy (1993) tells the story of that class and suggests those U.S. citizens who benefited most from the opportunities afforded by a comparatively prosperous American society were rarely its Vietnam-era citizen soldiers. Another challenge comes from Andrew Huebner (2008), who takes on the notion that the portrayal of the Vietnam-era citizen-soldier differed substantially from those of the Second World War and the Korean War; there may well have been fewer differences than the public has supposed. Despite the influence of the media and official propaganda, disenchantment and even cynicism toward war and wartime heroes may have been more pervasive since the mid-1940s than historians realized. These contributions cast doubt on the degree to which the separate concepts of citizen and soldier were ever actually intertwined in twentieth-century America.

Militarism, Antimilitarism, and Imperialism

Such views run counter to the arguments put forth by Andrew Bacevich (2005) and his many other books along the same line, which hold that America has become increasingly fascinated with its All-Volunteer Force. This fascination has led to an overuse of U.S. military power and the outright militarization of U.S. foreign policy. Can any society and its popular culture be both antimilitaristic and militaristic at the same time? Lisa Mundey (2012) answers in the affirmative, showing how depictions of the U.S. military in movies, television series, and comic strips created the intellectual space that made it possible for Americans to see other societies, especially those of Imperial Japan and Nazi Germany, as militaristic while at the same time overlooking the rise of military values and perspectives within their own society. Perhaps the answer is not simply one or the other, but rather that a dialectical or paradoxical oscillation occurs between them that is situationally contingent.

The meaning of militarism has evolved over time, but readers would do well to begin with Alfred Vagts' (1967) lengthy study, which defines militarism as privileging military institutions above civilian ones and allowing military ways of thinking to enter the "civilian sphere." The line between the civilian and military spheres has never been anything but gray and permeable, however. This is particularly true of American society, where people often used military service as a stepping stone toward a career in politics. Since Vagts' study, many nuanced as well as extremist interpretations of U.S. militarism have emerged. Examples of the former include "sentimental militarism" (Appy, 1993) and neo-militarism (Bacevich, 2005), while Stephen J. Whitfield (1996) represents the extremist claim that any non-negative reference to the U.S. military is evidence of militarism. In what we might call a form of reverse militarism in which military institutions copy or emulate civilian programs and values, Jennifer Mittelstadt (2015) sees the buildup of the soldier and family support programs that went into making volunteer service an attractive option for U.S. citizens as akin to creating a welfare state. More important than Mittelstadt's message that such a system is unfair and unwarranted is whether it has socially insulated the U.S. military from the populace it is sworn to protect.

Certainly, American culture—not unlike its European counterparts—has had its imperialist inclinations, as both Mary Renda (2001) and Max Boot (2002) attest. What Renda says about the powerful momentum of American ethnic and race-based paternalism and how it contributed to a culture of imperialism among leading policymakers, writers, and soldiers regarding Haiti can also be said of other U.S. interventions: Cuba, Puerto Rico, the Philippines, and elsewhere. Boot, on the other hand, does not see U.S. imperialism as negative, but rather as a natural outgrowth of America's erstwhile acceptance of its role as a global leader. In contrast, historians of the Canadian military have not grappled with whether Ottawa's foreign policies have been militarized, or whether its culture has unwittingly or too eagerly embraced military values. Mark Moss (2014) attempts to untangle the mixture of Victorian values and romanticized beliefs in the virtue of manliness that made up Canadian militarism in the decades before the First World War. An anthology coedited by Jody Berland and Blake Fitzpatrick (2010) traces the influence of militarism on the Canadian government's foreign and domestic policies; in the view of these editors, Canada's defense spending is unrealistically high compared to the reality of the threats their country actually faces—and the result is not unlike the style of militarism that Bacevich attributes to U.S. policymaking.

Race, Class, and Gender in Wartime

One glaring example of lost racial and cultural identities is the story of Native Americans. Robert M. Utley and Wilcomb E. Washburn (2002), Robert Wooster (1995), and Bill Yenne (2008) are just a few of many historians who have tried to arrive at objective portrayals of the U.S. government's wars against Native Americans during the late nineteenth century. Their efforts have led to a better appreciation of the diplomatic efforts, bravery, and stoicism of individual fighters on both sides, as well as the many myths and misunderstandings that emerged about those wars. Bruce Vandervort (2006) does the same for the Canadian side of the border. The question of genocide is explored by Alex Alvarez (2014), who carefully distinguishes between the intentional and unintentional annihilation of a group; he goes to some length describing the origins of the term "genocide" and its place in international law and the social sciences (26–31). Ultimately, readers must judge for themselves whether Alvarez's attempt to cast the question of Native American genocide as a complex matter succeeds, or comes uncomfortably close to an apology. What the study of Native Americans in U.S. society lacks is a more thorough treatment of how the various campaigns, battles, and massacres of the 1870s were viewed in urban areas in the east and northeast. What were the responses across the working and leisure classes, and how did they vary? As we know, images of Native Americans ranged from noble savages in romance novels to the posed photographs of "well-assimilated" young people at the Carlisle Indian School.

It is all but conventional wisdom to believe militaries both shape and are shaped by the societies they serve. But until the 1960s, the role of the U.S. military in putting down uprisings within the society it served had not been fully appreciated. Clayton D. Laurie and Ronald H. Cole (1997) correct that deficit for the period 1877 to 1945—that is, from the Great Railway Strike of 1877 to the plant seizures and race riots of the World War II era. A useful companion volume to this study is Jerry Cooper's (1980), which concentrates on the U.S. Army's interventions in the labor disputes of 1877–1900. Although Laurie and Cole's work is an official history, and thus tends to see military responses to civil disorders as reflective of the attitudes of the presiding civilian authorities at the time, it lays an impressive groundwork in terms of the sheer number of incidents it covers. To be sure, the U.S. military's interventions in such disorders never reached the magnitude of the mass bloodlettings that took place in many of Europe's modern counterrevolutions; yet incidents such as the Kent State shootings of 1970 cry out for more critical analyses of the role of America's citizen soldiers in stopping political and social change.

David Kennedy has produced several works covering the subject of mobilizing and motivating citizens more than soldiers; *Over Here* (1980) and *The American People in World War II* (1999) have set a high bar, describing how America's democracy was mobilized, both physically and psychologically, for each of its major wars. Jennifer Keene (2001) provides a more focused view of the politicization of U.S. "doughboys" and their experiences, and the impact of both on American attitudes toward war and military service. The First World War, as both Kennedy and Keene point out, raised the important question of whether a democratic government owed a debt to its citizen soldiers for their wartime sacrifices. For the United States, the answer was yes—but the scope of that debt and the ways in which it might be repaid, such as the GI Bill, have remained negotiable, and not always in favor of those to whom the debt is owed. Proper health care, for instance, has long been an issue for veterans who return from war with physical and emotional wounds.

Bernard Rostker (2013) casts the history of how the medical services have dealt with combat wounds and disease in a positive light—as a story of eventual if not always steady progress. Other scholarship, however, delves more deeply into how the casualties suffered in America's wars have become politicized, playing an influential role in domestic politics and presidential elections. In short, politicization is not limited to those citizen soldiers lucky enough to return home in one piece; it also includes those who were not so lucky, as well as the communities of friends and loved ones who must endure the loss of fathers, brothers, and—increasingly—mothers and sisters. Readers would do well to peruse Douglas Kriner and Francis Shen (2010), who address where and how wartime deaths are distributed across American society and the inequalities that result. Equally important is Stephen Casey (2014), who discusses how often American casualties have been obscured or outright manipulated—sometimes deflated, sometimes inflated—either to maintain popular support for a war or to undermine an opposing administration's strategies or policies. Contrary to popular belief, high casualties do not always weaken the American public's support for a war, but may in fact lead to the opposite by generating or regenerating anger or hostility. If war is too important to be left to the generals, in other words, it is surely too important to be left to political leaders as well. Kriner, Shen, and Casey underscore the growth of an activist element in American society, one determined to test the accuracy of official reports and to hold authorities accountable.

Such activism was not as evident in American society during the Second World War, which is still remembered as America's "good war." Indeed, this war is still the one all others are compared to, for better or worse. Although Morton Blum's *V Was for Victory* appeared more than forty years ago, it captures the mixture of opinions Americans had about the war; Blum gives us a sense of just how many "little Americas" made up the United States in the 1940s, and how each had to be appealed to by the presidency of Franklin Delano Roosevelt. Daniel Kryder (2000) examines Roosevelt's efforts with respect to race relations and war mobilization and demonstrates how important it was for the White House to retain the allegiance of African Americans, though only a small percentage of them were allowed to serve overseas. Of particular concern to FDR was the black press and how it represented the war to African Americans at home. Lee Finckle (1975) covers the history of the black press in the Second World War, whereas Patrick S. Washburn (1986) is more critical of the government's role. Nathan Brandt (1996) vividly describes the Harlem race riot of 1943 and its effects on Roosevelt's social policies. The picture of a racially variegated America gained further detail with Adriane Lentz-Smith (2009) and Chad L. Williams (2013), both of whom offer detailed examinations of race relations in the era of the Great War; ironically, while black Americans were being called to fight for democracy abroad, their struggle for full enfranchisement at home had yet to be won.

Neil A. Wynn (1993) considers the struggle for racial equality as it continued into the Second World War, as do the works of Jennifer Brooks (2004) and Christopher Parker (2009), both of whom deal with race relations in the Jim Crow South after the war. Kimberley Phillips (2005) takes the story from World War II to the campaign in Iraq, noting how the U.S. military's desegregation policies did not eliminate racial inequalities in military assessment and promotion, which persisted for decades. Bernard Nalty (1986) offers an overview of the black experience in America's militaries from 1775 to 1975, while Isaac Hampton (2013) focuses on African American officers from the Truman administration to the post-Vietnam era, paying particular attention to the tensions that emerged between the involvement of blacks in the antiwar movement and their growing representation and successful participation in the U.S. military. Two useful works on race relations in the Vietnam era are James Westheider (1999) and Herman Graham (2003); both show that definite racial divides existed, some subtle and some profound, between whites and blacks that sometimes affected morale and combat performance, despite the presumed bonding experience of war. One of Westheider's contributions is his distinction between perceived, personal, and institutional racism.

While much still remains to be learned about the black experience in the Vietnam conflict, even less is understood about the Chicano and Hispanic experience. Lorena Oropeza (2005) works to close that gap, but much more attention is needed here. A volume of essays edited by Maria Höhn and Seungsook Moon (2010) juxtaposes issues of race, sex, gender, and empire, and makes a much-needed contribution to our understanding of Asian-Caucasian relations. The same is true of Ji-yeon Yuh (2004), who relates the story of racial integration, or lack thereof, from the standpoint of many American military families.

Notions of gender and sexuality have long been conflated, and yet are powerfully persuasive. Kristin Hoganson (2000) attempts to show their coercive power in rhetoric and political debates that preceded the two armed conflicts that established America as a modern empire. How gender and sex coding was used to control female behavior in the Second World War is explored by Leisa Meyer (1998), Marilyn Hegarty (2008), and Meghan Winchell (2008). All grapple with the juxtaposition between the feminine role as nurturer and the masculine role of combatant. Heather Stur (2011) sidesteps Vietnam War literature and explores how gender norms were both reinforced and challenged during this tumultuous period. Kara Dixon Vuic (2009) traces the experiences of Army nurses in Vietnam, and captures gender roles in a state of flux in 1960s and 1970s America. Cynthia Toman (2008 and 2016) does the same for Canadian nurses during the Second World War and the Great War.

Of the many "masculine" images to emerge from America's experience in the Vietnam War, one of the most prevalent is that of an incompetent or duplicitous officer corps that sought only professional advancement. Ron Milam (2009) dashes that image to pieces by offering a counter-picture of a dedicated, albeit imperfect, U.S. officer corps struggling with the challenges of making war in a far-off land during an era of momentous cultural and social change. Perhaps no other war has provided as many negative images of the U.S. military as the Vietnam conflict. One of those—namely, that the post-Vietnam U.S. Army was all but crippled by drug abuse—has been corrected by Jeremy Kuzmarov (2009). Kuzmarov argues the prevalence of drug abuse was exaggerated by both the antiwar movement and conservative war hawks for their own political purposes; the former wanted to show the moral depravity of the war, and the latter tried to underscore the corrupting influence of the counterculture of the 1960s and 1970s. Demolishing still other myths, Brian M. Linn (2016) contends that "the US Army of the 1950s was the most diverse, representative, and in some ways egalitarian peacetime organization in the nation's history" (3).

Revisiting the American Way of War

Readers unfamiliar with the American way of war and its associated debates would do well to begin with the historiographical sections in Antulio J. Echevarria II (2014) and the historical survey by Matthew S. Muehlbauer and David J. Ulbrich (2018). Since America's loss in Vietnam and its near replication of that defeat in Iraq and Afghanistan in the opening decades of the twenty-first century, a great deal of literature has attempted to identify shortcomings in the American way of fighting. One of those purported shortcomings concerns the alleged American tendency to go into a conflict "log-heavy"—that is, with overwhelming logistical support, even to the point of turning U.S. forward-operating bases into "miniature Americas," replete with most of the comforts of home. Indeed, that tendency is one of the characteristics identified by Colin Gray (2006) as belonging to the American way of war. In Gray's view, the American way of war has been "logistically excellent" throughout most of its modern history. That observation is certainly supported by historical research, such as that by Meredith Lair (2011), who traces the American reliance on abundance in Vietnam to compensate for fighting against an irregular foe, for political objectives that were less than clear, and at a time when antiwar sentiment was steadily growing. Lair maintains this "log-heavy" approach proved counterproductive because it served to insulate U.S. soldiers from the people they were attempting to protect. Yet, Lair might well have explored the extent to which

American abundance acted like a magnet for many Vietnamese, and helped build bridges that spanned cultural differences. Nor is it true that this characteristic holds up to the test of history; plenty of U.S. operations were conducted with just barely enough logistical support.

Scholarly efforts to define an American way of war began, of course, with Russell Weigley's classic *The American Way of War* (1973). This magisterial work argued the American preference for annihilation strategies (which he misdefined) rendered it ill-suited for the kind of campaign it needed to wage in Vietnam. Successive interpretations, such as by Brian M. Linn (2002) and Echevarria (2014), have exposed the errors in Weigley's argument. Nonetheless, efforts to understand the American way of fighting and the reasons it seems to have come up short in two recent, major counterinsurgency campaigns have continued at a fast clip. Some, such as Adrian Lewis (2007), argued American "national culture" is to blame; in his view the United States prefers fighting traditional wars rather than "limited" ones. On the other hand, Benjamin Buley (2008) searched for explanations in American military culture. Others, such as Isaiah Wilson (2007), have examined U.S. civil-military relations. Still others, such as Reuben Brigety (2007) and Thomas Mahnken (2008), have tried to trace the American reliance on, or supposed love affair with, technology. These are not particularly American traits, however, especially given French advances in military technology from the late nineteenth century, with such state-of-the-art weaponry as the *chassepot* and *mitrailleuse*, to the mid-twentieth century, with that technological marvel called the Maginot Line. (Whether building the Maginot Line was the best strategic choice for France is another matter.)

Of the many contributions to the theme of the American way of war, two works deserve special mention. The first is Michael Lind's (2006), which suggests the American way of strategy reflects an historic balance, or dialectic, between the competing forces of the "legitimacy politics" of international liberalism versus the "power politics" of realism (23). Inconsistencies in American foreign policy and strategic performance are thus explained by the competing tensions created by upholding the values of self-determination and nonintervention, on the one hand, and addressing the threats posed by imperial powers or anarchy, on the other. In other words, America's way of war and its attendant way of strategy aptly reflect the political dynamics of American society. To perfect the former likely means harming the latter, perhaps irreparably.

The second is Hugh Rockoff's (2012), which offers a detailed analysis of the relationships between the American way of waging war and U.S. fiscal policies and economic realities. Rockoff shows how the fiscal costs of modern U.S. wars are either hidden or subtly transferred, but always come back to the public in some way. He also found a correlation between the rise of financial and human costs and how "cruel" the American prosecution of a war becomes (4). It is not likely that this correlation reveals a peculiarly American characteristic, however. Nonetheless, it does suggest certain characteristics may be more attributable to the types of wars being waged and their costs than to the alleged strategic cultures of the belligerents engaged. Paul A. C. Koistinen's *State of War* (2012), the final book of his five-volume *Political Economy of American Warfare*, shows how the military-industrial complex that President Dwight D. Eisenhower once warned about has become a reality—one that ultimately jeopardizes U.S. national security. Koistinen's conclusion that the "state of war" has become the American "state of mind" accords with the arguments of Andrew Bacevich regarding the militarization of U.S. foreign and domestic policies. Emulating the works of Rockoff and Koistinen, Rebecca Thorpe (2014) reveals how economic mobilization for the Second World War led to the establishment of a "permanent war economy"—a reflection of the military-industrial complex President Eisenhower warned about as he was leaving office (and ironically helped create)—that has since established economic dependencies in rural and semirural areas, and enabled numerous U.S. military interventions overseas. The military-industrial complex, readers ought to note, is something military professionals have frequently railed against because it means economic interests in the form of "big-ticket items"—as U.S. Army Lieutenant General H. R. McMaster remarked in April 2015—rather than true national security concerns or investments in the human side of conflict, are what drive defense expenditures (Altman, 2015). In other words, the U.S. armed forces are often stuck with expensive weapons and bases they neither want nor need, but are compelled to take to preclude economic downturns in individual congressional districts.

Studies of American strategic culture may provide grist for one's intellectual curiosity, but they ultimately assume too much. One example of assumptions taken too far is Stephanie Carvin and Michael Williams (2015), who assert

> Because the American mindset is one that demands total victory when confronted with war, and it suspends politics until after the war is complete, the country is willing to use whatever means possible to reach a victorious end state as quickly as possible.
>
> (86)

Yet, this assertion is patently false: many of America's small wars ended in negotiated settlements that were well short of total victory, and politics was never suspended during such conflicts or in America's major wars. If we think of a way of war as the search for historical patterns, then strategic culture is presumably the glue that holds such patterns together. However, as this essay has shown, America was and remains too deeply divided culturally, racially, and politically to have a single or uniform strategic culture. The quest to understand the American way of war and the associated search for a single American strategic culture became politicized shortly before 9/11, due in part to efforts to achieve "buy-in" for the U.S. Defense Department's attempt to realize a "revolution in military affairs," a thorough transformation based on information technologies. This politicization both helped and hurt the scholarship concerning how America has fought its wars. On the one hand, it stimulated greater interest in the American style of war, and encouraged research into underexplored areas. On the other hand, it prompted normative arguments that held that the American style of fighting is not what it should be: it ought to be more political, more strategic, less sensitive to casualties, and more open to technological change, for instance, while at the same time being less techno-centric (Echevarria, 2014).

Looking Ahead

Future generations of scholars studying war and society in North America after Reconstruction will surely add to the examination now underway regarding American and Canadian participation in the conflicts in Iraq and Afghanistan. Scholars will continue to study the major themes explored in this essay—the concept of the citizen-soldier, militarism and anti-militarism, race and gender, and the American way of war. Yet, what future scholars must do is build a broader base of comparison to add clarity to our understanding of North America's wars and societies. In other words, a common complaint about the American way of war is it lacks the cultural awareness and sensitivity needed to wage a "war amongst the people" successfully, as described by Sir Rupert Smith (2005, 3–4). If correct, it would be useful to determine to what extent such cultural ignorance is prevalent in other ways of war too.

The same holds true for such phenomena as the military-industrial complex. The United States has largely lost its industrial base; much of the manufacturing of war materiel—parts for aircraft and armored vehicles, antitank weapons, and so on—is done elsewhere. How has that affected the influence of the various components of the military-industrial complex? To what extent is there a military-industrial complex of sorts in Western Europe, or for that matter in Russia and China? And to what degree have the defense policies in those countries been affected by economic rather than security interests? (Indeed, to what extent are these interests separable?) Most European members of NATO, for instance, have not adhered to the agreed level of military spending in recent decades. Yet, the defense industry seems to thrive. Race, class, and gender studies would also benefit from greater cross-cultural or cross-national comparisons; are there some nations where integration proceeds with less friction, and why?

While we would surely expect to find many differences through such comparisons, the number of similarities we might discover could well surprise us.

Bibliography

Allen, Michael. *Until the Last Man Comes Home: POWs, MIAs, and the Unending Vietnam War.* Chapel Hill: University of North Carolina Press, 2009.

Altman, Howard. "General Dissects US Approach to War in Speech at University of South Florida." *Tampa Bay Times*, April 9, 2015; www.tbo.com/list/military-news/general-dissects-us-approach-to-war-in-speech-at-usf-20150408/

Alvarez, Alex. *Native America and the Question of Genocide.* Lanham, MD: Roman & Littlefield, 2014.

Ambrose, Stephen. *Citizen Soldiers: The U.S. Army From the Normandy Beaches to the Bulge to the Surrender of Germany,* June 7, 1944-May 7, 1945. 1997.

Appy, Christian. *Working-Class War: American Combat Soldiers and Vietnam.* Chapel Hill: University of North Carolina Press, 1993.

Bacevich, Andrew. *The New American Militarism: How Americans Are Seduced by War.* New York: Oxford University Press, 2005.

Backhouse, Constance. *Colour-Coded: A Legal History of Racism in Canada, 1900–1950.* Toronto: University of Toronto, 1997.

Bailey, Beth. *America's Army: Making the All-Volunteer Force.* Cambridge, MA: Belknap Press of Harvard University Press, 2009.

Barbeau, Arthur. *The Unknown Soldiers: African American Soldiers in World War I.* Philadelphia: Temple University Press, 1974.

Berland, Jody, and Blake Fitzpatrick, eds. *Cultures of Militarization.* Cape Breton: Cape Breton University Press, 2010.

Blum, John M. *V Was for Victory: Politics and American Culture During World War II.* New York: Harcourt Brace Jovanovich, 1976.

Boot, Max. *Savage Wars of Peace: Small Wars and the Rise of American Power.* New York: Basic Books, 2002.

Brandt, Nathan. *Harlem at War: The Black Experience in World War II.* Syracuse: Syracuse University Press, 1996.

Brigety II, Reuben E. *Ethics, Technology and the American Way of War: Cruise Missiles and US Security Policy.* London: Routledge, 2007.

Brokaw, Tom. *The Greatest Generation.* New York: Random House, 1998.

Brooks, Jennifer E. *Defining the Peace: World War II Veterans, Race, and the Remaking of Southern Political Tradition.* Chapel Hill: University of North Carolina Press, 2004.

Buley, Benjamin. *The New American Way of War: Military Culture and the Political Utility of Force.* London: Routledge, 2008.

Byers, Daniel. *Zombie Army: The Canadian Army and Conscription in the Second World War.* Toronto: UBC Press, 2016.

Carter, James. *Inventing Vietnam: The United States and State-Building, 1954–1968.* New York: Cambridge University Press, 2008.

Carvin, Stephanie, and Michael Williams. *Law, Science, Liberalism and the American Way of Warfare: The Quest for Humanity in Conflict.* Cambridge: Cambridge University Press, 2015.

Casey, Stephen. *When Soldiers Fall: How Americans Have Confronted Combat Losses From World War I to Afghanistan.* New York: Oxford University Press, 2014.

Chambers, John W. *To Raise an Army: The Draft Comes to Modern America.* New York: Free Press, 1987.

Cooper, Jerry M. *The Army and Civil Disorder: Federal Military Interventions in Labor Disputes, 1877–1900.* Westport: Greenwood, 1980.

Dalfiume, Richard M. *Desegregation of the Armed Forces: Fighting on Two Fronts, 1939–1953*. Columbia: University of Missouri Press, 1969.

Echevarria II, Antulio J. *Reconsidering the American Way of War: US Military Practice From the Revolution to Afghanistan*. Washington, DC: Georgetown University Press, 2014.

Erenberg, Lewis A., and Susan E. Hirsch, eds. *The War in American Culture: Society and Consciousness During World War II*. Chicago: University of Chicago Press, 1996.

Finkle, Lee. *Forum for Protest: The Black Press During World War II*. Cranbury, NJ: Associated University Presses, 1975.

Foley, Michael. *Confronting the War Machine: Draft Resistance During the Vietnam War*. Chapel Hill: University of North Carolina Press, 2003.

Fujitani, T. *Race for Empire: Koreans as Japanese and Japanese as Americans During World War II*. Berkeley: University of California Press, 2011.

Gaddis, John. *Surprise, Security, and the American Experience*. Cambridge, MA: Harvard University Press, 2005.

———. *We Now Know: Rethinking Cold War History*. New York: Oxford University Press, 1997.

Goedde, Petra. *GIs and Germans: Culture, Gender, and Foreign Relations, 1945–1949*. New Haven: Yale University Press, 2003.

Graham, Herman. *The Brothers' Vietnam War: Black Power, Manhood, and the Military Experience*. Gainesville: University Press of Florida, 2003.

Gray, Colin. *Irregular Enemies and the Essence of Strategy: Can the American Way of War Adapt?* Carlisle Barracks, PA: Strategic Studies Institute, 2006.

Hagen, Kenneth J. *This People's Navy: The Making of American Sea Power*. New York: Free Press, 1992.

Hampton, Isaac. *The Black Officer Corps: A History of Black Military Advancements From Integration Through Vietnam*. New York: Routledge, 2013.

Hegarty, Marilyn. *Victory Girls, Khaki-Wackies, and Patriotutes: The Regulation of Female Sexuality During World War II*. New York: New York University Press, 2008.

Hoganson, Kristin. *Fighting for the American Manhood: How Gender Politics Provoked the Spanish-American and Philippine-American Wars*. New Haven: Yale University Press, 2000.

Höhn, Maria. *GIs and Fräuleins: The German-American Encounter in 1950s West Germany*. Chapel Hill: University of North Carolina Press, 2002.

Höhn, Maria, and Seungsook Moon. *Over There: Living With the U.S. Military Empire From World War Two to the Present*. Durham, NC: Duke University Press, 2010.

Howard, Michael. "Total War in the Twentieth century: Participation and Consensus in the Second World War." In *War and Society*. Eds. Brian Bond and Ian Roy. New York: Holmes & Meier, 1975.

Huebner, Andrew. *The Warrior Image: Soldiers in American Culture From the Second World War to the Vietnam War Era*. Chapel Hill: University of North Carolina Press, 2008.

Jacobs, Seth. *America's Miracle Man in Vietnam: Ngo Dinh Diem, Religion, Race, and U.S. Intervention in Southeast Asia*. Durham, NC: Duke University Press, 2004.

Jeffords, Susan. *The Remasculinization of America: Gender and the Vietnam War*. Bloomington: Indiana University Press, 1989.

Keene, Jennifer. *Doughboys, the Great War, and the Shaping of Modern America*. Baltimore: Johns Hopkins University Press, 2001.

Kennedy, David. *The American People in World War II: Freedom From Fear, Part II*. New York: Oxford University Press, 1999.

———. *Over Here: The First World War and American Society*. New York: Oxford University Press, 1980.

Kindsvatter, Peter S. *American Soldiers: Ground Combat in the World Wars, Korea, and Vietnam*. Lawrence: University Press of Kansas, 2003.

Koistinen, Paul A.C. *State of War: The Political Economy of American Warfare, 1945–2011*. Lawrence: University Press of Kansas, 2012.

Kriner, Douglas, and Francis Shen. *The Casualty Gap: The Causes and Consequences of American Wartime Casualties*. New York: Oxford University Press, 2010.

Kruse, Kevin, and Stephen Tuck, eds. *Fog of War: The Second World War and the Civil Rights Movement*. New York: Oxford University Press, 2012.

Kryder, Daniel. *Divided Arsenal: Race and the American State During World War II*. New York: Cambridge University Press, 2000.

Kuzmarov, Jeremy. *The Myth of the Addicted Army: Vietnam and the Modern War on Drugs.* Amherst: University of Massachusetts Press, 2009.

Lair, Meredith H. *Armed With Abundance: Consumerism & Soldiering in the Vietnam War.* Chapel Hill: University of North Carolina, 2011.

Laurie, Clayton D., and Ronald H. Cole. *The Role of Federal Military Forces in Domestic Disorders, 1877–1945.* Washington, DC: U.S. Government Printing Office, 1997.

Lee, Ulysses. *The Employment of Negro Troops.* Special Studies: The United States Army in World War II. 1946. Reprint. Washington: The Center for Military History, 1963.

Lentz-Smith, Adriane. *Freedom Struggles: African Americans and World War I.* Cambridge, MA: Harvard University Press, 2009.

Lewis, Adrian. *The American Culture of War: The History of U.S. Military Force From World War II to Operation Iraqi Freedom.* New York: Routledge, 2007.

Lewis, Penny. *Hard Hats, Hippies, and Hawks: The Vietnam Antiwar Movement as Myth and Memory.* Ithaca: ILR Press of Cornell University Press, 2013.

Lind, Michael. *The American Way of Strategy: US Foreign Policy and the American Way of Life.* Oxford: Oxford University, 2006.

Linderman, Gerald. *The Mirror of War: American Society in the Spanish American War.* Ann Arbor: University of Michigan Press, 1974.

Linn, Brian M. "The American Way of War Revisited," *Journal of Military History* 66, no. 2 (April 2002): 501–530.

———. *The Echo of Battle: The Army's Way of War.* Cambridge: Harvard University Press, 2007.

———. *Elvis' Army: Cold War GI's and the Atomic Battlefield.* Cambridge, MA: Harvard University Press, 2016.

Logevall, Frederick. *Choosing War: The Lost Chance for Peace and the Escalation of the War in Vietnam.* Berkeley: University of California Press, 2001.

———. *Embers of War: The Fall of an Empire and the Making of America's Vietnam.* New York: Random House, 2013.

Lotchin, Roger W. *The Bad City in the Good War: San Francisco, Los Angeles, Oakland, and San Diego.* Bloomington: Indiana University Press, 2003.

MacPherson, Myra. *Long Time Passing: Vietnam and the Haunted Generation.* New York: Doubleday, 1984.

Madsen, Chris. *Another Kind of Justice: Canadian Military Law From Confederation to Somalia.* Vancouver: UBC Press, 1999.

Mahnken, Thomas. *Technology and the American Way of War Since 1945.* New York: Columbia, 2008.

Maraniss, David. *They Marched Into Sunlight: War and Peace Vietnam and America October 1967.* New York: Simon & Schuster, 2004.

Martini, Edwin. *Invisible Enemies: The American War on Vietnam, 1975–2000.* Amherst: University of Massachusetts Press, 2007.

Martini, Edwin, and Scott Laderman. *Four Decades On: Vietnam, the United States, and the Legacies of the Second Indochina War.* Durham, NC: Duke University Press, 2013.

McMaster, H. R. *Dereliction of Duty: Lyndon Johnson, Robert McNamara, the Joint Chiefs of Staff, and the Lies That Led to Vietnam.* New York: Harper Perennial, 1998.

Meyer, Leisa. *Creating GI Jane: Sexuality and Power in the Women's Army Corps During World War II.* New York: Columbia University Press, 1998.

Milam, Ron. *Not a Gentleman's War: An Inside View of Junior Officers in the Vietnam War.* Chapel Hill: University of North Carolina Press, 2009.

Mittelstadt, Jennifer. *The Rise of the Military Welfare State.* Cambridge: Harvard University Press, 2015.

Moon, Katherine H. S. *Sex Among Allies: Military Prostitution in U.S.-Korea Relations.* New York: Columbia University Press, 1997.

Morton, Desmond. *A Military History of Canada.* Toronto: MacClelland & Stewart, 1985.

Moss, Mark H. *Manliness and Militarism: Educating Young Boys in Ontario for War.* Toronto: University of Toronto Press, 2014.

Muehlbauer, Matthew S., and David J. Ulbrich. *Ways of War: American Military History From the Colonial Era to the Twenty-First Century.* 2nd ed. London: Routledge, 2018.

Mundey, Lisa. *American Militarism and Antimilitarism in Popular Media, 1945–1970.* New York: MacFarland, 2012.

Nalty, Bernard C. *Strength for the Fight: A History of Black Americans in the Military*. New York: Free Press, 1986.

Oropeza, Lorena. *Raza Si! Guerra No! Chicano Protest and Patriotism During the Viet Nam War Era*. Berkeley: University of California Press, 2005.

Ortiz, Stephen. *Beyond the Bonus March and GI Bill: How Veteran Politics Helped Shape the New Deal Era*. New York: New York University Press, 2012.

Parker, Christopher S. *Fighting for Democracy: Black Veterans and the Struggle Against White Supremacy in the Postwar South*. Princeton: Princeton University Press, 2009.

Phillips, Kimberley L. *War! What Is It Good For? Black Freedom Struggles and the U.S. Military From World War II to Iraq*. Chapel Hill: University of North Carolina Press, 2005.

Renda, Mary. *Taking Haiti: Military Occupation and the Culture of U.S. Imperialism, 1915–1940*. Chapel Hill: University of North Carolina Press, 2001.

Robertson, Linda. *The Dream of Civilized Warfare: World War I Flying Aces and the American Imagination*. St. Paul: University of Minnesota Press, 2005.

Rockoff, Hugh. *America's Economic Way of War: War and the US Economy From the Spanish-American War to the Persian Gulf War*. Cambridge: Cambridge University, 2012.

Rostker, Bernard. *Providing for the Casualties of War: The American Experience Through World War II*. Santa Monica, CA: RAND, 2013.

Sitkoff, Harvard. "African American Militancy in the World War II South: Another Perspective." In *Remaking Dixie: The Impact of World War II on the American South*, edited by Neil R. McMillen. Jackson: University Press of Mississippi, 1997.

Slotkin, Richard. *Gunfighter Nation: The Myth of the Frontier in Twentieth-Century America*. Norman: University of Oklahoma Press, 1998.

Smith, Rupert. *The Utility of Force: The Art of War in the Modern World*. London: Allen Lane, 2005.

Stur, Heather. *Beyond Combat: Women and Gender in the Vietnam War Era*. New York: Cambridge University Press, 2011.

Summerfield, Penny. *Reconstructing Women's Wartime Lives*. New York: Manchester University Press, 1998.

Terkel, Studs. *"The Good War": An Oral History of World War Two*. New York: Pantheon, 1984.

Thorpe, Rebecca U. *The American Warfare State: The Domestic Politics of Military Spending*. Chicago: University of Chicago Press, 2014.

Toman, Cynthia. *An Officer and a Lady: Canadian Military Nursing and the Second World War*. Toronto: UBC Press, 2008.

———. *Sister Soldiers of the Great War: The Nurses of the Canadian Army Medical Corps*. Toronto: UBC Press, 2016.

Utley, Robert M., and Wilcomb E. Washburn. *Indian Wars*. New York: American Heritage Press, 2002.

Vagts, Alfred. *A History of Militarism*. 2nd ed. New York: Free Press, 1967.

Vandervort, Bruce. *Indian Wars of Canada, Mexico, and the United States*. New York: Routledge, 2006.

Varon, Jeremy. *Bringing the War Home: The Weather Underground, the Red Army Faction, and Revolutionary Violence in the Sixties and Seventies*. Berkeley: University of California Press, 2004.

Vuic, Kara Dixon. *Officer, Nurse, Woman: The Army Nurse Corps in the Vietnam War*. Baltimore, MD: Johns Hopkins University Press, 2009.

Washburn, Patrick S. *A Question of Sedition: The Federal Government's Investigation of the Black Press During World War II*. New York: Oxford University Press, 1986.

Weigley, Russell F. *The American Way of War: A History of United States Military Strategy and Policy*. Bloomington: Indiana University Press, 1973.

Westheider, James. *Fighting on Two Fronts: African Americans and the Vietnam War*. New York: New York University Press, 1999.

Whitfield, Stephen J. *The Culture of the Cold War*. Baltimore: Johns Hopkins University Press, 1996.

Williams, Chad L. *Torchbearers of Democracy: African American Soldiers in the World War I Era*. Chapel Hill: University of North Carolina Press, 2013.

Wilson, Isaiah. *Thinking Beyond War: Civil-Military Relations and Why America Fails to Win the Peace*. New York: Palgrave Macmillan, 2007.

Winchell, Meghan K. *Good Girls, Good Food, Good Fun: The Story of USO Hostesses During World War II*. Chapel Hill: University of North Carolina Press, 2008.

Wooster, Robert. *The Military and United States Indian Policy, 1865–1903*. Lincoln: University Press of Nebraska, 1995.

Wu, Judy Tzu-yuan. *Radicals on the Road: Internationalism, Orientalism, and Feminism During the Vietnam Era.* Ithaca, NY: Cornell University Press, 2013.

Wynn, Neil A. *The Afro-American and the Second World War.* Revised ed. New York: Holmes and Meier, 1993.

Yenne, Bill. *Indian Wars: The Campaign for the American West.* Yardley, PA: Westholme, 2008.

Yuh, Ji-yeon. *Beyond the Shadow of Camptown: Korean Military Brides in America.* New York: New York University Press, 2004.

Zeiger, Susan. *Entangling Alliances: Foreign War Brides and American Soldiers in the Twentieth Century.* New York: New York University Press, 2010.

Part II

Thematic Approaches

14 Media and War

Jay Lockenour

Introduction

Media is an unwieldy term, in that it can refer to virtually any form of communication from any era. Our focus is sharpened only slightly by relating "media" to "war," an almost equally all-encompassing term that is used to describe conflicts small (Falklands War), large (Second World War), and even metaphorical (War on Poverty).

The relationship between media and war is an old one. Julius Caesar's commentaries on the Gallic Wars have been described as "political instruments" operating to elevate his friends, denigrate his enemies, and, most importantly, win favor among Roman citizens (Welch, Powell, et al., 2009; Taylor, 2003). The chroniclers of the Middle Ages in Europe wrote for courtly audiences, recording, often embellishing, the great military feats of kings, noblemen, and their ancestors, thus establishing their legitimacy and authority (Curry, 2015). Peter Paret (1997) has shown that visual media such as oil paintings and popular prints told stories about war that reflected some of the reality of battle, but also spoke of, reflected, and sometimes transformed the political and social realities of the day. In these works of art from the sixteenth to the nineteenth century one sees reflected not only the great deeds of kings and commanders but also the suffering of the common soldier and the victimization of civilians that speak so powerfully against war.

Most scholarly studies of media and war focus on the modern period, when rising literacy and the growing availability of print, as well as the inventions of radio, film, television, and now the Internet, allowed state and non-state actors, proponents or opponents of war alike, to communicate with a mass audience (Carruthers, 1999; Messinger, 2011; Seethaler, 2013). Not coincidentally, those technologies arose in an era of "mass" warfare, with larger and larger populations being mobilized not only to fight but also to support war in other ways. Using mass media, states urged their citizens to enlist, buy bonds, conserve resources, redouble their efforts in the factory or the laboratory, and preserve the political will to fight. By the same token, the enemy sought to undermine that will and with increasing violence attacked the nonmilitary sources of military power.

It has long been understood that wars are fought not only on the battlefield. Even before the "total wars" of the twentieth century, military commanders, politicians, and educated publics perceived that economic, intellectual, and social forces affected the origins, conduct, and outcome of war, and therefore those fields of endeavor could become contested arenas. In limited wars, the effort to preserve political will is arguably even more important, since the stakes of the conflict (for the side seeing the war as limited at least) are not as existential nor as obvious. It did not take the United States' war in Vietnam to make it obvious that the "hearts and minds" of one's own population, as well as the enemy's, were important targets in war.

It is vital to understand the role that media plays in telling the story of wars (whether impending, ongoing, or long past) in a way those populations can understand. Media help to "emplot" the events of war, give them narrative shape, and influence the understanding not only of participants but also of future generations and historians. The nature of that media changes over time and can include oral traditions, chronicles and official histories, newspapers, film, and today's "social media." Whether in print, on film, over the airwaves, or on the Internet, parties to war attempt to tell the story of a conflict in a way that, while perhaps not decisive for the outcome of the war, nevertheless gives that conflict meaning. Authors of these media messages strive for "authenticity" to heighten the impact and lend durability to their "war story." Victory and defeat are of course dependent on real power—battlefield results, diplomatic skill, economic might, and other factors. But the full impact of those realities is magnified, sometimes distorted, and sometimes undermined by their translation onto paper and screen.

The French Revolution unleashed the popular forces of "mass warfare." The revolutionary state's proclamation of the "levée en masse" in August 1793 created enormous, motivated bodies of soldiers that for over a decade brought France victory after victory and propelled Napoleon Bonaparte first to fame and eventually to the throne of a new French Empire. Though Napoleon controlled no truly mass media, he communicated regularly and directly to the people of France through official communiques, and Napoleon's image came to symbolize the might and authority of the French state and its armies (Holtman, 1941; Hanley, 2005).

The American Civil War is often credited with giving birth to modern journalism in the United States, with reporters aided, as British reporters had been in Crimea, by expansion of the telegraph network (Knightley, 1975). Technological advances in the 1880s also allowed newspapers to print in larger quantities at lower cost, making the press a truly mass medium. In the limited wars of the second half of the nineteenth century (addressed ahead), the public's thirst for news and the state's urge to control information grew and came into conflict.

Mass media was crucial to the conduct, if not necessarily the outcome, of the First World War that erupted in Europe in August 1914—arguably the first truly "total war." In Europe, newspapers immediately became dedicated almost exclusively to war news, reprinting government communiques as well as generating a great deal of (often equally biased) original reporting (Messinger, 1992). Nations viewed media as a powerful tool to influence not only their own but also enemy populations (Sanders and Taylor, 1982). Britain and Germany waged a relentless war for American opinion, in which the British always seemed to have the upper hand, thanks to both ham-fisted German tactics and genuine German atrocities, like the burning of the University of Leuven or the sinking of the *Lusitania* (Marquis, 1978; Welch, 2000). Opinion in other neutral nations mattered as well, and of course all nations directed propaganda at their own populations to boost morale.

A nascent film industry thrilled audiences during the First World War with moving images (often reconstructed) of battlefield events (Paris, 2000). So important did the German General Staff consider film propaganda that it fostered the creation of the Universum Film company (UFA), which dominated German cinema until 1945 (Kreimeier, 1996). Film became an important means not only of conveying information through newsreels but also of shaping the memory of the war in the decades that followed. The American film industry in particular threw itself into the war, offering entertaining films for the troops as well as selling war bonds at its domestic screenings (Isenberg, 1981; DeBauche, 1997; DeBauche, 2000).

So intense was the media war between 1914 and 1918 that scholarly study of propaganda, the effort to sway opinion through media, has its origins in the 1920s, which saw the publication of important works such as Walter Lippmann's *Public Opinion* (1922) and Harold Lasswell's *Propaganda Technique in the World War* (1927). It was Lasswell who pointed out the lack of subtlety in Germany's media policy in the First World War, directed as it was primarily by the military. British efforts, on the other hand, benefited from the social bonds, usually dating back to attendance at public school, of Britain's military and media elite. School ties gradually overcame military concerns over secrecy and ensured that British propaganda campaigns could successfully harness civilian creativity. Gradually in the 1920s, the sense arose that propaganda had gone too far in the First World War, poisoning the well of peace, which in some ways may have hamstrung more energetic reactions to German atrocities in the Second World War. After 1939, many Americans treated British efforts to draw the United States into the war warily, with the experience of 1914–1917 still relatively fresh in the national memory (Cull, 1995).

Radio came into its own during the 1930s and became a powerful force during the Second World War. The British Broadcasting Company (BBC) broadcast Winston Churchill's inspiring appeals to an anxious British public, which had an impact even when the original speech (often delivered in the House of Commons) was merely reread by an announcer (Schlesinger, 1987; Marquis, 1984; Briggs, 1985). Franklin Delano Roosevelt's "Fireside Chats," which had begun many years prior to the United States entering the war, carried important messages of reassurance and optimism on the airwaves to the American public. The Axis powers famously sought to undermine Allied morale with broadcasts in English by expatriates, such as "Lord Haw-Haw" and "Tokyo Rose" (Bergmeier and Lotz, 1997).

Cinema, including newsreels, came into its own in virtually all the combatant nations during the Second World War. Scores of books have been written about "Hollywood's World War," highlighting the important role that feature films played in boosting domestic morale, highlighting the threat to American democracy posed by the Axis dictatorships, and raising money through war bond drives hosted by popular stars (Doherty, 1993; Rollins and O'Connor, 1997; Shindler, 1979). The British film industry supported the war effort (Aldgate and Richards, 1986; Taylor and Museum, 1988), as did the Russian (Youngblood, 2007).

It is difficult to imagine an analysis of "media and war" that omits discussion of Nazi propaganda and Joseph Goebbels. The Reich minister for public enlightenment and propaganda for Adolf Hitler's National Socialist Germany, Goebbels is usually described as the master propagandist, whose understanding of the art and whose total control of German media enabled the Nazi state to manipulate a population of millions and drive them toward a war of aggression and genocide. Even if Nazi Germany ultimately lost the Second World War, Goebbels' propaganda campaigns remain for some "the war that Hitler won" (Herzstein, 1978).

That understanding is fraught with oversimplification, of course. We now have a more nuanced understanding of the nature of the National Socialist dictatorship, which was not as ironclad as is usually depicted in popular representation. The ability of Hitler or Goebbels to enact policy was powerfully shaped by a polycratic, even chaotic, system of confused and conflicting authorities, powerful vested interests, and corrupt, incompetent administration. Our awareness of the German people's combination of resistance and complicity, enthusiasm and fear has been heightened by the historiography of the last three or four decades such that historians no longer take for granted the Nazi state's ability (or need) to manipulate millions (Welch, 1993; Kershaw, 1987).

Nevertheless, Goebbels' use of media, and newsreels in particular, to tell the story of the Second World War is worth examination. Newsreels were the most important medium for communicating war news to the German people, and as such were critical to the narrative shape that the war took, especially during its early years. By 1938 four nominally private, independent production companies had been brought under centralized control, and Goebbels carefully vetted newsreels prior to release (Welch, 1983). These newsreels "emplotted" and justified German actions in Czechoslovakia and Poland, and touted the sturdiness of Germany's defenses in the West and the lack of preparation on the part of the French and British for war (Kallis, 2005). German newsreels reached the peak of effectiveness in 1941 and 1942. Skillful editing and the often grainy footage, unsteady camera work, and sometimes chaotic action that came from working under battle conditions seemed to place the viewer in the middle of combat at times, heightening the sense of realism so important to effective propaganda (Isenberg, 1981).

The 1942–1943 Battle of Stalingrad proved a turning point in the struggle for command of the war's "master narrative." Goebbels attempted to put a brave face on the defeat in his famous "total war" address at Berlin's Sportpalast on February 18, 1943. But the propaganda war was lost. Newsreel production and popularity declined. Early in the war, police reports on film attendance seemed to indicate that the newsreels were in fact the major audience draw. After 1943, Goebbels had to order the doors of the theater locked during the period between the newsreels and the feature, so that moviegoers could not sneak into the theater to see only the feature. Likewise Hitler's popularity, which had been the bulwark of Nazi power, began to decline after Stalingrad (Kershaw, 1987).

Especially early in the war, when his desire to glorify both Hitler and the feats of German arms corresponded to the reality of the *Wehrmacht's* successes against its enemies, Goebbels and other propagandists were able to produce effective propaganda, telling a story that excluded those being told by others from the media battlefield. Content (German superiority) corresponded to context (German victories) in such a way that contrary information or countervailing opinions, whether from domestic dissenters or foreign enemies, could gain no purchase (Kallis, 2005). By the end of the war, however, the bombing of German cities, the advance of the Soviet Army, and the mounting German casualties created a dissonance that undermined Goebbels' message. The realities of the battlefield made Goebbels' message inauthentic and less effective. Goebbels' suicide and Germany's surrender silenced the messenger but did not destroy the media, which survives in the miles of footage cataloged by Germany's military propaganda machine and used by today's documentarians.

Nor were the Germans the only ones manipulating information to their advantage. U.S. officials were especially concerned to control imagery of American casualties, for example. Tellingly, however, late in the war, when government officials were concerned that domestic commitment to the war might be lagging, they released more gruesome photos to the press to make clear the kinds of sacrifices that were necessary for victory (Roeder, 1993).

Allied propaganda enjoyed the advantage, at least after 1942, of being able to tell more of the truth on most fronts, reducing the dissonance faced by Goebbels or Japanese officials, who were forced to explain retreats or the presence of Allied bombers overhead while still proclaiming victory. In "total war" especially, real power (or the lack thereof) cannot simply be disguised with catchy slogans or carefully managed media. In the case of the Second World War, the "unconditional surrender" and military occupation of the Axis nations powerfully reinforced the victorious side's claim that democracy was good and dictatorship bad. How Germany, Italy, and Japan all became democracies after the war is of course a complicated question. The Americans in particular made great efforts to use media to "denazify" Germany and to reorient Japanese politics and society toward democracy (Fehrenbach, 1995; Smith, 1996; Tent, 1982; Dower, 1996; Dower, 1999).

Media is not only relevant in "total war," however. In the "limited wars" of the twentieth century the control or sympathy of the media has often been seen as critical to the successful prosecution of a war whose outcome (for one side at least) does not threaten national existence. In these situations, scholars and participants alike are more likely to posit a significant role for the media, as in Vietnam, which was dubbed "the living room war" while it was still being fought because of the allegedly profound impact of television coverage on American attitudes to the war (Arlen, 1969).

Long before Vietnam, media coverage of war influenced war itself. By 1890, a powerful, if informal, constituency had developed in the United States known as "jingoes" for their overweening, masculine, and belligerent patriotism. To this group belonged influential politicians, such as the future president Theodore Roosevelt and Senator Henry Cabot Lodge, the business magnate John Jacob Astor IV, and others (Thomas, 2010). A new sort of newspaper, the "yellow press" of editorial magnates, such as Joseph Pulitzer and William Randolph Hearst, fanned the imperial ambitions of the American public and America's interest in Cuba particularly.

Beginning in the 1860s, a series of independence movements challenged Spanish rule in Cuba. In 1895, José Marti and Máximo Gómez led a guerrilla uprising that was not particularly successful except in promoting harsh Spanish reprisals. In an effort to deny the rebels the support of the rural population on which they depended, Spain resettled hundreds of thousands of Cubans into camps where hundreds of thousands died of disease and malnutrition. The terrible conditions provoked international condemnation of Spain and provided newspapers, such as Pulitzer's *New York World* and Hearst's *New York Journal*, with ample material for this new brand of sensational news coverage (Thomas, 2010; Miller, 2011). Some historians have gone so far as to suggest that Hearst's coverage helped to drag a reluctant McKinley administration into war. Especially after the First World War, when the pernicious influence of propaganda was deemed by many to have impeded peace, such arguments found purchase (Wilkerson, 1932).

Two stories in particular epitomize Hearst's eagerness to parlay stories of Spanish barbarism and Cuban suffering into higher sales even at the risk of provoking war hysteria. The first involves a seventeen-year-old Cuban woman, Evangelina Cisneros, imprisoned by the Spanish for conspiring with the rebel forces. Hearst's paper splashed images of a demure young woman being strip-searched by leering Spanish authorities. The *Journal* spread unfounded rumors that she had been sentenced to a penal colony in Morocco and that she was subjected to torture, and predicted her death within the year. Hearst, in keeping with his newspaper's motto ("The Journalism That Acts"), himself sponsored a mission to free Cisneros and bring her to the United States. The daring rescue received extensive, imaginative coverage in the *Journal*. After her rescue, Hearst's paper crowed, "We have freed one Cuban girl—when shall we free Cuba?" referring to the "*Cuba libre*" movement to gain independence for the island (Hoganson, 1998).

Likewise the unexplained destruction of the American armored cruiser USS *Maine* in the harbor at Havana occupied the front page of Hearst's paper for over a week following the incident. Just two days after the explosion, Hearst's *Journal* proclaimed in a headline that the "Destruction of the war ship *Maine* was the work of an enemy," discounting the possibility of an accident (Miller, 2011, 58). Hearst was among the first to assert (without evidence) that a Spanish mine destroyed the vessel, and blaring headlines called for the U.S. government to avenge the more than 250 sailors killed when the *Maine* sank. Such "yellow press" coverage created an atmosphere in which President McKinley and House Speaker Thomas Reed found it increasingly difficult to negotiate a peaceful resolution of the crisis.

It is certain that other factors, such as economic interest and imperial ambition, and other actors, including Henry Cabot Lodge and Theodore Roosevelt, not to mention President McKinley, were decisive in pushing the United States over the brink into war with Spain in April 1898. But it was the ability of Hearst and others to use media to shape Americans' understanding of the conflict in Cuba and to tell the stories of Cisneros and the *Maine* incident in ways that provoked national outrage that made the war seem like the inevitable, even worthwhile, outcome of recent events. "Remember the *Maine*" became the rallying cry of pro-war "jingoes," and much of the American press was only too happy to preserve that memory on the front page.

The Cold War between the United States and Soviet Union after 1945 was also fought on a media battlefield. While the stakes in the Cold War hardly seem "limited," given the constant threat of nuclear annihilation, the conflict manifested itself in a series of limited and proxy wars in Korea, Vietnam, Afghanistan, and elsewhere. The role of the media has garnered particular attention in the Vietnam War. American officials of the era often saw the growing antiwar movement and the increasing public opposition to the war as a product of a hostile and unfettered press. By failing to manage press coverage through more polished briefings (Vietnam-era press conferences were dismissed as the "Five O'Clock Follies") or to control press access to battle zones more carefully, American officials believed they had allowed reporters too much leeway to tell their own story, to "emplot" the war in a way unfavorable to military goals.

Scholars are more skeptical. One of the most influential scholarly treatments, by Daniel Hallin (1986), suggests that the American press was remarkably sympathetic and sometimes even supportive in its early coverage of U.S. involvement in Southeast Asia. Rather, Hallin argues, the press turned hostile only after large sections of the American public began to oppose the war and after elite opinion also became divided in its assessment of the American mission in Vietnam.

Whether it worked the way U.S. officials feared or not, the North Vietnamese clearly hoped to use the media to influence American opinion. Their "Tet Offensive" at the beginning of 1968 envisioned attacks in Saigon, and on the U.S. embassy in particular, precisely because of the relatively dense population of American news crews in that city. Though the offensive can be considered to have failed in a military sense, the graphic images of fighting in Saigon and elsewhere seemed to belie official assurances that the U.S. Army and its South Vietnamese allies were winning the war and bringing the country under control. CBS News anchorman Walter Cronkite's visit to Vietnam after Tet was allegedly the occasion of his conversion to an antiwar stance, leading him famously to pronounce on air that the war was a stalemate and the United States should negotiate for peace (Halberstam, 1979).

Whatever the real impact of the media in Vietnam, scholarly interest in the topic of media and power surged in the 1960s and 1970s and produced a number of very influential works over the following twenty years. Though many of these works dealt only indirectly with war, writers like Marshall McLuhan and Jacques Ellul produced widely read works on the subject of propaganda and elite control (McLuhan and Fiore, 1968; McLuhan, Fiore, et al., 1967; Ellul, 1965). Noam Chomsky was motivated by his antiwar activism to write many insightful pieces on media and war. Chomsky's oeuvre is much too large to list, but works such as *Manufacturing Consent* and *Necessary Illusions* are staples of the literature (Chomsky, 1989; Herman and Chomsky, 1988). Another noted antiwar activist, Todd Gitlin (1980), has produced a number of very important works on the social impact of media, which are almost uniformly cited by scholars of media and war. One of Paul Virilio's earliest works, *War and Cinema* (1989), provides a foundation for our understanding of both the genre of war film and the interplay between cinema, war, and social norms.

Despite scholarly skepticism regarding the impact of media in Vietnam, policy makers around the world drew the conclusion that granting reporters unrestricted access to battle zones would lead to an erosion of public support as images of violence spilled across front pages and television screens. This "lesson of Vietnam" clearly influenced British planners as they prepared to respond to the Argentinian occupation of the disputed Falkland Islands (Harris, 1983; Morrison and Tumber, 1988). The occupation caught the British government by surprise, forcing them to improvise a hastily assembled task force to eject the Argentinians.

A mere twenty-nine reporters (a figure roughly triple what the military would have liked to send) had only a few hours to prepare to accompany the task force heading for the Falklands. The constraints on their freedom were numerous. They were dependent on shipboard communications in an era before satellite phones and were forced to submit their stories to onboard censors, who apparently interpreted the need for security more broadly than the reporters would have liked. Film and camera crews faced an even more laborious process of getting their material from the remote islands back to editors in the United Kingdom. The arrangement caused the media industry to bristle, but it seemed to please the government, which could more successfully control the message (Carruthers, 1999).

A similar media near-blackout obtained during the U.S. invasion of Grenada the following year and the "lessons of Vietnam" clearly shaped U.S. media policy during the First Gulf War in 1991. "Embedded" reporters traveled with certain units and the military carefully controlled their movements for the most part. New technologies, such as portable satellite uplinks and smaller video cameras, gave reporters some more freedom than they had in the Falklands, but most scholars agree that the war was very carefully packaged in favor of the official line (Taylor, 1992; Hallin and Gitlin, 1994; Bennett and Paletz, 1994; MacArthur, 1993). With the advent of CNN and 24/7 news coverage, war had also merged with entertainment. Numerous commentators compared the footage of coalition aircraft and cruise missiles striking their targets, sometimes even viewed from the perspective of the missile itself, to video games. Those similarities between war, games, and other forms of entertainment have only increased in the intervening decades, leading some scholars to speak of "Militainment" (Stahl, 2010; Schubart, 2009).

Now that the Cold War has ended, more general works on the role of media in the superpower confrontation have appeared (Shaw and Youngblood, 2010; Shaw, 2007; Aronson, 1990; Cull, 2008). More recently, scholarly attention has turned to the relationship between media and terrorism. In June 2014, when Abu Bakr al-Baghdadi issued his dubious call to establish a caliphate (an Islamic dictatorship with a theoretical claim to worldwide sovereignty), a contest was already underway that spanned new, global "social media" technologies. While these new technologies are notably more diffuse and less susceptible to centralized control than media such as newspapers or film (especially state-sponsored film), they nevertheless work in a quite similar way to "emplot," to give narrative form and consistency to a struggle known (by one side at least) as the Global War on Terror (abbreviated, unfortunately, as GWOT). Similarly, many jihadists call the campaign by the United States and its allies the "Global War on Islam" as they seek to attract followers to their holy cause.

This war is being fought on a media landscape as well as a physical one. "More than an armed confrontation, the war on terrorism is being played in the realm of narratives," writes Gabriel Weimann (Weimann, 2015, 18). Both sides endeavor to shape the story of the conflict in a way that encourages their citizens/followers to devote themselves to the cause, to submit to increased surveillance, to fund attacks on the other side, or to travel to the hotspots of this global war (whether smuggled across borders or landing in a Marine helicopter) to fight (Veer and Munshi, 2004).

There is an older scholarship on media and terrorism dealing with, for example, coverage of Irish Republican Army attacks in the United Kingdom or acts such as the murder of Israeli athletes at the 1972 Olympics (Schlesinger et al., 1983; Wardlaw, 1989; Schmid and Graaf, 1982). The scholarship on the media of today's war (GWOT) is obviously much smaller and of much more recent vintage than the work that has been done on previous conflicts. It appears in journals and conference proceedings of associations dedicated to security or terrorism studies as well as media studies. Those in the security field are often quite explicit in their aim of combating terrorism (as they define it) and increasing the security of the United States and its allies, from whence flows the money that funds their studies, in most cases. Their conclusions are more likely to be policy recommendations than arguments of particular historical relevance.

This scholarship contains important analyses of the content and character of jihadist propaganda. The media strategies of the Islamic State are certainly susceptible to an analysis similar to that provided for the Nazi state. Like Goebbels, the propagandists of the Islamic State seek to create a cohesive, encompassing explanation of the world surrounding its followers and potential recruits. Using social media technologies like Twitter, Facebook, and others, the Islamic State justifies its attacks, glorifies the heroes of its cause, and explains away its setbacks (Weimann, 2015; Winter, 2015; Veer and Munshi, 2004).

Social media has revolutionized the way that jihadists like the Islamic State communicate with their followers. Whereas radicals used to congregate in certain schools, mosques, or training camps, they can now disseminate information, issue online "fatwas" (religious rulings), and recruit new members using the Internet and social media. Social media has become a way in which believers in jihad can express their support virtually. "Any Muslim who intends to do jihad against the enemy electronically," claims the author of "Electronic Jihad," "is considered . . . a mujahid [fighter for Islam], as long as he meets the conditions of jihad," which include "sincere intention and the goal of serving Islam and defending it" (Weimann, 2015, 130). In 2008, jihadists even planned a "YouTube invasion" to "shame the Crusaders by publishing videos showing their losses" (Weimann, 2015, 142).

One scholar has argued that the extraordinary violence depicted in some of the Islamic State's productions blinds analysts in the West to the comprehensiveness of its overall message. The Islamic State does not just glorify the gruesome behavior of its fighters, whose atrocities, especially beheadings, it films and disseminates. It also offers more positive messages that create a sense of brotherhood and belonging among its (clearly mostly male) followers. Allah's mercy and the promise of utopia are among the themes the Islamic States media strategy pursues. It tells the story of Muslim victimization at the hands of Israel and the West, thereby implicitly, when not explicitly, justifying the atrocities it commits. And, perhaps just as important for the political outcome of the conflict, it portrays itself as possessing real power. Successful military action, captured American or Russian equipment, and the black flag flying on city rooftops create the impression of a legitimate state with a real army. We should remember that the Islamic State's struggle is not just with "the West," but also with many of its Shi'ite coreligionists, who have powerful state allies in places like Iran (Winter, 2015).

In fact, jihadist use of social media is quite sophisticated and conforms to many of the media and marketing strategies advocated in Western business literature. The official website of Hamas, maintained by the Palestinian Information Center, for example, offers visitors material in eight languages, including English and French. In each language, the site presents a different message, with more or less emphasis on violence, human rights, and antisemitism (Weimann, 2015).

Such a comprehensive media message seeks to surround its target with a singular vision of the world in order to increase the target's resistance to alternative viewpoints (Ellul, 1965). The strategy seems not all that different from the one practiced by William Randolph Hearst and Joseph Pulitzer, whose significantly fabricated news stories about Cuba eventually appear as fact in more conservative papers and even in the Congressional Record. Nor does it seem too different from the effort of Joseph Goebbels to tell a story of Aryan supremacy leading to military victory that ultimately foundered on the shoals of Stalingrad and the Combined Bomber Offensive.

It has been a feature of war throughout much of human history that combatants at all levels have sought to tell their "war story." In order to conduct a successful media policy, government ministries, newspaper editorial boards, and small terrorist cells alike have attempted to strike a fine balance between reality and representing their own interests, between context and content, in order to make their message effective. Authenticity is that combination of truth and perception for which messengers have striven.

Media also plays an important role in how wars are remembered, not just in how they are planned and fought. Readers may notice areas of overlap between this essay and the contribution "Memory and Memorialization" by Michael Dolski in this volume. Over time, literatures in the fields of memory studies and media studies will likely intersect even more explicitly, especially since media studies are already so thoroughly interdisciplinary. The works cited here are by historians, philosophers, sociologists, linguists, cultural critics, and more. Inspiration for this author's research in the field came from Anton Kaes (1989), professor of German and founder of the Department of Film and Media at my alma mater, the University of California (Lockenour, 2012). That interdisciplinarity and the sheer variety of approaches are daunting, but a source of great strength for this burgeoning field.

Media gives meaning to events, structures cause and consequence, and creates heroes and villains out of the raw material of occurrence. Messages conveyed by media, the stories that they tell about war, play an important role in shaping both contemporary and historical understanding of human conflict. If war is also at some level a contest of wills, then the power of media to influence that will can be important for the outcome as well. The foregoing historical examples demonstrated the limitations on that power. Goebbels' media policy could not overcome the Red Army; neither did William Randolph Hearst single-handedly start the Spanish-American War. Walter Cronkite was not the reason for the United States' defeat in Vietnam, nor will the perfect "Twitterstorm" establish a global caliphate. But studies of media and war will continue to provide powerful insights into the causes, conduct, consequences, and culture of war.

Bibliography

Aldgate, A., and J. Richards. *Britain Can Take It: The British Cinema in the Second World War*. Oxford: I.B. Tauris, 1986.

Arlen, M. J. *Living-Room War*. New York: Viking Press, 1969.

Aronson, J. *The Press and the Cold War*. New and expanded ed. New York: Monthly Review Press, 1990.

Bennett, W. L., and D. L. Paletz, eds. *Taken by Storm: The Media, Public Opinion, and U.S. Foreign Policy in the Gulf War*. American Politics and Political Economy Series. Chicago: University of Chicago Press, 1994.

Bergmeier, H.J.P., and R. E. Lotz. *Hitler's Airwaves: The Inside Story of Nazi Radio Broadcasting and Propaganda Swing*. New Haven: Yale University Press, 1997.

Briggs, A. *The BBC: The First Fifty Years*. Oxford: Oxford University Press, 1985.

Carruthers, S. L. *The Media at War: Communication and Conflict in the Twentieth Century*. New York: St. Martin's Press, 1999.

Chomsky, N. *Necessary Illusions: Thought Control in Democratic Societies*. Boston, MA: South End Press, 1989.

Cull, N. J. *The Cold War and the United States Information Agency: American Propaganda and Public Diplomacy, 1945–1989*. Cambridge: Cambridge University Press, 2008. doi:9780521819978.

———. *Selling War: The British Propaganda Campaign Against American "Neutrality" in World War II*. New York: Oxford University Press, 1995.

Curry, A. Agincourt. Great Battles. Oxford, UK: Oxford University Press, 2015.

DeBauche, L. M. *Reel Patriotism: The Movies and World War I*. Wisconsin Studies in Film. Madison: University of Wisconsin Press, 1997.

———. "The United States' Film Industry and World War One." In *The First World War and Popular Cinema: 1914 to the Present*. Ed. M. Paris. New Brunswick, NJ: Rutgers University Press, 2000, 138–161.

Doherty, T. *Projections of War: Hollywood, American Culture, and World War II*. New York: Columbia, 1993.

Dower, J. W. *Embracing Defeat: Japan in the Aftermath of World War II*. London: Allen Lane, 1999.

———. *Japan in War and Peace: Essays on History, Culture and Race*. London: Fontana, 1996.

Ellul, J. *Propaganda; the Formation of Men's Attitudes*. 1st American ed. New York: Knopf, 1965.

Fehrenbach, H. *Cinema in Democratizing Germany: Reconstructing National Identity After Hitler*. Chapel Hill: University of North Carolina Press, 1995.

Gitlin, T. *The Whole World Is Watching: Mass Media in the Making & Unmaking of the New Left*. Berkeley: University of California Press, 1980.

Halberstam, D. *The Powers That Be*. 1st ed. New York: Knopf, 1979.

Hallin, D. C. *The "Uncensored War": The Media and Vietnam*. New York: Oxford University Press, 1986.

Hallin, D. C., and T. Gitlin. "The Gulf War as Popular Culture and Television Drama." In *Taken by Storm: The Media, Public Opinion, and U.S. Foreign Policy in the Gulf War*. Eds. W. L. Bennett and D. L. Paletz. Chicago: University of Chicago Press, 1994, xvi, 308 pages.

Hanley, W. *The Genesis of Napoleonic Propaganda, 1796 to 1799*. New York: Columbia University Press, 2005.

Harris, R. *Gotcha!: The Media, the Government, and the Falklands Crisis*. London: Faber and Faber, 1983.

Herman, E. S., and N. Chomsky. *Manufacturing Consent: The Political Economy of the Mass Media*. 1st ed. New York: Pantheon Books, 1988.

Herzstein, R. E. *The War That Hitler Won: The Most Infamous Propaganda Campaign in History*. New York: Putnam, 1978.

Hoganson, K. L. *Fighting for American Manhood: How Gender Politics Provoked the Spanish-American and Philippine-American Wars*. Yale Historical Publications. New Haven: Yale University Press, 1998.

Holtman, R. B. "The Use of Propaganda by Napoleon." PhD dissertation, University of Wisconsin, 1941.

Isenberg, M. T. *War on Film: The American Cinema and World War I, 1914–1941*. Rutherford: Fairleigh Dickinson University Press, 1981.

Kaes, A. *From Hitler to Heimat: The Return of History as Film*. Cambridge, MA: Harvard University Press, 1989.

Kallis, A. A. *Nazi Propaganda and the Second World War*. Houndmills, UK: Palgrave Macmillan, 2005.

Kershaw, I. *The "Hitler Myth": Image and Reality in the Third Reich*. Oxford: Oxford University Press, 1987.

Knightley, P. *The First Casualty: From the Crimea to Vietnam: The War Correspondent as Hero, Propagandist, and Myth Maker*. 1st ed. New York: Harcourt Brace Jovanovich, 1975.

Kreimeier, K. *The UFA Story: A History of Germany's Greatest Film Company, 1918–1945*. 1st ed. New York: Hill & Wang, 1996.

Lasswell, H. D. *Propaganda Technique in the World War*. New York: Knopf, 1927.

Lippmann, W. *Public Opinion*. New York: Macmillan, 1922.

Lockenour, J. "Black and White Memories of War: Victimization and Violence in West German War Films of the 1950s," *Journal of Military History* 76, no. 1 (January 2012): 159–191.

MacArthur, J. R. *Second Front: Censorship and Propaganda in the Gulf War*. Berkeley: University of California Press, 1993.

Marquis, A. G. "Words as Weapons: Propaganda in Britain and Germany During the First World War," *Journal of Contemporary History* 13 (July 1978): 467–498.

———. "Written on the Wind: The Impact of Radio During the 1930s," *Journal of Contemporary History* 19, no. 3 (July 1984): 385–415.

McLuhan, M., and Q. Fiore. *War and Peace in the Global Village: An Inventory of Some of the Current Spastic Situations That Could Be Eliminated by More Feedforward*. New York: Bantam Books, 1968.

McLuhan, M., Q. Fiore, and J. Agel. *The Medium Is the Message*. New York: Bantam Books, 1967.

Messinger, G. S. *The Battle for the Mind: War and Peace in the Era of Mass Communication*. Amherst: University of Massachusetts Press, 2011.

———. *British Propaganda and the State in the First World War*. Manchester: Manchester University Press, 1992.

Miller, B. M. *From Liberation to Conquest: The Visual and Popular Cultures of the Spanish-American War of 1898*. Amherst: University of Massachusetts Press, 2011.

Morrison, D. E., and H. Tumber. *Journalists at War: The Dynamics of News Reporting During the Falklands Conflict*. London: SAGE, 1988.

Paret, P. *Imagined Battles: Reflections of War in European Art*. Chapel Hill: University of North Carolina Press, 1997.

Paris, M., ed. *The First World War and Popular Cinema: 1914 to the Present*. New Brunswick, NJ: Rutgers University Press, 2000.

Roeder, G. H. *The Censored War: American Visual Experience During World War II*. New Haven: Yale University Press, 1993.

Rollins, P. C., and J. E. O'Connor, eds. *Hollywood's World War I: Motion Picture Images*. Bowling Green, OH: Bowling Green State University Popular Press, 1997.

Sanders, M., and P. M. Taylor. *British Propaganda During the First World War, 1914–18*. London: Macmillan, 1982.

Schlesinger, P. *Putting "Reality" Together: BBC News*. London: Methuen, 1987.

Schlesinger, P., P. Elliott, and G. Murdock. *Televising "Terrorism": Political Violence in Popular Culture*. Comedia Series 16. London: Comedia, 1983.

Schmid, A. P., and J. D. Graaf. *Violence as Communication: Insurgent Terrorism and the Western News Media*. London: SAGE, 1982.

Schubart, R., ed. *War Isn't Hell, It's Entertainment: Essays on Visual Media and the Representation of Conflict*. Jefferson, NC: McFarland, 2009.

Seethaler, J., ed. *Selling War: The Role of the Mass Media in Hostile Conflicts From World War I to the "War on Terror"*, European Communication Research and Education Association Series. Bristol, UK: Intellect, 2013.

Shaw, T. *Hollywood's Cold War*. Culture, Politics, and the Cold War. Amherst: University of Massachusetts Press, 2007.

Shaw, T., and D. J. Youngblood. *Cinematic Cold War: The American and Soviet Struggle for Hearts and Minds*. Lawrence: University Press of Kansas, 2010.

Shindler, C. *Hollywood Goes to War: Films and American Society, 1939–1952*. Boston: Routledge, 1979.

Smith, A. L. *The War for the German Mind: Re-Educating Hitler's Soldiers*. Providence, RI: Berghahn Books, 1996.

Stahl, R. *Militainment, Inc.: War, Media, and Popular Culture*. New York: Routledge, 2010.

Taylor, P. M. *War and the Media: Propaganda and Persuasion in the Gulf War*. Manchester: Manchester Univ. Press, 1992. doi:Sbbt645817.

Taylor, P. M., and I. W. Museum. *Britain and the Cinema in the Second World War*. New York: St. Martin's Press, 1988.

Tent, J. F. *Mission on the Rhine: Reeducation and Denazification in American-Occupied Germany*. Chicago: University of Chicago Press, 1982.

Thomas, E. *The War Lovers: Roosevelt, Lodge, Hearst, and the Rush to Empire, 1898*. New York: Little Brown, 2010.

Veer, P.V.D., and S. Munshi, eds. *Media, War, and Terrorism: Responses From the Middle East and Asia*, Politics in Asia Series. London: Routledge, 2004.

Virilio, P. *War and Cinema: The Logistics of Perception*. London: Verso, 1989.

Wardlaw, G. *Political Terrorism: Theory, Tactics, and Counter-Measures*. 2nd ed. Cambridge, UK: Cambridge University Press, 1989.

Weimann, G. *Terrorism in Cyberspace: The Next Generation*. Washington, DC: Woodrow Wilson Center Press, 2015.

Welch, D. *Germany, Propaganda, and Total War, 1914–1918: The Sins of Omission*. New Brunswick, NJ: Rutgers University Press, 2000.

———. "Nazi Wartime Newsreel Propaganda." In *Film & Radio Propaganda in World War II*. Ed. K.R.M. Short. Knoxville: University of Tennessee Press, 1983, 201–219.

———. *The Third Reich: Politics and Propaganda*. London: Routledge, 1993.

Welch, K., A. Powell and J. Barlow, eds. *Julius Caesar as Artful Reporter: The War Commentaries as Political Instruments*. Pbk. ed. Swansea: Classical Press of Wales, 2009.

Wilkerson, M. M. *Public Opinion and the Spanish-American War: A Study in War Propaganda*. Baton Rouge: Louisiana State University Press, 1932.

Winter, C. "The Virtual 'Caliphate': Understanding Islamic State's Propaganda Strategy." In *Quillam Report* (2015). Published electronically July. www.quilliamfoundation.org/wp/wp-content/uploads/publications/free/the-virtual-caliphate-understanding-islamic-states-propaganda-strategy.pdf.

Youngblood, D. J. *Russian War Films: On the Cinema Front, 1914–2005*. Lawrence: University Press of Kansas, 2007.

15 War, the Body, and Health

Bobby A. Wintermute

Since the advent of the war and society specialization in military history, much of the emphasis has been placed on establishing the interconnectivity of armed conflict and broader social themes in the human experience, and to a lesser extent, the natural environment. As the specialization expands its reach to consider war's impact on our social institutions and awareness, it is only natural that it examines the most personal of physical and perceptual spaces—the connection and consequence to the human body. Again, this is in some ways a well-established subset of our collective literature on war. The experience of wounds and death has been recorded and contemplated since ancient times. Assyrian and Babylonian medical treatises detailed treatments for a variety of stabbing, cutting, and crushing wounds (Scurlock and Andersen, 2005). The most evocative descriptions of injury in ancient battle are found in *The Iliad*, including in graphic detail the injuries inflicted upon its participants. Roman accounts of battle are likewise rich in detail of combat's toll on the body. Such accounts, however, serve generally as descriptive narrative only, providing the reader an evocative sense of the ferocity of battle, but lacking any larger context. As a discursive and analytical lens, war and society methodology grants interested historians the ability to consider how other social institutions, including the medical profession, respond to conflict on multiple levels. These range from individual treatment of wounds to the creation of sophisticated support infrastructures dedicated to leveraging the coercive powers of the state to promote public health across the broader population as a whole, in the name of security.

For the purposes of this essay, the historiographic literature associated with war and society and the body is divided into three general categories. First are the works that consider the personal experience of combat and its personal effect on the body. These range from historians who consider battle injury as part of the larger experience of war and combat to those who focus explicitly on the issue of wounds and mortality. Second are the studies of the impact of wartime disease on armies, soldiers, and bystanders. Disease has often been called the handmaiden of war; this association extends to the medical advances of the nineteenth and twentieth centuries, new techniques and practices that have their basis in wartime military medicine. Third are histories devoted explicitly to the social and institutional development of military medical organizations. Finally are more broadly defined cultural studies that identify military medicine and the body as integral to processes of colonial and imperial expansion. Additionally, each of the categories will address relevant works that present the body itself as a socially defined construct, an object of contested control and negotiated participation, as well as address the role of military medicine as an arbiter of racial, ethnic, and gender-based social categorization.

Wounds and the Body

Wounds and injury are central to the reality of combat. All of the sanitized language associated with battle—"meeting the enemy," "defending/attacking" a fixed position or body of men, "inflicting casualties"—only conceals the essential violence of war and its tremendous toll on the individual. Since ancient times, fighting men have used ever more sophisticated, and yet efficient, tools to cut, stab, crush, dismember, and maim each other. As noted earlier, the physical toll of combat has always been accounted for in battle narratives, ranging from the poetic meter of the Bible and Homer to Robert Graves' and Tim O'Brien's more recent accounts. This is especially true since John Keegan's groundbreaking work *The Face of Battle: A Study of Agincourt, Waterloo, and the Somme* (1976). In his efforts to capture the essence of combat for the participant, Keegan devoted significant attention to the experience of injury and its immediate and lasting effects on the afflicted soldier. In doing this, Keegan set the tone for future historians interested in the relationship between war and the human body.

Such a rich trove of source material has given historians ample opportunity to describe the impact of wounds on the body. Some, like Victor Davis Hanson (1989), incorporate these primary sources to great effect to identify how brutal and effective hoplite combat was in killing the enemy. Hanson presents war between the ancient Greeks as a savage crush of men, butchering each other where they met on the field. For the ancient Greeks, wounds served as a virtual text that could describe the course of battle. Stabbing wounds to the front of the body described the collision of two phalanxes, and the frenetic push to gain dominance. Those in the rear of the body identified the point where decision turned the course of battle: in a culture where combat was a direct, face-to-face encounter, stabbing injuries from behind signaled the collapse of one phalanx to the other. In the ensuing rout, fleeing men became targets of opportunity for the victor.

Hanson's book influenced many other writers to reconsider the impact of hoplite warfare on the bodies of the participant, usually with similar results. There are some exceptions, however. Fellow classicist J. E. Lendon (2005) takes the nature of wounds in battle a step further. Here Lendon not only introduces and compares the differences and similarities in Greek and Roman approaches to war, but also considers the cultural imperative of wounds, not only for the dead but also for the survivor. He examines how battle injury was part of the individual Homeric soldier's quest for honor and memory—*kleos* and *kudos*—and how Greek and Roman alike venerated the example of personal combat even as their systems of war evolved tactically to emphasize group coordination and cohesion.

Sarah Covington (2009) describes the role of wounds for the surviving soldier and their transfiguration into physical symbols of personal martyrdom in seventeenth-century England. Accordingly the image of the battered, wounded soldier became a metaphor for the state under siege in the tumultuous events of the English Civil War and its aftermath. Employing primary accounts of surviving veterans, including well-known literary figures, like Daniel Dafoe and John Milton, Covington examines how a war-weary England drew larger analogies of its condition from the language of war wounds and the pain and suffering they evoked. In the aftermath of battle, "wounds could not help but become imbued with a larger aura of imposed significance, even at their most plainly represented" (6). For the survivor and his audience, the veteran's injuries and scars were evidence of an England undone and riven by disorder; just as the soldier's body was savaged by battle, so too was the nation broken by strife and chaos.

Such metaphysical approaches to the surviving soldier's wounds are part of a growing interest in how cultures engaged in war contextualize the human toll as a symbol of some larger flaw or disturbance in society. Joanna Bourke (1996) contextualizes the carnage of the First World War as an assault on British notions of masculinity. The challenge was not exclusively negative, as the shared experience of combat and injury on such a massive scale helped deconstruct class-based barriers that defined a distinct hierarchy of maleness in pre-war British society. A more egalitarian wartime aesthetic of the ideal male body helped to some degree but, as Bourke notes, the massive scale of injury among survivors, coupled with the awesome mortality of the war on the British and Imperial soldier, enabled the wounded survivor (save for the most extreme cases) to be accepted as part of the new normal, a virtual "'Everyman'" (15).

American conflicts reveal other postwar experiences of the wounded, as they and their families struggled to come to terms with their shattered bodies and minds. Eric T. Dean Jr. (1997) focuses primarily on the experiences of 291 Indiana veterans who were admitted after the American Civil War to the Indiana Hospital for the Insane. Broken emotionally by the war, these men and their stories reveal the sad denouement to the conflict that has otherwise been all but forgotten by historians. By stripping away the façade of normalcy that pervades most work on the Civil War, Dean reveals the universality of emotional trauma that underlies all wars. Brian Miller (2015) notes that for all of the public discourse of war and noble sacrifice in the postwar South, amputees were frequently shunned by family and neighbors, both celebrated and rejected as "incomplete and damaged 'heroes'" (8). Having sacrificed their masculinity many maimed veterans were perceived as doubly emasculated: unable to provide for themselves and their families, but also transformed by the war into both undesirable burdens on their communities and unwelcome reminders of the shame of defeat.

Disease and the Practice of Military Medicine

More common, however, are those narratives devoted explicitly to the subject of the treatment of wounds, injury, and disease in wartime. Until the 1990s, the field was dominated by a variety of traditional histories, many of them produced decades earlier. In the case of the American Civil War, for example, monographs like George Worthington Adams' (1952) and H. H. Cunningham's (1958) crafted the standard narrative of Civil War military medicine as practiced on both sides. The challenge was greater than expected. The Union Army's medical establishment was in a near-constant state of flux, thanks to the combination of political interference, philanthropic meddling, the erratic quality of volunteer medical officers, and the sheer scale of the casualties across such a wide area. Cunningham faced even greater challenges, as so many of the Confederate medical records were lost to an 1865 fire in Richmond. Nevertheless the two volumes retained significant influence for decades, largely because so few historians sought to follow their lead. A decade later Paul Steiner (1968) examined the prospects of infectious disease as a force multiplier (or reducer). Considering the impact of diseases on the war—over two-thirds of the war's dead fell victim to illness—Steiner concludes that not only was the toll unavoidable, given the state of mid-nineteenth-century medicine, but also this attrition dictated the terms of many campaigns over the course of the war.

Steiner's conclusion was recently revisited by Andrew McIlwaine Bell (2010). Malaria was a scourge for both sides, afflicting over 1.1 million Union soldiers and at least one-seventh of all Confederates in uniform during the first two years of the war alone. If the disease itself were not bad enough, its symptoms were a virtual seedbed for other lethal infections, including chronic diarrhea. Commanders on both sides were painfully aware of the consequences of campaigning in the "sickly season"—late summer and early fall—and strategic planning was adjusted accordingly. For Union soldiers in particular, the deep South was an alien land rife with pestilence, a place where men went to die without even sharing in the experience of combat. Bell's work is presented as a direct call to action for expanded scholarship in the field, especially of the nexus between epidemiology and military operations. "To date most medical histories of the conflict have given short shrift to [malaria and yellow fever]," he writes. "As a result, historians are left with an inchoate picture of daily life in the wartime South and an incomplete understanding of why various campaigns turned out the way they did" (6–7). Kathryn Shively Meier (2013) revisited the issue of disease and its impact on military operations in a detailed analysis of the Shenandoah and Peninsula campaigns. Meier considers the impact of illness from the perspective of common soldiers, noting that they usually had to determine preventive measures for themselves. Self-care took on many forms, ranging from proper shelters, identifying safe water supplies, basic sanitation, and tending to each other when disease struck. Accordingly most soldiers practiced the same palliative regimens they experienced in civilian life, and learned to distrust the frequently confusing, if not lethal, interventions of the army medical staff.

For all of its hyperbole, Bell's indictment of the field is ill-placed. While it is true that malaria and yellow fever have largely been ignored in traditional campaign narratives, suggesting that the medical histories of the war generally ignore the diseases is an exaggeration. Since the early 1990s, renewed interest in the belligerents' medical services has grown. Much of this literature is dominated by medical practitioners with an interest in Civil War history. Frank R. Freemon (1998), a physician with a doctorate in history, produced a detail-rich anecdotal survey of military surgery that presented the two military services as part of the same tradition of scientific-based medicine under assault by a Jacksonian-era emphasis on sectarian and nontraditional practices. His efforts have been followed by other physicians, including Alfred J. Bollet (2002) and Ira M. Rutkow (2005). Both narratives are in their own turn evocative and positivist, presenting the war as a transformative moment in medical and surgical practice.

More critical studies of the medical services, with special focus on disease, wounds, and the body, have followed as well. The most inclusive is Margaret Humphreys' (2013). According to Humphreys, the Civil War was the most challenging health crisis faced by the nation. Occurring just as medical science and practice were beginning to experience a transformative paradigmatic shift on the basis of germ theory, the war's tumult of disease, wounds, and injury challenged the boundaries of accepted therapeutics—but established the foundation for modern medicine and public health practice to come. Shauna Devine (2014) expands upon this theme by emphasizing the role of medical education before and during the Civil War. She shows how the Army Medical Department's antebellum culture of scientific data collection and reporting, along with the privileged status of educated licensed regular physicians, combined to establish a special cadre of Union senior medical officers, who were committed to preserving and learning from their experiences. Accusations that Civil War medicine was barbaric or unsophisticated miss the mark; by the standards of day (and given the sheer volume of cases), official medical care was quite advanced. Trained medical officers introduced new methods of patient care, battlefield evacuation and triage, surgery, and amputation. Most important was the tremendous collection of medical specimens and reports that became the core of the postwar Army Medical Museum and the six-volume work *The Medical and Surgical History of the War of the Rebellion*—establishing a legacy that would help advance medical science for decades to come.

While the subject of wartime health during the American Civil War has benefited from greater attention in recent years, subsequent conflicts and experiences have not been neglected. The toll of preventable camp diseases has been a central theme of the Spanish-American War since its end. Vincent J. Cirillo (2004) argues the U.S. Army's own culture of command stifled medical officer efforts to introduce basic camp sanitation and other advances associated with the recently accepted germ theory of disease and infection that could have prevented the epidemics of typhoid fever in the training camps. Another physician trained as a historian, Cirillo asserts that inherently conservative military culture tends to reject medical advice that challenges its legitimate authority, but fails to consider the outcome of the war and the subsequent validation of the Army Medical Department's actions. Bobby A. Wintermute (2010) revisits the fever camps episode, noting it makes up part of a larger quest for legitimacy by medical officers caught between two professional paradigms—the medical practitioner and the military professional. Wintermute notes that many of the problems within the training camps originated among the volunteer regiments and their resistance to centralized authority. Trained as scientifically oriented practitioners with a strong foundation in microbiology at the insistence of Surgeon General George Miller Sternberg, regular medical officers found their advice generally ignored by the volunteers. The ensuing disaster was mitigated before it could explode into a full-blown epidemic, in large part because of the scientific-based interventions of the regular and volunteer medical officers.

The problem of disease appeared again in the First World War, this time in the form of influenza. Carol R. Byerly (2005) describes a perfect storm of overconfident physicians, a deadly influenza virus taking shape in the trenches in France, and once again, a military establishment convinced of its own infallibility. Unable to stem the disease, medical officers could only look on in horror as the healthiest young men in their care died horribly. Wartime urgency to reach France prompted the Army, and the War and Navy Departments, to ignore recommendations against overcrowding on ships and trains, which only spread the disease.

The fact that the literature of subsequent conflicts focuses much less on disease is testament to the overall therapeutic success of the medical community and the impact of antibiotics. At the same time, however, aside from memoirs, the post-1918 literature has not enjoyed the same surge of appeal that has accompanied American Civil War studies. The field has been dominated by survey texts, like Mark Harrison's (2004) or Albert E. Cowdrey's (1994). Where Harrison presents a highly analytical focus on the Royal Army Medical Corps' success in helping the British Army maintain a high level of combat readiness, especially in the northwestern Europe campaign, Cowdrey's approach is more of an anecdotal and archival narrative. The latter, evocative and well-written, is a good introduction to a large and complicated subject that deserves more careful and detailed treatment.

The same can be said for more recent conflicts, including Korea, Vietnam, and the recent wars in Iraq and Afghanistan. While there is no shortage of official histories and memoirs related to these conflicts, the task of writing a comprehensive medical survey, let alone specific monographs related to different medical aspects of the wars, remains unfulfilled. Critics may point to the broad literature of post-traumatic stress and emotional trauma, following from the original start point of Robert Jay Lifton (1973; 2005) and Jonathan Shay (1995), but this subfield generally fails to account for the physicality of combat experience and the efforts to treat the wounded or diseased soldier afterward. There are also several general narratives, but many, including Ann Jones' (2013), are as much journalistic exposé as an attempt to produce historical meaning.

The Development of Military Medical Institutions

One of the immediate challenges for historians of military medicine is reconciling the two missions of the uniformed medical practitioner. Military physicians are expected to safeguard the health and wellness of soldiers whose primary obligation is to employ lethal force in the service of a (usually) legal authority. However, as some have noted, this places medical officers in direct opposition to the high ideal that as practitioners of the healing arts, they are empowered to do no harm to their patients, nor to participate in a system that is dedicated to armed coercion and violence. This debate, while not central to the task of historians studying military medicine institutions, is nevertheless addressed as they examine the broader culture of the military in society.

In her opening chapters, Byerly (2005) identifies the tensions physicians faced within their own professional community and American society at large as they balanced their medical responsibilities with their military duties. In the face of two simultaneous crises—the First

World War and the 1918 influenza pandemic—these professionals were torn between safe-guarding the health of soldiers under their charge and providing a reliable pool of manpower for service in France. She concludes that "Medical officers thus found themselves caught between their oath to the army and the nation and the Hippocratic oath" (5). This dilemma was not exclusive to the twentieth century. Bobby A. Wintermute (2010) presents it as a central facet of the nineteenth-century American medical officers' quest for dual legitimate authority. Since the beginning of the Jacksonian-era egalitarian rejection of elites, the medical profession was under assault by sectarians who deemed it to be an insult to the dignity of the common citizen. As practitioners with a formal traditional education—the "allopathic" tradition—aspiring medical officers were especially desperate to claim legitimacy on the basis of their privileged knowledge of the body and therapeutics. Simultaneously, they also were denied full status as U.S. Army officers by those on the line and in staff institutions. Throughout the antebellum period, and even after the Civil War, medical officers balanced their split identity as members of two professional groups. In the end, they reconciled the moral dilemma by emphasizing their perceived strategic institutional role by advancing a new construct, sanitary tactics, justifying their role in preserving the health of soldiers as a sort of force multiplier.

As historians consider the relationships between the formal institutions dedicated to military medicine and their parent societies and communities, other opportunities arise for crafting a well-informed social-based narrative that is at the core of the war and society field. Consider monographs on female nursing aides during the American Civil War. Many follow the traditional form of narrative-driven hagiography. Offering little social context, they present the story of women acting within their accepted roles as nurturers, neglecting the obstacles and resistance they faced from a male-dominated field. Prior to the Civil War, the task of nursing was the responsibility of male hospital stewards, the enlisted component of the Army Medical Department. Female participation in the hospital ward was considered a gender transgression. While women were expected to care for members of their own families, and in rare cases (lay sisters and nuns, in particular) within gender-segregated civilian hospital wards, the military hospital was a male-exclusive space. Their entry was subject to rigid oversight, lest their presence inflame emotions and create discord. Likewise, nineteenth-century gender ideology dictated that women were too frail and emotional to be of use in treating combat-oriented wounds and injuries. The inevitable hysterical response to blood and gore would only distract the male attendants and physicians from their jobs.

Jane E. Schultz (2004) examines the experiences of the over twenty thousand women who worked in Union and Confederate hospitals. She casts the hospital as a new front of contested identity and gender-based challenges to reach a state of shared legitimacy as partners in the recovery of men outside of the normal contours of domesticity. The entire hospital experience was an exercise in gender inversion for the male wounded and their female caregivers. Now rendered helpless, men were the object of pity and nurturing by strong women, imbued through their work with the same martial spirit that led their patients to the battlefield. Though denied access to combat, Schultz argues, female nurses and aides were warriors in their own right, challenging social norms and taboos against their presence in the military hospital to do their own part for the cause.

Despite their service in the Civil War, women were barred from military nursing in the United States through the nineteenth century. The camp crises of the Spanish-American War took place just as progressivism was taking shape as a reform-oriented movement, and provided the impetus for the incorporation of the Army Nurse Corps on February 1, 1901. The official history of the service, by Mary T. Sarnecky (1999), records in great detail the challenges in convincing the Army surgeon general, George Miller Sternberg, and a reluctant Congress to commission female nurses. Between the lobbying and organizational efforts of future nursing chief Anita Newcomb McGee and the untimely death of courageous contract nurses like Clara Maass, who volunteered as a test subject in the Medical Department's yellow fever study in Havana in 1901, the future of a female nursing service in uniform was secured. As Kimberly Jensen observes (2008), women were to become essential participants in the American military response in the Great War. And yet, while women served in a host of military roles, within and outside the Nurse Corps, they waged an internal struggle to achieve recognition and legitimacy in society at large through their participation in the war. Nurses played a significant role in this contest and, along with female ambulance drivers and orderlies, were placed directly in harm's way more than ever before. The cost they paid for challenging a deeply entrenched culture of gender separation and identity could be total.

By the Second World War, public attitudes toward the gendered aspects of nursing had been so totally transformed that now male nurses were seen as the intruder into a field recast as female-exclusive. Just as in the case of Civil War historiography, the story of female nurses in the Second World War is a very popular niche field. Most of the literature is very memoir-oriented, with more than a touch of positivist interpretation of the growing responsibilities and obligations of their contributions. Charissa J. Threat (2015) provides an uncommon example of a more analytically oriented treatment. By 1941 the occupation had become so dominated by white female nurses that both white male and African American female practitioners confronted significant obstacles in achieving acceptance in the service. During and after the Second World War, the Army Nurse Corps became a triple-contested space, as all three demographic groups vied for equal representation. Threat's book not only speaks to the issues surrounding army nursing but also serves as a reminder of the U.S. Army—and the Department of Defense—as a venue and advocate for civil rights in American society.

The role played by the U.S. Army Medical Department in establishing the contours of race and identity, particularly with regard to blackness, is ripe for new scholarship. This goes beyond the standard narratives of military medicine, with its emphasis on diagnostics and therapeutics. As Margaret Humphreys (2008) reveals, Union medical officers recognized the unique empirical opportunity of access to over 180,000 black soldiers presented to them for prolonged and broad study. Contemporary science—let alone crude stereotype—likened African Americans to beasts of burden, or biologically infantilized persons, incapable of expressing higher functions of reasoning and logic. As this rhetoric collided with the reality of battlefield performance, in which white officers became impressed with the nobility of the men under their command, Humphreys notes many medical officers sought to clarify the true biological standing of race and the black man in uniform. The implications were serious, as they would challenge the very notion of white exceptionalism in American culture. The work that ensued, however, was crafted to satisfy the racial status quo. While black soldiers were determined to be no different physiologically than their white counterparts (save perhaps for the consequences of poor diet as southern slaves and poor freemen in the North), the military medical establishment continued to find that blacks were morally unfit for soldiering, even if their bodies were sound. Humphreys quotes an 1866 report to Secretary of War Edwin M. Stanton: "'It is merely suggested that it is moral rather than physical; that the great susceptibility of the colored man to disease rose from lack of heart, hope, and mental activity, and that a higher moral and intellectual culture would diminish the defect'" (145). The role of medical officers and the Army Medical Department in asserting the legitimacy of racial hierarchies dependent upon white superiority is clear, and deserves greater attention by scholars.

One other area where the literature oriented around military medical institutions resides is the question of scientific and preventive medicine. A large part of this historiography is focused on epidemiology and military medicine. J. R. McNeill (2010) has written what in many ways is an environmental history of the Greater Caribbean, with disease at its core. The dilemma facing European armies in the region since the sixteenth century has been how to survive the area's tropical diseases. Until the nineteenth century, the Greater Caribbean was renowned as a fever pit, a place where white armies went to die. Numerous campaigns were waged over the islands' riches, most of them bitter failures, with thousands of English, French, Dutch, Spanish, and American soldiers left dead of malaria and yellow fever. McNeil shows how the European powers occupying these islands factored disease into their defenses. Spain in particular proved highly capable in leveraging yellow fever to its advantage as a force multiplier to offset its relatively small military commitment in defense of its possessions. Yet this would prove to be a mixed blessing, as local revolutions enjoyed the advantage of differential immunity against imperial forces sent there from Europe. "If they could avoid losing quickly on the battlefield, the revolutionaries could prevail in the long run thanks to the systematically partisan attacks of epidemics," McNeill wrote. "And prevail they did" (5). Only after military medical organizations identified the links between insect vectors and disease would the balance of control and power shift to a new American imperial hegemony in the Caribbean.

The role of military medical institutions in restricting the destructive potential of disease is the next logical step in charting the historiography. One of the earliest American examples of an organized response to infectious disease is recorded by Elizabeth A. Fenn (2001). The role of smallpox as a wartime scourge is well established; less understood is the significance of the Continental Army's adoption of inoculation against the disease. As early as July 1775, General George Washington balanced the long-term promise of inoculation against the short-term disadvantages associated with recovery from the procedure. The specter of infection, matched to military defeat, was realized in the abortive 1775–1776 Quebec Campaign, where the disease crippled the American forces under Generals Richard Montgomery and Benedict Arnold. After another two years of erratic campaigning, at Valley Forge in 1777, Washington finally authorized the inoculation of the entire army, at the urging of the Continental Army's chief physician, Benjamin Rush. Of course, the decision to adopt inoculation is only part of Fenn's masterful study, but it is central to the question raised here over the role of an army's medical support infrastructure in addressing disease and taking steps to remedy it.

A discussion of military medical initiatives in epidemiology would be incomplete without addressing the Army Medical Department's activities in Cuba after 1898 targeting yellow fever. Margaret Humphreys (1999) best describes the origins of the U.S. Army's interest in this disease. From Major General Benjamin Butler's successful quarantine of New Orleans to the 1897 Reed-Carroll commission's focus on testing Cuban doctor Carlos Finlay's mosquito vector theory (an effort that was interrupted by the Spanish-American War), Humphreys notes the Medical Department's activities were substantial. The subsequent 1901 Yellow Fever Board's activities—under the direction of Major Walter Reed—have been described by many historians and popular writers, with one of the best by John A. Pierce and Jim Writer (2005). While they uncover little new about the Yellow Fever Board and its work, the two coauthors capably describe the long-standing debates over the nature of the disease and the scientific challenges to validate Finlay's mosquito vector theory. Pierce and Writer also give due credit to Reed's peer, William Crawford Gorgas, for quickly transferring the confirmation of the role of *anopheles* mosquitos to the near immediate sanitation campaign in Havana that virtually eradicated the pest—and yellow fever—from the city. Reed would not survive to reap the recognition of his discovery, but Gorgas successfully transferred credit to his sanitary campaign to become the chief beneficiary of the ensuing acclaim.

Military Medicine, Race, and Imperialism

Military medicine is primarily associated with maintaining the health and fitness of those uniformed persons and their dependents in war and peace. This mission, however, is not exclusive. Since even before the Enlightenment, European army medics and navy surgeons were among the first whites to explore distant lands and classify the species, peoples, and diseases there. Along the way, military physicians helped establish the contours of racial exceptionalism and white superiority that enabled chattel slavery in the New World, crude and sophisticated schemas of racial hierarchies and classification, and the larger processes of coercion and control that underlie imperialism in Asia and Africa. These men—and in some rare cases, women—were not passive observers of the ideologized biology that took shape in the eighteenth and nineteenth centuries. They were active contributors, lending the gravity and legitimacy of their uniform, rank, and position to the creation of a racial order that would define Western responses to the rest of the world well into the twentieth century.

As in so many areas, the world historian William McNeill is a good starting point for understanding the connections between imperial expansion, the perceived hazards of the tropical world for Europeans, and military medical institutions. His landmark book *Plagues and Peoples* (1976) helped establish the centrality of disease as a historical force in popular and academic discourse. As a world historian, McNeill was a master in identifying and defining linkages across seemingly disparate topics and phenomena, presenting his conclusions in a very approachable fashion. As military history becomes more inclusive of nontraditional narratives, including the intersection of disease and empire, such works are a critical foundation for future study. Indeed, most of the historiography dealing with the activities of medical officers and other military personnel in the imperial sphere has come from outside the purview of military historians. Alfred W. Crosby (1986, 2004) considers the extent to which European global expansion was connected to biological factors, including disease and the importation of new plants and domestic animals, in an explanation that is in many ways an expansion of his Columbian exchange thesis (1973). It sets the standard for other *longue durée* attempts to link biology and geography with history, particularly that of Jared Diamond (1999).

More specifically, the theme of infectious disease assisting the conquest of indigenous populations and societies by outsiders is more carefully addressed by Ken De Bevoise (1995). The experience of war between the beginning of the Philippine revolution against Spain in 1896 through the end of the American–Philippine War in 1903 was exacerbated by repeated waves of disease, ranging from malaria and cholera to a dramatic upsurge in syphilis. The arrival of over 145,000 Spanish and American troops attracted thousands of prostitutes to Manila, who facilitated the spread of syphilis among the soldiers, who subsequently transferred infection across the islands and beyond. The arrival of smallpox and new strains of cholera in 1902 further decimated Filipinos, leading De Bevoise to conclude that the arrival of European and white troops at the turn of the nineteenth century not only was assisted by the diseases that accompanied them but also triggered an epidemiological disaster.

Interdisciplinary models provide a theoretical foundation for developing the context of military medical activity in support of the imperial project in the tropics. Environmental historian David Arnold (1996) lays out such a useful teleology, arguing the combination of physical environment, endemic disease, and the fecundity of flora and fauna in sub-Saharan Africa, South Asia, and Central and South America formed the basis for *tropicality*, a social construct that recast these regions as biological death traps for white Europeans.

> Landscapes, and other aspects of the physical environment, such as climate and disease, were endowed with great moral significance. Environments . . . were widely believed to have a determining influence upon cultures: savagery, like civilization, was linked to certain climatic or geographical features.
>
> (141–142)

Hence the tropics not only became a lethal space for whites but also was identified as a place for moral and physical degeneration. Military personnel, traders, administrators, and other expatriates risked corroding their own vitality and genetic superiority by prolonged sojourns there, making the physical transformation of the tropical space an essential object of imperialists.

The tropicality construct would become a critical feature of other works in this field. Philip D. Curtin (1998), though primarily concerned with the history of Africa and slavery, notes how the European imperial impulse was explicitly linked to questions of health and whiteness in the tropics. Even as the peacetime mortality rates of soldiers in Europe and the United States were improving during the nineteenth century, the disease toll among men engaged in imperial conquest and occupation in Africa was dramatically high. As white soldiers from temperate climates encountered malaria, yellow fever, typhoid fever, heat stroke, and other tropical diseases and conditions, they died by the scores. Among the many case studies Curtin employs is the French campaign in the Sudan. Between 1883 and 1888, he concludes, disease accounted for two hundred deaths per thousand, as compared with seventy-two per thousand in Senegalese garrisons (87). Similar tolls in other African imperial campaigns fed much of the anti-colonialist response—following the logic laid out by Arnold, the tropics were cast as a lethal space, the place where white men went to die.

The American response to the tropicality dilemma is described to varying extents by Warwick Anderson (2006) and by Bobby A. Wintermute (2010). Anderson describes the extent to which colonial administrators—with military medical officers at the fore—struggled to transform the Philippines into a more "American" space, safe from the archipelago's tropical hazards. The pressures of governing a strange and hostile land and people lent greater importance to crafting administrative systems that employed the new sciences of bacteriology, parasitology, and hygienic sanitation to maintain order. At the same time, military medicine proved critical in reinforcing American racial ideology in the Philippines, establishing the privileged status of whiteness over the indigene, while also transforming the social environment by using public health as part of a larger "nation-building" project. Wintermute further examines this theme of "making the tropics fit for white men." He argues the anti–yellow fever and anti-malaria activities of future surgeon general William Crawford Gorgas, and the discovery and successful remediation of hookworm infection in Puerto Rico by Lieutenant Bailey K. Ashford, were all part of an effort to effect a physical and social transformation of the tropics, removing the environmental hazards to successful white American commercial exploitation of the region. In the process, Army sanitarians and medical scientists achieved tremendous popular acclaim and status, giving Gorgas and his peers the cachet to reset the Medical Department as the nation's premier public health service. This emphasis remained until the United States entered the First World War.

The Future of the Fields

As this partial overview reveals, the subfield of military medical history is rich with opportunity for interdisciplinary social and cultural study. From the immediacy of combat to the impact of infectious disease, from empirical focus on the impact of wounds on the body to deconstructing the intersection of race, science, and empire, military medical practitioners and institutions provide salient insights into human nature, culture, and how we wage war and peace. War and those institutions devoted to its practice are considered as some of the oldest and most lasting social exchanges in recorded memory, provoking and responding to substantive transformations in those cultures, nations, and individuals taking part in it. Many historians and commentators have observed and sought to draw meaning from military affairs, but it is only within the last half century that a concerted effort to understand war as a social phenomenon, outside of the narrow range of political, strategic, and tactical studies, has occurred.

No doubt many students of military history will disagree with the broad statement that war is a social construct, albeit one with devastating material and humanitarian consequences. But as war and society scholarship indicates, there are few human endeavors that have such long-lasting and all-encompassing effects. Likewise, as a social and physical condition, the practice of armed defense and war necessitates—even in its lowest state of readiness—more total cooperation across society as a whole. War and society studies works because society dictates war (sadly) to be an essential function for defense, protection, and survival.

Bibliography

Adams, George Worthington. *Doctors in Blue: The Medical History of the Union Army in the Civil War.* Baton Rouge: Louisiana University Press, 1952, 1980.

Anderson, Warwick. *Colonial Pathologies: American Tropical Medicine, Race, and Hygiene in the Philippines.* Durham, NC: Duke University Press, 2006.

Arnold, David. *The Problem of Nature: Environment, Culture and European Expansion.* New Perspectives on the Past. Oxford: Wiley-Blackwell, 1996.

Bell, Andrew McIlwaine. *Mosquito Soldiers: Malaria, Yellow Fever, and the Course of the Civil War.* Baton Rouge: Louisiana State University Press, 2010.

Bollet, Alfred Jay. *Civil War Medicine: Challenges and Triumphs.* Tucson, AZ: Galen Press, 2002.

Bourke, Joanna. *Dismembering the Male: Men's Bodies, Britain, and the Great War.* Chicago: The University of Chicago Press, 1996.

Byerly, Carol. *Fever of War: The Influenza Epidemic in the U.S. Army During World War I.* New York: New York University Press, 2005.

Cirillo, Vincent J. *Bullets and Bacilli: The Spanish-American War and Military Medicine.* Piscataway, NJ: Rutgers University Press, 2004.

Covington, Sarah. *Wounds, Flesh, and Metaphor in Seventeenth-Century England.* New York: Palgrave Macmillan, 2009.

Cowdrey, Albert. *Fighting for Life: American Military Medicine in World War II.* New York: The Free Press, 1994.

Crosby, Alfred W. *The Columbian Exchange: Biological and Cultural Consequences of 1492.* Contributions in American Studies. Westport, CT: Greenwood Press, 1973.

———. *Ecological Imperialism: The Biological Expansion of Europe, 900–1900.* Second Edition. Cambridge, UK: Cambridge University Press, 1986, 2004.

Cunningham, H. H. *Doctors in Gray: The Confederate Medical Services.* Baton Rouge: Louisiana State University Press, 1958, 1986.

Curtin, Philip D. *Disease and Empire: The Health of European Troops in the Conquest of Africa.* Cambridge, UK: Cambridge University Press, 1998.

Dean, Eric T., Jr. *Shook Over Hell: Post-Traumatic Stress, Vietnam, and the Cold War.* Cambridge: Harvard University Press, 1997.

De Bevoise, Ken. *Agents of Apocalypse: Epidemic Disease in the Colonial Philippines.* Princeton: Princeton University Press, 1995.

Devine, Shauna. *Learning From the Wounded: The Civil War and the Rise of American Medical Science.* Chapel Hill: The University of North Carolina Press, 2014.

Diamond, Jared. *Guns, Germs, and Steel: The Fates of Human Societies.* New York: W.W. Norton, 1999.

Fenn, Elizabeth A. *Pox Americana: The Great Smallpox Epidemic of 1775–1782.* New York: Hill & Wang, 2001.

Freemon, Frank R. *Gangrene and Glory: Medical Care During the American Civil War.* Urbana: University of Illinois Press, 1998, 2001.

Hanson, Victor Davis. *The Western Way of War: Infantry Battle in Classical Greece.* New York: Knopf, 1989.

Harrison, Mark. *Medicine & Victory: British Military Medicine in the Second World War.* Oxford: Oxford University Press, 2004.

Humphreys, Margaret. *Intensely Human: The Health of the Black Soldier in the American Civil War.* Baltimore: The Johns Hopkins University Press, 2008.

————. *Marrow of Tragedy: The Health Crisis of the American Civil War.* Baltimore: The Johns Hopkins University Press, 2013.

————. *Yellow Fever and the American South.* Baltimore: The Johns Hopkins University Press, 1992, 1999.

Jensen, Kimberly. *Mobilizing Minerva: American Women in the First World War.* Urbana: University of Illinois Press, 2008.

Jones, Ann. *They Were Soldiers: How the Wounded Return From America's Wars—The Untold Story.* Chicago: Haymarket Books, 2013.

Keegan, John. *The Face of Battle: A Study of Agincourt, Waterloo, and the Somme.* New York: Viking, 1976.

Lendon, J. E. *Soldiers and Ghosts: A History of Battle in Classical Antiquity.* New Haven: Yale University Press, 2005.

Lifton, Robert Jay. *Home From the War: Neither Victims nor Executioner.* New York: Simon & Schuster, 1973; retitled: *Home From the War: Learning From Vietnam Veterans.* New York: Other Press, 2005.

McNeill, J. R. *Mosquito Empires: Ecology and War in the Greater Caribbean, 1620–1914.* New Approaches to the Americas. Cambridge, UK: Cambridge University Press, 2010.

McNeil, William. *Plagues and Peoples.* New York: Anchor, 1976.

Meier, Kathryn Shively. *Nature's Civil War: Common Soldiers and the Environment in 1862 Virginia.* Chapel Hill: The University of North Carolina Press, 2013.

Miller, Brian. *Empty Sleeves: Amputation in the Civil War South.* Athens: University of Georgia Press, 2015.

Pierce, John A., and Jim Writer: *Yellow Jack: How Yellow Fever Ravaged America and Walter Reed Discovered Its Deadly Secrets.* Hoboken, NJ: John Wiley & Sons, 2005.

Rutkow, Ira M. *Bleeding Blue and Gray: Civil War Surgery and the Evolution of American Medicine.* New York: Random House, 2005.

Sarnecky, Mary T. *A History of the U.S. Army Nurse Corps.* Philadelphia: The University of Pennsylvania Press, 1999.

Schultz, Jane E. *Women at the Front: Hospital Workers in Civil War Hospitals.* Chapel Hill: The University of North Carolina Press, 2004.

Scurlock, JoAnn, and Burton R. Andersen, translators and commentary. *Diagnoses in Assyrian and Babylonian Medicine.* Urbana: University of Illinois Press, 2005.

Shay, Jonathan. *Achilles in Vietnam: Combat Trauma and the Undoing of Character.* New York: Simon & Schuster, 1995.

Steiner, Paul. *Disease in the Civil War: Natural Biological Warfare in 1861–1865.* Springfield, IL: Charles C. Thomas, 1968.

Threat, Charissa J. *Nursing Civil Rights: Gender and Race in the Army Nurse Corps.* Urbana: University of Illinois Press, 2015.

Wintermute, Bobby A. *Public Health and the US Military: A History of the Army Medical Department, 1817–1917.* New York: Routledge, 2010.

16 The Combatant Experience

Eric W. Klinek

The "new military history" is no longer new. A shift away from the "drum and trumpet" approach began as early as the 1960s. The discipline now integrates diverse methodologies and is more conversant with fields like social, cultural, political, and economic history. Military historians have begun to appreciate how geography, demography, race, and gender impact the armed forces about which they write. They have increasingly equipped themselves with the intellectual tools required to assess the impact of war on society and, conversely, a society's role in determining how a nation prepares for and wages war. Modern narratives incorporate the stories of the individuals who fought for their hearth and home, for their small tribe or large nation-state, for territory or access to resources, or for abstract ideals, such as patriotism and nationalism. Combat history, an emergent subfield of the new military history, places primacy on the soldier's experience.

Source availability determines which combatant experiences historians can relate. From the eighteenth century to the present, as literacy rates increased and as wars grew larger in both geographic scope and number of men in arms, the documentary record accordingly expanded. Unit reports, diaries, letters, memoirs, and, more recently, oral histories have permitted historians to describe battle in greater detail. Scholarship, nevertheless, still tends to focus on the infantryman's war. There is a noticeable lacuna in the literature concerning the exploits of sailors, airmen, medics, support and logistics personnel, and engineers. Even less attention has been devoted to women, African Americans, minority combatants, and the non-Western battle experience. This essay will discuss the historiography of the combatant experience across time, conflict, and continent.

Bell I. Wiley was one of the forefathers of the new military history. Wiley's *The Life of Johnny Reb* (1943) and *The Life of Billy Yank* (1952) were groundbreaking social histories of Civil War soldiers' quotidian lives. He utilized diaries, letters, and memoirs to explain what these men thought and felt, why they fought, how they endured the ordeal of combat, and their opinions of the enemy. Wiley's influence on subsequent generations of scholars is well known. S.L.A. Marshall's *Men Against Fire* (1947) and John Keegan's *The Face of Battle* (1976), however, serve as the foundational texts on the combatant experience.

In many regards, Marshall's study is significant for the wrong reasons. While serving as an officer during World War II, Marshall conducted post-combat interviews to gather information on soldiers' experiences, training, unit cohesion, and small-unit leadership. He concluded that only 10 to 25 percent of GIs fired their weapons in combat because most individuals were psychologically reluctant to kill. The postwar American military widely accepted these findings, and many historians regurgitated Marshall's statistics as scripture. By the 1980s, though, scholars began challenging his assertions. Marshall did not offer statistical evidence to support his claims; in fact, he never questioned combatants about firing rates during his interviews. Despite its methodological flaws, *Men Against Fire* remains essential reading for the student of military history. Statistics and interviews—accurate ones of course—have become a crucial evidentiary base when writing about combat. By focusing on the individual combatant's thoughts and feelings, Marshall demonstrated the value of "bottom-up" history.

In *The Face of Battle*, Keegan pays homage to Marshall and even cites his predecessor's claims about firing rates. Keegan evaluates the common British soldier's experience in three battles—Agincourt (October 25, 1415), Waterloo (June 18, 1815), and the first day of the Somme (July 1, 1916). He describes the sights and sounds of battle, as well as soldiers' fears and wounds. The only true way to view the "face of battle" is by inserting death as a central feature of the narrative. Although the pre-nineteenth-century common soldiery was almost wholly illiterate, Keegan permits the combatants to relate their stories in their own words whenever such firsthand accounts exist.

Several authors have mimicked Keegan's case-study approach and compared the combatant experience across time and place. In *Men of War: The American Soldier in Combat at Bunker Hill, Gettysburg, and Iwo Jima* (2015), Alexander Rose attempts to write the American version of *The Face of Battle*. His aim is to describe the "*typical* soldier's *typical* combat experience" in three iconic battles. The dangers of such a reductionist framework should be self-evident. Who is a typical soldier? What is a typical combat experience? Do they even exist?

Peter Kindsvatter, in *American Soldiers: Ground Combat in the World Wars, Korea, and Vietnam* (2003), astutely avoids the pitfalls of defining a typical combatant or combat experience. He also challenges S.L.A. Marshall's assertions that soldiers were reluctant killers. By studying America's twentieth-century conscript-army wars, Kindsvatter benefits from the widely available records (e.g., memoirs, letters, and oral histories) that Army and Marine participants left behind. The author evaluates motivation, induction, training, life in the armed forces, and the African American soldier experience during the draft era.

Christopher Hamner and John C. McManus are also Keegan disciples. In *Enduring Battle: American Soldiers in Three Wars, 1776–1945* (2011), Hamner accounts for American soldiers' motivation in the Revolutionary War, Civil War, and World War II. He employs both sociological and psychological methodologies in his work. As warfare became more individualized and isolated, soldiers realized that their own actions could impact their chances of survival. Hamner challenges the "primary group thesis," largely attributable to S.L.A. Marshall, that suggests that social bonds influence men in combat. Hamner argues instead that strong group cohesion makes soldiers casualty-averse. In *Grunts: Inside the American Infantry Combat Experience, World War II Through Iraq* (2010), McManus examines ten engagements: four from World War II; three from Vietnam; one from Gulf War I; and two from Gulf War II. In this volume, Korea earns its nickname "the forgotten war." McManus de-emphasizes the role of technology in combat, finding instead that wars cannot be won without boots on the ground.

The aforementioned authors, on the whole, benefited from a significant written historical record on which to base their comparative studies of soldiers in combat. It is a far greater challenge to portray warfare in ancient times or the Middle Ages. Victor Davis Hanson's *Carnage and Culture* (2001) addresses the combatant experience in nine landmark battles over a 2,500-year period, from the Battle of Salamis (480 BCE) to the 1968 Tet Offensive. Hanson reasons that cultural factors, more so than environmental, geographical, or biological ones, have determined how Western powers wage war. Culture, then, accounts for Western military superiority over non-Western civilizations, and the battles he selects largely fit this paradigm. Hanson is not concerned with a Western power's morality or righteousness in waging war. Rather, taking a page from Russell Weigley, he contends that the Western way of war emphasizes annihilation of the enemy. Hanson admits that military historians and academics will likely find his generalizations to be chauvinistic.

John Lynn's *Battle: A History of Combat and Culture* (2003) is similar to Hanson's work in certain respects. Lynn examines various conflicts across the globe, from Ancient Greece to the current War on Terror, and posits that ideas and culture have a greater impact on warfare than technology. Culture is the lens through which Lynn evaluates how militaries and their corresponding ways of war have changed over time. But *Battle* is also very much a response to Hanson. Lynn attacks the notion of the "universal soldier." He challenges the monolithic concept of a Western way of war by stressing that combatants and combat must be understood within their specific historical contexts. Thus, he eschews providing a broad overview of warfare for the two-plus millennia under consideration in his book. Finally, Lynn differentiates between cultural conceptions of war (a society's discourse/s on war) and the reality of war.

Hanson traces the origins of Western civilization and a Western way of war to Ancient Greece. His edited volume *Hoplites* (1991) analyzes the Greek citizen soldiers' experiences, weapons, the battlefield environment, and tradition's influence on Greek warfare. The contributing authors relied on ancient literature, artifacts, iconography, epigraphy, and archaeological resources to depict the bloody and brutalizing nature of Greek battle. The final chapter of Adrian Goldsworthy's *The Roman Army at War, 100 BC–AD 200* (1996) highlights the individual soldier's experience in combat, while Gregory Daly's *Cannae: The Experience of Battle in the Second Punic War* (2002) details the feats of the Roman and Carthaginian commanders and soldiers who met on the battlefield in 216 BCE. Similar to Hanson, Daly portrays ancient warfare as horrific and frenzied.

It was during the Age of Reason, and the subsequent period of the American and French Revolutions, when firsthand written accounts of battle proliferated. Scholars of these and subsequent eras, consequently, have been able to compose more detailed studies. Christopher Duffy's *The Military Experience in the Age of Reason* (1987) is concerned primarily with the Seven Years' War and the professional soldier's life in the army and motivation in combat. Stephen Brumwell also analyzes the Seven Years' War, albeit in the American colonies. In *Redcoats: The British Soldier and War in the Americas, 1755–1763* (2002), Brumwell consults journals and diaries to recreate the battle experience for British regulars who served from Canada to Cuba.

In *A Devil of a Whipping: The Battle of Cowpens* (1998), Lawrence Babits utilizes Marshall's and Keegan's methodological approaches to explain how an amalgamated force of American Continentals and militia defeated British regulars in South Carolina in 1781. Babits draws on veteran pension applications and journals to create a comprehensive narrative of a victory that was pivotal to America's successful bid for independence. Caroline Cox's *A Proper Sense of Honor: Service and Sacrifice in George Washington's Army* (2004) demonstrates that the Continental Army mirrored colonial social stratification. The officers, on the whole, were wealthy landowners. They generally avoided socializing with lower-class enlisted men. These common soldiers were beholden to strict military obedience and harsh discipline, and misery and privation characterized their time in the army.

Matthew Spring and Don Hagist write about the British Army's experience in North America during the American Revolution. In *With Zeal and Bayonets Only* (2008), Spring reveals that, contrary to popular belief, the British Army was tactically adaptable on the battlefield. Hagist, in turn, has given voice to the common British soldier. In *British Soldiers, American War: Voices of the American Revolution* (2012), Hagist contextualizes nine firsthand accounts from British combatants. Rory Muir is also concerned with the British Army, but he studies its campaigns during the Napoleonic Era. In *Tactics and the Experience of Battle in the Age of Napoleon* (1998), Muir relies on memoirs, letters, and diaries to elucidate what combat was like and how soldiers acquitted themselves on the battlefield.

There are several excellent works on soldiers' experiences in the decades between the American Revolution and the American Civil War. In *For Liberty and the Republic: The American Citizen Soldier, 1775–1861* (2015), Ricardo Herrera contends that a "military ethos of republicanism" bound together the military and the society from which it was derived and that it defended. Republican culture provided the cohesive power necessary to overcome regional and social differences, and five principles influenced soldiers who took up arms for their country: virtue; legitimacy; self-governance; the interrelationship between religion and liberty; and glory, honor, and fame. Edward Skeen, in *Citizen Soldiers in the War of 1812* (1999), argues that state militias acquitted themselves poorly due to deficiencies in training, equipment, and discipline. Skeen also devotes some attention to morale, battle experiences, and camp conditions.

The Mexican War, 1846–1848, is fertile ground for additional research. In *Army of Manifest Destiny* (1992), James McCaffrey utilizes soldiers' letters and memoirs to portray their experiences, from enlistment to camp life, through postwar occupation in Mexico. His work is very much a social history of soldiers' everyday lives. Bruce Winders' *Mr. Polk's Army* (1997) is similar to McCaffrey's work in several respects. The author evaluates the American regulars and volunteers who fought in Mexico, as well as their daily routines and experiences, the weapons they used in combat, and their motivations for fighting.

There is no shortage of literature on the American Civil War. As the title of Gerald Linderman's *Embattled Courage: The Experience of Combat in the American Civil War* (1987) suggests, the author concludes that courage motivated and sustained men in combat. Though religion and honor also determined how a man comported himself on the battlefield, it was courage, more so than discipline, that accounted for unit cohesion. Yet, author Earl Hess believes that his book, *The Union Soldier in Battle* (1997), serves as a corrective to *Embattled Courage* in that it focuses on more than a soldier's mettle. Hess also comments on how Victorian values, ideology, religion, and camaraderie enabled northern soldiers to endure the strains of combat.

Reid Mitchell's *Civil War Soldiers: Their Expectations and Their Experiences* (1988) and James I. Robertson Jr.'s *Soldiers Blue and Gray* (1988) both recreate life in the Union and Confederate armies by exploiting soldiers' letters, diaries, and journals. Mitchell's aim is to understand a soldier's motivation for fighting and the impact of the war on his values. Though Mitchell concedes that Johnny Reb and Billy Yank were similar in certain respects, the men who composed the opposing forces were fundamentally different. In contrast, Robertson treats northern and southern soldiers as essentially the same. He investigates various aspects of military life during the Civil War, including enlistment, combat, casualties, privation, and emotions.

One of the most recent and unique books on the Civil War combatant experience is Mark M. Smith's *The Smell of Battle, the Taste of Siege: A Sensory History of the Civil War* (2015). Each of the book's five chapters concerns one of the senses during a specific episode of the war: the sound of the Confederate bombardment of Fort Sumter; the sight of opposing armies in a wide array of non-standardized uniforms at First Bull Run; the stench of decaying corpses in the hot summer sun at Gettysburg; taste and starvation during the siege of Vicksburg; and the sense of touch in the cramped, dark, and hot submarine *H.L. Hunley*. Similarly, Eric T. Dean Jr.'s *Shook Over Hell: Post-Traumatic Stress, Vietnam, and the Civil War* (1997) evaluates the nonphysical scars that warfare inflicts on participants. Dean challenges the notion that Vietnam veterans disproportionately suffered higher rates of post-traumatic stress disorder (PTSD) than combatants from other conflicts. He incorporates both psychological and historical methodologies in his analysis of pension records and asylum papers for 291 Union soldiers.

The period between the American Civil War and World War I is ripe for reexamination. There is a paucity of scholarship on the combatant experience during the Wars of German Unification, the "Indian" Wars, the various colonial wars (e.g., Boer War, Spanish-American War, Philippine Insurrection), and Japan, Russia, and China's activities at the turn of the twentieth century. Even less attention has been paid to the colonized peoples who struggled for sovereignty in these conflicts. John Laband has addressed, in part, this historical amnesia. In *Zulu Warriors: The Battle for the South African Frontier* (2014), the author studies four wars in the late 1870s between British troops and a handful of African states. Laband considers the impact of culture on military performance. By emphasizing the experiences of the states that resisted colonial rule, it becomes evident that the British simply outmatched their African counterparts.

While the colonial wars at the turn of the twentieth century redrew map lines on the "periphery," the events of 1914 to 1918 shattered Europe, the Middle East, and centuries-old empires. Edward M. Coffman's *The War to End All Wars: The American Military Experience in World War I* (1968) is essential reading for any scholar of that conflict. The book was at the cutting edge of the new military history when it appeared. Coffman interviewed veterans and utilized collections of personal papers to paint a broad picture of the American wartime experience. Jennifer Keene's *World War I: The American Soldier Experience* (2011) situates American involvement in the Great War in 1917–1918 in a broader geographical and temporal context; in the latter regard, her study begins with American military intervention in Mexico in 1914 and concludes with the passage of the Bonus Act in 1924. She not only considers life and death in combat but also assesses recruitment and conscription, soldier training, and doughboy interactions with French civilians. In *Doughboys of the Great War: How American Soldiers Viewed Their Military Service* (2014), Edward Gutierrez relates what members of the American Expeditionary Forces (AEF) said about their combat experiences upon returning to the United States in 1919. Four states—Utah, Minnesota, Connecticut, and Virginia—provided returning veterans with questionnaires that covered all aspects of military life, including time in camp, camaraderie on the line, weapons and equipment, and life and death in the trenches.

John Ellis' *Eye-Deep in Hell: Trench Warfare in World War I* (1976) discusses daily life in the mud on the front lines and in no-man's-land. By drawing from both official sources and personal letters and diaries, Ellis provides a comprehensive account of trench structures, daily routines, disease and vermin, casualties, motivation and morale, and death and decay. In *No Man's Land: Combat and Identity in World War I* (1979), Eric Leed addresses why men eagerly went to war in 1914, what motivated them, and what sustained them in combat. He relies on anthropological, sociological, and psychoanalytical methodologies to evaluate modern industrialized warfare's impact on the lives—and minds—of the men who fought in the trenches. Two studies in particular focus on the Great War's psychological toll on the British "Tommy." Considered together, Peter Leese's *Shell Shock: Traumatic Neurosis and the British Soldiers of the First World War* (2002) and Fiona Reid's *Broken Men: Shell Shock, Treatment and Recovery in Britain, 1914–30* (2010) contemplate prewar notions and treatment of mental illness, early attempts to diagnose and cure "shell shock," and postwar care for the men seemingly caught in perpetual battle.

The majority of English-language books on combat experience in World War I are about the Anglo-American forces on the Western Front. In contrast, George Mosse's *Fallen Soldiers: Reshaping the Memory of the World Wars* (1990) is primarily concerned with the German experience. He links the two world wars by tracing the rise of the mythos of the fallen soldier, a concept that enabled warriors and societies to derive meaning and purpose from modern warfare and its attendant death and suffering. In the decades following World War II, scholars propagated the myth that the *Wehrmacht* was innocent of perpetrating crimes against humanity. In *Hitler's Army* (1992), however, Omer Bartov accounts for the *Wehrmacht*'s complicity in committing atrocities on the Eastern Front. He finds that de-modernization of the front, the primary group and discipline's impact on unit cohesion and morale, and ideology and indoctrination brutalized and radicalized the German soldiers. Wolfram Wette's *The Wehrmacht: History, Myth, Reality* (2006) similarly reveals the German armed forces' involvement in war crimes on the Eastern Front. Wette's study further shatters the commonly accepted argument that it was the *Schutzstaffel* (SS) alone that implemented Hitler's exterminationist policies. In *Frontsoldaten* (1995), Steven Fritz employs a bottom-up approach to depict what the average German soldier (*Landser*)—as opposed to a member of the *Waffen-SS*—experienced during World War II (primarily on the Eastern Front). His account is based largely on letters, diaries, and oral histories; he neglects to consult official sources and memoranda. Fritz concludes that ideology (i.e., Nazism) motivated *Landsers* in combat.

Limited access to Russian and former Soviet archives has precluded scholars from examining the everyday experiences of Soviet soldiers on the Eastern Front. Catherine Merridale's *Ivan's War: Life and Death in the Red Army, 1939–1945* (2006), therefore, is a pioneering study. The author culled never-before-seen material from military and secret police archives, mined personal letters and diaries, and interviewed men and women who fought in the East. Merridale is one of the first scholars to evaluate warfare in the East from the perspective of the common Soviet soldiers who fought and died in the mud and snow.

While the historiography of the combatant experience on the Eastern Front is limited, the literature on the American GIs' experience is quite extensive. Interestingly, though, it was the country's worst military defeat—Vietnam—that led to a renewed historical focus on the American experience in World War II. Following a period of military, political, and social turmoil, scholars turned their attention to the "good war" to determine, in part, what went wrong in the 1960s and 1970s. Studs Terkel's Pulitzer Prize–winning *The Good War* (1985), a collection of oral histories from Americans from all walks of life, initiated this reevaluation of the World War II war effort.

Harold Leinbaugh and John Campbell's *The Men of Company K: The Autobiography of a World War II Rifle Company* (1985) was one such post-Vietnam study. Leinbaugh and Campbell were officers in Company K, 333rd Infantry Regiment, 84th Infantry Division, a unit that saw action from the Battle of the Bulge to V-E Day. Based largely on their own memories and those of the men with whom they served, the authors tell one rifle company's story in the war against Germany. But even more significantly, Leinbaugh and Campbell were some of the first authors to challenge S.L.A. Marshall's claims about fire ratios. They saw firsthand that far greater than 25 percent of their company fired in combat.

Stephen Ambrose similarly related the experiences of a small unit during World War II. *Band of Brothers* (1992) follows the men of E Company, 506th Parachute Infantry Regiment, 101st Airborne Division, from stateside formation and training in Georgia, through combat jumps into Normandy and the Netherlands, to the Battle of the Bulge, and to the surrender of Germany. Though plagued later in his career by accusations of plagiarism and exaggerating his relationship with Dwight D. Eisenhower, Ambrose's books have had a significant impact on the genre of combat history. John Ellis' *The Experience of War Through the Eyes of the Allied Soldiers in World War II* (1991) and Lee Kennett's *G.I.: The American Soldier in World War II* (1987) paint broader pictures of life in combat during World War II. Ellis, in particular, made effective use of oral history in his work. Masayo Umezawa Duus' *Unlikely Liberators: The Men of the 100th and 442nd* (1987) focuses on the now well-known exploits of the Nisei (Japanese American) soldiers who fought for a country that interned many of their families. The author benefited from access to personal papers and veteran interviews. The 442nd Regimental Combat Team achieved fame in the brutal campaign on the Italian Peninsula and in the Vosges Mountains of Eastern France. It would become one of the most highly decorated U.S. Army units of World War II.

The fiftieth anniversaries of D-Day, V-E Day, and V-J Day in 1994 and 1995 sparked renewed popular and scholarly interest in American participation in World War II. As aging veterans began passing away, there was a dramatic increase in the publication of memoirs and books about their experiences. Furthermore, the Smithsonian's controversial 1995 *Enola*

Gay exhibit reintroduced the war into the popular consciousness in an unintended way by forcing historians to question the "good war" thesis. Scholars also began reexamining the military's performance and effectiveness.

Three works simultaneously evaluate American combat efficiency in the European theater and the soldier experience: Michael Doubler's *Closing With the Enemy: How GIs Fought the War in Europe, 1944–1945* (1994); Peter Mansoor's *The GI Offensive in Europe: The Triumph of American Infantry Divisions, 1941–1945* (1999); and Robert Rush's *Hell in Hürtgen Forest: The Ordeal and Triumph of an American Infantry Regiment* (2001). Doubler concludes that, despite prewar neglect with regard to funding, equipping, and training, the Army was still effective in battle. He depicts a process of innovation, growth, and improvement in combat, which culminated in a greater understanding of combined-arms operations. His focus is the tactical level and the common soldier, and he draws upon a variety of official documents, interviews, memoirs, personal papers, and oral histories. Mansoor similarly identifies a process of innovation and adaptation, albeit at the divisional level. He finds that flexibility and the ability to learn from previous experience resulted in increased effectiveness. Finally, Robert Rush highlights the engagements of the 22nd Infantry Regiment, 4th Infantry Division, in November and December 1944 in the meat-grinder that was the Hürtgen Forest. He argues that the American infantry was more effective than its German counterpart and that the Army's replacement system permitted units to sustain "battle worthiness." Rush echoes Omer Bartov's claims about the importance of group and organizational cohesion, thus contradicting S.L.A. Marshall's "primary group" thesis.

Other significant works on the Army in the European Theater of Operations (ETO) include Gerald Linderman's *The World Within War: America's Combat Experience in World War II* (1997) and Peter Schrijvers' *The Crash of Ruin: American Combat Soldiers in Europe During World War II* (1998). John McManus' *The Dead and Those About to Die* (2014) highlights the 1st Infantry Division's exploits on D-Day and puts readers on Omaha Beach with the soldiers. There are several important works on ground combat in the Pacific theater as well. Craig Cameron's *American Samurai* (1994) is an innovative study of how "imaginary constructs" and myths impacted the Marine Corps' ethos and self-image. He evaluates how the warrior elite's conceptions of "us versus them" impacted the First Marine Division's campaigns on Guadalcanal, Peleliu, and Okinawa. Patrick O'Donnell's *Into the Rising Sun: In Their Own Words, World War II's Pacific Veterans Reveal the Heart of Combat* (2002) offers a detailed look at Marine and Army infantrymen as they island-hopped across the Pacific. In *The GI War Against Japan* (2004), Schrijvers utilizes soldiers' personal papers and military documents to describe the American combat experience in the Pacific and Asia. Like Cameron, Schrijvers demonstrates that warfare in the Pacific was particularly brutal. For Schrijvers, the nature of the conflict in this theater was the result, in part, of the intersection of GIs' preconceived notions of the region and people there, the terrain the soldiers encountered, and the enemy that Americans engaged.

In general, scholarship tends to equate "combatant" with "ground soldier." There are relatively few books on the sailor or airman's experience during World War II. Mark K. Wells' *Courage in Air Warfare: The Allied Aircrew Experience in the Second World War* (1995) compares how the U.S. Eighth Air Force and Royal Air Force Bomber Command formed their crews, utilized their manpower, and addressed the psychological strain of air warfare. Donald Miller's *Masters of the Air: America's Bomber Boys Who Fought the Air War Against Nazi Germany* (2006) offers a gripping narrative that recreates the life of an air crewman, from downtime on base in England between missions to terrifying moments in flak- and fighter-filled skies over Nazi-occupied Europe.

Literature on the American experience in World War II has largely ignored African Americans. Though written between 1947 and 1951, Ulysses Lee's *The Employment of Negro Troops*

(1965) has long stood as the foundational work on black soldiers' experiences in a segregated military. The language in the title, however, clearly indicates the time period in which it was published. Subsequent scholars have employed various methodologies to provide a fuller account of what African Americans endured. Mary Penick Motley's edited volume, *The Invisible Soldier: The Experience of the Black Soldier, World War II* (1975), consists of more than fifty oral histories with officers and enlisted men who served in all branches of the Jim Crow military. Bryan D. Booker's *African Americans in the United States Army in World War II* (2008) expands on Lee's work even further. He not only addresses both combat and non-combat personnel but also evaluates their experiences during World War I and the interwar period.

Two authors offer unique perspectives on combat, as both men served in World War II and then enjoyed postwar careers as prominent academics. Jesse Glenn Gray, a professor of philosophy, and Paul Fussell, a cultural and literary historian and critic, analyzed their own experiences in an attempt to derive meaning from battle. Their books are not mere memoirs. Rather, they provide a larger analytical framework for understanding soldiers' motivations and actions on the battlefield. In *The Warriors: Reflections on Men in Battle* (1959), Gray, who served as an intelligence officer in North Africa, Italy, France, and Germany, reflected on his wartime letters and journals in an effort to heal his emotional wounds. He philosophized on how warfare—particularly total warfare—impacted man. Fussell was an infantry lieutenant in the European theater. In *The Boys' Crusade: The American Infantry in Northwest Europe, 1944–1945* (2003), he demolishes the "good war" and "greatest generation" theses by portraying war in all its gruesomeness. The American citizen soldiers were merely scared young men and boys who witnessed carnage, death, and decay firsthand. For Fussell and his comrades, combat was neither glorious nor heroic.

The Korean War not only is the "forgotten war" in the American popular consciousness but also is overlooked in the historiography. The extant literature focuses heavily on the Marine Corps in the November 1950 Chosin Reservoir campaign. Eric M. Hammel's *Chosin: Heroic Ordeal of the Korean War* (1981) depicts the First Marine Division's fight against the tens of thousands of Chinese soldiers who crossed the Yalu River and swarmed into North Korea. Rocky terrain, inhospitable weather, and subzero temperatures proved almost as deadly as the communist forces. Veteran Martin Russ' *Breakout: The Chosin Reservoir Campaign, Korea 1950* (2000) evaluates how twelve thousand encircled Marines and GIs endured the Chinese onslaught and broke through enemy lines. Though *Breakout* is mainly concerned with the First Marine Division, the author incorporates all belligerents' perspectives: American, Chinese, and Korean.

Bob Drury and Tom Clavin's *The Last Stand of Fox Company: A True Story of U.S. Marines in Combat* (2009) relates how one company of the First Marine Division held out on an isolated hill at the "Frozen Chosin" for the better part of a week in November 1950. This vivid account relies, in part, on survivor interviews. Patrick K. O'Donnell's *Give Me Tomorrow: The Korean War's Greatest Untold Story—the Epic Stand of the Marines of George Company* (2010) follows one company that landed at Inchon in September 1950 and fought its way through the streets of Seoul. It then found itself as part of an Allied force at the Chosin that was outnumbered ten to one by Chinese troops.

There is a rich historiography of American involvement in Vietnam, but the majority of these works deal with strategy, tactics, and political and military leadership. John Wood's *Veteran Narratives and the Collective Memory of the Vietnam War* (2016) offers a literary and cultural analysis of how Vietnam veterans' memoirs have impacted collective memory and understanding of the war. The book should serve as a useful companion to two of the most well-known Vietnam memoirs: Philip Caputo's *A Rumor of War* (1977) and Lieutenant Colonel Hal Moore and war correspondent Joseph Galloway's *We Were Soldiers Once . . . and Young: Ia Drang, the Battle That Changed the War in Vietnam* (1992). Caputo, a Marine who served in Vietnam in 1965 and 1966, examines men's actions in war and warfare's impact on soldiers. Combat quickly disabused him of his idealized notions of serving his country, and he and his comrades became engaged in a bloody, savage, unconventional war of attrition in an inhospitable climate and terrain. *We Were Soldiers* recounts the 1st Battalion, 7th Cavalry's experiences in one of the first, and one of the few conventional, Vietnam battles. Moore and Galloway interviewed hundreds of veterans, including the commanders of the opposing Vietnamese forces.

Andrew Wiest has given long overdue attention to America's primary ally in the war: the Army of the Republic of Vietnam (ARVN). In *Vietnam's Forgotten Army* (2008), which is based on interviews with two ARVN officers, the author explores the ARVN's reputation as an ineffective combat force. Wiest attributes the ARVN's performance to poor leadership and an American strategy that utilized it to wage conventional—as opposed to counterinsurgency—warfare.

James Ebert's *A Life in a Year: The American Infantryman in Vietnam, 1965–1972* (1993) is derived from interviews with more than fifty Marine and Army veterans. Their recollections reveal that there was no "one" Vietnam War; the rice-paddy war was very different from the jungle war, and fighting in central Vietnam was distinct from combat in the Mekong Delta. The Vietnam War of 1965 was not the same as the Vietnam War of 1972. Kyle Longley's *Morenci Marines* (2013) relates the experiences of the "Morenci Nine," a group of men from an Arizona mining camp who served in Vietnam. Six of them died in combat. Longley not only addresses war's impact on the survivors but also considers the relationship between the military and the communities from where its soldiers came.

Scholars and the public have long believed that the uniqueness of the Vietnam War negatively affected combatants' mental welfare. Though Dave Grossman's *On Killing* (1995) does not focus solely on Vietnam, it argues that most soldiers are reluctant killers. Consequently, the objective of military training is to teach recruits to overcome this abhorrence of bloodletting. Grossman believes that the result has been a rise in reported cases of PTSD and a society that has become desensitized to violence. In *Achilles in Vietnam* (1994), Jonathan Shay, a psychiatrist, simultaneously conducts a literary analysis of Homer's *Iliad* and Vietnam veterans' uncensored accounts of combat to identify the underlying social, psychological, and political origins of PTSD.

The Vietnam experience forever altered the American military, particularly with the 1973 implementation of the all-volunteer armed forces. Beth Bailey's *America's Army: Making the All-Volunteer Force* (2009) examines the Army and society's relationship from the Vietnam-era draft protests to the wars in Iraq and Afghanistan. In addition to analyzing the concepts of duty and service, she also evaluates the Army's efforts to market itself to the public and lure potential recruits to its ranks. Ground warfare in the era of the All-Volunteer Force has been characterized by small-unit actions and the employment of special and elite forces. A small portion of the citizenry has borne the burden of fighting seemingly perpetual warfare, and soldiers are now subject to successive deployments to different combat zones. Journalist Sebastian Junger followed the soldiers of 2nd Platoon, Battle Company, 173rd Airborne Brigade during their fifteen-month deployment in 2007 and 2008 in Afghanistan's Korengal Valley. In *War* (2010), Junger describes the sights, sounds, and smells of warfare on an isolated mountain outpost. He also addresses how the compartmentalized and isolated nature of combat in the "War on Terror" impacts the soldiers.

Though there exists a rich literature on the combatant experience, this essay has revealed that scholars still continue to conflate "combatant" and "infantryman." There is ample room for additional research on the battle experiences of airmen, tankers, medics, artillerists, and sailors, to name a few. New scholarship should explore the experiences of oft-overlooked combatants—women, African Americans and other minority groups, and soldiers in non-Western and non-first-world countries and territories. There is also a heavy bias toward relating the Western battle experience, especially during the nineteenth and twentieth centuries. This phenomenon is due, in part, to the abundance of written sources and oral histories that soldiers in these eras generated. There is a paucity of books on the Japanese battle experience during World War II, for instance. There also remains much work to be done on the Korean War from both American and Korean perspectives. Future works on the combatant experience, particularly those that focus on non-Western participants, should consider how different societies prepare for and wage war, distinct notions of service and sacrifice, what motivates warriors and soldiers, and the physical and psychological impact of warfare on the body and mind. Scholars must be cognizant of the fact that battle is very much influenced by the particular era and geographical location in which it occurs, as well as by the people who participate in it. The "universal soldier" does not exist and, hence, neither does the universal combat experience.

Bibliography

Ambrose, Stephen E. *Band of Brothers: E Company, 506th Regiment, 101st Airborne From Normandy to Hitler's Eagles Nest*. New York: Simon & Schuster, 1992.

Babits, Lawrence E. *A Devil of a Whipping: The Battle of Cowpens*. Chapel Hill: University of North Carolina Press, 1998.

Bailey, Beth. *America's Army: Making the All-Volunteer Force*. Cambridge: The Belknap Press of Harvard University Press, 2009.

Bartov, Omer. *Hitler's Army: Soldiers, Nazis, and War in the Third Reich*. New York: Oxford University Press, 1992.

Booker, Bryan D. *African Americans in the United States Army in World War II*. Jefferson, NC: McFarland, 2008.

Brumwell, Stephen. *Redcoats: The British Soldier and War in the Americas, 1755–1763*. Cambridge, UK: Cambridge University Press, 2002.

Cameron, Craig M. *American Samurai: Myth, Imagination, and the Conduct of Battle in the First Marine Division, 1941–1951*. New York: Cambridge University Press, 1994.

Caputo, Philip. *A Rumor of War*. New York: Holt, Rinehart and Winston, 1977.

Coffman, Edward M. *The War to End All Wars: The American Military Experience in World War I*. New York: Oxford University Press, 1968.

Cox, Caroline. *A Proper Sense of Honor: Service and Sacrifice in George Washington's Army*. Chapel Hill: University of North Carolina Press, 2004.

Daly, Gregory. *Cannae: The Experience of Battle in the Second Punic War*. New York: Routledge, 2002.

Dean, Jr., Eric T. *Shook Over Hell: Post-Traumatic Stress, Vietnam, and the Civil War*. Cambridge: Harvard University Press, 1997.

Doubler, Michael D. *Closing With the Enemy: How GIs Fought the War in Europe, 1944–1945*. Lawrence: University Press of Kansas, 1994.

Drury, Bob, and Tom Clavin. *The Last Stand of Fox Company: A True Story of U.S. Marines in Combat*. New York: Atlantic Monthly Press, 2009.

Duffy, Christopher. *The Military Experience in the Age of Reason*. London: Routledge and Kegan Paul, 1987.

Duus, Masayo Umezawa. Translated by Peter Duus. *Unlikely Liberators: The Men of the 100th and the 442nd*. Honolulu: University of Hawaii Press, 1987.

Ebert, James R. *A Life in a Year: The American Infantryman in Vietnam, 1965–1972*. Novato, CA: Presidio Press, 1993.

Ellis, John. *Eye-Deep in Hell: Trench Warfare in World War I*. New York: Pantheon Books, 1976.

———. *On the Front Lines: The Experience of War Through the Eyes of the Allied Soldiers in World War II*. New York: Wiley, 1991.

Fritz, Stephen G. *Frontsoldaten: The German Soldier in World War II*. Lexington: University of Kentucky Press, 1995.

Fussell, Paul. *The Boys' Crusade: The American Infantry in Northwestern Europe, 1944–1945*. New York: Modern Library, 2003.

Goldsworthy, Adrian Keith. *The Roman Army at War, 100 BC–AD 200*. New York: Oxford University Press, 1996.

Gray, J. Glenn. *The Warriors: Reflections on Men in Battle*. New York: Harcourt Brace, 1959.

Grossman, Dave. *On Killing: The Psychological Cost of Learning to Kill in War and Society*. Boston: Little, Brown, 1995.

Gutierrez, Edward A. *Doughboys of the Great War: How American Soldiers Viewed Their Military Service*. Lawrence: University Press of Kansas, 2014.

Hagist, Don N. *British Soldiers, American War: Voices of the American Revolution*. Yardley, PA: Westholme, 2012.

Hammel, Eric M. *Chosin: Heroic Ordeal of the Korean War*. New York: Vanguard Press, 1981.

Hamner, Christopher H. *Enduring Battle: American Soldiers in Three Wars, 1776–1945*. Lawrence: University Press of Kansas, 2011.

Hanson, Victor Davis. *Carnage and Culture: Landmark Battles in the Rise of Western Power*. New York: Doubleday, 2001.

———, ed. *Hoplites: The Classical Greek Battle Experience*. London: Routledge, 1991.

Herrera, Ricardo A. *For Liberty and the Republic: The American Citizen Soldier, 1775–1861*. New York: New York University Press, 2015.

Hess, Earl J. *The Union Soldier in Battle: Enduring the Ordeal of Combat*. Lawrence: University Press of Kansas, 1997.

Junger, Sebastian. *War*. New York: Twelve, 2010.

Keegan, John. *The Face of Battle*. London: J. Cape, 1976.

Keene, Jennifer D. *World War I: The American Soldier Experience*. Lincoln: University of Nebraska Press, 2011.

Kennett, Lee. *G.I.: The American Soldier in World War II*. New York: Charles Scribner's Sons, 1987.

Kindsvatter, Peter S. *American Soldiers: Ground Combat in the World Wars, Korea, and Vietnam*. Lawrence: University Press of Kansas, 2003.

Laband, John. *Zulu Warriors: The Battle for the South African Frontier*. New Haven, CT: Yale University Press, 2014

Lee, Ulysses. *The Employment of Negro Troops*. United States Army in World War II, Special Studies. Washington, DC: Center of Military History, United States Army, 1994. First published 1966 by U.S. Government Printing Office.

Leed, Eric. *No Man's Land: Combat and Identity in World War I*. Cambridge, UK: Cambridge University Press, 1979.

Leese, Peter. *Shell Shock: Traumatic Neurosis and the British Soldiers of the First World War*. New York: Palgrave Macmillan, 2002.

Leinbaugh, Harold P., and John D. Campbell. *The Men of Company K: The Autobiography of a World War II Rifle Company*. New York: W. Morrow, 1985.

Linderman, Gerald F. *Embattled Courage: The Experience of Combat in the American Civil War*. New York: Free Press, 1987.

———. *The World Within War: America's Combat Experience in World War II*. New York: Free Press, 1997.

Longley, Kyle. *The Morenci Marines: A Tale of Small Town America and the Vietnam War*. Lawrence: University Press of Kansas, 2013.

Lynn, John A. *Battle: A History of Combat and Culture*. Boulder, CO: Westview Press, 2003.

Mansoor, Peter R. *The GI Offensive in Europe: The Triumph of American Infantry Divisions, 1941–1945*. Lawrence: University Press of Kansas, 1999.

Marshall, S.L.A. *Men Against Fire: The Problem of Battle Command*. Washington: Infantry Journal, 1947.

McCaffrey, James M. *Army of Manifest Destiny: The American Soldier in the Mexican War, 1846–1848*. New York: New York University Press, 1992.

McManus, John C. *The Dead and Those About to Die: D-Day: The Big Red One at Omaha Beach*. New York: NAL Caliber, 2014.

———. *Grunts: Inside the American Infantry Combat Experience, World War II Through Iraq*. New York: New American Library, 2010.

Merridale, Catherine. *Ivan's War: Life and Death in the Red Army, 1939–1945*. New York: Metropolitan Books, 2006.

Miller, Donald L. *Masters of the Air: America's Bomber Boys Who Fought in the Air War Against Nazi Germany*. New York: Simon & Schuster, 2006.

Mitchell, Reid. *Civil War Soldiers: Their Expectations and Their Experiences.* New York: Viking Press, 1988.

Moore, Harold G., and Joseph L. Galloway. *We Were Soldiers Once … and Young: Ia Drang, the Battle That Changed the War in Vietnam.* New York: Random House, 1992.

Mosse, George L. *Fallen Soldiers: Reshaping the Memory of the World Wars.* New York: Oxford University Press, 1990.

Motley, Mary Penick, ed. *The Invisible Soldier: The Experience of the Black Soldier, World War II.* Detroit, MI: Wayne State University Press, 1975.

Muir, Rory. *Tactics and the Experience of Battle in the Age of Napoleon.* New Haven, CT: Yale University Press, 1998.

O'Donnell, Patrick K. *Give Me Tomorrow: The Korean War's Greatest Untold Story—the Epic Stand of the Marines of George Company.* Cambridge, MA: Da Capo Press, 2010.

———. *Into the Rising Sun: In Their Own Words, World War II's Pacific Veterans Reveal the Heart of Combat.* New York: Free Press, 2002.

Reid, Fiona. *Broken Men: Shell Shock, Treatment and Recovery in Britain, 1914–30.* London: Continuum, 2010.

Robertson, Jr., James I. *Soldiers Blue and Gray.* Columbia: University of South Carolina Press, 1988.

Rose, Alexander. *Men of War: The American Soldier in Combat at Bunker Hill, Gettysburg, and Iwo Jima.* New York: Random House, 2015.

Rush, Robert Sterling. *Hell in Hürtgen Forest: The Ordeal and Triumph of an American Infantry Regiment.* Lawrence: University Press of Kansas, 2001.

Russ, Martin. *Breakout: The Chosin Reservoir Campaign, 1950.* New York: Penguin Books, 2000.

Schrijvers, Peter. *The Crash of Ruin: American Combat Soldiers in Europe During World War II.* New York: New York University Press, 1998.

———. *The GI War Against Japan: American Soldiers in Asia and the Pacific During World War II.* New York: New York University Press, 2002.

Shay, Jonathan. *Achilles in Vietnam: Combat Trauma and the Undoing of Character.* New York: Simon and Schuster, 1994.

Skeen, Carl Edward. *Citizen Soldiers in the War of 1812.* Lexington: University Press of Kentucky, 1999.

Smith, Mark W. *The Smell of Battle, the Taste of Siege: A Sensory History of the Civil War.* New York: Oxford University Press, 2015.

Spring, Matthew H. *With Zeal and Bayonets Only: The British Army on Campaign in North America, 1775–1783.* Norman: University of Oklahoma Press, 2008.

Wells, Mark K. *Courage and Air Warfare: The Allied Aircrew Experience in the Second World War.* London: Frank Cass, 1995.

Wette, Wolfram. *The Wehrmacht: History, Myth, Reality.* Translated by Deborah Lucas Schneider. Cambridge, MA: Harvard University Press, 2006.

Wiest, Andrew. *Vietnam's Forgotten Army: Heroism and Betrayal in the ARVN.* New York: New York University Press, 2008.

Wiley, Bell I. *The Life of Billy Yank: The Common Soldier of the Union.* Indianapolis: Bobbs-Merrill, 1952.

———. *The Life of Johnny Reb: The Common Soldier of the Confederacy.* Indianapolis: Bobbs-Merrill, 1943.

Winders, Bruce. *Mr. Polk's Army: The American Military Experience in the Mexican War.* College Station: Texas A&M University Press, 1997.

Wood, John A. *Veteran Narratives and the Collective Memory of the Vietnam War.* Athens: Ohio University Press, 2016.

17 Prisoners of War and Repatriation

Paul J. Springer

In every military conflict, there has been the opportunity to capture members of the opposing forces. Over more than two millennia of organized warfare, a common understanding of the necessity of taking prisoners, and of the ways in which they may be treated by their captors, has gradually arisen, particularly in the West; and Western assumptions about the proper conduct of warfare have generally dominated international political and legal agreements for the past several centuries. As such, the historiography of prisoners of war (POWs) and their repatriation is largely an examination of Western policies and practices, although there are some exceptions to that rule. However, before commencing a study of modern nation-state POW operations, a brief examination of older approaches will illustrate the need for the modern codification of the expected treatment of POWs.

In one of the earliest works of military history, Greek philosopher Thucydides repeatedly referred to the mass torture, starvation, and ritual execution of POWs taken in the Peloponnesian War (Strassler, 1998). The Roman Republic allowed captives to live long enough to be paraded through the streets in a triumphal procession, but then typically ordered the prisoners strangled once the excitement had passed. A select few war captives might be allowed to fight in the gladiatorial games, but this only put off their inevitable demise and prolonged the enjoyment of the citizen observers (Polybius, 2010). For more than a thousand years, war captives were subjected to ritual executions, killings for convenience, or grisly murders as a means to terrorize the enemy. One of the most notorious instances of brutality toward the enemy came after the Battle of Kleidion in 1014. The victorious Byzantine army captured fourteen thousand Bulgarians, and rather than enslaving or killing them, Emperor Basil II devised a fiendish plan to use his captives as a means to compel Bulgarian czar Samuel's complete surrender. The prisoners were separated into groups of one hundred and ordered to draw lots. The "winner" of the drawing was blinded in one eye, and the remaining ninety-nine of each group were blinded in both eyes. Basil then ordered the partially blind prisoners to lead their comrades on the march home, where their arrival heralded the terrible fate that might befall any who resisted the empire's expansion (Runciman, 1930).

A medieval code of chivalry emerged that offered some protection to fellow Europeans, particularly those of noble birth, but it did little or nothing to protect non-European populations. Such protection included the accepting of surrender, the negotiation of ransoms, and the release of captives as part of a peace treaty. When the Crusades brought European Christians into conflict with the Muslims of the Middle East, neither side showed much interest in protecting the enemy. Ronald Finucane (1983) argued that prisoners might be taken as a means to coerce the enemy, either through threats of retaliation against hostages or to compel a ransom, but there was little compunction about selling the enemy into slavery to finance the war effort. Lev Yaacov (2001) noted that enemy captives were compelled by force to build fortifications, and at times were used as human shields to protect the fruit of their labors.

The chivalric code's effect was perhaps best exemplified by the capture of French king Jean II at the Battle of Poitiers. He was held for a ransom of three million crowns; when he proved unable to raise the sum, he voluntarily returned to British captivity (Rogers, 1993). Chivalry might offer some protection between the elites of European society, but it did not create a guarantee of safety even after an enemy accepted one's surrender. If prisoners became more of an encumbrance than they were worth, or if the tide of a battle threatened to turn against a captor, they might be executed out of necessity or convenience. When Henry V's army captured several hundred knights at the Battle of Agincourt in 1415, they disarmed their captives and bound them behind the English lines. However, seeing the numerically superior French army readying for another charge, Henry ordered the prisoners killed, lest they be freed and able to attack the English from behind. English nobles might have quietly disagreed with the order, but the common foot soldiers wasted no time in slaughtering the hapless captives where they lay (Curry and Mercer, 2015).

The genesis of modern international law governing POWs can be traced back to the writings of Hugo de Grotius, who observed the horrors of the Thirty Years' War, and took particular umbrage at the treatment of troops who surrendered. Geoffrey Parker (1997) argues that the war constituted a European civil war, in which any generosity toward the enemy quickly dissolved in the face of unmitigated atrocities. Peter Wilson (2009) considers the Thirty Years' War a systematic devolution in the conduct of warfare, in which atrocities became commonplace and enemies were slaughtered without regard for retaliation or consequences. In such a conflict, refusal of quarter became the norm, in part because most of the combatants claimed a divine right to utterly destroy their opponents.

Writing in the context of seventeenth-century Europe, Hugo Grotius (1925) argued that prisoners should not be considered the property of their individual captors, as had been the case prior to the rise of national forces operating at the behest of the state. Rather, Grotius considered POWs to be in the possession of the state employing troops who had received a surrender. Further, Grotius noted that the ancient practices of slaughtering or enslaving captives should be ended and replaced by merely holding captives until the end of a conflict, and then returning them to their home states. If nations wished to exchange captives on an equal basis, Grotius did not object, though he considered it more humane to simply remove captured troops from the battlefield to await the end of the conflict. Doing so would mitigate some of the worst carnage of state conflicts, and would also provide an incentive for troops in an impossible position to surrender rather than fight to the death. Later, Emmerich de Vattel (1758) built upon these concepts, arguing that common decency and chivalry should remain a part of even the bloodiest fields of battle.

The historiography of prisoners of war can largely be divided into four categories—namely, consolidation efforts that examine multiple periods of conflict; studies that examine a single war; analyses of smaller regions within a larger conflict, to include studies of a single captivity location; and personal memoirs of prisoners. Each of these types can offer significant enhancement to the understanding of POW affairs, but each also suffers from inherent weaknesses. Works that attempt to compare conflicts over a long period of time are typically forced to limit coverage of any given individual conflict, but can identify enduring patterns of behavior. Books and articles that study the captivity conditions of a single conflict often miss the broader context of earlier practices, but can focus upon the unique conditions germane to one war. Regional studies have a tendency to be hyper-focused upon a small area, and often fail to contextualize how that area either represents or significantly departs from the larger narrative. Individual POW memoirs provide a primary source for understanding the prisoners' dilemma, but are often rife with exaggerations and excoriations of the captors.

The earliest broad treatment of prisoners of war throughout history was provided by Herbert C. Fooks in 1922. His topical organization allowed the reader to understand the gradual changes in POW experiences over time, but much of his work was heavily influenced by the very recent experiences of World War I, in which millions of soldiers became captives. To Fooks, the most important question of the day was whether the Hague Convention of 1907 had supplied adequate protections for POWs. Although mortality rates in European prison camps were lower than had been the case in previous conflicts, Fooks and his contemporaries pushed for a more comprehensive international convention to protect captured troops. Later William E. S. Flory's *Prisoners of War* (1942) tackled many of the same subjects as Fooks' earlier work, but did so in the middle of World War II. Flory focused almost entirely upon the United States and Western Europe, despite numerous reports of mistreatment of prisoners in the Pacific theater and the capture of millions of German and Soviet forces on the Eastern Front. Flory castigated war planners for giving insufficient thought to designing a system that might safely intern millions of enemy troops, and even potentially put them to work on behalf of their captors.

In the postwar era, U.S. Army Lieutenant Colonel George G. Lewis and Captain John Mewha (1955) examined the means by which the U.S. military had employed POWs for labor purposes in previous conflicts. The vast majority of the work was dedicated to World War II, when nearly five hundred thousand enemy prisoners were utilized for nonmilitary labor, contributing their efforts to agricultural harvests, pulpwood logging, and industrial production of nonmilitary goods. Although they rated the labor program an overwhelming success, they noted that it was largely improvisational, and that if the United States and other belligerents had planned ahead for how to effectively employ their captives, the camp labor system would have proven far more efficient.

At the height of the Cold War, A. J. Barker (1974) offered a comprehensive examination of POW treatment in recent conflicts. He believed that modern warfare was most accurately characterized as a struggle of survival between competing ideologies, which in turn reduced the emphasis upon humanitarian behavior toward captured enemies. Essentially, belligerent forces came to see enemy troops more as numbers than as individuals, and meted out care to captives with an emphasis upon efficiency more than anything else. Barker also noted that guerrilla forces had become far more common on modern battlefields, rendering previous definitions of POWs obsolete.

Richard Garrett (1981) sought to illustrate that the behavior of captors did not markedly change from the Hundred Years' War to the Vietnam War. Garrett argued that POW operations have almost always been an afterthought or an irritant to captors. The moment of surrender was the most dangerous point for any combatant seeking captivity, but once accepted, the chance of surviving until repatriation was relatively high, although an individual prisoner's lot tended to be characterized by filthy conditions, meager food allowances, minimal medical care, and above all, continual boredom punctuated by periods of extreme terror.

Pat Reid and Maurice Michael's *Prisoner of War* (1984) took the unique approach of breaking down all POW experiences into six phases: surrender; march; camp life; work; captivity effects; and escape. They then compared the life of the average POW in different time periods within each phase to demonstrate the common aspects of POW experiences regardless of the era. Unfortunately, they supply no citations and a poor bibliography, making it difficult to determine how they reached their conclusions regarding the captivity experience.

After almost three decades without a single work purporting to examine the broad issues of POWs, three separate works analyzed the treatment of enemy captives by the U.S. government in the same year. Paul Springer's *America's Captives* (2010) argued that the United States tended to cede the initiative regarding prisoners to the enemy, and essentially rely upon reciprocity and retaliation to govern acceptable treatment of captives. Although the American military could be counted upon to hold itself to a higher standard than the enemy regarding the treatment of POWs, its tendency to improvise policies led to inconsistent practice toward enemy captives. Further, the relative leniency offered by the United States extended only to international conflicts—wars with Native American populations typically resulted in few prisoners, and those taken could expect few protections.

Robert Doyle's *The Enemy in Our Hands* (2010) relied upon anecdotal examples to illustrate the typical treatment of enemy captives held by the United States. Like Springer, he found that the United States tended to be relatively humane, but also could have offered much better standards of treatment to the enemy at minimal cost. Unlike Springer, however, Doyle chooses to include civilian prisoners and internees in his analysis, thus including personnel who are not traditionally treated in the same fashion as enemy troops.

Although Stephanie Carvin's *Prisoners of America's Wars* (2010) also purports to examine more than two centuries of U.S. POW policy, her book disproportionately focuses on the captivity practices of the George W. Bush administration. As such, Carvin essentially equates individuals captured fighting on behalf of al-Qaeda with those captured while serving in conventional national, uniformed military organizations. In this way, she breaks with convention by extending recognition of POW status to groups explicitly excluded by international law. The standard definition of POWs requires that they are uniformed troops of a state; are bearing arms openly; are under a discernible command structure; and follow the laws of war; four conditions entirely unmet by al-Qaeda.

In addition to the works examining broad periods of POW treatment, there have also been hundreds of works dedicated to the treatment of prisoners in individual conflicts. These range from overarching treatments of an entire war to individual accounts based upon a single prisoner's experiences. Thousands more works have touched upon captives as one element in a larger conflict, or as part of a discourse upon the nature of war. The historiography of works examining POWs in a single conflict is best examined via the chronology of the conflicts under study, because few of the authors sought to develop a broad interpretation of POW history.

In the sixteenth through nineteenth centuries, a hybrid warfare developed in the Americas, with European regulars, colonial militias, and native auxiliaries all fighting on the same battlefields, each with different objectives. European-descended colonists often viewed natives as an existential threat to be annihilated rather than an honorable enemy to be treated with respect. Captured enemies might be sold into slavery, tortured, or killed out of hand, but they were rarely given the opportunity for ransom or exchange (Nash, 1974). As Michael Fickes (2000) noted, at the conclusion of the Pequot War in 1637 English colonists forced three hundred Pequots to bound labor as servants. Conversely, Native American peoples often sought to integrate captives into their societies, though interactions with colonists strained and altered such practices over time (Brooks, 2002; Fickes, 2000; Richter, 1992)

At times, such as the fighting between the British and Continental Armies during the American Revolutionary War, European norms prevailed, and surrenders were generally accepted without incident. However, the customs of chivalry to a captured enemy provided scant protection for prisoners condemned to the floating hulks used to hold captives in New York Harbor. The British chose to minimize their expenditures on prisoner upkeep by condemning most military and civil prisoners to the hulks, relying upon their absolute control of the sea to prevent any attempt at rescue. High-ranking officers might be allowed a limited parole or to be quartered with Loyalists, but for the enlisted, the hulks were the only option (Metzger, 1971; Dring, 2010). American captors also tried to maintain their prisoners with limited success due to a fundamental lack of resources and poor planning (Burroughs, 2008; Borick, 2012; Krebs, 2015). In the case of General John Burgoyne's surrender, the U.S. Congress simply reneged on the terms of the capitulation and refused to allow Burgoyne's troops to take ships home, condemning them to a long captivity in the interior (Sampson, 1995). In the same conflict, though, back-country partisans often refused quarter to a prostrate enemy, triggering an unending wave of retaliatory actions (Lee, 2001).

In Europe, the French Revolution triggered a new wave of nationalism and a series of wars as the conservative monarchies sought to contain unleashed popular passions. Under the Reign of Terror, the French government decreed that all counterrevolutionaries fighting against the French should be immediately executed (Lynn, 1984). This led to a policy of no quarter and the slaughter of thousands of British, Hanoverian, and Spanish troops. When the French Directory assumed power from the National Convention, mass prison camps and the use of floating hulks became the norm for enemy captives (Forrest, 1989). Napoleon Bonaparte's attempts to dominate the Continent produced the largest battles seen in Europe to date. The thousands of captives taken as a result faced years of confinement when enemy governments refused to negotiate treaties of exchange (Smith, 2001). When convenient, Napoleon preferred to hold captives as hostages against their nation's behavior, but when those prisoners proved too much of an encumbrance, wholesale massacres might ensue. This was particularly true during Napoleon's invasion of Russia in 1812—troops on both sides quickly learned that surrender led only to mass executions (Zamoyski, 2004).

The Industrial Revolution, while it enabled nations to raise enormous forces, posed challenges for the composition and conduct of armies, including the treatment of large masses of enemy prisoners. During the American Civil War, for example, legal scholar Francis Lieber attempted to codify the limits of modern warfare. His efforts resulted in President Abraham Lincoln issuing General Orders No. 100 on April 24, 1863. These orders, commonly called the Lieber Code, prescribed a specific set of expectations upon Union troops operating in the field. Of the 157 articles, 38 were concerned with the capture, treatment, and repatriation of enemy prisoners (Springer and Robins, 2014). Reiterating ideas propounded by Grotius, Lieber noted that POWs belong to the captor nation, not the individual units that accept a surrender; they must be removed from danger as much as practicable and given sufficient food and medical care. Captives could not be executed out of hand or mistreated, and must be held safely until the end of the war or the creation of an exchange cartel (Friedel, 1947; Hartigan, 1983).

The Lieber Code supplied a principled set of regulations. But even with this new set of expectations, more than 15 percent of American Civil War POWs perished in captivity, making surrender actually more dangerous than service in the field. Writing in the 1920s and 1930s, William Best Hesseltine (reprint, 1964) argued that neither the North nor the South deliberately attempted to harm their captives, but both sides devoted little effort or resources to their upkeep. In the aftermath of the war, the Union effectively won the information campaign to blame the South for the high mortality rate in its POW camps, while absolving itself for any similar responsibility for an almost equally high mortality rate. Lonnie Speer (1997) undertook a comprehensive study of all of the major POW sites in the Union and Confederacy, and demonstrated that a significant difference in the treatment of prisoners at each location could be traced to the leadership of the site and the resources available in the immediate region to assist in their upkeep. Charles Sanders (2005) challenged the prevailing view of benign neglect, and instead claimed that Union and Confederate political and military leaders deliberately ordered the mistreatment of captives as a coercive means of affecting the behavior of the enemy.

Lieber's arguments supplied the foundation for a series of international agreements, including the Brussels Declaration of 1874; the Hague Conventions of 1899 and 1907; and the Geneva Conventions of 1929 and 1949 (Witt, 2012). Each of these agreements sought to ameliorate the conditions of wartime captivity in the name of common humanity. While no agreement could guarantee the safety and survival of all POWs, these accords reminded belligerents that limits exist in warfare, and are a necessary aspect of conflict. At the same time these agreements were implemented, prisoners came to be valued for more than exchanges or any information they might possess. In the twentieth century, POWs became recognized as potential laborers to keep the industrial economies of the Europeans states churning forward. While each agreement banned forcing enemy captives to engage in work directly related to the war effort, especially the production and transportation of munitions, allowing POWs to provide labor facilitated accepting surrenders and incentivized good behavior toward potential captives. The use of POWs for agricultural labor also helped offset the costs of their upkeep, reducing the burden to the captor.

In World War I, more than two million troops fell into enemy hands. In Europe, for the most part, captives were relatively well treated, and neutral international agencies, such as the Red Cross, conducted POW camp inspections on behalf of the combatants to ensure that legal norms were being followed (Yarnall, 2011). For the first three years of the war, the United States acted in the role of a protecting power, and after conducting camp inspections, sent reports across battle lines to reassure each side that their prisoners were receiving adequate care (Speed, 1990). Only in the last year of the war, when the British blockade of German ports began to exhaust the Central Powers' supplies, did the camp system begin to collapse. Chronic shortages in POW camps led to disease outbreaks and malnutrition—but the same was true within the civilian population, and there is little evidence of a deliberate attempt to mistreat enemy captives (Feltman, 2015). Elsewhere in the world, conditions varied, from the benign treatment of German captives by the Japanese (Burdick and Moessner, 1984) to the brutal treatment of Ottoman Turks captured by the Russians, many of whom were not repatriated until five years after the war ended (Yanikdağ, 1999).

When World War II erupted, few observers realized how catastrophic the conflict might become. Japan did not sign the 1929 Geneva Convention, and thus had no legal obligation to uphold any of its statutes. Japanese troops terrorized the civilian populations of conquered territories (Chang, 1997), and Western troops who surrendered to the Japanese faced starvation, enslavement, torture, and possibly execution before the end of the conflict. On the Asian mainland, the advancing Japanese Army engaged in a series of atrocities, behavior it considered entirely justified according to its own cultural norms (Tanaka, 1996). In particular, the Japanese bushido code reviled any warriors who surrendered to enemy forces, and essentially allowed the captor to mete out punishment to captives at will.

As American and Allied forces advanced across the Pacific, they faced a Japanese enemy unwilling to surrender due to their assumptions that surrender carried a permanent stain of dishonor, and that captives would be mistreated or killed out of hand. Gregory Urwin (2010) traced the experiences of American prisoners captured on Wake Island, and found that unit cohesion and the establishment of survival and resistance networks contributed to higher survival rates. Kelly Crager (2008) followed a single battalion of U.S. troops captured on Java and forced to work on the construction of the Burma-Thailand railway, one of the deadliest activities for captured Allied troops due to the brutal conditions. Like Urwin, he found that members of the unit supported one another and increased the likelihood of survival when compared to prisoners not taken as part of a unit.

Allied forces soon stopped demanding capitulation after a series of faked submissions. The deliberate effort by the leadership of both sides to dehumanize the enemy made atrocities far more likely, as did assumptions about the behavior and motivations of the opponent (Dower, 1986). In turn, the Japanese fostered racial resentment among Asian populations as a means to coopt their support in an anti-colonial war against Europeans (Horne, 2004). Alison Gilmore analyzed Allied efforts to induce Japanese surrenders, particularly examining the reliance upon captured Japanese troops to create more effective propaganda in the hopes of inducing further surrenders. Ulrich Strauss (2003) examined the role of Japanese culture upon the decision to surrender or to fight to the death, and found that when Japanese troops believed their cause held the possibility of victory, they placed little emphasis upon their own survival. As they lost faith in Japan's chances of winning the war, the previously unthinkable option of surrender became more palatable.

In Europe, the fight between Nazi Germany and the Western Allies usually followed the laws of war, such that mortality rates among POWs averaged less than 5 percent. Interviews of survivors who ended the war in captivity found a mutual respect and a common understanding of the laws of warfare (Carlson, 1997). Aerial forces, in particular, seemed to return to chivalric principles toward one another, including after capture, although angry civilians who captured enemy fliers might extract revenge before turning prisoners over to authorities. Prisoners on both sides reported relatively benign conditions in the camps, although the situation deteriorated on the Continent after the Normandy landings. John Nichol and Tony Rennell (2003) followed the saga of Allied prisoners held at Colditz, who were marched across Central Europe in an attempt to prevent their recapture, through interviews of survivors of the ordeal. Captured aerial personnel held by the Germans noted that they maintained a privileged status, and received better food and medical care in the prisons run by the *Luftwaffe* (Richard, 2000). Robin Quinn (2015) analyzed the British decision to retain German prisoners until 1948, using them for forced labor to rebuild areas damaged by German aerial attacks.

Those Axis prisoners shipped to the United States found ample food, relatively comfortable living conditions, and a respite from the dangers of warfare. By far the best overview of their experiences is Arnold Krammer's *Nazi Prisoners of War in America* (1979), which argues that the United States far exceeded any expectations of its responsibilities as a captor. One of the most thorough individual German memoirs of captivity found that the American guard personnel seemed favorably disposed toward their charges, and seemed to hold little ill will against them (Pabel, 1955). General officers captured by the enemy were treated almost as visiting dignitaries, albeit ones who remained in confinement for most of their visit (Mallett, 2013). In the last year of the war, American authorities attempted a massive reeducation project, hoping to create a group of repatriated prisoners who could assist occupying troops in administration and law enforcement duties in postwar Germany. The program violated the spirit if not the letter of the Geneva Convention, but also provided a much-needed liaison force that mitigated the worst aspects of occupation (Gansberg, 1977; Robin, 1995; Smith, 1996).

When Germany surrendered, the Western Allies accepted responsibility for millions of Axis prisoners, overwhelming the capacity of the camp system and creating the potential for a humanitarian crisis. Canadian journalist James Bacque examined official reports of the era and misinterpreted them in a fashion that suggested General Dwight Eisenhower had systematically murdered millions of German POWs (1991). A collection of very prominent historians thoroughly debunked his position (Bischof and Ambrose, 1992) and demonstrated that the POW camps, while hardly a model of efficiency, saved countless lives and provided a mechanism to deliver much-needed supplies to war-torn Europe. Far from murdering helpless prisoners, Eisenhower had simply ordered that the very oldest and youngest prisoners be released from captivity to make their way home and help feed their families. Returning POWs who had been held in North America and instructed in the tenets of democracy and capitalism moved into key liaison functions in Western Germany, bringing key skills into the administration of the occupied zone.

On the Eastern Front, World War II proved to be a titanic struggle of national survival between Germany and the Soviet Union, two societies with political systems that could not coexist. German racial policies called for the extermination of "subhuman" races, a fact that created a racial hierarchy within the German POW system that placed Soviet prisoners at the very bottom (Megargee, 2006). German soldiers internalized the racial attitudes of their leadership, and used them to justify atrocities against their Soviet enemies, including the widespread slaughter of surrendered enemy troops (Bartov, 1991). Like the Japanese, and the Germans on the Eastern Front, the Soviets did not adhere to the Geneva Convention, and when German and Soviet forces captured one another, the unfortunate captives faced mass executions or enslavement into war industries. Both Adolf Hitler and Josef Stalin believed that a policy of annihilation would intimidate the enemy, and a fear of retaliation would motivate their own troops to fight to the death (Wette, 2006). German war planners expected to capture millions of Soviet troops, but had little interest in preserving their health. Timothy Snyder (2010) argues that much of the machinery of the Holocaust was tested on Soviet prisoners before being used against civilian victims. Those Germans who were not killed out of hand or worked to death in Siberian gulags were retained for up to twelve years after the war's end (Hirschfeld, 1986).

The Korean War also pitted two diametrically opposed political systems against one another, and was complicated by lingering racial antipathy. POWs captured from the United Nations Command faced a 40 percent mortality rate due to deliberate mistreatment by their captors, who were later accused of attempting to "brainwash" their prisoners (Carlson, 2002). As fighting devolved into a stalemate, negotiations for an armistice dragged on for two years, largely over the issue of POW repatriations (Biderman, 1963). Communist nego-tiators utilized a strict interpretation of the Geneva Convention, while the West sought to win a propaganda victory by allowing thousands of captured enemies to remain in South Korea or transfer to Taiwan, escaping their communist governments (Hermes, 1966). In the end, a Neutral Nations Repatriation Commission, composed of five neutral states and led by India, ruled that POWs could choose to refuse repatriation. This option was exercised by approximately fifteen thousand Chinese and eight thousand North Korean POWs. More surprisingly, one British and twenty-three American soldiers also initially refused repatria-tion, although most eventually elected to voluntarily return to their home countries. While a small number, the fact that any Western troops chose to remain under communist control rather than be repatriated proved embarrassing to their governments (MacKenzie, 2012; McKnight, 2014).

During the Vietnam War, captivity behavior followed many of the norms set in Korea. Captured Viet Cong and North Vietnamese troops were largely held at the island prison of Phu Quoc, with much the same results as those obtained in Korea (Davis, 2000). Red Cross delegations routinely accused the South Vietnamese government, and by extension the United States, of mistreating prisoners. Captured U.S. troops faced torture, starvation, and execution in enemy hands, with some facing years in captivity before repatriation (Roch-ester and Kiley, 1999; Philpott, 2001). The most notorious prison in North Vietnam, the "Hanoi Hilton," held approximately six hundred aircrew in its walls. Survivors reported harrowing conditions characterized by torture, wanton cruelty, and occasional executions (McCain, 1999).

When the Cold War ended, so did Soviet protection of puppet regimes and client states that mistreated captives. The United States returned to its habit of meting out relatively excellent treatment to enemy captives, particularly as demonstrated in the Persian Gulf War (Department of Defense, 1992)—though this was not reciprocated by the Iraqi govern-ment under Saddam Hussein (Cornum, 1992). But when al-Qaeda operatives killed nearly three thousand U.S. citizens on September 11, 2001, they triggered a renewed wave of U.S. interventions into the Middle East and Southwest Asia. The wars in Afghanistan and Iraq, with major underpinnings rooted in religious ideology, undermined many of the advances in POW treatment. American president George W. Bush announced that the United States did not view members of al-Qaeda or the Taliban as lawful combatants, and thus would not consider them POWs. Karen Greenberg (2010) noted that this led to the creation of a prison compound at Guantanamo Bay that fell outside the jurisdiction of U.S. courts, creating a recipe for mistreatment. Soon, allegations of mistreatment of captives by American authori-ties began to emerge, first from Camp X-Ray at Guantanamo Bay, and next from Abu

Ghraib in Iraq. In both cases, photographs showed enemy captives being subjected to acts of sexual humiliation, torture, and possibly murder. The images set off a firestorm of protest and served as a major recruitment tool for budding insurgencies throughout the conflict region. They forced American citizens to reconsider the role of the United States in the world, and how it should conduct itself during warfare (Greenberg and Dratel, 2005).

Despite the enormous amount of scholarship that has been dedicated to POWs, captivity, and repatriation, there are still a number of areas that remain understudied. In particular, the treatment of enemy captives in unconventional conflicts has been largely overlooked. In 1977, international negotiators (not including the United States) ratified a series of additional protocols to the Geneva Convention Relative to Prisoners of War. The protocols extended the legal status of POWs to guerrilla forces, a consideration the United States adamantly opposed. Very little scholarship has examined the effects of the protocols, which threatened to blur the lines between national forces and non-state actors, a key component of twenty-first-century warfare. Key case studies might include any conflict since 1977 that has included significant participation by irregular forces. In particular, the conflicts between Israel and non-state actors, such as Hamas and Hezbollah; the fight between Boko Haram and the Nigerian government; the Russian engagement in Chechnya; and the Balkan Wars of the 1990s all might offer opportunities to examine the function of the protocols.

Modern academics searching for topics within the field of POWs would do well to consider the effects of race and gender upon the decision to accept a surrender and how to treat the resulting captives. Examples here include Lyle Adair, an African American Union soldier captured by Confederate cavalry and sent to Andersonville (Robins, 2011). Private Jessica Lynch, serving in Iraq, became the most well-known U.S. prisoner of war in a generation—her story illustrates not only the dangers of the modern battlefield but also the special challenges presented by incorporating women into combat zones and eventually combat units, including the use of rape as a terrorizing element (Bragg, 2003). Likewise, the effect of ideology, whether religious, cultural, or political, upon one's behavior toward enemies, offers an enormous avenue of new scholarship to enterprising academics.

Another major avenue of study that has been almost completely overlooked by scholars to date is the examination of POWs from the captors' perspectives. There are almost no analyses of camp life from the perspective of guard personnel, camp officials, or even the local population living in the vicinity of POW camps, with the exception of formal reports rendered to military superiors. One rare exception, provided by Erik Saar and Viveca Novak (2005), illustrated the challenges of detaining enemy prisoners who do not fall under the traditional definition of POWs—specifically forces captured in the War on Terror—and questioned whether the United States should be holding such captives on an indefinite basis. The lack of coverage on this vital aspect of POW operations is undoubtedly partly due to numbers and available records, as there are always more prisoners than guards. An enterprising graduate student might do well to search for such accounts, though, as they would go a long way toward completing our understanding of captivity in different conflicts.

Of course, as more conflicts occur, more POW experiences will accumulate. Also, as time passes, previously classified material becomes available, allowing scholars to compare the generally accepted narratives with new sources. Thus, the analysis of captivity in previous wars is always subject to new interpretations and examinations, and there is always a new generation of scholars ready and willing to undertake the task. Likewise, the norms and expectations of capturing enemy forces continue to evolve, and new policies and practices will draw upon our collective understanding of what has functioned and what has failed in the past, making the continual study of POW issues a vital aspect of military history.

Bibliography

Bacque, James. *Other Losses: The Shocking Truth Behind the Mass Deaths of Disarmed German Soldiers and Civilians Under General Eisenhower's Command.* New York: Prima, 1991.

Barker, A. J. *Prisoners of War.* New York: Universe Books, 1974.

Bartov, Omer. *Hitler's Army: Soldiers, Nazis, and War in the Third Reich.* New York: Oxford University Press, 1991.

Biderman, Albert D. *March to Calumny.* New York: Macmillan, 1963.

Bischof, Gunter, and Stephen E. Ambrose, eds. *Eisenhower and the German Prisoners of War.* Baton Rouge: Louisiana State University Press, 1992.

Borick, Carl P. *Relieve Us of This Burthen: American Prisoners of War in the Revolutionary South, 1780–1782.* Columbia: University of South Carolina Press, 2012.

Bragg, Rick. *I Am a Soldier, Too: The Jessica Lynch Story.* New York: Knopf, 2003.

Brooks, James F. *Captives & Cousins: Slavery, Kinship, and Community in the Southwest Borderlands.* Chapel Hill: University of North Carolina Press, 2002.

Burdick, Charles, and Ursula Moessner. *The German Prisoners of War in Japan, 1914–1920.* Lanham, MD: University Press of America, 1984.

Burroughs, Edwin G. *Forgotten Patriots: The Untold Story of American Prisoners During the Revolutionary War.* New York: Basic Books, 2008.

Carlson, Lewis H. *Remembered Prisoners of a Forgotten War: An Oral History of Korean War POWs.* New York: St. Martin's Press, 2002.

———. *We Were Each Other's Prisoners.* New York: Basic Books, 1997.

Carvin, Stephanie. *Prisoners of America's Wars: From the Early Republic to Guantanamo.* New York: Columbia University Press, 2010.

Chang, Iris. *The Rape of Nanking.* New York: Basic Books, 1997.

Cornum, Rhonda. *She Went to War: The Rhonda Cornum Story.* Edited by Peter Copeland. Novato, CA: Presidio Press, 1992.

Crager, Kelly. *Hell Under the Rising Sun: Texas POWs and the Building of the Burma-Thailand Death Railway.* College Station: Texas A&M University Press, 2008.

Curry, Anne and Malcolm Mercer, ed. *The Battle of Agincourt.* New Haven, CT: Yale University Press, 2015.

Davis, Vernon E. *Long Road Home: U.S. Prisoner of War Policy and Planning in Southeast Asia.* Washington, DC: Office of the Secretary of Defense, 2000.

Department of Defense. *Final Report to Congress on the Conduct of the Persian Gulf War.* Washington, DC: Government Printing Office, 1992.

Dower, John W. *War Without Mercy: Race and Power in the Pacific War.* New York: Pantheon Books, 1986.

Doyle, Robert C. *The Enemy in Our Hands: America's Treatment of Enemy Prisoners of War, From the Revolution to the War on Terror.* Lexington: University Press of Kentucky, 2010.

Dring, Thomas. *Recollections of Life on the Prison Ship Jersey.* Edited by David Swain. Yardley, PA: Westholme, 2010.

Feltman, Brian K. *The Stigma of Surrender: German Prisoners, British Captors, and Manhood in the Great War and Beyond.* Chapel Hill: University of North Carolina Press, 2015.

Fickes, Michael L. "'They Could Not Endure That Yoke': The Captivity of Pequot Women and Children After the War of 1637," *New England Quarterly* 73 (March 2000): 58–81.

Finucane, Ronald C. *Soldiers of the Faith: Crusaders and Moslems at War.* New York: St. Martin's Press, 1983.

Flory, William E. S. *Prisoners of War: A Study in the Development of International Law.* Washington, DC: American Council on Public Affairs, 1942.

Fooks, Herbert C. *Prisoners of War.* Federalsburg, MD: J. W. Stowell, 1924.

Forrest, Alan. *Conscripts and Deserters: The Army and French Society During the Revolution and Empire.* New York: Oxford University Press, 1989.

Friedel, Frank B. *Francis Lieber, Nineteenth-Century Liberal.* Baton Rouge: Louisiana State University Press, 1947.

Gansberg, Judith M. *Stalag USA: The Remarkable Story of German POWs in America.* New York: Thomas Y. Crowell, 1977.

Garrett, Richard. *POW.* London: David & Charles, 1981.

Gilmore, Allison B. *You Can't Fight Tanks With Bayonets: Psychological Warfare Against the Japanese Army in the Southwest Pacific.* Lincoln: University of Nebraska Press, 1998.

Greenberg, Karen J. *The Least Worst Place: Guantanamo's First 100 Days.* New York: Oxford University Press, 2010.

Greenberg, Karen J., and Joshua L. Dratel. *The Torture Papers: The Road to Abu Ghraib.* New York: Cambridge University Press, 2005.

Grotius, Hugo. *De Jure Belle ac Pacis.* Trans. Francis W. Kelsey. Oxford: Clarendon Press, 1925.

Hartigan, Richard Shelly. *Lieber's Code and the Laws of War.* Chicago: Precedent, 1983.

Hermes, Walter G. *Truce Tent and Fighting Front.* Washington, DC: Center of Military History, 1966.

Hesseltine, William Best. *Civil War Prisons: A Study in War Psychology.* Reprint. New York: F. Ungar, 1964.

Hirschfeld, Gerhard. *The Policies of Genocide: Jews and Soviet Prisoners of War in Nazi Germany.* Boston: Allan & Unwin, 1986.

Horne, Gerald. *Race War! White Supremacy and the Japanese Attack on the British Empire.* New York: New York University Press, 2004.

Krammer, Arnold. *Nazi Prisoners of War in America.* New York: Stein & Day, 1979.

Krebs, Daniel. *A Generous and Merciful Enemy: Life for German Prisoners of War During the American Revolution.* Norman: University of Oklahoma Press, 2015.

Lee, Wayne. *Crowds and Soldiers in Revolutionary North Carolina: The Culture of Violence in Riot and War.* Gainesville: University Press of Florida, 2001.

Lewis, George G., and John Mewha. *History of Prisoner of War Utilization by the United States Army, 1776–1945.* Washington, DC: Government Printing Office, 1955.

Lynn, John. *The Bayonets of the Republic: Motivation and Tactics in the Army of Revolutionary France, 1791–1794.* Urbana: University of Illinois Press, 1984.

MacKenzie, S. P. *British Prisoners of the Korean War.* New York: Oxford University Press, 2012.

Mallett, Derek R. *Hitler's Generals in America: Nazi POWs and Allied Military Intelligence.* Lexington: University Press of Kentucky, 2013.

McCain, John. *Faith of My Fathers.* New York: Random House, 1999.

McElroy, John. *Andersonville: A Story of Rebel Military Prisons.* Toledo: D. R. Locke, 1879.

McKnight, Brian D. *We Fight for Peace: Twenty-Three American Soldiers, Prisoners of War, and "Turncoats" in the Korean War.* Kent, OH: Kent State University Press, 2014.

Megargee, Geoffrey P. *War of Annihilation: Combat and Genocide on the Eastern Front.* Lanham, MD: Rowman & Littlefield, 2006.

Metzger, Charles H. *The Prisoner in the American Revolution.* Chicago: Loyola University Press, 1971.

Nash, Gary B. *Red, White, and Black: The Peoples of Early America.* Englewood Cliffs, NJ; Prentice-Hall, 1974.

Nichol, John, and Tony Rennell. *The Last Escape: The Untold Story of Allied Prisoners of War in Europe, 1944–45.* New York: Viking, 2003.

Pabel, Reinhold. *Enemies Are Human.* Philadelphia, PA: Winston, 1955.

Parker, Geoffrey. *The Thirty Years' War.* London: Routledge, 1997.

Philpott, Tom. *Glory Denied.* New York: Norton, 2001.

Polybius. *The Histories.* Translated by Robin Waterfield. New York: Oxford University Press, 2010.

Quinn, Robin. *Hitler's Last Army: German POWs in Britain.* Stroud, UK: The History Press, 2015.

Ransom, John. *John Ransom's Diary.* New York: P.S. Eriksson, 1963.

Reid, Pat, and Maurice Michael. *Prisoner of War: The Inside Story of the POW From the Ancient War to Colditz and After.* New York: Beaufort Books, 1984.

Richard, Oscar G., III. *Kriegie: An American POW in Germany.* Baton Rouge: Louisiana State University Press, 2000.

Richter, Daniel K. *The Ordeal of the Longhouse: The Peoples of the Iroquois League in the Era of European Colonization.* Chapel Hill: University of North Carolina Press, 1992.

Robin, Ron Theodore. *The Barbed-Wire College: Reeducating German POWs in the United States During World War II.* Princeton: Princeton University Press, 1995.

Robins, Glenn, ed. *They Have Left Us Here to Die: The Civil War Prison Diary of Sgt. Lyle Adair, 111th U.S. Colored Infantry.* Kent, OH: Kent State University Press, 2011.

Rochester, Stuart I., and Frederick Kiley. *Honor Bound: American Prisoners of War in Southeast Asia, 1961–1973.* Annapolis: Naval Institute Press, 1999.

Rogers, Clifford J. "The Military Revolutions of the Hundred Years' War," *Journal of Military History* 57 (April 1993): 241–278.

Runciman, Steven. *A History of the First Bulgarian Empire*. London: G. Bell and Sons, 1930.

Saar, Erik, and Viveca Novak. *Inside the Wire: A Military Intelligence Soldier's Eyewitness Account of Life at Guantanamo*. New York: Penguin, 2005.

Sampson, Richard. *Escape in America: The British Convention Prisoners, 1777–1783*. Chippenham: Picton, 1995.

Sanders, Charles W. *While in the Hands of the Enemy: Military Prisons of the Civil War*. Baton Rouge: Louisiana State University Press, 2005.

Smith, Arthur L. *The War for the German Mind: Re-Educating Hitler's Soldiers*. Providence, RI: Berghahn Books, 1996.

Smith, Denis. *The Prisoners of Cabrera: Napoleon's Forgotten Soldiers, 1809–1814*. New York: Four Walls Eight Windows, 2001.

Snyder, Timothy. *Bloodlands: Europe Between Hitler and Stalin*. New York: Basic Books, 2010.

Speed, Richard B. *Prisoners, Diplomats, and the Great War: A Study in the Diplomacy of Captivity*. New York: Greenwood Press, 1990.

Speer, Lonnie. *Portals to Hell: Military Prisons of the Civil War*. Mechanicsburg, PA: Stackpole Books, 1997.

Springer, Paul J. *America's Captives: Treatment of POWs From the Revolutionary War to the War on Terror*. Lawrence: University Press of Kansas, 2010.

Springer, Paul J., and Glenn Robins. *Transforming Civil War Prisons: Lincoln, Lieber, and the Politics of Captivity*. New York: Routledge, 2014.

Strassler, Robert B., ed. *The Landmark Thucydides: A Comprehensive Guide to the Peloponnesian War*. Translated by Richard Crawley. New York: Free Press, 1998.

Strauss, Ulrich. *The Anguish of Surrender: Japanese POWs of World War II*. Seattle: University of Washington Press, 2003.

Tanaka, Yuki. *Hidden Horrors: Japanese War Crimes in World War II*. Boulder, CO: Westview Press, 1996.

Urwin, Gregory J. W. *Victory in Defeat: The Wake Island Defenders in Captivity*. Annapolis, MD: Naval Institute Press, 2010.

Vattel, Emmerich de. *The Law of Nations or the Principles of Natural Law*. 1758. Eds. Bela Kapossy and Richard Whatmore. Indianapolis: Liberty Fund, 2008.

Wette, Wolfram. *The Wehrmacht: History, Myth, Reality*. Translated by Deborah Lucas Schneider. Cambridge, MA: Harvard University Press, 2006.

Wilson, Peter H. *The Thirty Years' War: Europe's Tragedy*. Cambridge: Harvard University Press, 2009.

Witt, John Fabian. *Lincoln's Code: The Laws of War in American History*. New York: Free Press, 2012.

Yaacov, Lev. "Prisoners of War During the Fatimid-Ayyubid Wars with the Crusaders." In *Tolerance and Intolerance: Social Conflict in the Age of Crusades*. Eds. M. Gervers and J. M. Powell. Syracuse, NY: Syracuse University Press, 2001.

Yanikdağ, Yücel. "Ottoman Prisoners of War in Russia, 1914–22," *Journal of Contemporary History* 34, no. 1 (1999): 69–85.

Yarnall, John. *Barbed Wire Disease: British & German Prisoners of War, 1914–19*. Stroud, UK: The History Press, 2011.

Zamoyski, Adam. *1812: Napoleon's Fatal March on Moscow*. New York: Harper Collins, 2004.

18 Home Fronts

Bianka J. Adams

The term "home front" had its origin in World War I. In that global conflict all societal assets and groups mobilized to support the war effort. Men volunteered for military service, women left their traditional station and volunteered to work outside the house in factories or as nurses, and children volunteered their time to collect household items or scrap material that might aid war production. Domestic industries converted from peacetime to military production. Society at large accepted deprivation and sacrifice as a patriotic duty. In World War II, large-scale mobilization of the home front was essential for almost every combatant nation. As in the previous war, industries converted from peacetime to war production and, owing to technological advances in mobile warfare, industries sprang up to produce the new war machines. Again, women marched into factories in unprecedented numbers to work in jobs formerly performed by men gone off to war.

The definition of home front as essential to the war effort applies mostly to the two world wars in the twentieth century. For better or worse, home fronts in this sense are linked to the definition of "total war," in which the whole population and all resources of the warring nation are committed to complete victory and therefore become legitimate military targets (Manfred Boemeke, 1999). While the fronts and home fronts of the American Civil War, 1860–1865, displayed many of the traits later attributed to total war, eminent Civil War historians debate the validity of this term for the war. Stephen Ash and Mark Grimsley argued that Union generals' and soldiers' restraint in dealing with Confederate civilians and their property ruled out classification as a total war. In contrast James M. McPherson, Edward Hagerman, and Stanley L. Engerman and J. Matthew Gallman hold against these interpretations given that the consequences, level of destruction, and the Confederacy's coercive direction of manpower and resources meant that the South came close to waging total war (Stig Förster, 1997). Based on this interpretation, the Civil War home front is included in this survey.

Through extensive study of personal papers and municipal documents, William Rogers (1999) retraces life on the home front in Montgomery, Alabama, during the Civil War. Also closely examining a local community, Thomas O'Connor (1997) reveals how events on the front lines affected the lives of citizens living on the northern home front in Boston, Massachusetts. Taking a wider geographical approach that encompasses the whole northern home front, J. Matthew Gallman (1994) demonstrates how individuals experienced the Civil War and how northern society collectively reacted to the challenges of the war. His thorough analysis of the northern economy and war financing shows that while the northern economy remained prosperous, its growth slowed due to the war. The role of women has become a popular topic, and is the subject of one anthology in particular (Giesberg, 2009). Another asserts that multiple regions composed "the North," and explores the distinctiveness of the "Midwest" (Aley and Anderson, 2013).

Works analyzing the home fronts of belligerents in World War I largely concentrate on how governments mobilized their populations, what kind of an impact the war had on women or—as in the United States—on minorities, and on civilian wartime experiences. John Horne (1997) and Gerald Feldman (2000) examine the belligerent states' ability to mobilize and "remobilize" their entire populations and economies. Comparing Britain, France, Germany, Italy, and Austria-Hungary, contributors to Horne's volume find that the governments of Great Britain and France were not only more capable of mobilizing their nation but also better prepared to deal with the consequences of the Great War. Feldman agrees with this assessment, arguing that democratically elected governments deemed legitimate by their populations were in a better position to ask for sacrifices in return for special privileges and increased rights. Giovanna Procacci (1999) claims, however, that this was not the case in Italy. Even though governed by a liberal regime, the way Italians reacted to the war and its demands was similar to how Germans, Russians, and Central Europeans reacted to their authoritarian governments, because they felt betrayed by them.

Belinda Davis (2000), Ute Daniel (1989), and Maureen Healy (2004) investigate the plight of women on the home fronts in the Central Powers, Germany and Austria. Davis examines the German government's efforts to deal with food shortages in Berlin and argues that food riots amounted to political acts of women who had yet to gain suffrage. In contrast, Daniel disputes that the war had any emancipatory effect for women because those who filled factory jobs had already worked outside the house and were thus merely redeployed into the war industry. Healy studies the complex interrelation of women's citizenship and the Austro-Hungarian monarchy's efforts to prevent food riots in Vienna.

Arthur Marwick (1965) saw World War I as a crucible for social change in Great Britain. He describes life on the home front during the war, arguing that the social changes such as women's work outside the home translated into lasting progress for women and transformed the country. While many historians adopted his interpretation at the time, two decades later it came under increasing criticism. Basing their analysis on oral history interviews, diaries, and letters, Gail Braybon and Penny Summerfield (1987) convincingly argue that women's station in society remained unchanged by the war.

Jay Winter and Jean-Louis Robert (1997; 2007) explore life in Berlin, Paris, and London during World War I. In their first volume contributors investigate material living conditions of inhabitants in the three capital cities. The perceived and real uneven distribution of hardships led in some cases to hatred and suspicion among classes, urban and rural dwellers, and against ethnic and religious minorities. Contributions to the second volume focus on diverse aspects of urban life during the war and on contemporary reactions to the conflict. In Winter and Richard Wall (2005), sixteen scholars inspect the different ways in which family life was affected and changed by the war. In addition to exploring the social, economic, and ideological dimensions of the war, the volume also sheds light on the political forces that helped restore traditional gender roles and family life at the end of the conflict. The essays pay attention to living standards in wartime Europe and how the war affected women's work and disturbed family life. They also investigate how political and social groups, supported by the state, restored the traditional social order. David Kennedy (1980) examines the effects of America's comparatively short military involvement in the fighting in Europe on U.S. society, in general, and women, African Americans, and labor, in particular. Similar to developments in Europe, women, blacks, and labor groups gained some concessions and privileges during wartime, but at the end of it long-standing class, gender, and racial traditions remained intact.

While historians have studied the home fronts of the main antagonists in World War I— namely, Germany, Great Britain, France, and latecomer America in great detail—home fronts in Central Europe, Russia, but also occupied Belgium and France have received little or no attention. Richard Stites and Aviel Roshwald's (1999) volume on home fronts in "forgotten" countries is a notable exception. Contributors examine among others Poland, the Balkans, and Russia. Their findings are that while some countries, such as Croatia and Hungary, remained relatively untouched by the war, in others, particularly Poland, Russia, and also Germany, resentment against the upper classes and urban dwellers went hand in hand with growing antisemitism.

Twenty years after the end of the Great War, advances in tactics and technology made World War II the bloodiest conflict in recorded history. The upheaval it caused in the lives of ordinary people on the home fronts of all belligerent states was unprecedented. Similar to World War I, works that address Germany, Great Britain and its dominions, and the United States dominate the historiography on home fronts.

The first multivolume history examining the struggles on the German home front was Martin Broszat's (1977–1983) pioneering *Alltagsgeschichte* [history of everyday life] in Bavaria. In studying the social and political behavior of Bavarians, one major conclusion the series reached was that in everyday life resistance to and compliance with the National Socialist (Nazi) regime were not mutually exclusive. While Bavarians, for instance, might have resisted the Nazi Party's pressure to have their child participate in the Hitler Youth organization, some may have complied with the regime to keep their jobs.

More conventional historical treatments of the German home front such as Earl Beck (1986) and Richard Evans (2009) mainly concentrated on the effects of the Allied bombing on civilian morale. Beck disputed the conventional wisdom about the strengthening of resolve and instead emphasized the Nazi Party's struggle to control the increasing disillusionment and dissolution of the *Volksgemeinschaft* [people's community]. Evans places the home front in the context of Germany's military campaigns, the Nazi ideology, and its implementation by the armed forces and everyday Germans. He reiterates the profoundly devastating effect the Allied bombing had on the economy and people's lives.

Yet, none of these publications portrayed the German population as innocent victims of catastrophic destruction wrought by vindictive Allied bombardments. In *Der Brand*, however, Jörg Friedrich (2002) takes up exactly this line of argument. Based mainly on city chronicles and local history projects, Friedrich's voluminous account of lives lost and towns demolished stays so close to the local level that he hardly even mentioned the larger context of Nazi Germany's aggressive warfare on all fronts. Publication of the book caused an immediate emotional public debate about Germans as victims of the Allied bombing and of brutal expulsions (similar, in the way they were carried out, to what is nowadays termed ethnic cleansing) from formerly German occupied territories after the end of the war. As leading German and British historians weighed in, the debate became a *Historikerstreit* [Historians' Dispute], of which Lothar Kettenacker (2003) collected and published the most important contributions.

In 2004 and 2005, the Militärhistorisches Forschungsamt of the German Armed Forces concluded its magisterial official history of World War II with two volumes on the German home front. Edited by Jörg Echternkamp, the volumes contain essays about nearly every aspect of the German home front, excepting the roles of religion, churches, and gender. These topics are at the center of Michael Burleigh's (2000), Claudia Koontz's (1987), and Karen Hagemann and Stefanie Schüler-Springorum's (2002) works. Burleigh examines the quasi-religious character of the Nazi ideology. He argues that the hold the Nazis had on Germans—even as the war became hopeless—stemmed from the popular belief in the national community and Adolf Hitler's messianic role. Koontz showed that women were at once victims, followers, and resisters of Nazism. Hagemann and Schüler-Springorum's collection of essays focuses on the changes in gender roles and the blurring of the distinction between front and home front before, during, and after the two world wars.

Early British histories about the World War II home front tended to be optimistic about the positive impact total mobilization of all societal groups had for formerly disadvantaged groups and also women. Richard Titmuss' official history (1950) argued that social policies had reduced inequalities and had raised women's status in society. A younger generation of scholars attacked this overly optimistic view of the war's impact in the 1990s. Contributors to Harold Smith's edited anthology (1990) examined the effects of World War II on British society, concluding that surprisingly few wartime changes lasted beyond the war.

Home fronts in the British dominions have also received scholarly attention. Nancy Taylor's two-volume official history of the home front in New Zealand (1986), based exclusively on published official records and newspapers, provides a comprehensive account of the effects of World War II on the civilian population. Kate Darian-Smith's (1990) work on wartime Melbourne, Australia, studies social dislocation and increased government interference. It also scrutinizes the challenges to women's traditional role as homemakers and the influence of the presence of thousands of American GIs on Australian nationalism and identity.

Entering World War II as a belligerent nearly two and a half years after its outbreak in Europe, the United States mobilized its population, all sectors of its economy, and its military to fight the Axis powers on two fronts thousands of miles from its shores. In an early account Richard Polenberg (1972) writes about the transformative power of that wartime mobilization on American society. He claims that the wartime experience radically altered the character of American society and shaped Americans' political expectations lastingly.

The optimistic view of the socially and culturally progressive impact of World War II in the 1970s gave way to revisionist interpretations in the 1980s and 1990s, when Studs Terkel (1984), Allan Winkler (1986), and Michael Adams (1993) led the revisionist charge and challenged the "myth of the good war." Winkler concludes that Americans embraced wartime changes but, at the same time, clung to their prewar values. In his study, William O'Neill (1993) concentrates on the limits of American commitment to democracy at home for blacks, immigrants, workers, women, and Japanese Americans.

To supply a two-front war, the American economy underwent unprecedented and far-reaching changes. Harold Vatter (1985) describes the major trends, policies, and government controls in wartime industry. While he also includes the agricultural sector in his analysis, Walter Wilcox's (1947) study of the American farmer in World War II is still considered the standard work on American agriculture during the war. The war's impact on organizations of the labor movement, in general, and the Congress of Industrial Organizations (CIO), in particular, is the subject of Nelson Lichtenstein's (1982) seminal work. He claims that tensions within the labor movement and with the American business community had far-reaching consequences that lasted beyond the war years. Andrew Kersten (2006) analyzes the CIO's rival organization, the American Federation of Labor (AFL). He contends that the AFL, though neglected in labor historiography, was a "proponent of federal intervention in the economy and in labor relations" [ix] and therefore at the center of the New Deal's liberal coalition.

In her landmark study on women's roles at the American home front, Karen Anderson (1981) examines the types of jobs women held in war plants, their earnings, and trade unions' response to them. Susan Hartmann (1982) agrees with Anderson that any changes in women's roles during war time were temporary. She argues that white women's roles in society remained unchanged by the war, and so did barriers for employment of black women. Ronald Takaki (2000) arrives at a similar conclusion through examination of nine ethnic American cultures during World War II. Similar to World War I, the changes the war caused in the lives of women and African Americans proved to be ephemeral.

Discrimination and mistreatment in wartime America was not limited to African Americans. Immigrants of Japanese, German, or Italian descent were suspected of having loyalties to their countries of birth, and many of them, chief among them Japanese Americans, were "evacuated" from the West Coast and other regions to internment camps in places in the interior. Peter Irons' (1983) history of the internment of Japanese Americans claims that the government even engaged in a campaign of altering and suppressing evidence to prevent the U.S. Supreme Court from declaring the camps unconstitutional. Arnold Krammer (1997) sheds light on less studied internments of German and Italian Americans during World War II.

Home fronts in other major belligerent states receive attention in Jeremy Noakes' (1992) collection of essays. Besides Britain, Germany, and the United States, contributors examine the Soviet Union, Japan, Italy, Poland, France and the Netherlands. For each country, the essays explore the impact of World War II through the examination of characteristics such as propaganda, labor mobilization, and the role of women. Noakes' collection is probably the most wide-ranging of a very small group of Anglophone studies about Central and Southern Europe. The home front in Japan is another field that has not received much attention in the English-speaking world. Hence, Haruko and Theodore Cook's (1992) work on wartime Japan is still one of the very few. The Cooks based their study on hundreds of oral history interviews with men and women to offer insights into the life of ordinary Japanese on the home front.

The study of the Soviet home front has benefited from the end of the Cold War when some of the formerly secret archives were opened to historians. John Barber and Mark Harris (1991) examine how the Soviet political and economic systems braved the German invasion and the burdens of total war, and how the experience changed the social and economic order. Rodric Braithwaite (2006) takes a different, more personal approach in his analysis of how Muscovites experienced the defense of their city against the German invaders in November 1941. He based his book on personal remembrances of soldiers, politicians, writers, artists, workers, and even schoolchildren.

In conflicts and wars during the Cold War, the home front in the "classical" sense of total mobilization of the population and economic resources to sustain the war ceased to exist. On both sides of the "iron curtain," but particularly in the United States and in the Soviet Union, war industries, built up during World War II, continued to develop and produce vast amounts of arms and military equipment. Armies, downsized immediately after the end of the war, soon grew again and remained large. Containment of the Soviet Union's perceived expansionist ambitions and nuclear proliferation became the goal of American foreign and national defense policy. In this standoff between two super powers, divergent social forces battled it out on the American home front. In her seminal work, Elaine Tyler May (1994) shows that "containment" also applied to American domestic life. The home became the fortress of traditional values and gender roles in the defense against dangerous new feminist and civil rights movements that threatened the old order in the Vietnam War era.

While social upheaval and antiwar protests continued throughout the Vietnam War years, the war directly touched only the everyday lives of those American families whose sons fought in the war or whose daughters served as nurses or in other roles near the front lines. It was, however, the first war almost every average American family experienced on television. That kind of exposure to the fighting and dying at the front right in their living rooms transformed the American public into a new kind of virtual and sometimes real home "front." In his masterful study, William Hammond (1998) investigates how the media become the battleground for the war at the home front. He reveals how the relationship between the government and the media soured over the years as hopes of victory faded and public sentiment shifted against the war. Lawrence Eldridge (2011) concentrates on how the African American press handled the Vietnam War. He dissects how the black press reported the war while, at the same time, weighing its impact on the civil rights movement.

As the war dragged on year after year, public opinion became increasingly critical, questioning the reasoning for the seemingly endless fighting. Some parts of society even turned openly hostile toward the war and those that were fighting it. Pacifists, conscientious objectors, and those wanting to avoid the draft were joined by political opponents to the war and civil rights groups, fighting against racism and for equal rights for African Americans. Charles DeBenedetti and Charles Chatfield (1990) offer a comprehensive history of the anti-Vietnam War movement. They show the movement as a social and cultural phenomenon that energized people but failed to develop into a political force. Maureen Ryan (2008) examines the aftereffects of the war on the early "Baby Boomers" who came of age during the war and their attitudes about the military and foreign policy. She bases her findings on how that period is reflected in memoirs, short stories, films, and novels by veterans of the war, and contrasts them with those written and produced by women, antiwar activists, and Vietnamese refugees among others. Heather Stur (2011) studies Cold War America's idea of masculinity and how men and women returning from Vietnam challenged and also reinforced gender roles in America.

Similar Anglophone investigations of the home front in Vietnam or in the Soviet Union during its occupation of Afghanistan or in Iraq or Iran during their war against each other are unfortunately hard to find. There are, however, some studies about another evolution of the American "home front" during the War on Terror in Afghanistan and Iraq. Ole Holsti (2011) examines public attitudes toward U.S. involvement in the war and how public opinion influenced policy making. In a second edition of his *The Politics of Protest*, David Meyer (2007) includes chapters about new and old media and the impact of online activism. Andrew Carroll's (2006) volume gives voice to service members and their family through their eyewitness accounts, private journals, short stories, and letters. What sets this book apart is that the soldiers and their loved ones write about their experiences while the conflict is ongoing. In another analysis of the "new home front" in the War on Terror in Great Britain, Gillian Youngs (2010) points out that there are many links between "home" (national or domestic) and "front" (Iraq and Afghanistan). Her particular emphasis is on the complex relationship between the ethical contexts of policies and politics related to the war, in view of new media that have replaced the traditional top-down linkages between political institutions, the mass media, and the public, with horizontal and real-time online connections. Thus, the Vietnam War set the stage for the war playing out inside American homes. Advances in communication technologies and almost universal access to personal computers and cellular telephones subsequently increased the scope of the lens.

What is apparent in this wide-ranging bibliographical essay about home fronts in total wars, and their later iterations during the Vietnam War and the War on Terror, is that there is room for expansion of the scholarship into more regions of the globe and more topic areas. Thus far, scholarship on home fronts outside the English-speaking world is few and far between. While there are translations of major works for some countries, there is a lot of room for Asia, Eastern Europe, and Russia specialists to expand the field. Lastly, only a handful of historians and political scientists have investigated some aspects of the home fronts during the War on Terror. For the most part, though, developments regarding home fronts in countries that participated in the wars in Afghanistan and Iraq have only received attention in national and international media outlets. Their scholarly examination in professional journals and book-length treatments is the next step.

Bibliography

Aley, Ginette, and Joseph L. Anderson, eds. *Union Heartland: The Midwestern Home Front During the Civil War.* Carbondale: Southern Illinois University Press, 2013.

Anderson, Karen T. *Wartime Women: Sex Roles, Family Relations, and the Status of Women During World War II.* Westport, CT: Greenwood Press, 1981.

Barber, John, and Mark Harrison. *The Soviet Home Front, 1941–1945: A Social and Economic History of the USSR in World War II.* London: Longman, 1991.

Beck, Earl R. *Under the Bombs: The German Home Front, 1942–1945.* Lexington: University Press of Kentucky, 1986.

Boemeke, Manfred F., Roger Chickering, and Stig Förster, eds. *Anticipating Total War: The German and American Experiences, 1871–1914.* Cambridge, UK: Cambridge University Press, 1999.

Braithwaite, Rodric. *Moscow 1941: A City and Its People at War.* New York: Knopf, 2006.

Braybon, Gail, and Penny Summerfield. *Out of the Cage: Women's Experiences in Two World Wars.* London: Pandora Press, 1987.

Broszat, Martin, Elke Fröhlich, and Falk Wiesermann, eds. *Bayern in der NS-Zeit. Soziale Lage und Politsches Verhalten der Bevölkerung im Spiegel vertraulicher Berichte.* 6 Vols. Munich: R. Oldenbourg Verlag, 1977–83. [Bavaria in the Nazi Era. Social and Political Behavior of the Population as Reflected in Confidential Reports.]

Burleigh, Michael. *The Third Reich: A New History.* New York: Hill and Wang, 2000.

Carroll, Andrew, ed. *Operation Homecoming: Iraq, Afghanistan, and the Home Front, in the Words of U.S. Troops and Their Families.* Chicago: The University of Chicago Press, 2006.

Cook, Haruko Taya, and Theodore Cook. *Japan at War: An Oral History.* New York: New Press, 1992.

Daniel, Ute. *The War From Within: German Working-Class Women in the First World War.* Transl. Margaret Ries. Oxford: Berg, 1997. [German edition 1989. Arbeiterfrauen in der Kriegsgesellschaft.]

Darian-Smith, Kate. *On the Home Front: Melbourne in Wartime, 1939–1945.* Melbourne: Melbourne University Press, 1990.

Davis, Belinda J. *Home Fires Burning: Food, Politics and Everyday Life in World War I Berlin.* Chapel Hill: University of North Carolina Press, 2000.

DeBenedetti, Charles, and Charles Chatfield. *An American Ordeal: The Antiwar Movement of the Vietnam Era.* Syracuse, NY: Syracuse University Press, 1990.

Echternkamp, Jörg, ed. *Das Deutsche Reich und der Zweite Weltkrieg, Bd. 9/1 and 9/2: Die Deutsche Kriegsgesellschaft 1939 bis 1945.* Stuttgart: Deutsche Verlags-Anstalt, 2004 and 2005. [*Germany and the Second World War, Vol. 9/1: German Wartime Society 1939 to 1945.* Transl. Derry Cook-Radmore, Ewald Osers, Barry Smerin, Barbara Wilson. Oxford: Clarendon Press, 2008, and Oxford University Press, 2014]

Eldridge, Lawrence Allen. *Chronicles of a Two-Front War: Civil Rights and Vietnam in the African American Press.* Columbia: University of Missouri Press, 2011.

Evans, Richard J. *The Third Reich at War.* New York, Penguin Press, 2009.

Feldman, Gerald. "Mobilizing Economies for War." In *The Great War and the Twentieth Century.* Eds. Jay M. Winter, Geoffrey Parker, and Mary Habeck. New Haven, CT: Yale University Press, 2000, 166–186.

Förster, Stig, and Jörg Nägler, eds. *On the Road to Total War. The American Civil War and the German War of Unification, 1861–1871.* Cambridge: Cambridge University Press, 1997.

Friedrich, Jörg. *Der Brand. Deutschland im Bombenkrieg 1940–1945.* Munich: Propyläen Verlag, 2002. [*The Fire: The Bombing of Germany, 1940–1945.* Transl. Allison Brown. New York: Columbia University Press, 2008]

Gallman, J. Matthew. *The North Fights the Civil War: The Home Front.* Chicago: Ivan R. Dee, 1994.

Giesberg, Judith Ann. *Army At Home: Women and the Civil War on the Northern Home Front.* Chapel Hill: University of North Carolina Press, 2009.

Hagemann, Karen, and Stefanie Schüler-Springorum. *Home/Front: The Military, War, and Gender in Twentieth-Century Germany.* Oxford: Berg, 2002.

Hammond, William M. *Reporting Vietnam: Media and Military at War.* Lawrence: University Press of Kansas, 1998.

Hartmann, Susan M. *The Home Front and Beyond: American Women in the 1940s.* Boston: Twayne, 1982.

Healy, Maureen. *Vienna and the Fall of the Habsburg Empire: Total War and Everyday Life in World War I.* Cambridge: Cambridge University Press, 2004.

Holsti, Ole R. *American Public Opinion on the Iraq War.* Ann Arbor: University of Michigan Press, 2011.

Horne, John, ed. *State, Society and Mobilization in Europe During the First World War.* Studies in the Social and Cultural History of Modern Warfare, vol. 3. Cambridge: Cambridge University Press, 1997.

Irons, Peter. *Justice at War: The Story of the Japanese Internment Cases*. New York: Oxford University Press, 1983.

Kennedy, David M. *Over Here: The First World War and American Society*. New York: Oxford University Press, 1980.

Kersten, Andrew. *Labor's Home Front: The American Federation of Labor During World War II*. New York: New York University Press, 2006.

Kettenacker, Lothar, ed. *Ein Volk von Opfern? Die neue Debatte um den Bombenkrieg 1940–1945*. Berlin: Rowohlt Verlag, 2003. [A Nation of Victims? The New Debate About the Bombing of Germany.]

Koontz, Claudia. *Mothers in the Fatherland: Women, the Family and Nazi Politics*. New York: Routledge, 1987.

Krammer, Arnold. *Undue Process: The Untold Story of America's German Alien Internees*. London: Rowman & Littlefield, 1997.

Lichtenstein, Nelson. *Labor's War at Home: The CIO in World War II*. New York: Cambridge University Press, 1982.

Marwick, Arthur. *The Deluge: British Society and the First World War*. New York: Atlantic—Little Brown, 1965.

May, Elaine Tyler. *Pushing the Limits: American Women 1940–1961*. New York: Oxford University Press, 1994.

Meyer, David S. *The Politics of Protest: Social Movements in America*. 2nd Edition. New York: Oxford University Press, 2007.

Noakes, Jeremy, ed. *The Civilian in War: The Home Front in Europe, Japan and the USA in World War II*. Exeter: University of Exeter, 1992.

O'Connor, Thomas H. *Civil War Boston: Home Front and Battlefield*. Boston: Northeastern University Press, 1997.

O'Neill, William. *A Democracy at War: America's Fight at Home and Abroad in World War II*. New York: Free Press, 1993.

Polenberg, Richard. *War and Society: The United States, 1941–1945*. Philadelphia: J.B. Lippincott, 1972.

Procacci, Giovanna. *Dalla Rassegnazione Alla Rivolta: Mentalita' e Comportamenti Popolari Nella Grande Guerra* [From Resignation to Revolt: Popular Mentality and Behavior in the Great War.], Storia e documenti, vol. 30. Rome: Bulzoni, 1999.

Rogers, William Warren, Jr. *Confederate Home Front: Montgomery During the Civil War*. Tuscaloosa: University of Alabama Press, 1999.

Ryan, Maureen. *The Other Side of Grief: The Home Front and the Aftermath in American Narratives of the Vietnam War*. Amherst: University of Massachusetts Press, 2008.

Smith, Harold L., ed. *War and Social Change: British Society in the Second World War*. Manchester: Manchester University Press, 1990.

Stites, Richard, and Aviel Roshwald, eds. *European Culture in the Great War: The Arts, Entertainment, and Propaganda, 1914–1918*, Studies in the Social and Cultural History of Modern Warfare, vol. 6. Cambridge, UK: Cambridge University Press, 1999.

Stur, Heather Marie. *Beyond Combat: Women and Gender in the Vietnam War Era*. New York: Cambridge University Press, 2011.

Takaki, Ronald. *Double Victory: A Multicultural History of America in World War II*. New York: Little, Brown, 2000.

Taylor, Nancy M. *The Home Front*. Volumes I and II. Wellington, NZ: Historical Publications Branch, Department of Internal Affairs (Official History of New Zealand in the Second World War, 1939–45), 1986.

Terkel, Studs. *The "Good War": An Oral History of World War Two*. New York: Pantheon Books, 1984.

Titmuss, Richard M. *Problems of Social Policy*. London: HMSO and Longmans, Green, 1950.

Vatter, Harold G. *The U.S. Economy in World War II*. New York: Columbia University Press, 1985.

Wilcox, Walter. *The Farmer in the Second World War*. Ames: Iowa State College Press, 1947.

Winkler, Allan M. *Home Front U.S.A.: America During World War II*. Arlington Heights: Harlan Davidson, 1986.

Winter, Jay, and Jean-Louis Robert, eds. *Capital Cities at War: Paris, London, Berlin 1914–1919*. 2 vols. Cambridge, UK: Cambridge University Press, 1997 and 2007.

Winter, Jay, and Richard Wall, eds. *The Upheaval of War: Family, Work and Welfare in Europe, 1914–1918*. Cambridge, UK: Cambridge University Press, 2005.

Youngs, Gillian. "The New Home Front and the War on Terror," *International Affairs* 86, no. 4 (2010): 925–937.

19 War, Atrocity, and Genocide

Hilary Earl

What is genocide? How do we define it? Where does it come from? Is it a modern phenomenon? What is the role of state building in genocide? What is its relationship to war, colonization, and imperialism? These are the questions that are at the heart of scholarly debates and research in the burgeoning and multidisciplinary field of genocide studies, which encompasses history, political science, sociology, psychology, social psychology, law and legal studies, anthropology, health sciences, and others. If the questions seem simple enough, the answers are anything but and depend on *who* is asked. Regardless of questions or answers, the historical study of genocide and atrocity is well suited to the study of war and society, encompassing interdisciplinary factors that extend beyond traditional methodologies in military history.

Part of the definitional conundrum comes from the imprecise legal definition of genocide, codified as law in the 1948 United Nations *Convention on the Prevention and Punishment of Genocide*. The law was a compromise document, signed as the Cold War was heating up, and which defines genocide as the "intent to kill, in whole or in part, a national, ethnical, racial, or religious group, as such." That political and other groups were excluded from the definition was a compromise to the Soviets, who never would have signed the *Convention* if it had included them. Another imprecision of the *Convention* is that it does not define what constitutes "part of the group," leaving open for interpretation the number of dead required to constitute "genocide." Cultural destruction is also not mentioned in the 1948 definition, although other ways to destroy the group are, including preventing births, transferring children from the group, and inflicting untenable life conditions (Earl, 2009, 91–92; *Genocide Convention*, 1948, Article II).

The imprecise definition has led to raucous scholarly debates about the precise meaning of the term, compounded by disciplinary differences. Social scientists tend to work comparatively and look for similarities, whereas historians mostly do contextual, case-specific research. As such, the definition of "genocide," its origins, and relationship to war are *all* contested (Moses, 2002; Straus, 2001; Weitz, 2003; Fein, 1991; Charny, 1994; Shaw, 2003). There is no single element of the legal, historical, or sociological category we call "genocide" that scholars agree on (Straus, 2001). In fact, genocide scholars have been heavily criticized for spending too much time defining the term and not enough time analyzing and comparing genocidal events (Straus, 2001, 359). Not only do academics from various fields define and conceptualize the term differently but also scholars from the *same* discipline often disagree about the core attributes of this (relatively) newly coined term, and therefore what *is* and *is not* genocide. Not even the most obvious example, the Holocaust—the deliberate and systematic murder of six million European Jews between 1933 and 1945—is universally identified as genocide.

This division is reflected in the main journals in the field, *Holocaust and Genocide Studies*, which almost exclusively publishes articles on the Holocaust, and the *Journal of Genocide Research*, which seldom publishes articles dealing with the Holocaust. Some very notable scholars, including Yehuda Bauer (2001) and Stephen Katz (1981, 1996), have argued that the Holocaust is unlike other mass murders, atrocities, and genocides in history. The distinction, as articulated by Bauer (2001, 12), is that while genocide is planned, it somehow is not as "bad" as "Holocaust" because it is selective killing of members of the group, whereas "Holocaust," he believes, entails the "radicalization of genocide" where the perpetrators intend to murder all members of the targeted group (Bauer, 2001, 12; Rosenfeld, 1999; Rosenbaum, 1996; Bloxham, 2009). For instance, the murder of the Sinti and Roma during World War II was not a "Holocaust" but rather a genocide because the Nazis had no intention of killing every last "Gypsy." In this interpretation the Holocaust is believed to be a more complete or total form of genocide. The irony, of course, is that Raphael Lemkin, a Polish Jewish lawyer whose entire family except one brother was killed by the Nazis, originally coined the term "genocide" in 1944 as a way to describe and categorize the crimes committed by the Germans during World War II—that is, the Holocaust (Lemkin, 1944/2008; Earl, 2013; Moses, 2004). Not surprisingly, Holocaust and genocide studies have had a troubled academic relationship (Mattaus et al., 2011), which was reaffirmed at a recent conference of genocide scholars (Ahren, 2016).

The United Nations defined genocide as an international crime on December 9, 1948, in its *Convention for the Prevention and Punishment of Genocide*, which came into force on January 12, 1951, and which 147 states have now ratified as law. In its modern incarnation, genocide is almost always linked to forms of state violence (Kren and Rappoport, 1980; Bartrop, 2002). However, Article I of the *Convention* affirmed that genocide can also "be committed during times of peace," not just during "times of war," allowing for the organizational and planning period that precedes such murder to be considered part of the genocidal process. It also recognizes there may be situations where war and genocide are not intertwined, such as with the Ukrainian famine of 1932–1933, the Indonesian massacres of 1965–1966, and the Chinese takeover of Tibet since 1950 (Bartrop, 2002, 526–527). The *Convention* also defines genocide as an intentional act. Importantly, this means that genocide is not spontaneous military aggression or atrocity—what international law today refers to as "war crimes"—but rather a planned process that almost always begins before armed conflict commences.

The British sociologist Martin Shaw (2003) disagrees with the legal definition that requires the "intent to destroy" to be proved in order for the crime to be considered "genocide." Rather, he sees modern war where civilians are targeted for destruction as indistinguishable from genocide, regardless of intention. As Paul Bartrop notes (2002), this should not be surprising since genocide is almost always seen to have something to do with war (519)—though the exact relationship is debatable. Traditional views regard war as an ancient, socially accepted, and legitimate state practice carried out between two or more armed belligerents to advance their political aims; and, while killing is central to war, it has rules and limits. Over the centuries the act of killing civilians in the course of war has escalated to the point where during contemporary state conflicts, civilian casualties often outweigh military ones. Nearly thirty-three million people died in World War I, of which only 5 percent were civilians, whereas 66 percent of those killed in World War II were unarmed noncombatants; today the number of civilian casualties has escalated to nearly 80 percent (Bartrop, 2002, 529–530). This fact complicates the relationship between war and genocide.

Is Martin Shaw right—are war and genocide indistinguishable (Shaw, 2003, 5–6)? Certainly war and genocide are linked, and at the very least scholars note that war affords perpetrating states an opportunity to carryout genocidal policies against long-standing enemies, such as the case in World War II (Bergen, 2003). Moreover, modern military conflicts are not the exclusive purview of professional soldiers. Rather more often than not, paramilitary forces and nonmilitary personnel carry out state aims against targeted enemies, such as against Armenians in World War I, Jews in World War II, and Bosnian Muslims during the wars of succession in the former Yugoslavia. The increased reliance on paramilitary forces, explains Michael Ignatieff, "may be one reason why postmodern war is so savage, why war crimes and atrocities are now integral to the very prosecution of war," and why Shaw believes modern conflicts are more like genocide than traditional limited war (Ignatieff, quoted in Bartrop, 2002, 530).

Lt. Col. David Grossman (1995/2009, 201) labels this development "the atrocity spectrum." At one end is the killing of one armed combatant by another armed combatant, which is universally considered a legal act. At the opposite end of the spectrum is "atrocity," which is universally considered criminal. In the military context atrocity literally means breaking the rules of war. Atrocity is the act of killing combatants who have ceased fighting or unarmed civilians who pose no immediate threat to the soldier (Grossman, 1995/2009, 195–204). Legally, this act becomes genocide when the state targets entire civilian groups as enemies to be destroyed. In the end, aside from Shaw, few scholars dispute that there is a distinction between war, war crimes or atrocities, and genocide. Unlike atrocity, as historian Erich Haberer (2001) has noted, perpetrating genocide requires both planning and intention. Genocide might seem like a sudden eruption of violence; however, it is not. Rather it requires preliminary steps to carry out a "final solution" to a nation's problem (Bartrop, 2002, 524). Raul Hilberg (1961) explained this development best in his masterful and groundbreaking book *The Destruction of the European Jews*, when he described genocide as a four-step process that includes the identification of the group targeted for destruction, their oppression and concentration, their deportation, and finally the last and decisive phase: extermination (Hilberg, 1961).

Thus far, the twentieth century seems to be historically unprecedented and the most deadly (Hobsbawm, 1994). It has been called at various times the age of extremes, moral atrocity, and mass violence. In his now famous essay "Why Is the Twentieth Century the Century of Genocide?" Mark Levene (2000) answered his own question when he noted that 187 million people died from political violence during the century, including the victims from at least seven genocides: the genocide against the Herero by the Germans (1904–1907), the Armenian genocide by the Young Turks (1915), the Nazi Holocaust against the Jews (1933–1945), the Maya genocide in Guatemala (1966–1968), the Cambodian genocide committed by the Khmer Rouge against the Vietnamese and other groups of Cambodians (1975–1979), the Rwandan genocide of Tutsi and moderate Hutu (1994) by the extremist Hutu, and finally the Bosnian genocide by the Serbs (1995). It remains to be seen if the century will live up to its name, given early twenty-first-century developments, such as the ongoing and escalating violence in Sudan, Syria, Congo, Central African Republic, and Yemen, to name but a few examples.

Because of the definitional muddle, the scope of the field, and the myriad of debates, this chapter limits itself to three key historiographical discussions. First it examines the question of temporality: whether historians and other scholars have categorized genocide as a modern phenomenon or as a practice repeated for centuries and across cultures. Second, it assesses scholarship on the historical context of colonial violence, particularly that of Dirk Moses. Finally, this chapter explores how historians and other scholars have scrutinized motives of the individual and participant groups in genocide.

Is Genocide Modern?

The preamble to the 1948 *Convention on the Prevention and Punishment of Genocide* begins by noting that the crime is timeless, having caused "great losses" to humanity as an "odious scourge," and for that reason genocide requires international attention to identify it and prevent it from happening again. Leo Kuper, one of the founding fathers of genocide studies, begins his path-breaking book *Genocide* with a chapter that elucidates various examples of the ancient crime from biblical times to the present (1981, 11–18). *The Oxford Handbook of Genocide Studies* (Bloxham and Moses, 2013) has five essays devoted to what it calls "premodern" genocide. Even the father of the term "genocide," Rafael Lemkin (1944, 79), tells us that genocide is a new name applied to "an old practice." But is it? Are the medieval crusades, to take one of Kuper's examples (12–13), the same genus as the murder of Bosnian Muslims by Serbian nationalists in 1995? Like all elements of genocide, its temporality is contested. Part of the confusion around whether genocide is a "modern" phenomenon, or an action that transcends time and place, comes from Lemkin himself, whose definition of genocide is elucidated in Chapter 9 of his book *Axis Rule in Occupied Europe* (1944, 79–95).

While Lemkin did say that genocide was an old practice, he also said that the context of violence had changed over time. Specifically he said that genocide "denote[s] an old practice in its modern development" (Lemkin, 1944, 79). Does that mean that genocide is new, or rather that the new context has changed the meaning of the old violence? Lemkin is not clear on this. He states that genocide is a crime committed first and foremost against a "national group as an entity"—in other words as a corporate crime of one group targeting another group (Lemkin, 1944, 79; Earl, 2009, 91–95). There is a caveat, though: it is not enough to simply target the group; the intention to wholly destroy the group must also be present, perhaps distinguishing it from earlier forms of violence (Lemkin, 1944, 79; *Genocide Convention*, 1948, Article 2). With the exception of Moses, most scholars agree that what distinguishes genocide from other forms of atrocity committed over the years and through epochs, including the newly defined "crimes against humanity," is the intent to destroy the group. With the exception of the thirteen Nuremberg trials held between 1945 and 1949, where the intention to destroy the group was assumed based on the outcome of six million dead European Jews, proving intention in a court of law (e.g., what defines first-degree murder) is quite difficult.

Today most scholars agree that one of the defining features of the modern period is that interstate conflict involves the murder of increasingly larger numbers of civilians (Levene, 2000; Mazower, 2002). The problem has been and continues to be in naming these episodes. The mounting civilian deaths during times of state violence have been defined by various scholars in various ways over time, as total war, atrocity, mass murder, mass violence, and terrorism. As the scope of violence increases scholars have found it necessary to create more new terms, such as ethnic cleansing (the removal of entire groups from geographic areas that often leads to genocide), crimes against humanity (murder of the parts of the group without the intention to destroy the entire group), and genocide (the deliberate attempt to destroy the group in whole or in part) (Naimark, 2001). The question is what distinguishes historical events and different types of atrocities over the centuries?

A good example of how historical events have been reinterpreted using new historical and legal frameworks such as genocide is the Holocaust, or the deliberate murder of six million European Jews. The Holocaust used to be examined almost exclusively within the framework of premodern religious hatred. The argument went something like this: the Nazi destruction of almost all European Jews was the culmination of hundreds of years of Christian persecution and hatred that were characterized in earlier centuries by exclusion, periodic violence, and murder, and that in the modern period manifested as total destruction or genocide (Hilberg, 1961). Put simply, Jews had always been the target of Christian Europeans; the Nazis simply took that hatred to its logical conclusion and killed them all (Reuther, 1977; Yerushalmi, 1977).

Zygmunt Bauman (2000) was one of the first scholars to move beyond the "hatred as explanation" interpretation. In his now famous book he argues that the Holocaust was a modern phenomenon, a product of Enlightenment ideas. Bauman is a sociologist by training, and not surprisingly defines modernity in terms of the structures of the state, bureaucracy, and industrialization. His basic premise is that the Holocaust is uniquely modern largely because it harnessed modern technology (gas chambers), bureaucracy (the SS), and ideas (racism) to bring it to fruition. The murder of six million European Jews was the result of industrial society and not ancient religious hatreds, as previous scholarship has argued. In Bauman's thinking the Holocaust was a unique and unprecedented genocide precisely because the state harnessed its bureaucratic functions to carry out the murders. For historians and social scientists alike, the problem with Bauman's argument as an explanatory model is that it renders the Holocaust unique and thus makes it incomparable to other modern genocides, and modernity itself—in his definition—is decontextualized (Birken, 1995).

Mark Levene, one of the most prolific scholars working in the field today, rejects Bauman's (2000) theory in that it reduces the genocide discussion to "the form genocidal killing takes," and that form does not apply to all genocides equally (Levene, 2000, 307). For instance, Levene notes that the Rwandan genocide was carried out by machete, and not gas chambers, not to mention the Herero and Armenian examples, where the victims died largely as a result of displacement (Levene, 2000). Levene certainly believes genocide is modern, but his definition of modernity is understood in historical context—similar to Moses'—as part of a larger process intertwined with the international system that created the modern nation-state. In Levene's words, "genocide is a by-product of very general drives toward . . . political power" that "in contemporary history can only effectively succeed through control of modernized or at least modernizing nation-states operating within a global community of such states" (2002, 65). In other words, genocide is linked to state building. Levene's thesis situates genocide in a world-historical context, not just in Europe, as other scholars have done.

Donald Bloxham, one of a group of contemporary historians who have taken the field by storm, employs a similar "modern state" theory as Levene. Rather than situating the Holocaust in a world-historical context, though, he positions it in the European context. He sees the murder of the Jews as part of wider European practices that since the 1870s had increasingly used violence against various civilian groups, including ethnic "minorities . . . indigenous peoples in European colonies . . . opponents of the economic order. . . [and] the socially marginal" (2009, 2). Hence violence against civilians was not a uniquely German practice. Rather, the state-directed violence that culminated in the Holocaust was the result of modern conceptions of race and the state that Germany happened to practice more efficiently than others.

Intention and Outcome: The Colonial Context

The field of genocide studies was born in the 1980s by a group of pioneering social scientists, mostly sociologists and political scientists who worked comparatively. As social scientists, they engaged in model building and worked mostly with secondary sources. Then in the 1990s the field exploded with a second wave of genocide scholarship, sparked in part by the Rwandan and Bosnian genocides, which many scholars lamented could have been prevented (Power, 2002; Dallaire, 2003). There is a plurality of scholarship during this period that can be loosely classified in three groups. First were those scholars who tried to understand genocide using law. Second was a group that consisted mostly of Americans who focused on how to prevent or detect genocide before it happened and, finally, a third group of postcolonial scholars consisting mostly of historians, such as Dirk Moses, Isabel Hull, Donald Bloxham, and Mark Levene. The latter engaged in archival research to try to understand genocide historically, as part of larger and longer historical processes, such as colonialism, frontier violence, and the dynamic of power in settler societies that they defined globally, rather than focusing exclusively on individual nations in Europe.

Among these postcolonial scholars, Dirk Moses is one of the most important in the field of genocide studies. He is prolific and influential, with eight books to his credit, including *The Oxford Handbook of Genocide Studies* (2013), which he coedited with Donald Bloxham. He has been the editor-in-chief of the International Network of Genocide Scholars' (INoGS) *Journal of Genocide Research* since 2011, a multidisciplinary forum for scholarly research in the field. Moses is a historian by training and perhaps best known for his work on "settler" and "colonial" genocide. At the heart of his research is the view that the Holocaust should not be held up as *the* paradigm for all genocides, as it is a flawed and Eurocentric argument that creates stasis in the field as well as clouds our understanding of non-paradigmatic (i.e., Holocaust) violence, what Moses (2004, 2005) cleverly refers to as "the Gorgon effect." Genocide, he forcefully argues, is not new; rather it is an old racist, colonial practice that sometimes includes "non-lethal forms of racial and ethnic persecution," especially of indigenous peoples (2004, 535). He notes violence may take on different forms than total physical annihilation, particularly culturally, where entire indigenous groups have lost their language and culture to forced assimilationist practices, including the forced removal of indigenous children from their families.

The 1948 *Convention* has very little to say about colonial and cultural genocide, notes Moses. He especially takes issue with Article II, which emphasizes "*the intention* [of the perpetrator] to destroy, in whole or in part, a national, ethnical, racial, or religious group, as such[.]" Proving that the Nazis intended to kill the Jews is relatively easy, he notes, although *when* the decision was made is significantly more contentious. *Mein Kampf* (1924), Hitler's blueprint for power, identifies "the Jews" as a mortal and supernatural enemy that Germany must destroy or be destroyed themselves (Birken, 1995; Jäckel, 1981); that the Nazis saw Jews as such was enough for the Nuremberg prosecutors to make a case for intentional destruction. But rarely, notes Moses, can intention be discerned so easily in cases of settler violence, which is his main area of study. Moreover, in the colonial context, most deaths are the result of disease and displacement, not from "intentional" acts of violence or an intentional policy decision. How indigenous people die should not matter, argues Moses (2005), because the outcome is the same—the cultural and physical destruction of the group, which *is* the core of the genocide concept.

Moses (2005) makes an important point here. Genocidal intention, or what law refers to as the guilty mind or *mens rea*, does not need to be explicitly stated for the act to be considered genocide. Rather it can be found in the outcome itself, because the likely result, for example, of settling a geographic area that does not belong to your state is the death of the indigenous population. Even contemporaries knew that, as Charles Darwin observed in the mid-nineteenth century: "We can see that the cultivation of the land will be fatal in many ways to savages for they cannot, or will not, change their habits" (Moses, 2005, 29). Moses' paradigmatic shift away from the intention of "total physical annihilation" as the *only* prototype for genocide may be one of the most important new interpretations of genocide that the field has seen since its inception.

Genocidal Motives

The final question this survey will address is also the most morally vexing. What explains the behavior of the perpetrators of genocide? How do scholars explain the willingness of individuals to shoot, stab, hack, rape, or gas women and children, the elderly, and their neighbors at close range? While the behavior of the perpetrator is unquestionably abhorrent—as social psychologist James Waller (2007) has so aptly noted—states have never had difficulty recruiting people to carry out the task. Of course, genocide is not an individual crime; rather it is a corporate endeavor that requires a variety of types of perpetrators, from those firsthand killers who pull the trigger or slit the throat to those indirect participants who schedule the trains, man the barricades, and carry out propaganda (Earl, 2009). As early observers of the crime have noted, genocide is conducted like a "large-scale industrial enterprise" (Kren and Rappoport, 1980, 8; see also Hilberg, 1961).

In the 1990s the perpetrator debate exploded with the publication of Christopher Browning's (1992) much-lauded study of Reserve Police Battalion 101, a group of reserve policemen who operated in Poland during the early phase of the Final Solution against Jews, and Daniel Goldhagen's (1996) reinterpretation of the motives of the same group of perpetrators. At the heart of the controversy was Goldhagen's thesis that Germans killed Jews because they possessed a particular type of murderous antisemitism he called "eliminationist." Germans, he argued, were born to hate Jews so much that when given the chance, as was the case during World War II, they willingly and eagerly killed them. Browning's argument, on the other hand, was equally as disturbing. He observed that the men of Police Battalion 101 were quite "ordinary," by which he meant that under similar circumstances any non-ideological person would have behaved in the same fashion as the policemen had in the field, because the factors that motivated them were neither innate hatred nor structural, but rather situational. In other words, Browning (1992) believes that killing is dependent on context. In the case of the men of Police Battalion 101, he noted these policemen killed for banal reasons, such as peer pressure and career advancement.

Browning's (1992) "ordinary men" thesis turned the field of genocide studies on its head. The problem of course is that we want to believe that only bad people do bad things, that people kill because they hate or because they are somehow mentally ill, psychopathic, or vengeful. Accepting the possibility that human beings kill for banal reasons means we must accept Browning's (1992) conclusion that under the right circumstances, we too could kill. That fact alone, argues Donald Bloxham, makes "understanding . . . perpetration. . . *the* essential element to understanding genocide" (2004, 414). The question of what motivates otherwise ordinary men—whether soldiers or paramilitary personnel—to commit atrocities and genocide is the focus of the last section of this chapter.

Killing large numbers of civilians requires planning, resources, and manpower. Scott Strauss (2004) has estimated there were as many as 214,000 direct perpetrators in the Rwandan genocide, and they killed approximately 800,000 people in a hundred days. No one has attempted such a count for the Nazi genocide of the Jews, largely because scholars cannot agree on the temporal parameters of the event. However, they generally acknowledge that between 1939 and 1945, it probably took as many as two hundred thousand people from a host of European countries to carry out the genocide that killed six million people.

Who were the killers? Research has shown that they came from all walks of life, and were not limited to specific classes, professions, or geographic regions. My own research, for example, looks at twenty-four commanders of the paramilitary SS-*Einsatzgruppen*, who killed approximately 1.1 million Jewish men, women, and children in open-air shootings in the Soviet Union between 1941 and 1943 (Earl, 2009). They were mostly well-educated professionals, whereas their men came mostly from the German working and lower classes (Earl, 2009). This is not an unusual finding. Research on other genocides demonstrates quite clearly that perpetrators can come from all classes. The only commonality between perpetrators across time and space is that men are more likely than women to become *direct* perpetrators, and this may have more to do with historical context and opportunity than anything innately different between the sexes. Early research on this subject suggests that while some women have directly engaged in the murder process, their roles tend to be indirect, as enablers and helpers, rather than firsthand killers (Lower, 2013).

Who exactly should be considered a perpetrator is another important historiographical debate. Is it only the person who pulls the trigger? Since genocide is a group crime that requires significant planning and organization, not just murder, many different types of jobs—and thus perpetrators—are required depending on the context and timing of the process. Two broad classifications of killers are identified in the historiography on perpetrator motivation: firsthand or direct killers, and indirect murderers. Firsthand killers are those individuals who directly participate in the killing process, including soldiers, members of paramilitary organizations, policemen, and civilians. Firsthand murderers are told—at varying times during the process—that their task is to kill enemies of the state. This is not always the case with indirect perpetrators, some of who may not have realized at the time they were directly or indirectly contributing to genocide.

Within the historiography, the category of indirect perpetrators was made famous by Hannah Arendt's 1961 study of desk murderer Adolf Eichmann, *Eichmann in Jerusalem: A Report on the Banality of Evil*. Desk murderers are indirect perpetrators who kill using their pens. For example, Adolf Eichmann planned the deportation of millions of Europe's Jews to the death camps in Poland, but who, as far as we know, never killed anyone directly himself. In his study of the Final Solution, Raul Hilberg (1961) explains that the mass murder of Europe's Jews was in fact made possible by fastidious administrators, such as Eichmann, who may have been removed from the direct killings, but who nonetheless were responsible for sending millions of Jews to their death.

Recent research has shown that the direct-indirect binary may be too rigid. For instance, my research on the *Einsatzgruppen* leaders reveals there were men whose behavior and proximity to the killing process placed them firmly in neither camp—dubbed "hybrid killers" (Earl, 2009). They are not in an office removed from the killing process, but rather operate in the field with paramilitary units, giving orders to murder and witnessing executions as they occur. They see firsthand and close up the impact of their decisions and know full well that their orders mean certain death for the victims.

A final group of perpetrators is the elites, who in some ways occupy a category all their own. These are the individuals whose ideas and ideology govern the state. They are the highest-ranking political figures responsible for conceptualizing, organizing, and carrying out violence. They are almost always men. Sometimes they see the results of their work close up, as did Heinrich Himmler in the autumn of 1941, when he visited an *Einsatzgruppen* shooting, or Albert Speer, whose presence at Dora labor camp is well documented. This is a shockingly large group that includes such historical figures as Talat Pasha, Talat Bey, Enver Pasha, Pol Pot, Adolf Hitler, Reinhard Heydrich, Heinrich Himmler, Albert Speer, Radovan Karadžić, and Slobodan Milošević, to name but a few. We will not concern ourselves here with the decision makers, the ideologues, and those in charge of the state. More important for a volume on war and society is the men who participate in genocide directly as members of the military and paramilitary organizations operational on the ground.

What motivates these men to kill? In the historiography on genocide, there are two basic motivational categories for genocidal murder. The first is internal, the belief that what one is doing—killing for the state—is somehow right. In the literature on perpetrator motivation, this is often referred to as ideology. The second category is external, what some scholars call situational factors, which may include coercion, obedience to orders, peer pressure, professional advancement, and even economic gain. Of course, direct and indirect perpetrators can be motivated by more than one of these at any time. Seldom do human beings act for the same reason in every situation. For this reason, perpetration must be understood in the context in which it takes place (Waller, 2007; Westermann, 2005, 2016; Hatzfeld, 2003).

Research on the question of perpetrator motivation is immense, and began immediately after World War II with psychiatrists and psychologists who were hired by the Allies to interview the major Nazi perpetrators in custody at Nuremberg. These early interpretations promoted the idea that perpetrator behavior was somehow abnormal and extraordinary. For instance, Gustav Gilbert (1963), one of the prison psychologists at the 1945 Nuremberg International Military Tribunal (IMT) proceedings, examined the defendants and others in custody at length and concluded that the *Allgermeine Schutzstaffel* ("General SS") units— the German paramilitary stormtroopers who were Nazi Party members and carried out the genocide on the Eastern Front—took normal individuals and transformed them into "murderous robots." Gilbert (1963) sees the SS men as machines, devoid of conscience, and programmed to be obedient. Brainwashed and lacking agency, this group was seen to do the bidding of their leaders, who Gilbert identified as the most important perpetrators.

Alongside this perspective of the SS perpetrators as automatons is the social-psychological view that they possessed "authoritarian personalities." The nature of this personality type is bifurcated, in that the individual has the desire both to hold power over others and to submit to a higher authority. The classic statements on the authoritarian personality are Erich Fromm's *Escape From Freedom* and Theodor Adorno et al.'s *The Authoritarian Personality*. In the 1960s and 1970s, Yale social psychologist Stanley Milgram (1974) officially (and unethically, some would argue) conducted more than two dozen obedience to authority experiments. The most well-known of these was the one in which 65 percent of Milgram's "ordinary" participants obediently and willingly followed the orders of an experimenter in a white coat who instructed them to administer lethal electric shocks on a total stranger. Milgram's experiment illustrated the willingness of adults to obey authority. As Grossman has aptly noted, and what Browning's 1992 research has demonstrated, is that "if this kind of obedience could be obtained with a lab coat and a clipboard by an authority figure who has been known for only a few minutes, how much more would the trappings of military authority and months of bonding accomplish?" (Grossman, 1995/2009, 141–143). Of course, what is also significant about Milgram's (1974) findings is that virtually every Nazi war criminal argued that he too was "obeying orders." Even Otto Ohlendorf, the leader of *Einsatzgruppe* D and one of the most committed Nazi ideologues, told the court at Nuremberg that "it is inconceivable that a subordinate leader should not carry out orders given by the leaders of the state" (Ohlendorf, quoted in Alvarez, 2010, 120).

Outside of the social-psychological realm, not much in the way of critical analysis or interpretive frameworks have been developed to explain perpetrator motivation, let alone to assess their self-representation in a legal arena. Exceptions include Hannah Arendt's 1963 controversial *Eichmann in Jerusalem*, which she wrote in 1961 during Adolf Eichmann's trial in Jerusalem and which made famous the prototype of the "banal" desk murderer. Arendt depicted Eichmann as a relatively ambitious man who openly admitted he was neither a hater of Jews nor a firsthand killer, yet whose conscientious bureaucratic activities helped facilitate the murder of hundreds of thousands of innocent people. Because of these personality traits, Arendt (1963) concluded that Eichmann was neither mad nor a monster, but rather an ordinary albeit ambitious man who committed extraordinary acts of evil.

Arendt's (1963) thesis is inapplicable in the context of shootings on the Eastern Front where the perpetrators killed their victims face-to-face. My own research on *Einsatzgruppen* leaders demonstrates that without exception, these men were of strong character (Earl, 2009). Many possessed charismatic traits; they were, for the most part, well-educated professionals with minds and wills of their own, anything but "insane" or automatons. Some were ideologically motivated, and others came to the killing process through their work. They were all career-driven and some were even likeable. Others were the epitome of the nasty, jackbooted SS man. There are literally thousands of pages of testimony and documentation from the trial of the *Einsatzgruppen* leadership that attest to the fact that these men were anything but puppets (Earl, 2009).

Bloxham labels this shift in the historiography that encompasses my research as the "'voluntarist turn,'" which for historians means "cutting through the courtroom apologia" to get at the real motive for perpetration (2009, 265). Sparked by Browning's *Ordinary Men*, it now includes the work of Andrej Angrick, Waitman Boern, Hilary Earl, Ulrich Herbert, Klaus-Michael Mallmann, Edward Westermann, and Michael Wildt, and they offer more complicated depictions of the SS and policemen in the field (Paul and Mallmann, 2000; Wildt, 2003; Heinemann, 2001). Unfortunately, though, this picture does not allow for easy characterizations. Instead, it yields a myriad of personality types and variable factors that more accurately reflect the complex human reality of events on the ground.

What about soldiers—where do they fit into this discussion about atrocity and genocide? In the case of the Third Reich, until recently it was believed that the German Army had clean hands and was not involved in the genocide against Europe's Jews; if soldiers did participate it was not willingly but rather because they were forced to obey the regime or they would be punished and terrorized. Omer Bartov was one of the first scholars to argue that, far from opposing the Third Reich, the German Army was "an integral part of rather than a separate entity from the regime" (1991, 10). Bartov's conclusions that the German Army was a willing and effective tool of the Nazis set the field on fire, and has led to a reexamination of the military in genocide that includes important work by Isabel Hull on the German Army and its operations against the Herero during 1904–1907 (2005) and Waitman Boern's (2014) research on the German Army on the Eastern Front.

Outside of the German context, many have tried to explain the motives of perpetrators from a variety of different disciplinary models. This is a vibrant field and one of the most important questions to answer. If we come up with a satisfactory-enough one, perhaps one day, as the most hopeful scholars suggest, genocide will be predicted and prevented entirely.

Bibliography

Ahren, Raphael. "Academics Go to War Over the Study of Mass Killing," *The Times of Israel*, June 26, 2016. Accessed October 5, 2016 www.timesofisrael.com/academics-go-to-war-over-the-study-of-mass-killings/

Alvarez, Alex. *Genocidal Crimes*. New York: Routledge, 2010.

———. *Native America and the Question of Genocide*. Lanham, MD: Rowman & Littlefield, 2014.

Apsel, Joyce, and Ernesto Verdeja, eds. *Genocide Matters: Ongoing and Emerging Perspectives*. New York: Routledge, 2013.

Arendt, Hannah. *Eichmann in Jerusalem: A Report on the Banality of Evil*. New York: Penguin, 1963.

Bartov, Omer. *Hitler's Army: Soldiers, Nazis, and War in the Third Reich*. New York: Oxford University Press, 1991.

Bartrop, Paul. "The Relationship Between War and Genocide in the Twentieth Century: A Consideration," *Journal of Genocide Research* 4, no. 4 (2002): 519–532.

Bauer, Yehuda. *Rethinking the Holocaust*. New Haven: Yale University Press, 2001.

Bauman, Zygmunt. *Modernity and the Holocaust*. Ithaca: Cornell University Press, 2000.

Beorn, Waitman. *Marching Into Darkness: The Wehrmacht and the Holocaust in Belarus*. Cambridge, MA: Harvard University Press, 2014.

Bergen, Doris L. *War and Genocide: A Concise History of the Holocaust*. New York: Rowman & Littlefield, 2003.

Birken, Lawrence. *Hitler as Philosophe: Remnants of the Enlightenment in National Socialism*. Westport, CT: Praeger, 1995.

Bloxham, Donald. *The Final Solution: A Genocide*. Oxford: Oxford University Press, 2009.

———. "From Streicher to Swoniuk: The Holocaust in the Courtroom." Dan Stone, editor. *The Historiography of the Holocaust*. New York: Palgrave Macmillan, 2004.

Bloxham, Donald, and A. Dirk Moses. *The Oxford Handbook of Genocide Studies*. New York: Oxford University Press, 2013.

Brantlinger, Patrick. *Dark Vanishings: The Discourse on the Extinction of Primitive Races, 1800–1930*. Ithaca: Cornell University Press, 2003.

Browning, Christopher. *Ordinary Men: Police Battalion 101 and the Final Solution in Poland*. New York: Harper Perennial, 1992.

Carmichael, Cathie. *Genocide Before the Holocaust*. New Haven: Yale University Press, 2009.

Chalk, Frank, and Kurt Jonassohn, ed. *The History and Sociology of Genocide*. New Haven: Yale University Press, 1990.

Charny, Israel. "Toward a Generic Definition of Genocide." In *Genocide: Conceptual and Historical Dimensions*. Ed. George J. Andreopoulos. Philadelphia: University of Pennsylvania Press, 1994.

Chhay, Chanda. *War and Genocide: A Never-Ending Cycle of Human Brutality*. Amazon Digital Services, 2011.

Dallaire, Roméo. *Shake Hands With the Devil: The Failure of Humanity in Rwanda*. Toronto: Random House, 2003.

Earl, Hilary. *The Nuremberg SS-Einsatzgruppen Trial: Atrocity, Law, and History*. New York: Cambridge University Press, 2009.

———. "Prosecuting Genocide Before the Genocide Convention: Raphael Lemkin and the Nuremberg Trials, 1945–1949," *Journal of Genocide Research* 15 (September 2013): 317–338.

Fein, Helen. *Genocide: A Sociological Perspective*. London: SAGE, 1991.

Gerlach, Christian. *Extremely Violent Societies: Mass Violence in the Twentieth-Century World*. Cambridge, UK: Cambridge University Press, 2010.

Gilbert, Gustav. "The Mentality of the SS Murderous Robots," *Yad Vashem Studies on the European Jewish Catastrophe and Resistance* 5 (1963): 35–41.

Gocek, Fatma Muge. *Denial of Violence: Ottoman Past, Turkish Present, and Collective Violence Against the Armenians 1789–2009*. Oxford: Oxford University Press, 2015.

Goldhagen, Daniel Jonah. *Hitler's Willing Executioners: Ordinary Germans and the Holocaust*. London: Little, Brown, 1996.

Gourevitch, Philip. *We Wish to Inform You That Tomorrow We Will Be Killed With Our Families: Stories From Rwanda*. New York: Farrar, Straus and Giroux, 1998.

Grossmann, Dave. *On Killing: The Psychological Cost of Learning to Kill in War and Society*. New York: Back Bay Books, 1995/2009.

Guichaoua, Andre, and Scott Straus. *From War to Genocide: Criminal Politics in Rwanda, 1990–1994*. Madison: University of Wisconsin Press, 2015.

Haberer, Erich. "Intention and Feasibility: Reflections on Collaboration and the Final Solution," *East European Jewish Affairs* 31, no. 2 (2001): 64–81.

Hatzfeld, Jean. *Machete Season: The Killers in Rwanda Speak*. New York: Farrar, Straus and Giroux, 2003.

Heinemann, Isabel. "'Another Type of Perpetrator': The SS Racial Experts and Forced Population Movements in the Occupied Regions," *Holocaust and Genocide Studies* 12, no. 3 (Winter, 2001): 387–411.

Hilberg, Raul. *The Destruction of the European Jews*. Chicago: Quadrangle Books, 1961.

Hobsbawm, Eric J. *The Age of Extremes: The Short Twentieth Century*. London: Vintage, 1994.

Hull, Isabel V. *Absolute Destruction: Military Culture and the Practices of War in Imperial Germany*. Ithaca: Cornell University Press, 2005.

Jäckel, Eberhard. *Hitler's World View: A Blueprint for Power*. Cambridge, UK: Cambridge University Press, 1981.

Katz, Steven. "The 'Unique' Intentionality of the Holocaust," *Modern Judaism* (September 1981): 161–183.

———. "The Uniqueness of the Holocaust: The Historical Dimension." In *Is the Holocaust Unique? Perspectives on Comparative Genocide*. Ed. Alan S. Rosenbaum, 19–38. Boulder: Westview Press, 1996.

Kiernan, Ben. *Blood and Soil: A World History of Genocide and Extermination From Sparta to Darfur*. New Haven: Yale University Press, 2007.

Korb, Alexander, and Dieter Pohl. *Mass Violence and Genocide in Eastern Europe and the Balkans: The Second World War and Its Aftermath*. London: Bloomsbury Academic, 2017.

Kren, George M., and Leon Rappoport. *The Holocaust and the Crisis of Human Behavior*. New York: Holmes & Meier, 1980.

Kuper, Leo. *Genocide: Its Political Use in the Twentieth Century*. New Haven: Yale University Press, 1981.

Lemkin, Raphael. *Axis Rule in Occupied Europe: Laws of Occupation, Analysis of Government, Proposals for Redress*. Washington, DC: Carnegie Endowment for International Peace, 1944, reprint 2008.

Levene, Mark. "Genocide: A Twentieth Century Phenomenon?" In *Will Genocide Ever End?* Eds. Carol Rittner et al. St. Paul: Paragon House, 2002.

———. *Genocide in the Age of the Nation State, vol. 1: The Meaning of Genocide*. London: I.B. Tauris, 2005.

———. "Why Is the Twentieth Century the Century of Genocide?" *Journal of World History* 11, no. 2 (Fall, 2000): 305–336.

Lower, Wendy. *Hitler's Furies: German Women in the Nazi Killing Fields*. Boston: Houghton Mifflin Harcourt, 2013.

Magargee, Geoffrey P. *War of Annihilation: Combat and Genocide on the Eastern Front, 1941.* Lanham, MD: Rowman and Littlefield, 2007.

Mamdani, Mahmood. *When Victims Become Killers: Colonialism, Nativism, and the Genocide in Rwanda.* Princeton: Princeton University Press, 2001.

Mattäus, Jügen, et al. "The Final Solution: A Genocide," *Journal of Genocide Research* 13, nos. 1–2 (2011).

Mazower, Mark. "Violence and the State in the Twentieth Century," *The American Historical Review* 107 (October, 2002): 1158–1178.

Milgram, Stanley. *Obedience to Authority: An Experimental View.* New York: Harper and Row, 1974.

Moses, A. Dirk. "Conceptual Blockages and Definitional Dilemmas in the 'Racial Century': Genocides of Indigenous Peoples and the Holocaust," *Patterns of Prejudice* 36 (2002).

———, ed. *Empire, Colony, Genocide: Conquest, Occupation, and Subaltern Resistance in World History.* New York: Berghahn Books, 2008.

———, ed. *Genocide and Settler Society: Frontier Violence and Stolen Indigenous Children in Australian History.* New York: Berghahn Books, 2005.

———. "The Holocaust and Genocide," Dan Stone, editor, *The Historiography of the Holocaust.* New York: MacMillan, 533–555, 2004.

Naimark, Norman M. *Fires of Hatred: Ethnic Cleansing in Twentieth-Century Europe.* Cambridge, MA: Harvard University Press, 2001.

Paul, Gerhard, and Klaus-Michael Mallmann, eds. *Die Gestapo im Zweiten Weltkrieg: "Heimfront" und besetztes Europa.* Darmstadt: Wissenschaftliche Buchgesellschaft, 2000.

Power, Samantha. *A Problem From Hell: America and the Age of Genocide.* New York: Basic Books, 2002.

Raben, Remco. "On Genocide and Mass Violence in Colonial Indonesia," *Journal of Genocide Research* 14, nos. 3–4 (2012): 485–502.

Reuther, Rosemary. "Antisemitism and Christian Theology." In *Auschwitz: Beginning of a New Era? Reflections on the Holocaust.* Ed. Eva Fleischner. New York: KATV, 1977.

Rosenbaum, Alan S., ed. *Is the Holocaust Unique? Perspectives on Comparative Genocide.* Boulder: Westview Press, 1996.

Rosenfeld, Gavriel D. "The Politics of Uniqueness: Reflections on the Recent Polemical Turn in Holocaust and Genocide Scholarship," *Holocaust and Genocide Studies* 13 (Spring, 1999): 28–61.

Scheffer, David. "Genocide and Atrocity Crimes," *Genocide Studies and Prevention* 1, no. 3 (2006).

Scheper-Hughes, Nancy. "The Genocidal Continuum: Peace-Time Crimes." In *Power and the Self.* Ed. Jeanette Mageo. Cambridge, UK: Cambridge University Press, 2002.

Shaw, Martin. *War & Genocide: Organized Killing in Modern Society.* Cambridge: Polity Press, 2003.

Sofsky, Wolfgang. *Violence: Terrorism, Genocide, War.* London: Granta Books, 2004.

Stone, Dan, ed. *The Historiography of Genocide.* New York: Palgrave MacMillan, 2010.

Straus, Scott. "Contested Meanings and Conflicting Imperatives," *Journal of Genocide Research* 3, no. 3 (2001): 349–375.

———. "How Many Perpetrators Were There in the Rwandan Genocide? An Estimate," *Journal of Genocide Research* 6 (March 2004): 85–98.

Travis, Hannibal. "Extremely Violent Societies: Mass Violence in the Twentieth-Century World," *Journal of Genocide Research* 14, no. 1 (2012): 99–104.

Waller, James. *Becoming Evil: How Ordinary People Commit Genocide and Mass Murder.* 2nd ed. Oxford: Oxford University Press, 2007.

Weiss-Wendt, Anton. "Problems in Comparative Genocide Scholarship," in Dan Stone (ed.), *The Historiography of the Genocide.* New York: Palgrave Macmillan, 2010, 42–70.

Weitz, Eric D. *A Century of Genocide: Utopias of Race and Nation.* Princeton: Princeton University Press, 2003.

Westermann, Edward. *Hitler's Police Battalions: Enforcing Racial War in the East.* Lawrence: University Press of Kansas, 2005.

———. "Stone Cold Killers or Drunk With Murder? Alcohol and Atrocity During the Holocaust," *Holocaust and Genocide Studies* 30, no. 1 (Spring 2016): 1–19.

Wildt, Michael. *Generation des Unbedingten: Das Führungskorps des Reichssicherheitshauptamtes.* Hamburg: Hamburger Editions HIS Verlagages, 2003.

Yerushalmi, Yosef Hayim. "Response to Rosemary Ruether," Eva Fleischner, editor, *Auschwitz: Beginning of a New Era? Reflections on the Holocaust.* New York: KATV, 1977, 97–107.

20 War and Environment

Jason W. Smith

If we are to believe Napoleon that armies march on their stomachs or Walt Whitman that war was "about nine hundred and ninety-nine parts diarrhea to one part glory," then military historians must seriously consider the place of the natural world in their scholarship (Smidgall, 2001, 187). And if armies march on their stomachs, they also march over mountains and through valleys, forests, and plains, rivers, swamps, and jungles, riding horses and swatting mosquitos. Navies operate in so-called blue water, brown water, and even green water. Implicit in such terms is the recognition that the marine environment fundamentally shapes warfare at sea and the naval forces that wage it. Militaries and the historians who study them have long understood that there is a connection between war and the environment. The "most marked if . . . not the most important . . . specialty of military activity," the Prussian military theorist Carl von Clausewitz wrote in *On War*, is the "connexion which exists between War and country or ground," and one finds similar environmental and geographical awareness from Sun Tzu to Alfred Thayer Mahan (Clausewitz, 1873, 32).

In recent years, environmental historians, whose subfield emerged out of the French *Annales* School, the history of the American West, and the environmental movement of the 1960s and 1970s, have studied the relationship among nature, war, and militaries in new and more complex ways. Both the American Society for Environmental History and the Society for Military History have sponsored conference panels, pressing their members to more deeply examine the ways in which environments influence militaries and vice versa. Environmental historians, in particular, have pushed their peers in military history to see the natural world not just in terms of its tactical, operational, and strategic value but also through its material and imagined significance to influence the causes, conduct, and consequences of military affairs in war and peace. These collaborations have generated a number of historiographical questions, which have been fruitful for both fields and enriched what had been, until quite recently, a no-man's-land in between.

Scholars have offered at least four distinct but interrelated conceptual frameworks for examining the relationship between militaries and the natural world. The first, long recognized by military historians even if it has not commonly been central to their analysis, is the environment's tactical, operational, and strategic significance. Military historians acknowledge that weather, terrain, natural resources, and logistical systems are central to war. This frame might be termed "the military gaze," to borrow a term that Judith Bennett briefly employs in her book *Natives and Exotics* (Bennett, 2009). Perhaps drawing on similar terms charged with cultural and historiographical currency, such as "the male gaze" and "the imperial gaze," Bennett suggests that militaries, their leaders, officers, and personnel perceived the natural world in particularly instrumental ways. Bennett's "military gaze" reminds us that militaries often interpret the natural world through a tactical, operational, and strategic lens.

The second frame, posited by Richard Tucker and Edmund Russell in their anthology *Natural Enemy, Natural Ally*, presses historians to think of the environment in dualistic terms in which "the very usefulness of nature to one side of a conflict has often made it the enemy of another" (Tucker and Russell, 2004, 5). Seen in this way, territory, disease, or natural resources became central objectives or factors on which conflicts turned. This perspective also forces historians to think outside narrowly conceived national histories that privilege Euro-American perspectives to understand the ways in which environmental forces acted upon all belligerents. Moreover, military histories tend to emphasize their subjects' destructive effect on the environment, conjuring, for example, images of the Somme (Hupy, 2008; Closmann, 2009). While acknowledging war's environmental destruction, Tucker and Russell suggest this relationship is more complex than simply a story of environmental, biological, and ecological declension. In fact, it has sometimes led to greater ecological diversity, the regrowth of forests, or rebounding of once endangered species. Finally, implicit in the concept of natural enemy, natural ally is the long-held belief that the natural world has agency, which, together with human contingency, accounts for historical change. In Tucker and Russell's estimation and, indeed, that of nearly all environmental historians, the environment is an actor.

Third, Lisa Brady argues that environmental histories of war might adopt Clausewitz's notion of friction as a category of analysis useful to both military and environmental historians (Drake, 2015). Just as Clausewitz identified friction as a key factor in battle, so too might historians see it as a useful lens through which to view the convergence of human and natural forces that exist on the battlefield. While never explicit in Clausewitz's treatise, Brady suggests that historians might intuit nature as a form of combat friction that "can provide a linguistic and conceptual bridge between military and environmental history" (147).

Finally, much recent scholarship has focused on what Chris Pearson has come to call "militarized landscapes," which he defines as "simultaneously material and cultural sites that have been partially or fully mobilized to achieve military aims" (Pearson, 2012, 1). Transcending the "reductive and instrumental" ways in which terrain and weather have influenced the course and outcomes of war, militarized landscapes push historians to consider the relationship between militaries and nature within larger social contexts not just in war but also in peacetime. The concept has influenced a number of studies, particularly examining military bases (Martini, 2015). It has proven useful for challenging the often artificial boundaries between military and civilian worlds (as well as between wartime and peacetime) and for examining the larger questions of war and society, social history, and memory (Pearson et al., 2010). As Pearson and others suggest, battlefields can sometimes be only the starkest—and most ephemeral—examples of militarized landscapes. Military bases and proving grounds have otherwise had a more complex, pervasive, and sustained influence on environments and the societies that surround them.

For those new to the subfield, two terms are worth further elaboration: nature and agency. Both are fraught, and not all scholars in the subfield or its parent fields of military and environmental history agree on the definition of these terms or, for that matter, their utility. What is nature? This is a complex question that cannot be resolved here. Environmental historians tend to be critical of the way the term is used, loaded as it is with notions of pristine and unchanging environments (Cronon, 1995). Such constructions are themselves ahistorical. Rather, environmental historians have argued that the natural and the human are inextricable. There are very few places in the world—either now or in the past—that are an Eden untouched or unchanged by human presence. "The unmade and the made are everywhere mixed," observes Angela Gugliotta (2005, 37). The important point for scholars of war and environment is to clearly define how they intend to use the term "nature" in their work. Despite its inherent problems, most historians in the subfield have defined it simply as all that is not human.

Yet, environmental historians, including those who study war and militaries, press us to see humans as part of—rather than apart from—the natural world. After calling attention to the role of disease in the fall of empires in his seminal *Plagues and Peoples*, the environmental historian William McNeill went so far as to "discern patterns of microparasitism among human kind," and particularly in the relations among nations in *The Pursuit of Power*, a book that otherwise took up questions quite familiar to military, political, and diplomatic historians (McNeill, 1982, vii). While McNeill was careful not to push this metaphor too far, he was nevertheless suggesting that rather than seeing the roles of humans and other organisms as dichotomous, perhaps historians ought to think about humans themselves as part of the natural world.

The question of natural agency, then, is doubly problematic, and scholars outside environmental history have generally been suspicious in perceiving agency in anything other than human thoughts and actions. Agency suggests intent to imagine change and then to execute it. Environmental historians, however, understand environmental forces as important agents of historical change. It is a central tenet of their field. Eschewing critiques of environmental determinism, environmental historians understand historical change to be the outcome of both human and natural agency (Tucker and Russell, 2004; Brady, 2012; Crosby, 1986; McNeill, 2010). "It is worth considering how our stories might be different," argues Linda Nash, "if human beings appeared not as the motor of history but as partners in a conversation with a larger world, both animate and inanimate, about the possibilities of existence" (Nash, 2005, 69). Indeed, in the interstices between military and environmental history, the question of natural agency may well become one of the central historiographical debates. It must be addressed if military and environmental historians are to have fruitful debate with one another.

Long before environmental historians engaged the field, however, military historians were noting the importance of environmental factors in battle. As early as the 1950s, Samuel Eliot Morison highlighted the Americans' ignorance of "dodging tides" in the costly amphibious assault on Tarawa (Morison, 1958). Other historians of the Pacific War have continued to stress the influence of island environments in that theater. "Of all the unpleasant islands the Marines saw in World War II," writes Allan Millett, "Iwo Jima was the nastiest, prepared by nature and the Japanese armed forces as a death trap for any attacker" (Millett, 1980, 427). More generally, John Keegan argues that historians must "account . . . as a prologue to consideration of individual experiences" what he terms "the physical circumstances of battle." His portrayal of the Somme, for example, begins almost poetically with "the slow-moving river, winding its way through a peat-bottomed valley below beech woods and bare chalk downland" (Keegan, 1976, 134, 204). Following Keegan, social histories of combat could not ignore the elemental relationship between humans and nature in battle (Meier, 2013). Yet, for Keegan, Millett, and Morison, the environment was not central to their work. While they acknowledged environmental factors, the natural world served them primarily as an indispensable background to set the stage for the course of dramatic human events.

In recent years, scholars outside military history—almost all of them geographers, environmental historians, or scholars in environmental studies and policy—have approached questions of war and environment by placing the environment at the center, emphasizing its agency in the process and outcome of war. Harold Winters concludes that the natural world "is neither passive nor presumptive," but a dynamic force with the "potential to shape conflict" and "sometimes . . . a decisive factor in its outcome" (Winters, 1998, 3). He ranges widely, from the thirteenth-century cyclones that turned back Kublai Khan's two abortive invasions of Japan to the fog enshrouding the American base at Khe Sanh during the Vietnam War, not just referring to the environment in the abstract but also, for example, examining the processes of monsoonal circulations and high atmospheric pressure. Implicit in Winters' argument is that historians must understand these larger natural processes, from rain, climate, and glacial geology to the ultisol soils of tidewater Virginia that thwarted Ambrose Burnside's initial attack at Fredericksburg during the American Civil War. Just as weather and long-term atmospheric and climatological processes turned back the Mongols in 1274 and 1281, such forces also decided the timeline for the Allies' cross-channel invasion of Normandy in 1944.

Other recent works from environmental historians illustrate how an environmental lens can deepen or revise our understanding of battle, war, nation building, and the rise and fall of states and empires. Megan Kate Nelson, for example, has shown how the desert environment of the American Southwest led to poor Union command decisions that military historians had previously attributed to cowardice. David Biggs examines the ways in which Vietnamese hydrological practices and knowledge of the Mekong Delta undermined French and American imperial and operational designs in the region (Drake, 2015; Biggs, 2010). Meanwhile, environmental studies and policy scholars have placed in historical context their concerns about the connections between war and climate change, access to water and other resources, and other ecological changes.

Amid growing concerns about environmental change in our own time, it is no surprise that historians, scientists, and policy experts are interested in studying old conflicts through new lenses (Stockholm International Peace Research Institute, 1980; Hillel, 1994). In his recent book *Global Crisis*, Geoffrey Parker finds in patterns of seventeenth-century global cooling the seeds of "revolutions and state breakdowns around the world," from Ming China and the Mughal Empire to Russia and the Spanish Monarchy. Parker, a military historian, identifies two distinct categories of data—a "human archive" and a "natural archive"—in other words, concepts already familiar to environmental historians (Parker, 2013, xvi–xvii).

A second line of inquiry long established by military historians and augmented recently by environmental historians is the role of natural resources, in particular, within the larger natural, logistical, agricultural, and industrial systems that make war possible. Pioneering works on the role of salt shortages in the defeat of the Confederacy (Lonn, 1933) and the significance of North American timber resources to the Royal Navy in the Age of Sail (Albion, 1926) represented, in their times, new awareness for the fundamental linkages between large military organizations, societies at war, and the natural world. Lonn and Albion suggested that wars were won or empires defended not just on the battlefield but also in the larger fight for access to and denial of natural resources. It is also worth noting just how forward-thinking these works were. Albion's *Forests and Sea Power*, for example, contains a remarkable illustration of various species of trees appropriate for use as ship timbers, with the various kinds of timbers—from futtocks to ships' knees—splayed into the very branches of the trees themselves. Here was a kind of imagined commodification of nature long before pioneers in environmental history, like William Cronon, brought historiographical currency to the idea (Cronon, 1991). For military historians, these books formed the antecedents to more recent works (van Creveld, 1977; Lynn, 1993), which set environmental factors within a broader analysis of military logistics.

Here, again, environmental historians have added depth to the discussion of military logistics, showing that ways of knowing, controlling, and managing the environment influenced war making and the operations of military organizations. Scholarship by Richard Tucker, Simo Laakonen, A. Joshua West, J. R. McNeill, and Greg Bankoff has shown that forestry, conservation practices, and timber resources from the Philippines and Finland to Indonesia and India were central to war making, imperial expansion, and the subordination of people around the world to Euro-American colonialism (Tucker and Russell, 2004; West, 2003; McNeill, 2004; Closmann, 2009). Historians of the American Civil War have traced Confederate secession and the Army of Northern Virginia's Gettysburg campaign to the South's agricultural systems, which both wedded the South to the institution of slavery and in 1863 pushed Robert E. Lee to take his hungry army away from war-ravaged Northern Virginia into Pennsylvania's rich farmland (Gates, 1965; Tucker and Russell, 2004). Similarly, the production of food for militaries and home fronts has attracted the attention of several historians who have seen in it the impulse for territorial expansion in Nazi Germany and Imperial Japan during the 1930s. Others have traced the long shadow of war across a postwar Pacific world forever altered by Allied logistical systems, which, among other things, introduced Spam to Pacific islanders and thus irrevocably changed indigenous foodways, diets, and health (Collingham, 2012; Bennett, 2009).

Historians of environment and war have called attention not only to the material landscapes of farms and battlefields, but also to imagined environments, illustrating that the ways in which policy makers, officers, and soldiers perceived or constructed mental understandings of nature were inextricably linked to military strategy and the daily experience of combat. Lisa Brady, for example, contends that the Union Army consciously implemented a strategy that would ravage southern fields, forests, and cities, turning a previously ordered and improved landscape into a chaotic wilderness, thereby hastening the Confederacy's defeat (Brady, 2012)—an argument that Megan Kate Nelson elaborates in her examination of ruination of southern landscapes, houses, and bodies (Nelson, 2012). Together, Brady's and Nelson's work constitutes part of a larger, coherent body of scholarship focusing on the environmental history of the Civil War (Drake, 2015; Meier, 2013).

Historians of Austria and Germany during the Great War, meanwhile, have pointed to the Alps and the Eastern Front as landscapes that fundamentally influenced the ways soldiers interpreted their experience and the ways these postwar societies understood their service. The Alps became a symbolic bulwark of German nationalism as the East became fixed in the German mind as a place of mystery and uncivilized people ripe for present and future conquests (Liulevicius, 2000; Keller, 2009). As Gabriel Vejas Liulevicius argues in his excellent *War Land on the Eastern Front*, "the dynamic mindscape turned description of the land into a prescription for how it was to be faced, confronted, and approached" (156). As these scholars demonstrate, military historians must account for the ways in which armed forces and those in their ranks made sense of their often profound encounters with nature.

These works engage larger historiographies that seek to place militaries within broader national or cultural understandings of nature and the ways in which centralized governments attempted to engineer environments to serve the nation-state. The debate has been particularly intense in the German context, centered on questions about the degree to which German conservation efforts were tied directly to national socialism or whether they were more deeply rooted in German culture through the notion of *Heimat*, or homeland (Blackbourn, 2006; Brüggemeier et al., 2005). In the French case, Kieko Matteson and Chris Pearson have noted the critical question of state control and management of forest environments in the coming of the French Revolution and of Napoleon III's desire to change so-called marginal lands into militarized landscapes, such as the Châlons Camp. These are part of a longer French history that linked state-sponsored environmental change to notions of progress and modernity associated with the growth of the nation-state (Matteson, 2015; Pearson, 2012). As David Blackbourn argues in the German context, "the human domination of nature has a lot to tell us about the nature of human domination" (Blackbourn, 2006, 7). Still, as David Biggs has shown, French efforts to engineer nature in colonies such as Indochina as part of its so-called *mission civilisatrice* often proved more complicated, rendered largely ineffective by the very nature of the environment itself (Biggs, 2010). Whether in German, French, or American contexts, environmental "improvements" were often the purview of militaries and carried out for military purposes. Each evoked the militarized rhetoric of conquest and domination over nature even as their activities elicited anxieties about environmental change grounded in cultural ideas that linked a pristine, primordial nature to national consciousness and identity.

Just as environmental historians have placed questions of war and nature at the center of the growth of the modern nation-state, they have also offered significant contributions to the scholarship on the rise of the West. In fact, the ways in which natural agents explain the European conquest of the Americas and, indeed, much of the world formed one of the pioneering avenues of inquiry for the field of environmental history. Beginning with seminal works like William McNeill's *Plagues and Peoples* and Alfred Crosby's *The Columbian Exchange* and *Ecological Imperialism*, environmental historians have long pointed to the decisive and often unintentional role that biological actors—namely, plants, animals, and disease—played in European expansion (W. McNeill, 1976; Crosby, 1972, 1986; Fenn, 2000). For Crosby and others, the sixteenth-century Spanish conquest of Mexico was made possible not so much by superior European weaponry, military discipline, leadership, or indigenous alliances, but by an epidemic of smallpox and other diseases unknowingly brought by Europeans. Crosby's "portmanteau biota" are the agents of European expansion. While humans "benefited from the great majority of these [biological] changes," he argues, "their role was less often a matter of judgment and choice than of being downstream of a bursting dam" (Crosby, 1986, 192).

Critics of Crosby and subsequent works (Diamond, 1999) see in this line of argument an ecological determinism that largely dismisses human contingency (McAnany and Yoffee, 2009). Questions such as these underscore the central debate that frames the field of war and environment relative to military history more generally. How much agency should historians attribute to environmental factors versus human choices in explaining change over time? Nevertheless, Crosby's work has been extraordinarily influential. We can no longer account for the fall of Tenochtitlán, for example, without crediting disease.

More nuanced in its balance of biological and human factors is J. R. McNeill's *Mosquito Empires*, which examines the role of the mosquitos *Aedes aegypti* and *Anopheles quadrimaculatus*, the vectors for yellow fever and malaria, as "key actors . . . if not. . . *dramatis personae*" central to the defense and, later, the collapse of European empires in the tropical Caribbean (J. McNeill, 2010). "Yellow fever," along with malaria, McNeill writes, "formed a crucial part of Spanish Imperial defense," evinced most infamously by the disastrous siege of Cartagena in 1741, in which disease claimed the vast majority of the more than eight thousand British soldiers and sailors killed (3–4). Military historians have long acknowledged the irony that illness among concentrations of men in camps, prisons, ships, and fortifications crippled many more armies than enemy swords and bullets at least before the end of the nineteenth century (Wintermute, 2011). Indeed, the ways soldiers coped with these diseases—including "straggling"—constituted an important part of their wartime experience in the era before quinine and penicillin, and, of course, it was a significant factor in an army's ability to wage war at all (Meier, 2013). Later, McNeill contends, these diseases aided revolutions that undermined European empire in the Caribbean from the 1820s to the turn of the twentieth century. Yet, McNeill is careful to acknowledge that ecological factors alone did not account for these changes. "Strictly speaking, they did not determine the outcomes of struggles for power," he admits, "but they governed the probabilities of success and failure in military expeditions and settlement schemes" (2).

In studying the mosquito, McNeill is one of many environmental historians who have identified nonhuman organisms as important agents of historical change. There is an entire subfield of environmental history, for example, devoted to the significance of domesticated and feral animals in narratives of empire and settlement and to agricultural and urban history. Along with mosquitos, military historians must, of course, acknowledge the horse in questions of war and environment. Long the mainstay of an army's mobility and the backbone of its logistical networks, horses were fundamental components of preindustrial militaries. On the battlefield, they died in droves. More horses and mules died in the American Civil War than humans, and Ann Greene argues that the advent of railroads, in fact, only expanded the importance of horses in the American Civil War, bringing supplies from depots and railroad nodes to armies in the field (Greene, 2004).

While Crosby's conquistadors and McNeill's mosquitos unwittingly hastened biological agents to the premodern battlefield, modern war has been characterized by the threat of chemical, radiological, and nuclear weapons consciously employed toward destructive ends. In *War and Nature*, Edmund Russell examines the cooperative and complementary relationship among the American chemical industry, the federal government, and the U.S. military from the Great War through the publication of Rachel Carson's *Silent Spring*. The latter raised critical awareness of the use of DDT in American agriculture and ultimately culminated in the United States' ban on the use of Agent Orange in Vietnam and the emergence of the American environmental movement. Most fascinating is Russell's analysis of the ways in which the American military, federal government, and chemical manufacturers constructed agricultural pests and wartime enemies—particularly the Japanese—as objectives to be exterminated through the application of chemical agents (Russell, 2001). As Russell puts it, "war and nature coevolved: the control of nature expanded the scale of war, and war expanded the scale on which people controlled nature" (2). For Russell, manipulation of the natural world is one of the hallmarks of modern war. His work reveals the ways in which the environmental dimensions of modern warfare permeated American society, science, and business.

The Cold War only intensified these dimensions as the United States, the Soviet Union, and their allies and satellites took the threat of war and annihilation to new environments, manipulated nature in new ways, and left military-industrial footprints on the Earth that will remain for many centuries (McNeill and Unger, 2010). Among questions of Cold War military history and the environment, the herbicide Agent Orange and its various modifications have long loomed large, but until recently, environmental historians have overlooked it. Two recent studies have deepened our understanding. Edwin Martini has cast its shadow in a much larger transnational lens, looking at its effect on American veterans and American society, the environmental consequences of its disposal in places like Johnston Atoll and Gulfport, Mississippi, and American and Vietnamese attempts at restitution (Martini, 2012). Throughout, he highlights the medical and scientific uncertainties surrounding the herbicide's effects. For Martini, "herbicidal warfare was simply one more failed attempt . . . to impose control over a nation, a people, and a landscape—indeed, over nature itself—all of which refused to accept the dictates of American power" (6). David Zierler, meanwhile, links Agent Orange to the notion of "ecocide," which gained currency during the Vietnam War and amid growing concerns in the United States that the Cold War and decades of industrial activity had left or would soon leave the natural world in ecological ruin (Zierler, 2011).

Implicit in concepts like ecocide was the fear that the Cold War state's intersection of modern science, technology, and war making might leave the natural world fundamentally and irrevocably altered on a global scale. In Jacob Hamblin's estimation, meanwhile, the United States attempted nothing less than to use "nature itself as a weapon," enlisting the environment in the Cold War struggle, for instance, in attempts to poison Soviet agriculture, to melt the polar ice caps through nuclear explosions, and to engineer rain in drought-stricken India in hopes of sealing an American victory in the Cold War ideological struggle (Hamblin, 2013, 36). "The collaboration between scientists and the armed services created a scientific worldview obsessed with environmental change, manipulation, and vulnerability," Hamblin writes. One senses his growing horror at the ways in which "catastrophic war" led to "catastrophic environmental change," and, indeed, the sometimes casual way in which scientists and military planners considered these possibilities (11, 158). As Hamblin shows, climate change is not a phenomenon reserved only for present debates. Rather, it had roots in a postwar world in which the American military-scientific-industrial complex intended to win the Cold War by fundamentally altering the natural world.

Not surprisingly, environmental historians have identified in the Cold War the roots of the emerging environmental movement in the United States and Europe. But environmentalism itself was not immune from geopolitical and geostrategic posturing as the Soviet Union and the United States used environmental protection—aside, ironically, from the simultaneous threat of catastrophic environmental destruction—as a weapon in Cold War ideological battles (McNeill and Unger, 2010). Nevertheless, fears about ecocide, the production of military-agricultural chemicals that culminated in DDT, and revelations over the effects of Agent Orange in Vietnam were catalysts for the modern environmental movement in the United States.

The proliferation of Cold War military bases and proving grounds, including those at which atomic and thermonuclear bombs were tested, also proved important sites of growing environmentalist consciousness. Chris Pearson, for instance, has shown that protests surrounding the proposed French base at Larzac in the 1970s were historical moments in which competing uses of nature and the broader interaction between the military and French society were negotiated (Pearson, 2012). Indeed, military bases were often flashpoints of Cold War environmentalism. With subsequent environmental regulations as well as shifts in cultural attitudes toward the environment, several historians have also noted the ways in which militaries have adopted "military environmentalism"—to borrow Pearson's phrase—as a central element of civil-military relations. While historians remain suspect of the underlying motives of this "greening" of the military, many have also pointed out the positive ecological effects of federal land reserved for military use, such as Fort Stewart, Georgia, or conserved by decades of standoff, like Korea's demilitarized zone. In these places, ecosystems and their natural flora and fauna have thrived, protected from encroaching outside development (Martini, 2015; Brady, 2008). Furthermore, the process of commemorating and memorializing military sites and battlefields has also raised questions about environment, preservation, and memory (Pearson et al., 2010).

While the breadth of scholarship in war and environment has focused on land warfare and armies, historians have written comparatively little about military operations spanning the seven-tenths of the world covered by salt water. Yet, navies perhaps more than any other military organization are creatures of their environments. Ship design, technology and weapons systems, administration, tactical, strategic, and intellectual debates, and institutional culture all bear marks of the sea's influence. "Without the ocean," Gary Weir has observed, "words like *ship*, *navy*, and *oceanography* have no meaning" (Weir, 2001, xi). For scholars interested in questions of environmental agency, moreover, the sea offers one of the most dynamic and inhospitable environments to human activity of all kinds, including marine navigation, hydrographic and oceanographic science, exploration, communication or command and control, and naval combat. Rising sea levels, melting polar ice, and Chinese claims to human-constructed islands built on half-submerged reefs suggest that marine environments from the littorals to the deep sea and the resources within and beneath them present pressing questions that beg historical context.

Works that touch on navies and the marine environment have done so largely through the lens of the history of science. Hydrographic and later oceanographic science in support of navigation, exploration, cartographic claims, and later undersea warfare stimulated naval science since the eighteenth century. In her pioneering work on marine science and the exploration of the deep sea in Great Britain and the United States, Helen Rozwadowski argues that by the mid-nineteenth century the sea had become a destination for marine scientists working within or closely with European and American navies (Rozwadowski, 2005). She highlights the work of Lieutenant Matthew Fontaine Maury at the U.S. Naval Observatory and the Royal Navy's Challenger Expedition as well as the sometimes stilted ways in which civilian scientists interacted with naval scientists in the confines of a ship. Michael Reidy, meanwhile, examines the study of the tides in Great Britain as central to its commercial and imperial ascendance from the eighteenth to the nineteenth century. "The machinery of empire required as its lubricant a science of the sea," Reidy writes, "essential for any overseas expansion of trade or successful military campaign" (Reidy, 2008, 6). Both Gary Weir and Jacob Hamblin have studied the relationship between civilian oceanographers and the U.S. Navy during the Cold War (Weir, 2001; Hamblin, 2005). Implicit in this historiography are the challenges of conducting scientific work at sea, not least due to the vastness and dynamism of the marine environment itself.

My own work builds on this scholarship, arguing that hydrographic science in the U.S. Navy was central to the expansion of American commercial and military-strategic empire during the nineteenth century. The surveyors and scientists of the United States Exploring Expedition, for instance, hoped to turn Pacific islands and their coasts into ordered environments for antebellum American maritime commerce by replacing an exotic wilderness with the empiricism of cartographic hydrography, a process that included the application of devastating military force on the Fijian island of Malolo in 1838 (Smith, 2013). By the turn of the century, the expansion of American sea power had become inextricable from hydrographic knowledge of the marine environment. Following hydrographic difficulties in establishing the inshore blockade on the coast of Cuba during the war with Spain, the U.S. Navy embarked on comprehensive surveys and chart-making to consolidate its new empire in the Caribbean and the Pacific. These percolated to the highest levels of strategic debate as the General Board of the Navy discussed the comparative advantages of islands, coasts, and bays for the defense of the American empire (Smith, 2014). My forthcoming book traces the emergence of an American cartographic empire in the nineteenth century, which rested on knowledge of the marine environment represented by the Navy's chart-making and its larger material and cultural encounter with the sea.

More work must be done in order to more deeply understand the relationship among navies, naval warfare, and the marine environment. Naval historians have not been quick to embrace environmental history. For their part, environmental historians have long remained terrestrial in focus, only recently beginning to embrace marine environmental history. Nevertheless, ocean history is rapidly gaining historiographical and methodological interest across other subfields. Three recent books dealing with the history of the ocean have won the Bancroft Prize. The most recent, Andrew Lipman's *The Saltwater Frontier*, recasts the Anglo-Dutch-Native American encounter during the seventeenth century from a watery perspective, showing the significance of armed conflict in the littoral where Native Americans had a decided advantage (Lipman, 2015). Lipman shows how reframing one's environmental lens can shed new light on old wars.

The most glaring need is for more scholarship that is comparative or transnational, and whose focus is outside Europe and the United States, as well as before the modern era. Of course, this is a challenge in the larger fields of environmental and military history as well. Judith Bennett has shown how a transnational environmental synthesis of war should be done in her wide-ranging book *Natives and Exotics*, in which she sets disease, logistics, the effects of atomic and conventional weapons, memory, and many other questions in comparative perspective. We must have for other wars what Bennett has done here for the South Pacific during and immediately after World War II (Bennett, 2009). Building on the work of William Tsutsui on wartime Japan or Roger Levine on Zulu and Xhosa-speaking groups in colonial Africa, there must be more work dealing with environments and conflicts outside the Euro-American context (Tucker and Russell, 2004). Additional research on wars in the

Middle East over water or oil resources will prove important if not timely. There has been virtually nothing written on environmental histories of warfare in the Ancient World and very little in the premodern (Slavin, 2014; Zappia, 2016). In fact, military environmental history is particularly well situated to push its larger parent fields in these directions since armed conflict encompasses so much of the globe over nearly all of its human history and transcends environments from land to sea, air, and space.

Environmental histories of guerilla warfare and counterinsurgencies, conflicts in which the natural world is deeply intertwined with the conduct of war, require further analysis. David Biggs' work on Vietnam offers an excellent model for how this can be done by examining the ways the environment of the Mekong Delta underscored and undermined French colonialism in Indochina. The Philippine Insurrection and the Mujahidin's war against the Soviets are two examples of the many wars whose histories might be deepened by adopting a similar environmental lens. This extends to urban warfare as well. If natural terrain is fundamental to tactical, operational, and strategic decision-making, why not the urban landscape, which environmental historians have interrogated through the concept of "second nature"? Built environments have their own unique character and bearing on armed conflict. The Algerian War for Independence from 1954 to 1962, for example, which saw brutal urban warfare in the streets of Algiers among other places, begs an environmental analysis along these lines. As Andrew Lipman's work suggests, historians of colonial encounters with Native Americans or the U.S. Army's wars against Native Americans in the nineteenth century would benefit from closer attention to the environmental contexts of those conflicts.

A final avenue of prospective study must acknowledge the place of armies and navies in scientific studies of the natural world and as engineers of environmental change in both war and peace. Among other things, my own scholarship argues that the U.S. Navy may well have been the most important scientific organization in the United States before the Civil War even as many of its officers eschewed scientific expertise for command at sea and glory in battle. More scholarship on American naval science is needed in addition to the work of the British, French, and Russian navies in matters of science and exploration. Of course, armies are also instruments of science and exploration, particularly in exploring and surveying the American West in the early nineteenth century. Armies have also been the reserve of engineering knowledge and professionalism, from Sébastien Le Prestre de Vauban to Joseph Totten. Military engineering fundamentally changed the natural world in war and, perhaps more significantly, in peace, linking military history to larger historiographical debates about the rise of the nation-state, consolidation of its borders, boundaries, and territory, and the attendant spirit of environmental conquest that so often pervaded the rhetoric of these efforts.

Indeed, there is need for more work in questions of war and environment generally. In many ways, this is still a very new subfield. The following bibliography shows that much of the work in this subfield dates from the last ten years. As Richard Tucker and Edmund Russell noted when they sought contributors for *Natural Enemy, Natural Ally* in 2004, the pool seemed shallow. Much has changed in the last decade, which speaks to the current vitality of the subfield. Yet, much remains to be done, particularly if we are to move beyond claiming an important place for the environment in war and peace to more nuanced debates about the ways nature affects and is affected by militaries. Furthermore, the vast majority of the work in war and environment today comes from the pens of environmental historians. Questions of war have permeated that field such that warfare occupies a prominent place among the most important synthetic works of environmental history (Steinberg, 2002; Fiege, 2012). The same cannot be said for surveys of military history. Decades ago now, the new military history expanded the parameters of the field, generally embracing studies focusing on class or social questions, race, and—more recently—gender. It is time the field more fully embraced environmental history as well.

Bibliography

Albion, Robert G. *Forests and Sea Power: The Timber Problem of the Royal Navy, 1652–1862*. Cambridge, MA: Harvard University Press, 1926.

Bennett, Judith A. *Natives and Exotics: World War II and Environment in the Southern Pacific*. Honolulu: University of Hawai'i Press, 2009.

Biggs, David. *Quagmire: Nation-Building and Nature in the Mekong Delta*. Seattle: University of Washington Press, 2010.

Blackbourn, *The Conquest of Nature: Water, Landscape, and the Making of Modern Germany*. New York: W.W. Norton, 2006.

Brady, Lisa M. "Life in the DMZ: Turning a Diplomatic Failure into an Environmental Success," *Diplomatic History* 32 (September 2008): 585–611.

———. *War Upon the Land: Military Strategy and the Transformation of Southern Landscapes During the American Civil War*. Athens: The University of Georgia Press, 2012.

Brüggemeier, Franz-Josef, Mark Cioc, and Thomas Zeller. *How Green Were the Nazis? Nature, Environment, and Nation in the Third Reich*. Athens: Ohio University Press, 2005.

Clausewitz, Carl von. *On War*. Trans. J.J. Graham. London: N. Trübner, 1873.

Closmann, Charles E. *War and the Environment: Military Destruction in the Modern Age*. College Station: Texas A&M University Press, 2009.

Coates, Peter, Tim Cole, Marianna Dudley, and Chris Pearson. "Defending Nation, Defending Nature?: Militarized Landscapes and Military Environmentalism in Britain, France, and the United States," *Environmental History* 16 (July 2011): 456–491.

Collingham, Lizzie. *The Taste of War: World War II and the Battle for Food*. New York: The Penguin Press, 2012.

Cronon, William. *Nature's Metropolis: Chicago and the Great West*. New York: Norton, 1991.

———, ed. *Uncommon Ground: Toward Reinventing Nature*. New York: Norton, 1995.

Crosby, Alfred W. *The Columbian Exchange: Biological and Cultural Consequences of 1492*. Westport, CT: Greenwood, 1972.

———. *Ecological Imperialism: The Biological Expansion of Europe, 900–1900*. New York: Cambridge University Press, 1986.

Degroot, Dagomar. "'Never Such Weather Known in These Seas:' Climatic Fluctuations and the Anglo-Dutch Wars of the Seventeenth Century, 1652–1674," *Environment and History* 20 (May 2014): 239–273.

Diamond, Jared. *Guns, Germs, and Steel: The Fates of Human Societies*. New York: Norton, 1999.

Drake, Brian Allen, ed. *The Blue, the Gray, and the Green: Toward an Environmental History of the Civil War*. Athens: The University of Georgia Press, 2015.

Eden, Lynn. *Whole World on Fire: Organizations, Knowledge, & Nuclear Weapons Devastation*. Ithaca: Cornell University Press, 2004.

Evenden, Matthew. "Aluminum, Commodity Chains, and the Environmental History of the Second World War," *Environmental History* 16 (January 2011): 69–93.

Fenn, Elizabeth. "Biological Warfare in Eighteenth-Century North America: Beyond Jeffrey Amherst," *Journal of American History* 86 (March 2000): 1552–1580.

Fiege, Mark. *The Republic of Nature: An Environmental History of the United States*. Seattle: University of Washington Press, 2012.

Gates, Paul W. *Agriculture and the Civil War*. New York: Knopf, 1965.

Greene, Ann N. "War Horses: Equine Technology in the Civil War," in *Industrializing Organisms: Introducing Evolutionary History*. Edited by Susan Schrepfer and Philip Scranton. New York: Routledge, 2004.

Gugliotta, Angela. "Environmental History and the Category of the Natural," *Environmental History* 10 (January 2005): 37–39.

Hamblin, Jacob Darwin. *Arming Mother Nature: The Birth of Catastrophic Environmentalism*. New York: Oxford University Press, 2013.

———. *Oceanographers and the Cold War: Disciples of Marine Science*. Seattle: University of Washington Press, 2005.

———. *Poison in the Well: Radioactive Waste in the Oceans at the Dawn of the Nuclear Age*. New Brunswick, NJ: Rutgers University Press, 2008.

Hillel, Daniel. *Rivers of Eden: The Struggle for Water and the Quest for Peace in the Middle East*. New York: Oxford University Press, 1994.

Hupy, Joseph P. "The Environmental Footprint of War," *Environment and History* 14 (August 2008): 405–421.

Keegan, John. *The Face of Battle*. New York: Viking Press, 1976.

Keller, Tait. "The Mountains Roar: The Alps During the Great War," *Environmental History* 14 (April 2009): 253–274.

Lipman, Andrew. *The Saltwater Frontier: Indians and the Contest for the American Coast*. New Haven: Yale University Press, 2015.

Liulevicius, Vejas Gabriel. *War Land on the Eastern Front: Culture, National Identity and German Occupation in World War I*. New York: Cambridge University Press, 2000.

Lonn, Ella. *Salt as a Factor in the Confederacy*. New York: W. Neale, 1933.

Lynn, John A., ed. *Feeding Mars: Logistics in Western Warfare From the Middle Ages to the Present*. Boulder, CO: Westview Press, 1993.

Martini, Edwin. *Agent Orange: History, Science, and the Politics of Uncertainty*. Amherst: University of Massachusetts Press, 2012.

———, ed. *Proving Grounds: Militarized Landscapes, Weapons Testing, and the Environmental Impact of U.S. Bases*. Seattle: University of Washington Press, 2015.

Matteson, Kieko. *Forests in Revolutionary France: Conservation, Community, and Conflict, 1669–1848*. New York: Cambridge University Press, 2015.

McAnany, Patricia A., and Norman Yoffee, ed. *Questioning Collapse: Human Resilience, Ecological Vulnerability, and the Aftermath of Empire*. New York: Cambridge University Press, 2009.

McNeill, J. R. *Mosquito Empires: Ecology and War in the Greater Caribbean, 1620–1914*. New York: Cambridge University Press, 2010.

———. "Woods and Warfare in World History," *Environmental History* 9 (July 2004): 388–410.

McNeill, J. R., and Corinna R. Unger, eds. *Environmental Histories of the Cold War*. New York: Cambridge University Press, 2013.

McNeill, William H. *Plagues and Peoples*. Garden City, NY: Doubleday, 1976.

———. *The Pursuit of Power: Technology, Armed Force, and Society Since 1,000 A.D.* Chicago: University of Chicago Press, 1982.

Meier, Kathryn Shively. *Nature's Civil War: Common Soldiers and the Environment in 1862 Virginia*. Chapel Hill: The University of North Carolina Press, 2013.

Millett, Allan R. *Semper Fidelis: The History of the United States Marine Corp*. New York: MacMillan, 1980.

Morison, Samuel Eliot. *History of United States Naval Operations in World War II*. Vol. 7. Boston: Little, Brown, 1958.

Nash, Linda. "The Agency of Nature or the Nature of Agency?" *Environmental History* 10 (January 2005): 67–69.

Nelson, Megan Kate. *Ruin Nation: Destruction and the American Civil War*. Athens: The University of Georgia Press, 2012.

Parker, Geoffrey. *Global Crisis: War, Climate Change, & Catastrophe in the Seventeenth Century*. New Haven: Yale University Press, 2013.

Pearson, Chris. *Mobilizing Nature: The Environmental History of War and Militarization in Modern France*. Manchester, UK: Manchester University Press, 2012.

Pearson, Chris, Peter Coates, and Tim Cole, eds. *Militarized Landscapes: From Gettysburg to Salisbury Plain*. New York: Bloomsbury, 2010.

Reidy, Michael S. *Tides of History: Ocean Science and Her Majesty's Navy*. Chicago: University of Chicago Press, 2008.

Rozwadowski, Helen M. *Fathoming the Ocean: The Discovery and Exploration of the Deep Sea*. Cambridge, MA: Belknap Press of Harvard University Press, 2005.

Russell, Edmund. *War and Nature: Fighting Humans and Insects With Chemicals From World War I to Silent Spring*. New York: Cambridge University Press, 2001.

Slavin, Philip. "Warfare and Ecological Destruction in Early Fourteenth-Century British Isles," *Environmental History* 19 (July 2014): 528–550.

Smidgall, Gary, ed. *Intimate With Walt: Selections From Walt Whitman's Conversations With Horace Traubel, 1882–1892*. Iowa City: University of Iowa Press, 2001.

Smith, Jason W. "The Bound[less] Sea: Wilderness and the United States Exploring Expedition in the Fiji Islands," *Environmental History* 18 (October, 2013): 710–737.

———. "'Twixt the Devil and the Deep Blue Sea: Hydrography, Sea Power, and the Marine Environment," *Journal of Military History* 78 (April 2014): 565–604.

Steinberg, Ted. *Down to Earth: Nature's Role in American History*. New York: Oxford University Press, 2002.

Stockholm International Peace Research Institute. *Warfare in a Fragile World: Military Impact on the Human Environment*. London: Taylor & Francis, 1980.

Sunseri, Thaddeus. "Reinterpreting a Colonial Rebellion: Forestry and Social Control in German East Africa, 1874–1915," *Environmental History* 8 (July 2003): 430–451.

Szasz, Ferene M. "The Impact of World War II on the Land: Guinard Island, Scotland, and Trinity Site, New Mexico as Case Studies," *Environmental History Review* 19 (Winter, 1995): 15–30.

Tucker, Richard, and Edmund Russell. *Natural Enemy, Natural Ally: Toward an Environmental History of War*. Corvallis, OR: Oregon State University Press, 2004.

Van Creveld, Martin. *Supplying War: Logistics From Wallenstein to Patton*. New York: Cambridge University Press, 1977.

Weir, Gary. *An Ocean in Common: American Naval Officers, Scientists, and the Ocean Environment*. College Station: Texas A&M Press, 2001.

West, A. Joshua. "Forests and National Security: British and American Forestry Policy in the Wake of World War I," *Environmental History* 8 (April 2003): 270–293.

Wintermute, Bobby A. *Public Health and the US Military: A History of the Army Medical Department, 1818–1917*. New York: Routledge, 2011.

Winters, *Battling the Elements: Weather and Terrain in the Conduct of War*. Baltimore: The Johns Hopkins University Press, 1998.

Zappia, Natale. "Revolutions in the Grass: Energy and Food Systems in Continental North America, 1763–1848," *Environmental History* 21 (January 2016): 30–53.

Zierler, David. *The Invention of Ecocide: Agent Orange, Vietnam, and the Scientists Who Changed the Way We Think About the Environment*. Athens: The University of Georgia Press, 2011.

21 War and Terrorism

Stephen Connor

In many ways, the search for a historiography (or perhaps even an understanding) of terrorism runs counter to the nature of Clio. Historians it seems are poorly suited to provide the kinds of answers, the kinds of generalizations so sought after in the growing field of terrorism studies. History addresses questions and provides answers that are always rooted in the specific context of the past, in essence in a time and under conditions that have come and gone. As Yale historian Beverly Gage (2011) notes, there is for terrorism at best only a developing historiography. Further, writing in the 1970s, Walter Laqueur (1986; 1999) argued that agreement on a definition had proven impossible and was unlikely to be remedied anytime soon. Histories of terrorism can tell us about characteristics, conditions, methods, and means but, akin to Laqueur's reflection on definition risks, when generalized, tend to be either imprecise or, worse, misleading. As he insightfully noted, there is "no such thing as terrorism pure and unadulterated, specific and unchanging . . . rather there are a great many terrorisms" (1986, 88–89). This then is the rich and fallow ground where historians have and must continue to ply their trade to tell us how we got here but not what to do next.

This essay considers some of the fundamental challenges facing historians examining modern terrorism and considers some of the ways that they have engaged and contributed to deeper understandings of the phenomenon. The essay first begins with a discussion of the search for definition and the challenges facing historians (and the humanities) engaging in a field largely dominated by the social sciences. Second, the essay considers the two primary ways that historians have addressed terrorism and constructed their narratives, suggesting that, at present, a historiography of terrorism remains underdeveloped. Finally, I turn to a more specific focus, by way of example, and examine the current historical perspective on terrorism, war, and insurgency, which of course are intimately linked.

The Definition Debate

Analysis of terrorism, historical or otherwise, proves a daunting task. Indeed, the word itself, while so common a lexical currency, seems to defy definition. Defining terrorism remains a thorny, contested undertaking, and there is little consensus. Given the wide range of theories and frameworks through which the phenomenon is studied and analyzed by academics, not to mention its use by media and general populations, the possibility (and perhaps value) of a single definition remains hard to pin down (Schmid, 2013; Hoffman, 2006). Given the myriad of meanings, some have argued that the politicized and often rhetorical nature of the word's use ensures that defining it is entirely futile or simply a catalyst for endless controversies (Laqueur, 1977).

Randall Law (2009) notes that terrorism's very suffix proves a complicating factor in understanding it. Confusion results from an assumption that terrorism, akin to familiar "isms," such as liberalism or communism, contains an inherent ideological or world view and attendant political program. In short, he argues, terrorists are always "something else, be they communists, nationalists or fascists" (3). Akin to efforts to define other contested "isms," such as fascism, rather than establishing a universal classification, the lure of embracing a "check-list" definition based on terrorism's most common attributes remains strong (Law, 2009). Yet as Alex Schmid (2013) notes, definition is not found in merely listing similar or recurrent features. Political scientist Bruce Hoffman succinctly summed up the challenge, stating, "the most compelling reason perhaps [for the difficulty of defining terrorism] is because the meaning of the term has changed so frequently over the last two hundred years" (2006, 2). In the end, just as there are a great many terrorisms, so too are there a great many meanings of it, which, as Beverly Gage (2011) asserts, have changed in accordance with time and political context. Indeed, even if social scientists did produce a consensus meaning, a puzzle that has yet to be solved, historians are, as Randall Law (2015) reminds, unlikely to universally embrace it but rather problematize the definition. Perhaps, akin to Angelo Tasca's reflection on fascism, to define terrorism is to write its history.

Regardless of such challenges historians must begin by defining their terms. Indeed, virtually every historical study of terrorism begins first by noting the challenges of finding a definition and then, however conditionally, providing one. Walter Laqueur (1986) asked the question outright: What is terrorism? If the goal was to develop a single, general definition, then he argued the undertaking was as impossible as it was particularly meaningful (Hoffman, 2006). If consensus meaning could be established at all, he reasoned, it was found in describing features of terrorism rather than defining it. Yet as Ariel Merari (2015) rightly points out, a working definition of terrorism is a required beginning point.

Hoffman again proves helpful as he too allows that while a definition remains elusive, it is possible to detail and differentiate it from other forms of violence and ultimately to "identify the characteristics that make terrorism the distinct phenomenon of political violence that it is" (Hoffman, 2006, 2–3; Gage, 2011). In effect, historical distinction allows for both specific and more general, global discussions, if not a cohesive historical narrative of the phenomenon. In general, terrorists employ violence as publicly as possible to engender fear as a means to achieve specific goals. This, however, tells us *about* terrorism, and from this rather general understanding, only further generalization (and definition) often follows. From the outset, specific detail, context, and the careful consideration of historical experience come to define, to return again to Laqueur, not terrorism in general but a single historically contextualized terrorism (1986, 88).

Richard English (2010), an expert on political violence and Irish terrorism, offers a useful if not flawless classification that contends that terrorism, put simply, is the sum of particular parts (characteristics). He asserts that

> Terrorism involves heterogeneous violence used or threatened with a political aim; it can involve a variety of acts, of targets, and of actors; it possesses an important psychological dimension, producing terror or fear among a directly threatened group and also a wider implied audience in the hope of maximizing political communication and achievement; it embodies the exerting and implementing of power, and the attempted redressing of power relations; it represents a subspecies of warfare, and as such it can form part of a wider campaign of violent and non-violent attempt at political leverage.
>
> (24)

English's definition highlights terrorism's essential political nature but also sets out central markers or distinctions that historians can investigate in specific contexts. Historians can then search for distinction and characteristics, which, after all, is the meat and potatoes of the profession. Yet, however complete a checklist, Law (2015) warns academics that turning to definitions that rely on an endlessly expanding set of criteria risks shifting focus away from terrorism's essential nature, the creation and celebration of violent spectacle. A way forward through the definitional labyrinth may lie, as he contends, in ceasing to "pigeonhole" terrorism with an artificially precise definition (2009, 3). Understanding the phenomenon historically necessitates that political violence can "only be understood as terrorism and its significance appreciated within and because of a particular context" (6). Perhaps, as Darren Mulloy put it, as historians "the best we can do is provide carefully constructed contextualized studies" (2011, 113).

The primacy of the search for consensus, coupled with definition as a starting point, threatens to send histories of terrorism in the wrong direction from the outset. In such cases, historical inquiry serves to uncover and detail characteristics that reinforce the definition. To avoid such myopic outcomes, historians do the opposite: namely, investigate the different dimensions and different angles of the historical manifestations of terrorism. At their intersection, Law (2015; 2009) contends, terrorism itself is found.

Writing History and Histories of Terrorism

The purpose of this section is to show how historians have written about terrorism in light of disciplinary conventions and the challenges of definition. We cannot even get to a historiography until we have a set of specifics by which to evaluate terrorism. Historians examining terrorism turn to the case study, usually with a level of specificity and conditionality that all but defies generalization and abstraction. Modern historians (at least initially) establish themselves based on a level of specialization. Only later career historians, and then only rarely, produce the kinds of works of topical synthesis best represented by the likes of Laqueur in the late 1970s, or Law writing three decades later. Two further considerations only heighten the problem. First, historians consider terrorism within a time frame that traditionally is linked to a specific place, meaning that—as others have noted—studying the phenomenon historically requires a command of the broader society in which it manifested. Consequently, academic studies of the history of terrorism tend to provide depth rather than breadth. Second, this predisposition toward specificity and contextual mastery is further narrowed by the limitations imposed by other factors, such as the availability of archival resources and even language fluency. Clearly, historians do not study terrorism as understood in the social sciences, but the phenomenon embedded in a particular and often exclusive context. Put another way, historians consider terrorism *and*, *as*, and *in*. The cumulative effect has been that historians can tell us about terrorism in a specific place and time rather than about it more broadly, and come only slowly to more general narratives.

Historians have detailed many different terrorisms rooted in particular periods, contexts, and often unique circumstances, necessitating unique categorizations (Laqueur, 1986; English, 2010). This, as we have seen, confounds efforts to establish a comprehensive or universal definition, but it also profoundly shapes the ways in which historians have researched and taught about terrorism (English, 2010). On the one hand, to explore a particular terrorism—among many—historians necessarily consider it deeply and on its own terms. On the other hand, there remains a need to consider terrorism, if not as a single phenomenon, more broadly and generally in scope.

Histories of terrorism enter an increasingly crowded field of scholarship and have been, and largely remain, on the periphery of what has become known as terrorism studies (Schmid, 2013; Ross, 2006; Schmid and Jongman, 1988). While Laqueur overstates the point with his contention that "history mostly did not figure in at all" (1977, 146), Gage's view that "professional historians kept the subject at arm-length, rarely engaging in the policy centered world of terrorism studies" is, on the whole, accurate (2011, 79). This reality, of course, should not be surprising given the historian's disciplinary toolbox focused on particularity and context. As the political scientist Dipak Gupta remarked, "academic historians have had . . . much to say about individual rebellion and terrorist activity but remained reticent about addressing the question of 'terrorism'" (2011, 99).

A final point bears consideration. The challenge of identifying a terrorism historiography, or even perhaps even a subfield, is further compounded by the nature of historical debate and discussion. Case studies do indeed spark it, but discussions tend to progress by digging deeper into the specifics of time and place. For example, Peter Hart's (1999) controversial *The IRA and Its Enemies* unleashed an acrimonious debate that ultimately spilled beyond the rather collegial, if often testy, border of the academy and into the public domain. However, the debate itself had less to do with terrorism and rather more with allegations of historical revisionism deeply focused on the specifics of the Irish experience. In effect, key debates around terrorism are always conditional, always rooted in the specifics of context. Considering a historiography or subfield of terrorism is, in the end, akin to painting with smoke.

Reflective of disciplinary conventions, historical studies focused on terrorism can broadly be divided into two approaches: the specific case study and the global/general history. The first approach details a particular regional, national experience and explores movements, conditions, methods, responses, and outcomes rooted in a specific historical context, such as the People's Will in Imperial Russia (Offord, 1986) or the Shining Path in Peru (Stern, 1998). The second encompasses histories broader in scope, perspective, and chronology, examples of which include Randall Law's *Terrorism: A History* and Walter Laqueur's seminal *Terrorism* (Carr, 2007; Miller, 2012; Burleigh, 2010; Laqueur, 2004; Whittacker, 2012; Horgan and Braddock, 2011).

This division of approach presents its own set of unique challenges. First, a case study's specificity risks missing the forest for the trees, a problem that global narratives have attempted to address. Yet such forests still remain, even at their best, a collection of trees, a reality quickly confirmed by cursory examination of various tables of contents. Second, both species can prove problematic, particularly when taken to inform the discussion and analysis of contemporary manifestations of terrorism. Specific case studies can reinforce supposed links between terrorism and a particular ideology and region while global and general treatments risk producing, as Gage asserts, "a teleological narrative in which all previous acts of terrorism build dramatically toward 9/11" (2011, 82).

Engaging these challenges has, on the whole, been greatly aided by David Rapoport's "four waves" theory (2004). For Rapoport, modern terrorism encompasses waves—anarchist, anti-colonial, New Left, and religious—that reflect its dominant (but not sole) feature. Each wave, he contends, represents a cycle of activity, expressed internationally and "driven by a common predominant energy that shapes the participating groups[']" characteristics and mutual relationships (2004, 47). Rapport's theory proves a useful framework for both considering the historiography of terrorism and teaching about it. First, virtually every academic, historical study of modern terrorism either can be placed within a wave or reflects its periodization. Second, as case studies can more or less be situated within a particular wave, the forest-trees pitfall is lessened by considering various terrorist manifestations both within and across a particular wave. Similarly, more recent global/general narratives largely reflect Rapport's theory as a means to effectively organize and engage specific focuses (Chaliand and Blin, 2015; Law, 2009, 2015).

While Gage's concerns of forests and trees are not entirely unmerited, the emerging historical literature is starting to change. Historians have begun to engage core problems, such as definition, lack of historiography, and the need to generalize. In effect, each species promotes, challenges, informs, and builds upon the other. Equally significant is the role careful histories serve in challenging seemingly dominant assumptions about the phenomenon both within the academy and without. Gage is certainly right to note that within terrorism studies, the Middle East and Islam continue to receive the lion's share of attention, yet historians are certainly active in investigating a wide range of places and times (Law, 2015; Hanhimaki and Blumenau, 2013). Further, the alleged novelty or indeed the notion of "new terrorism" is challenged from a number of perspectives, ranging from a focus on the premodern world to considering modern terrorism within "waves" often far removed from contemporary political violence (Laqueur, 2004; Rapoport, 2004; Sageman, 2004; Mockaitis, 2007; Jackson and Sinclair, 2012).

Ultimately such efforts can counter the dangers posed by policy makers who, in search of lessons learned, adopt decontextualized histories as capable of informing the present. Finally, akin to the training of historians rooted in specialization, the university setting presents its own set of organizational and pedagogical problems. For those offering courses about terrorism, there is a challenge to master each specific context, and by necessity asking students to do the same. Put bluntly, academics *teaching* about terrorism turn to both species of history as required reading.

War, Insurgency, and Terrorism

What is the relationship between war and terrorism? Terrorism fits along the spectrum of political violence and is seen by many as a subspecies of warfare (Law, 2009; Polk, 2007; English, 2010). Charles Townshend (2011) believes that war and terrorism are intimately related, noting that at a most fundamental level, war and terrorism generate fear, whether as an objective or an outcome (Schmid, 2004). Further, given that in the post-1945 world, wars within states are more common than between states, the need to consider the relationship between war and terrorism is both relevant and pressing (Hughes, 2015; Duyvesteyn and Fumerton, 2010). In its modern expression, terrorism is also linked to insurgency as a form of warfare (O'Neill, 2005). But here again the problem of definition emerges, strongly evidenced by the attractive yet inaccurate conflation of terrorism and guerrilla warfare. Given the challenge of defining terrorism, compounded by its pejorative and immoral connotations, guerrilla warfare as synonymous with terrorism offered a semantic alternative. Indeed, beginning in modern terrorism's "third wave," the lure of adopting the "one man's terrorist is another's freedom fighter" perspective proved particularly inviting (Rapoport, 2004). As early as the 1970s, however, Laqueur (1976, 1977) warned against such a conflation, arguing that the terrorist strategy of urban guerrilla warfare inaccurately linked the two (English, 2010; Hughes, 2015; Ganor, 2002; Joes, 2007). Akin to the challenges of defining the phenomenon in toto, academics have yet to present a "clear and objective definition that distinguishes between 'terrorism' and 'insurgency' as a whole" (Hughes, 2015, 384).

Insurgency, at its heart, is the effort to challenge state-based adversaries (Hughes, 2015; Cronin, 2008). To that end, as Bard O'Neil (2005) asserts, insurgent violence "is manifested in different forms of warfare"—namely, "terrorism, guerrilla war and conventional warfare" (33). Terrorism is a mode of warfare, a tactic adopted within the context of a wider insurgency, and often motivated by necessity and the particular challenges many states face in countering it. Based largely on tactical characteristics, terrorism exhibits key differences from other forms of the political violence, such as guerilla and conventional war. For example, conventional warfare seeks to establish control over geographic space, whereas terrorism seeks to exert psychological influence largely free of the material elements of guerilla warfare. Indeed, even at the tactical level, terrorism differs considerably in unit size, armaments, and operations. Further, as a mode of struggle, terrorism neither recognizes specific war zones nor enjoys international legality, characteristics common to both conventional and guerrilla conflicts.

In general, whether considered strategically or tactically, terrorism shares little common ground with either conventional or guerrilla warfare (Law, 2009; Merari, 2015; Laqueur, 1976; Clausewitz, 1989). As Ariel Merari puts it, "terrorism is the easiest form of insurgency" (2015, 12). Richard English argued that terrorism was "a mechanism among others within a broad and varied campaign—for exerting influence over an opponent towards political ends" and often an element in a "wider violent and political repertoire of complex actions" (2010, 12, 25–26). If then terrorism is a mode of insurgent warfare, the question remains: what methods do terrorists embrace, how, and why? Terrorism necessarily and opportunistically targets symbols and constituencies for psychological coercion (Law, 2009; Richards, 2014). This focus on the psychological is so central that it is often cited as a defining characteristic of terrorism itself.

Counter to contemporary conventional warfare notions of "shock and awe" to rapidly degrade enemy forces, terrorism embraces the notion of the "long war" (Galula, 2006). Further, given that terrorism is most often adopted as a strategy by groups limited in size and political and material resources, victory can be achieved only through psychological rather than physical compulsion over time. In effect, terrorism aims to induce behavioral modification of the audience, to both "coerce a government and to intimidate its population into acceding to political demands" (Hughes, 2015, 384; O'Neill, 2005, 33). While surely similar to other forms of warfare in that insurgent terrorism is purposeful violence, psychological coercion remains the primary goal even if operations often seek to achieve a range of specific ends—often concurrently. Terrorists look to spread their message, intimidate, provoke, sow chaos, and wear down the state. Even the most well-supplied terrorist organization cannot conventionally defeat a targeted regime. Operations seek to spread the word, to transmit tenets, and to bring attention to perceived grievances. Whereas for nineteenth-century terrorists "propaganda by the deed" tended to target symbolic individuals within the regime, twentieth-century and contemporary terrorist attacks often seek to gain widespread media coverage and reach multiple audiences through large-scale, shocking, spectacular attacks (Saunders, 2015). In short, such propaganda transforms the witnesses, however remotely, into participants.

Closely linked to this notion is the terrorist goal of shaping social behavior. More specifically, attacks can, whether targeting specific constituencies or the general population, serve as means to intimidate. Whether intended to compel or end particular behaviors, in effect terrorists simultaneously seek to demonstrate and project their own power. In some cases, operations are intended to coerce populations to accept and engage a "shadow," parallel government established by the terrorists themselves (Race, 2010; Hunt, 1995). In effect, attacks act as a means establishing both legitimacy and social control. Faced with spectacular and influential attacks, regimes are compelled to respond. Indeed, promoting reaction is often specifically the intention of operations. Importantly, particularly in liberal democracies, it is most often the violence itself, rather than simply the presence of anti-regime movements and ideologies, that prompts state intervention. For terrorists, the goal is to provoke harsh retaliations, primarily conducted by the state or allied foreign coalition security forces (Marighella, 1971). In this way, the regime faces a dilemma in which failure to respond jeopardizes legitimacy while responses, particularly violent ones, risk alienating populations, creating fresh grievances, and empowering the desire for re-violence and revenge within previously pacific constituencies (Kilcullen, 2011). In short, repressive actions by the government not only empower terrorist rhetoric but also ensure that elements of the population begin to see their interests best represented (or at least protected) by the terrorists rather than the regime.

Beyond alienating and potentially surrendering specific populations to terrorists, regime intervention and reprisal also risk propelling expected operational friction into a broader sense of tension and destabilization. More specifically, security forces can find themselves caught in a dilemma in which inaction (or ineffectiveness) can be as damaging as disproportionate repression. In these cases, a failure to respond to terrorist operations or prevent further attacks empowers the notion that the regime has lost effective control over the state. In some cases, this can lead to demands for regime change. For terrorists, such calls and changes do not signal imminent victory but rather a new and generally harsher state more willing to act violently and punitively, thus risking the alienation of specific constituencies. For Hughes, the danger can be even more profound as he asserts that "states can themselves erode the foundations of their constitutions by their own reactions to internal violence" (Hughes, 2015, 392; Koch, 2016; Boyle, 2010).

This strategy of forcing an ungovernable state is closely related to that of attrition. This principle is predicated on the notion that the military and political endgame is directly linked to the insurgent's capacity to endure rather than inflict violence. Again understanding that conventional engagement with security forces would prove disastrous, terrorist persistence and perseverance over the course of a "long war" look to push the state toward utilitarian, generally political-economic calculation. In short, terrorism is not a means to win, in a conventional sense, but rather an effort to insure insurgents survive and push the regime to a point where the cost of continuing proves unacceptable and compels the state to end the conflict (O'Brien, 1995).

A final point bears consideration. Wars, Clausewitz famously reminds, are fought for political objectives, and violence used for strategic ends (Hughes, 2015). Yet, as we have seen, terrorism as a strategy of insurgency differs from conventional warfare (Freedman, 2007). Terrorism profoundly blends the military and political and intertwines both, engendering both a conceptual and practical murkiness that challenges and confounds academics and policy makers alike. As a strategy of insurgence, terrorism is shaped by objective conditions rather than by strategic conceptions of the insurgents and represents a means to square "meager means and grandiose objectives" (Merari, 2015, 31). Such efforts, both historically and indeed contemporarily, have proven unsuccessful. As a sole method of political violence, terrorism has failed to depose a government. Indeed, only in cases in which terrorists have transitioned their campaign to full insurgency has regime change been achieved (Hughes, 2015). Terrorism then represents a particular, perhaps transitional set of insurgent choices framed by circumstance (Metz, 2012). Historians have much to add to our understanding of such interaction. Terrorism and insurgency, as Hughes (2015) states, encompass a convergence of war, politics, and crime demanding that policy makers and academics acknowledge its practical and conceptual complexity. To that end, historians have a great deal of work to do—namely, to analyze the context, characteristics, and conditions of each case on their own terms (Hughes, 2015).

In the aftermath of 9/11 and the subsequent Global War on Terror, scholarly interest in terrorism increased dramatically. So too did the tempo of academic *production*. In short, terrorism studies was supercharged by an infusion of stakeholders, ranging from academics to policy makers, militaries, the media, and private industry. As a result of this frenzy of publication, gaining command of the literature is like trying to drink from a fire hose (Gupta, 2011). As we have seen, it is within this context that historians are building an emerging historiography.

The Future Study of War and Terrorism

A historical understanding of terrorism reveals two key features. First, terrorism is not in and of itself an ideology. Indeed historical case studies labor to identify the worldview and ideologically motivated aspirations that underpin terrorist violence (Chaliand and Blin, 2015; Law, 2009, 2015). Terrorism, strategically and tactically, is a means to an end. Consequently, wars can be fought to combat terrorists and deny their goals but not against terrorism itself. Second, historians are adept at revealing and detailing the context from which a particular terrorist manifestation emerged. This process highlights the reality that a decision to engage in terrorism is a conscious and deliberate one rather than an act of madness. In both features, analyzing case studies requires historians to grapple with factors often outside the purview of traditional military history, and thus posit the systematic study of this terrorism firmly in the war and society milieu.

Terrorism has been and continues to be examined from a myriad of academic perspectives, and certainly historians are well placed to contribute to the discussion. They cannot, however, provide definitive answers to the challenges and dangers it presents. What historians can do is highlight that terrorism has a history, rooted in context—a truth often as elusive to the general public as to national policy makers. These studies are likewise so slippery because conflicting social, cultural, religious, political, regional, and other factors need to be disentangled or at least recognized. The lure of providing "lessons learned" is a powerful one, reinforced by both the nature of terrorism studies and a lingering tradition within military history writing to seek them out. Historians of terrorism face significant challenges, ranging from the quest for definition to navigating the discipline's place in the ever-growing field of terrorism studies. Finally, histories of terrorism, the warp and weft of the historiography, are only beginning to move significantly toward global and general narratives and to embrace careful comparative studies that will allow for continuities to be exposed and examined critically. To this end, writing the history of the phenomenon, like terrorism itself, shows no sign of abating.

Bibliography

Boyle, Michael. "Do Counterterrorism and Counterinsurgency Go Together?," *International Affairs* 86, no. 2 (2010): 333–353.

Burleigh, Michael. *Blood and Rage: A Cultural History of Terrorism*. New York: Harper Perennial, 2010.

Carr, Matthew. *The Infernal Machine: A History of Terrorism*. New York: The New Press, 2007.

Chaliand, Gérard, and Arnaud Blin, eds. *The History of Terrorism: From Antiquity to ISIS*. Berkeley: University of California Press, 2015.

Clausewitz, Carl von. *On War*. Princeton: Princeton University Press, 1989.

Cronin, Audrey Kurth. "Ending Terrorism: Lessons From Defeating al-Qaeda." Adelphi Paper 394. Abingdon, UK: Routledge for International Institute for Strategic Studies, 2008.

Duyvesteyn, Isabella, and Mario Fumerton, "Insurgency and Terrorism: Is There a Difference?" In *The Character of War in the 21st Century*. Eds. Caroline Holmqvist-Jonsater and Christopher Coker. Abingdon, UK: Routledge, 2010, 27–41.

English, Richard. *Terrorism: How to Respond*. Oxford: Oxford University Press, 2010.

Freedman, Lawrence. "Terrorism as a Strategy," *Government & Opposition* 42, no. 3 (2007): 314–339.

Gage, Beverly. "Terrorism and the American Experience: A State of the Field," *The Journal of American History* 98, no. 1 (2011): 73–94.

Galula, David. *Counterinsurgency Warfare: Theory and Practice*. Westport: Greenwood, 2006.

Ganor, Boaz. "Defining Terrorism: Is One Man's Terrorist Another Man's Freedom Fighter?" *Police Practice and Research* 3, no. 2 (2002): 287–304.

Gupta, Dipak. "Terrorism, History and Historians: A View From a Social Scientist," *The Journal of American History* 98, no. 1 (2011): 95–100.

Hanhimaki, Jussi, and Bernhard Blumenau. *An International History of Terrorism: Western and Non-Western Experiences*. Abingdon, UK: Routledge, 2013.

Hart, Peter. *The IRA and Its Enemies: Violence and Community in Cork*. Oxford: Oxford University Press, 1999.

Hoffman, Bruce. *Inside Terrorism*. New York: Columbia University Press, 2006.

Horgan, John, and Kurt Braddock, eds. *Terrorism Studies: A Reader*. Abingdon, UK: Routledge, 2011.

Hughes, Geraint. "Terrorism and Insurgency." *Routledge History of Terrorism*. Ed. Randall Law. Abingdon, UK: Routledge, 2015, 383–395.

Hunt, Richard. *Pacification: The American Struggle for Vietnam's Hearts and Minds*. Boulder: Westview Press, 1995.

Jackson, Richard and Samuel Sinclair, eds. *Contemporary Debates on Terrorism*. Abingdon, UK: Routledge, 2012.

Joes, Anthony. *Urban Guerrilla Warfare*. Lexington: University of Kentucky Press, 2007.

Kilcullen, David. *The Accidental Guerrilla: Fighting Small Wars in the Midst of a Big One*. Oxford: Oxford University Press, 2011.

Koch, Bettina, ed. *State Terror, State Violence: Global Perspectives*. Wiesbaden: Springer VS, 2016.

Laqueur, Walter. *Guerrilla*. Boston: Little, Brown, 1976.

———. *The New Terrorism: Fanaticism and the Arms of Mass Destruction*. Oxford: Oxford University Press, 1999.

———. "Reflections of Terrorism," *Foreign Affairs* 65, no. 1 (1986): 86–100.

———. *Terrorism*. Boston: Little, Brown, 1977.

———. *Voices of Terror: Manifestos, Writings and Manuals of Al Qaeda, Hamas, and Other Terrorists From Around the World and Throughout the Ages*. New York: Reed Press, 2004.

Law, Randall, ed. *The Routledge History of Terrorism*. Abingdon, UK: Routledge, 2015.

———. *Terrorism: A History*. Cambridge: Polity, 2009.

Marighella, Carlos. "Manual of Urban Guerrilla Warfare," in Robert Moss, *Urban Guerrilla Warfare*, London: International Institute for Strategic Studies, 1971.

Merari, Ariel. "Terrorism as a Strategy of Insurgency." In *The History of Terrorism: From Antiquity to ISIS*. Eds. Gérard Chaliand and Arnaud Blin. Berkeley: University of California Press, 2015, 12–51.

Metz, Steven. "Re-thinking Insurgency." In *The Routledge Handbook of Insurgency and Counterinsurgency*. Eds. Paul Rich and Isabella Duyvesteyn. Abingdon, UK: Routledge 2012, 32–44.

Miller, Martin. *The Foundations of Modern Terrorism: State, Society and the Dynamics of Political Violence*. Cambridge, UK: Cambridge University Press, 2012.

Mockaitis, Thomas. *The "New" Terrorism: Myths and Reality*. Westport, CT: Greenwood, 2007.

Mulloy, D. J. "Is There a 'Field'? And If There Isn't, Should We Be Worried About It?" *The Journal of American History* 98, no. 1 (2011): 111–114.

O'Brien, Brendan. *The Long War: The IRA and Sinn Féin*. Dublin: O'Brien Press, 1995.

Offord, Derek. *The Russian Revolutionary Movement in the 1880s*. Cambridge, UK: Cambridge University Press, 1986.

O'Neill, Bard. *Insurgency and Terrorism: From Revolution to Apocalypse*. Dulles, VA: Potomac Books, 2005.

Polk, William. *Violent Politics: A History of Insurgency, Terrorism, and Guerrilla War, From the American Revolution to Iraq*. New York: HarperCollins, 2007.

Race, Jeffrey. *War Comes to Long An: Revolutionary Conflict in a Vietnamese Province*. Berkeley: University of California Press, 2010.

Rapoport, David. "The Four Waves of Modern Terrorism." In *Attacking Terrorism: Elements of a Grand Strategy*. Eds. Audrey Kurth Cronin and James Ludes. Washington: Georgetown University Press, 2004.

Rich, Paul, and Isabella Duyvesteyn, eds. *The Routledge Handbook of Insurgency and Counterinsurgency*. Abingdon, UK: Routledge 2012.

Richards, Anthony. "Conceptualizing Terrorism," *Studies in Conflict and Terrorism* 37, no. 3 (2014): 213–236.

Ross, Jeffrey Ian. *Political Terrorism: An Interdisciplinary Approach*. New York: Peter Lang, 2006.

Sageman, Marc. *Understanding Terror Networks*. Philadelphia: University of Pennsylvania Press, 2004.

Saunders, Robert. "Media and Terrorism." In *Routledge History of Terrorism*. Ed. Randall Law. Abingdon, UK: Routledge, 2015, 428–441.

Schmid, Alex. "Frameworks for Conceptualizing Terrorism," *Terrorism and Political Violence* 16, no. 2 (2004): 197–221.

————, ed. *The Routledge Handbook of Terrorism Research*. Abingdon, UK: Routledge, 2013.

Schmid, Alex, and Albert Jongman. *Political Terrorism: A New Guide to Actors, Authors, Concepts, Databases, Theories, and Literature*. Piscataway: Transaction, 1988.

Stern, Steve, ed. *Shining and Other Paths: War and Society in Peru, 1980–1995*. Durham: Duke University Press, 1998.

Thorup, Mikkel. *An Intellectual History of Terror: War, Violence and the State*. Abingdon, UK: Routledge, 2010.

Townshend, Charles. *Terrorism: A Very Short Introduction*. Oxford: Oxford University Press, 2011.

Whittacker, David, ed. *The Terrorism Reader*. Abingdon: Routledge, 2012.

22 Religion, Ethics, and War

Jacqueline E. Whitt

If two subjects can lay claim to universality in the human experience across time and space, war and religion top many lists. In many times and places, the history of the two was (and perhaps is) deeply intertwined. Some scholars and popular writers have claimed that religion is inherently violent—or at least that it exacerbates violent conflict—because it arouses emotions and passions that are not subject to reason (Avalos, 2005; Hitchens, 2007; Dawkins, 2006). On the other hand, critics of this point of view (often religious apologists) claim, with equal conviction, that religion is nothing of the sort; rather, empire and politics are the key source of conflict in the world (Armstrong, 2014). They note the secular ideologies of Nazism, Stalinism, and Maoism in particular to rebut the claim that religion is a key motivator of violence. A third group argues, more dispassionately, for a more neutral relationship, arguing that religion is neither inherently violent nor pacifistic, but has been used to justify and animate a variety of positions in relation to the organized violence of war—in other words, "religion and violence are clearly compatible, but they are not identical" (Eller, 2010, 236). Charles Selengut writes that "the history and scriptures of the world's religions tell stories of violence and war as they talk of peace and love" (2003, vii). William Cavanaugh further argues that even the attempt to separate "religious" from "secular" violence is itself analytically problematic, as what counts as "religious" or "secular" is historically specific and culturally constructed (2009).

Though these fundamentally philosophical and theological questions are interesting and important, they are not of immediate concern for most military historians. Instead of entering this debate, this chapter uses the third interpretive position described earlier as a starting point. This chapter will focus on the empirical, historical relationship between the realms of religion and war—recognizing that in rhetoric and practice, the history of war and the history of religion are often linked, but not deterministically so.

Having acknowledged these assertions and assumptions, we can identify four major lines of inquiry about the relationship between war, religion, and ethics. First, if "religion" is neither inherently violent nor pacifistic, how can we best understand the ways in which religious ideas have been used to justify or prohibit war and to encourage religious people to fight or abstain from it? Second, how have religious, ethical, and moral considerations affected the conduct of war, especially as multipliers or limiters of violence? Third, how has war affected the beliefs and practices of religious people and institutions—both during and after conflict? And fourth, how have military and religious institutions interacted with each other? Regardless of which framework or question is the focus, scholars must recognize that it is impossible to look at the "religious" aspects of such conflicts without also carefully considering the social, political, and economic factors also at play.

Perhaps no conflicts better underscore the complicated relationship between religion and warfare than the Crusades. By the eleventh century, Christendom was divided between East (centered in Byzantium) and West (centered in Rome), and the political landscape was even more variegated, as monarchs and nobles battled for control. In the Muslim world, a series of invasions and political infighting had left its cities and rulers vulnerable to attack. When Pope Urban II called for the First Crusade to provide military support for Byzantium, in response to incursions by Turks from Anatolia, his motivations were likely a mix of political (to unify the church with himself as the head) and religious (to recapture Jerusalem from Muslim control). In Europe, the response was rapid and enthusiastic: Urban preached that the Crusade was God's will and promised indulgences (absolution from sin) for participants. Thus, in many historians' views, the Crusades must be viewed in light of this penitential purpose, where Crusaders were often motivated by egoistic reasons as opposed to corporate ones. And though the Crusades often targeted non-Christian populations, they also occasionally turned inward, seeking to root out heretical Christian practices as well (Riley-Smith, 2009). In each of the Crusades that followed, there emerged a complex mix of religious, political, economic, and social justifications and consequences (Tyerman, 2006).

The geopolitical and religious legacies of the Crusades have been long-lasting. For the Christian Church, the Crusades inured clergy and laypeople alike to the idea of warfare for holy causes, so the language and ideology of crusading (combating the "enemies of God") permeate much of the Western Christian tradition. T. Walter Herbert has been particularly incisive on this point, arguing that the U.S. response to the terrorist attacks of 9/11 represented only the latest in a long trend of an "imperialist version of Christian Americanism" that blurred the lines between religious war and contemporary nationalism (2009, 3).

Closely related to the question of "holy war" is the idea of a "just war," and in the Christian tradition, much of the foundational just war writing originated in the aftermath of the Crusades and during the consolidation of European monarchies. If war requires men and women to act outside of the norms of acceptable behavior, then it follows that an intellectual tradition of moral reasoning about war should also emerge as people write about its causes, conduct, and consequences to define the boundaries of what is permissible amid hostilities. As might be expected, much of this reasoning about war has developed within the world's religious traditions, both within sacred texts (scriptures) and in the interpretation of those texts by scholars and clerics, though there are nonreligious conventions and norms that have governed the rules of war as well (Corey and Charles, 2012). On the whole, Christianity and Islam have the most developed theological and intellectual traditions exploring the questions of holy war and just war, and the relationships between the two (Fine, 2015). Although there are distinctive features within each tradition, there are significant commonalities in substantive coverage and structure that also suggest some exchange of ideas and writing between contemporaries (Hashmi, 2012). Further, these just war traditions provide further insight into how religion can be understood to both limit and exacerbate the violence of war (Johnson, 1975 and 1981).

Broadly, there are two key questions considered by the literature on just war. First, under what conditions is it permissible to go to war (*jus ad bellum*), and second, what constitutes right or proper conduct during war (*jus in bello*); some modern thinkers have added a just peace (*jus post bellum*) to the rubric for determining the just nature of a war. Scholars generally agree that just war norms evolved in order to regulate and limit war between two culturally similar adversaries; when, however, an adversary or enemy is regarded as an alien "other," usually in racialized or religious terms, these norms and conventions are frequently ignored. The Crusades against Islamic enemies may be seen as prime examples of this phenomenon (Hartigan, 1974).

The Christian tradition provides much of the vocabulary used in analyzing just war theory (both religious and secular). Saint Augustine is credited with the first commentary on the ethics of war, but Thomas Aquinas developed the most systematic articulation of a Catholic just war doctrine in the thirteenth century. Aquinas' formulation for *jus ad bellum*, in *Summa*

Theologica, proposes a three-part test to determine whether a war is just. First, the war must be declared by a proper authority; second, the cause of the war must be just; and third, the intention must be right (i.e., to promote good and avoid evil). These three conditions form the basis for most subsequent Western and Christian explorations of just war theory, as these have expounded upon (and argued with) Aquinas (Reichberg, 2016). Contemporary theologians (Protestant, Catholic, and Orthodox) have added other considerations or requirements for the legitimate use of force, including that war must be in response to a grave or existential threat; that war should be a means of last resort; that there must be the genuine prospect of achieving objectives; and that the use of force must not produce greater evil than that which is to be eliminated.

Within the realm of *jus in bello*, Catholic doctrine—historical and contemporary—has centered around the ideas of human reason, proportionality, discrimination, and accountability. Respectively these norms suggest that war does not necessitate an absence of reason or morality, that responses in war must be proportional to the damage caused by an attack, that actions in war should discriminate whenever possible between combatants and noncombatants or between legitimate and illegitimate targets, and that individuals remain accountable for their actions in war (J. T. Johnson, 1981; Reichberg et al., 2014; Stoyanov, in Popovski, 2009).

If the Christian tradition must come to terms with the historical Crusades and a crusading ideology writ large, the Islamic tradition must similarly grapple with the concept of jihad and its historical and contemporary interpretations and applications. Translated simply as "struggle" or "striving," jihad can be understood historically, theologically, and pragmatically in two ways: first and most importantly, as a category of personal religious struggle unrelated to warfare or violence (the greater jihad), and second, as a call to organized violence against the enemies of Islam (the lesser jihad). The call to jihad could be defensive or offensive.

Jurisprudence has varied significantly between Sunni and Shi'ite sects, and scholars, theologians, and practitioners have differed on who, if anyone, in the Muslim community has the authority to order jihad. Shi'ite Muslims generally believe that only the Infallible Imam has the authority to call for offensive jihad, while the Sunni tradition historically held that it must be a permanent condition for all adherents. The classical Sunni view, usually seen as encouraging offensive jihad, must be understood in light of the growth and consolidation of Islamic empires in the eighth to thirteenth centuries. Contemporary Sunni literature on jihad (and war and peace more generally) is typically seen as more measured, reflecting the loss of power of Islamic empires at the hands of expanding European ones in the fourteenth through eighteenth centuries. Most importantly, scholars argue, jihad must be understood in historical and theological context as a doctrine that has evolved over time, and for which there have been multiple interpretations (Kelsay, 1993; Kelsay, 2007; Sonbol, in Popovski, 2009; Feirahi, in Popovski, 2009; Bonner, 2006).

Of the world's major religious traditions, and particularly within the Abrahamic faith groups, Judaism has the least developed just war tradition, although both biblical and rabbinical sources speak to the ethics of war (Niditch, 1993). Reuven Firestone has explored the idea of Jewish Holy War and argued that it disappeared for nearly two thousand years within Jewish discourse before reappearing with the advent of Zionism and the founding of Israel (2012). Since the second-century Bar Kochba rebellion, Jews have been the victims of violence and war, rather than the primary perpetrators of it. The advent of the Jewish State of Israel has complicated this view: the Israel Defense Forces have been embroiled in near-continuous war and conflict with its neighbors and Palestinian nationalist groups since its formation in 1948, insinuating a religious element to most of these conflicts (Afterman and Afterman, in Reichberg, 2014).

While much of the language for just war theory originates within the Christian tradition and is most easily analyzed within the Abrahamic traditions, the world's oldest religious faiths also have texts and intellectual traditions that offer normative and ethical prescriptions for the declaration of and conduct during war. In the case of India, many scholars have argued that there was no clear tradition of military philosophy or strategic thinking before the colonial period, while others claim such ideas existed and had deeply influenced Hindu thought for more than two thousand years (Brekke, 2006; Roy, 2012). This is a complicated project because Hinduism is not monotheistic and does not have a single authoritative sacred text, but most sources approve of violence and force as a way to advance order and for defense (Morkevicius, in Hensel, 2010). In his book *Hinduism and the Ethics of Warfare in South Asia* (2012), Kaushik Roy explores the tradition of military philosophy in India, focusing on the tension between the ideal types of *Dharmayuddha* (just war) and *Kutayuddha* (unjust war). Furthermore, Roy examines the ways in which religious heterogeneity within India has affected ethical thinking about war, examining not only Hinduism but also the influences of Buddhism, Jainism, Sikhism, Christianity, and Islam.

The global tradition of Buddhism, which adapted to local conditions after also originating in the Indus Valley and subsequently spreading through much of Asia, offers many interesting ways for scholars to engage questions about just war theory and the ethics of war. Peace and nonviolence, both internal and external, are central tenets to the many practices of Buddhism. Yet there is a long history of predominantly Buddhist states undertaking war. Some scholars and practitioners argue that when Buddhists participate in war, they are acting outside of the norms of the tradition (Degalle in Jerryson, 2010; and Victoria in Jerryson, 2010). But Michael Jerryson notes that Buddhists—using Buddhist texts—have justified war against coreligionists, against external enemies of different faiths, and for explicitly political aims, such as conquest and rebellion (2010).

In the late twentieth century, there began a concerted effort to develop a secular philosophy of just war. Michael Walzer in *Just and Unjust Wars* (in its fifth edition as of 2015) has articulated a well-developed and extensively argued secular philosophy of just war. Placing most of the moral burden on the aggressor in a war, he argues that there are universal moral principles both in the decision to go to war and in the conduct of war that undermine realist views in which decisions about war are made from material and utilitarian perspectives. In his work, Walzer also addresses the question of individual conduct and concludes that combatants are still morally culpable for their actions during war (i.e., that atrocities committed during war are the responsibility of those who commit them), but also that they must adhere to the "war convention," which states (1) who soldiers may kill in war and (2) when and how they may do so (Walzer, 1977).

Other secular studies of ethics and war exist as well. Writing in the utilitarian tradition of John Stuart Mill and Jeremy Bentham (which Walzer rejects, but which reaches many similar conclusions about restraint in war), William H. Shaw argues in favor of a utilitarian war principle in which war is morally right only if no other available course of action has a greater expected good attached to it. Ultimately, Shaw (in a somewhat interesting turn) puts forward a list of general principles that reveal significant similarities to traditional, religious just war thinking (Shaw, 2016).

Within most religious traditions that have developed theological and legal positions on the norms of war, there are alternative traditions that reject such norms in favor of religiously motivated pacifism (Dyck, 1996). The standard work on this topic remains John Howard Yoder's *Nevertheless: The Varieties and Shortcomings of Religious Pacifism* (1973). Some religious traditions, such as Jainism, are entirely pacifistic in their orientation, while other traditions, such as Buddhism and Hinduism, also have significant pacifist strains. An ethic of virtue and nonviolence has been a key characteristic of many religious leaders, from Buddha to Jesus to

Gandhi to Martin Luther King (Gier, 2003). Duane Cady offers a wide-ranging overview of the "moral spectrum" on the ethics of war in *From Warism to Pacifism: A Moral Continuum*. Cady identifies himself as a "reluctant pacifist," unable to defend just war theory, and argues that pacific ends are unlikely to be won from violent means. Cady's work is notable also for its argument that warism, the idea that war is morally justifiable and sometimes required, is the default paradigm for Western societies (Cady, 2010).

Just war theory is not the only useful framework for understanding the relationship between war and religion. A number of important works seek to understand religiously motivated violence, often in the form of terrorism, at the margins or even outside of mainstream religious practice (Selengut, 2003; Juergensmeyer, 2000; Stern, 2003). Here, violence perpetrated by Islamic extremist groups, sometimes under the banner of jihad, is juxtaposed not with the systemic and institutionally sanctioned conflict of the Crusades but alongside that of other marginalized groups that use (for some, abuse) religious ideas to justify bloody means to accomplish a political objective. In studying this sort of religiously motivated violence, scholars might focus on the organization, theology, and operations of specific groups—the Ku Klux Klan, the Lord's Resistance Army, Abu Sayeff, Da'esh (ISIL)—rather than assigning brutal motives, rhetoric, and actions to a religion writ large.

In the search for understanding the motivations for violent religious extremists, specifically terrorists and terrorist organizations, the theme of dangerous ideologies and radicalization has been a key source of debate. Authors such as Mary Habeck have argued that if the actors themselves understand their motives as fundamentally *religious* ones, scholars and policy makers ought to take those pronouncements seriously—in other words, "we" should take "them" at their word (Wiktorowicz, 2005; Habeck, 2006). Others have argued, just as strongly, that the real causes of terrorism are economic disfranchisement, lack of education, and meager economic prospects, rather than religion or ideology (Mackinlay, 2010; Kundnani, 2014). The debate matters in large part because the policy prescriptions for governments waging counterterrorism campaigns are often quite different, depending on which view is accepted.

Other kinds of religiously motivated violence fall outside the purview of traditional military history, but which may provide fruitful avenues for research under the broader construct of organized violence. Such groups often have a particular violent and apocalyptic eschatology, and a fascination with technology and weaponry that may pit them against state authorities. Aum Shinrikyo, the perpetrators of the sarin gas attack on the Tokyo subway in 1995, believed, for example, that "salvation could be achieved only by bringing about the deaths of just about everyone on this earth" and that killing others would "provide them with a favorable rebirth" (Lifton, 1999, 8). These violent groups (or radicalized individuals) may be of interest to military historians primarily in the ways they interact with the state and its military forces. In other words, military historians might examine the ways in which military forces respond to sects and other marginalized religious groups as well as military responses to religiously motivated terrorist or paramilitary groups.

The particular religious beliefs and practices of individuals who participate in war (both in combat and on the homefront) are also studied by military historians trained in using the methodologies and sources of social and cultural history. A model of this sort is Jonathan Ebel's *Faith in the Fight* (2010), which explores the religious lives of American service members during the First World War. Ebel argues against the view that World War I ushered in rapid and widespread secularization in the twentieth century. Instead, Ebel suggests that Americans frequently saw the war as a fundamentally religious contest that held both individual and national significance and further that Americans' religious worldviews shaped both their experiences in war and their interpretations of it after the fact. Michael Snape has contributed significantly to this body of work for the United States in the Second World War (2015), as has Steven Woodworth for the American Civil War (2001).

In a non-U.S. context, several studies have considered the religious worlds of Christians in Nazi Germany. Doris Bergen's *Twisted Cross* explores the German Christian movement, which sought to expunge all Jewish elements from Christianity (to include rejecting the Jewish heritage of Jesus and throwing out the Old Testament) and, Bergen argues, occupied key positions within the German Christian church even after the Nazi regime rejected their agenda (1996). More recently, Lauren Faulkner Rossi has written about Catholic priests who served willingly in the Third Reich, examining their moral and theological justifications, their attempts to retain independence from the regime, and their failures in response to the genocidal Nazi project (2015).

In addition to the lived religion of wartime participants, historians have also begun to explore the world of religious ideas that have motivated leaders and participants in war in specific contexts. The most thorough examinations of this sort have addressed the American Civil War, which offers historians of religion and the military significant grist for the mill. Both sides were steeped in deeply similar cultural and theological worlds, yet used religion to justify perfectly opposite positions on slavery, war aims, and American identity. These ideas have been explored in depth in an edited collection, *Religion and the American Civil War*, which was the result of an interdisciplinary conference on the subject (Miller et al., 1998). More recently, Mark Noll (2006, 3) has characterized the Civil War as a "theological crisis," while Harry Stout has asserted that, as the conflict "progressed onto increasingly eroded moral ground, something transformative simultaneously took place that would render the war the defining phenomenon in American history. Patriotism itself became sacralized to the point that it enjoyed coequal or even superior status to conventional denominational faiths." (Stout, 2006, xviii).

The interactions of religion and religious people with political institutions has been a persistent theme across time and space, in part because organizing and deploying military forces require significant bureaucratic and organizational power. Even as secular states have developed and consolidated military power, people of faith continue to populate their military forces and, as we have seen, religious norms continue to inform many of the discussions related to ethics and warfare. Military organizations interact with religious people and institutions by employing military chaplains and allowing for religious accommodations; providing religiously influenced ethical training to their troops; and employing religious rhetoric in support of war efforts. One of the key books introducing these themes is Anne Loveland's *American Evangelicals and the U.S. Military*. Loveland argues that between the end of the Second World War and the early 1990s, conservative evangelicals in the United States, through concerted and coordinated effort, gained significant influence within the military and political institutions. Like others, Loveland concludes that the Vietnam War marked a major turning point in the trajectory of the relationship between military and religious institutions—with the influence of mainline Protestantism on the decline and evangelical culture on the rise. Evangelical Christians influenced the military by providing an increasing number of military chaplains, forming parachurch organizations that provided alternative religious communities to the official chaplain program, focusing on the military as a mission field, and supporting military interventions abroad (Loveland, 1996).

These institutional connections did not disappear after the 1990s. Writing in the aftermath of 9/11, Kelly Denton-Borhaug, a theologian, has written an incisive critique of some of these institutional-cultural connections in her 2010 work, *U.S. War-Culture, Sacrifice and Salvation*. Denton-Borhaug argues that sacrificial rhetoric, rooted in evangelical Protestant culture, has furthered militarism (Denton-Borhaug uses the term "war-culture") in the United States, distorting both individual and institutional attitudes to the point that war, and its effect on American culture and discourse, is regarded as normal, rather than anomalous (2010). And writing against the narrative of secularization in the twentieth century, Tom Lawson and Stephen Parker have offered an edited collection that explores the relationship between the Church of England and war. Overall, this collection demonstrates the complicated relationship between an established church, a secular state, and a culturally religious but not particularly devout population (Parker and Lawson, 2012). All of these works suggest that the relationship between secular states and religious institutions is more complicated than popularly imagined.

The American war in Vietnam provides a good case study for examining many (though not all) of the aforementioned topics in some more depth, and in relation to one another. The war can be understood in a multitude of ways: as a proxy war embedded within the framework of the broader Cold War; as a war for national liberation or decolonization; as an insurgency and counterinsurgency contest; or as a civil war. Though none of the motivations for the Vietnam War were explicitly religious in the sense of it being a crusade or holy war, there were religious undertones from the beginning of the conflict. The Cold War itself was frequently understood in ideological terms that relied on civil-religious and Judeo-Christian language and imagery—particularly in the United States, which saw itself as a moral bulwark against atheistic communism (Kirby, 2003; Inboden, 2008). Furthermore, the nation of Vietnam was fractured on religious lines, thanks to the legacies of indigenous religious traditions, Buddhist influences from China (in the North) and India (in the South), Catholic missions and French colonial rule, and communism. Particularly in South Vietnam, the political and religious contest between Catholics and Buddhists was intense and often violent (Duiker, 1995).

For Americans—citizens, soldiers, and policymakers alike—religion played a central role in both rhetoric and policy toward Vietnam. American political support for Ngo Dinh Diem was based at least in part on racial and religious thinking, as Kennedy administration officials came to trust and believe in Diem's ability to lead Vietnam because of his devout Catholic faith (Jacobs, 2004). Even after Diem's assassination in 1963 and the introduction of a series of Buddhist leaders, U.S. policy makers often struggled to understand and manage the various religious groups vying for power in South Vietnam. As the war expanded to include significant American involvement, and as U.S. fortunes in Vietnam seemed less than optimistic, religious opposition to the war also increased. American religious communities—Protestant and Catholic, Christian and non-Christian, mainline and evangelical, white and black, liberal and conservative—vigorously debated the morality, justness, and righteousness of the Vietnam War, its conduct, and whether devout citizens should serve in it (Settje, 2011, Gill, 2011, Hall, 1990). They reached no consensus positions, though within popular imagination and to some extent in scholarship, the position of religious groups vis-à-vis the Vietnam War has come to be seen as primarily one of dissent and pacifism. Somewhat relaxed policies about conscientious objection also contributed to this perception. In reality, though, there were many religious organizations that zealously supported the war (or at least those serving in it) over its duration.

Even though the war sparked intense and often vitriolic debates within and among American religious groups, these homefront debates rarely affected the lived experience of service members actually stationed in Vietnam. Here, we see a stark difference between institutional religious concerns and the "foxhole" faith of military personnel. It is important, however, to recognize that the lived religious experience of the Vietnam War varied widely—some found war damaged their faith, others said war left their faith unchanged, and still others found it strengthened (Kindsvatter, 2003). To put it simply: there *were* atheists in foxholes—just as there were Christians, Jews, Muslims, Buddhists, and a host of others—and each of them practiced his or her religion in different ways. Some relied on institutional support from chaplains or lay leaders; others prayed or recited scripture on their own; some carried talismans and good-luck charms; still others abandoned or modified traditional religious practices to suit the peculiarities of combat. The diversity of soldiers' religious practice in the Vietnam War is primarily established through the extensive memoir and fiction literature coming out of the war (Johnson, 2001; Whitt, 2014).

Military chaplains, then, served at the intersection of many of these competing institutional expectations and personal experiences. As full members of both the military and of their religious organizations, military chaplains had to navigate a host of relationships that left them in liminal spaces between sacred and secular, civilian and military, officer and enlisted, and faith groups and denominations (Bergen, 2004). Some have argued that the military chaplain's liminal position induces severe role conflict because they cannot reconcile conflicting expectations without subordinating one role to another or without significant cognitive dissonance. They generally argue that chaplains forsake their religious values in favor of military ones (Zahn, 1969; Cox, 1973). With the passage of time and access to a variety of new sources, later scholars have argued that chaplains' roles were more contingent and variable than these earlier studies suggested (Abercrombie, 1977; Whitt, 2014).

The aftermath of the Vietnam War on American military and religious institutions has been significant. The Vietnam War did not *cause* a rift between conservative and liberal factions of American religious groups, but it did exacerbate and accelerate an already-existing divide that was made plain in the so-called culture wars of the 1980s. Institutionally, the military chaplaincy faced critiques from both the political and religious left and right in the years after Vietnam. Liberal critics of the chaplaincy declared it unconstitutional on First Amendment establishment grounds and decried the increasing conservatism of chaplains who filled its ranks. Though its constitutionality was eventually upheld, defending itself against this charge took a toll (Drazin and Currey, 1995). The divide between conservative and liberal religious institutions over the war also exacerbated trends within the chaplaincy as it became both more demographically diverse and more theologically and politically conservative and homogenous in the decades following the war. These trends had significant effects on the institution's vision of its own mission and chaplains' First Amendment protections; its response to the implementation and eventual repeal of the "Don't Ask, Don't Tell" policy about the service of gay military members; and its response to ongoing questions about the chaplaincy's relevance in a country that is becoming less religious and more diverse over time (Whitt, 2014; Loveland, 2014).

The number of works that explore the relationship between war, religion, and ethics has grown considerably in the past decade, but there is still ample room for rigorous and innovative scholarship, particularly regarding the late twentieth century, and in both Western and non-Western contexts. This is also a field in which scholars have much to say to practitioners and policy makers within military and religious organizations.

For Americans, the 9/11 terrorist attacks on the United States prompted interest in violent Islamic jihad in particular, and in violent religious extremism more generally. Extended American involvement in Iraq highlighted the significance of understanding, for example, the history of conflict between Sunni and Shi'a Muslims in the region, or the complications of Western countries using "crusading" language to discuss their strategic aims. At the same time, concerns within the American military about resiliency, post-traumatic stress (disorder), moral injury, and an alarming suicide rate promoted the services to look seriously at comprehensive fitness programs. These included a "spiritual" component, even as demographic surveys suggested the force was becoming less traditionally religious (Mahedy, 1986; Tick, 2005; Hauerwas et al., 2014). The latter trend has also renewed discussions about the role of chaplains, the necessity and extent of religious accommodations, and balancing First Amendment concerns about free exercise and establishment within the military (Hansen, 2012). All of these policy-relevant discussions offer historians the opportunity to contextualize the debate by examining recent issues regarding the U.S. military.

On a global stage, the field remains quite open for scholars to tackle many of the questions outlined earlier, particularly in contemporary, non-Western contexts. There is ample room, for example, for the study of the religious lives of ordinary military personnel—how have military personnel prayed, thought about fate and luck, worshipped, and interpreted combat through a religious lens—or not? To what extent is the lived religion of combat universal, and to what extent is it tied to particular beliefs in particular times and places? Another path would be the study of institutional relationships between governments and religious organizations— for example, how have secular states made allowances for diverse religious practices while maintaining the uniformity and cohesion thought to be required in a military setting? Some models for this type of work can be found in the edited collection *Religion in the Military Worldwide*, which includes chapters on Japan, India, Iran, Israel, and Turkey in addition to analyses of several Western militaries (Hassner, 2013). Alternately, historians could offer detailed rhetorical and cultural analysis of how religious language and imagery are used in justifying and motivating war, examining, for example, religious language and imagery used in recruiting materials or in public statements about a group's strategy.

One key challenge for historians wishing to write for audiences in both military and religious history is that both fields require a specialized lexicon, which can be daunting to master and often requires translating from one field to another—a scholar of religion may not be familiar with military organizations or operations, and military historians may likewise be unfamiliar with theological ideas or the traditions and texts of world religions. The recent proliferation of academic studies that combine the two themes should help break some of these barriers, and the possibilities for fruitful study and innovative scholarship are sufficient to make the effort worthwhile.

Bibliography

Abercrombie, Clarence L. *The Military Chaplain*. Beverly Hills: SAGE, 1977.

Armstrong, Karen. *Fields of Blood: Religion and the History of Violence*. New York: Knopf, 2014.

Avalos, Hector. *Fighting Words: The Origins of Religious Violence*. Amherst, NY: Prometheus Books, 2005.

Bergen, Doris L. *The Sword of the Lord: Military Chaplains From the First to the Twenty-First Century*. Notre Dame, IN: University of Notre Dame Press, 2004.

———. *Twisted Cross: The German Christian Movement in the Third Reich*. Chapel Hill: University of North Carolina Press, 1996.

Bonner, Michael David. *Jihad in Islamic History: Doctrines and Practice*. Princeton: Princeton University Press, 2006.

Brekke, Torkel. *The Ethics of War in Asian Civilizations: A Comparative Perspective*. London: Routledge, 2006.

Cady, Duane. *From Warism to Pacifism. A Moral Continuum*. 2nd ed. Philadelphia, PA: Temple University Press, 2010.

Cavanaugh, William T. *The Myth of Religious Violence: Secular Ideology and the Roots of Modern Conflict*. Oxford: Oxford University Press, 2009.

Corey, David D., and J. Daryl Charles. *The Just War Tradition: An Introduction*. Wilmington, DE: ISI Books, 2012.

Cox, Harvey. *Military Chaplains: From Religious Military to a Military Religion*. New York: American Report Press, 1973.

Dawkins, Richard. *The God Delusion*. Boston: Houghton Mifflin Harcourt, 2006.

Denton-Borhaug, Kelly. *U.S. War-Culture, Sacrifice and Salvation*. London: Equinox, 2010.

Drazin, Israel, and Cecil B. Currey. *For God and Country: The History of a Constitutional Challenge to the Army Chaplaincy*. Hoboken, NJ: KTAV, 1995.

Duiker, William J. *Sacred War: Nationalism and Revolution in a Divided Vietnam*. New York: McGraw-Hill, 1995.

Dyck, Harvey L., and Peter Brock. *The Pacifist Impulse in Historical Perspective*. Toronto: University of Toronto Press, 1996.

Ebel, Jonathan H. *Faith in the Fight: Religion and the American Soldier in the Great War*. Princeton, NJ: Princeton University Press, 2010.

Eller, Jack David. *Cruel Creeds, Virtuous Violence: Religious Violence Across Culture and History*. Amherst, NY: Prometheus Books, 2010.

Fine, Jonathan. *Political Violence in Judaism, Christianity and Islam: From Holy War to Modern Terror*. Lanham, MD: Rowman & Littlefield, 2015.

Firestone, Reuven. *Holy War in Judaism: The Fall and Rise of a Controversial Idea*. New York: Oxford University Press, 2012.

Gier, Nicholas F. *The Virtue of Nonviolence: From Gautama to Gandhi*. Albany: State University of New York Press, 2003.

Gill, Jill K. *Embattled Ecumenism: The National Council of Churches, the Vietnam War, and the Trials of the Protestant Left*. DeKalb: NIU Press, 2011.

Habeck, Mary R. *Knowing the Enemy: Jihadist Ideology and the War on Terror*. New Haven: Yale University Press, 2006.

Hall, Mitchell K. *Because of Their Faith: CALCAV and Religious Opposition to the Vietnam War*. New York: Columbia University Press, 1990.

Hansen, Kim Philip. *Military Chaplains and Religious Diversity*. New York: Palgrave Macmillan, 2012.

Hartigan, Richard Shelly. "War and Its Normative Justification: An Example and Some Reflections," *The Review of Politics* 36, no. 4 (1974): 492–503.

Hashmi, Sohail H. *Just Wars, Holy Wars, and Jihads: Christian, Jewish, and Muslim Encounters and Exchanges*. New York: Oxford University Press, 2012.

Hassner, Ron E. *Religion in the Military Worldwide*. New York: Cambridge University Press, 2013.

Hauerwas, Stanley, Jonathan Shay, and Robert Emmet Meagher. *Killing From the Inside Out: Moral Injury and Just War*. Cascade Books, 2014.

Hensel, Howard M. *The Prism of Just War: Asian and Western Perspectives on the Legitimate Use of Military Force*. Burlington, VT: Ashgate, 2010.

Herbert, T. Walter. *Faith-Based War: From 9/11 to Catastrophic Success in Iraq*. London: Equinox, 2009.

Hitchens, Christopher. *God Is Not Great: How Religion Poisons Everything*. New York: Twelve, 2007.

Inboden, William. *Religion and American Foreign Policy, 1945–1960: The Soul of Containment*. Cambridge, UK: Cambridge University Press, 2008.

Jacobs, Seth. *America's Miracle Man in Vietnam: Ngo Dinh Diem, Religion, Race, and U.S. Intervention in Southeast Asia, 1950–1957*. Durham: Duke University Press, 2004.

Jerryson, Michael K., and Mark Juergensmeyer, eds. *Buddhist Warfare*. Oxford: Oxford University Press, 2010.

Johnson, James D. *Combat Chaplain: A Thirty-Year Vietnam Battle*. Denton: University of North Texas Press, 2001.

Johnson, James Turner. *Ideology, Reason, and the Limitation of War*. Princeton, NJ: Princeton University Press, 1975.

———. *Just War Tradition and the Restraint of War*. Princeton, NJ: Princeton University Press, 1981.

Juergensmeyer, Mark. *Terror in the Mind of God: The Global Rise of Religious Violence*. Berkeley: University of California Press, 2000.

Kelsay, John. *Arguing the Just War in Islam*. Cambridge, MA: Harvard University Press, 2007.

———. *Islam and War: A Study in Comparative Ethics*. Louisville, KY: Westminster/John Knox Press, 1993.

Kindsvatter, Peter S. *American Soldiers: Ground Combat in the World Wars, Korea, and Vietnam*. Lawrence: University Press of Kansas, 2003.

Kirby, Dianne. *Religion and the Cold War*. Houndmills, UK: Palgrave, 2003.

Kundnani, Arun. *The Muslims Are Coming!: Islamophobia, Extremism, and the Domestic War on Terror*. New York: Verso, 2014.

Lifton, Robert Jay. *Destroying the World to Save It: Aum Shinrikyō, Apocalyptic Violence, and the New Global Terrorism*. New York: Henry Holt, 1999.

Loveland, Anne C. *American Evangelicals and the U.S. Military, 1942–1993*. Baton Rouge: Louisiana State University Press, 1996.

———. *Change and Conflict in the U.S. Army Chaplain Corps Since 1945*. Knoxville: University of Tennessee Press, 2014.

Mackinlay, John. *The Insurgent Archipelago: From Mao to Bin Laden*. New York: Columbia University Press, 2010.

Mahedy, William P. *Out of the Night: The Spiritual Journey of Vietnam Vets*. New York: Ballantine Books, 1986.

Miller, Randall M., Harry S. Stout, and Charles Reagan Wilson. *Religion and the American Civil War*. New York: Oxford University Press, 1998.

Niditch, Susan. *War in the Hebrew Bible: A Study in the Ethics of Violence*. New York: Oxford University Press, 1993.

Noll, Mark A. *The Civil War as a Theological Crisis*. Chapel Hill: University of North Carolina Press, 2006.

Parker, Stephen, and Tom Lawson. *God and War: The Church of England and Armed Conflict in the Twentieth Century*. Burlington, VT: Ashgate, 2012.

Popovski, Vesselin, Gregory M. Reichberg, and Nicholas Turner, eds. *World Religions and Norms of War*. Tokyo, Japan: United Nations University Press, 2009.

Reichberg, Gregory M. *Thomas Aquinas on War and Peace*. New York: Cambridge University Press, 2016.

Reichberg, Gregory M., and Henrik Syse, eds. *Religion, War, and Ethics: A Sourcebook of Textual Traditions*. New York: Cambridge University Press, 2014.

Riley-Smith, Jonathan Simon Christopher. *The First Crusade and the Idea of Crusading*. Philadelphia: University of Pennsylvania Press, 2009.

Rossi, Lauren Faulkner. *Wehrmacht Priests: Catholicism and the Nazi War of Annihilation*. Cambridge, MA: Harvard University Press, 2015.

Roy, Kaushik. *Hinduism and the Ethics of Warfare in South Asia: From Antiquity to the Present*. New York: Cambridge University Press, 2012.

Selengut, Charles. *Sacred Fury: Understanding Religious Violence*. Walnut Creek, CA: Altamira Press, 2003.

Settje, David E. *Faith and War: How Christians Debated the Cold and Vietnam Wars*. New York: New York University Press, 2011.

Shaw, William H. *Utilitarianism and the Ethics of War*. New York: Routledge, 2016.

Snape, M. F. *God and Uncle Sam: Religion and America's Armed Forces in World War II*. Woodbridge, UK: Boydell Press, 2015.

Stern, Jessica. *Terror in the Name of God: Why Religious Militants Kill*. New York: Ecco, 2003.

Stout, Harry S. *Upon the Altar of the Nation: A Moral History of the American Civil War*. New York: Viking, 2006.

Tick, Edward. *War and the Soul: Healing Our Nation's Veterans From Post-Traumatic Stress Disorder*. Wheaton, IL: Quest Books, 2005.

Tyerman, Christopher. *God's War: A New History of the Crusades*. Cambridge, MA: Belknap Press of Harvard University Press, 2006.

Walzer, Michael. *Just and Unjust Wars: A Moral Argument With Historical Illustrations*. New York: Basic Books, 1977.

Whitt, Jacqueline E. *Bringing God to Men: American Military Chaplains and the Vietnam War*. Chapel Hill: University of North Carolina Press, 2014.

Wiktorowicz, Quintan. *Radical Islam Rising: Muslim Extremism in the West*. Lanham, MD: Rowman & Littlefield, 2005.

Woodworth, Steven E. *While God Is Marching On: The Religious World of Civil War Soldiers*. Lawrence: University Press of Kansas, 2001.

Yoder, John Howard. *Nevertheless: A Meditation on the Varieties and Shortcomings of Religious Pacifism*. Scottdale, PA: Herald Press, 1971.

Zahn, Gordon C. *The Military Chaplaincy: A Study of Role Tension in the Royal Air Force*. Toronto: University of Toronto Press, 1969.

23 Race and Ethnicity

David J. Ulbrich and Alexander M. Bielakowski

Race and ethnicity have intertwined but not identical meanings. Race designates groups of people according to their biological characteristics, such as skin pigment, facial features, or skeletal frame. These differences can carry with them culturally based prejudices—racism—that assume physical differences signal inferior or superior intellectual, emotional, or other traits. Ethnicity refers to culturally determined identities attributed to population groups, such as custom, ancestry, region, history, religion, cuisine, physical appearance, language, or other characteristics. The belief in the inherent superiority of one ethnicity relative to other groups is called ethnocentrism. Like racism, ethnocentrism creates artificial hierarchies based on differences among groups. Both play on xenophobia of the "others" to strengthen identity and cohesion in the "self" group. The "self" group also denigrates the "others" as barbarians or subhumans (Brace, 2005; Eliav-Feldon et al., 2009).

Although racism and ethnocentrism arouse vitriolic hatreds during conflicts, much of the "traditional" historiography of warfare dismissed these factors altogether or treated them as prosaic and normative. In these interpretations, race and ethnicity rarely affected the decisions of generals, the outcomes of battles, or the planning of campaigns. Indeed, this focus on elites dominated most fields in history, including military history, for several centuries. Then, in the United States, revisionist histories started appearing in the 1960s amid the social upheavals during the Vietnam War and movements for civil rights and women's rights. Elsewhere in the world, long-term effects of the Second World War, such as decolonization, also heralded changes in those respective historiographies. In both cases, scholars wanted to add groups of society to their new histories. To do so, they sought answers to key questions about hierarchy, power, and stereotyping that the traditional histories had possessed no language to ask, let alone answer.

Studies of race and ethnicity started to carve niches in the historiography of warfare during the 1980s. Since then, other historical and interdisciplinary methodologies have steadily grown in number and influence in the field of military history. This chapter traces this process as interpretations of race and ethnicity entered the field as part of a larger "war and society" historiography.

Treatments of European conquest of the western hemisphere after 1492 exemplify how recent histories have challenged the more traditional ones. Until the late twentieth century, this domination interpretations pointed to some combination of greed, opportunity, territorial expansion, glory, and religion as motivations for Spain's conquests in Central and South America (Morison, 1942). When mentioned, race as a factor was treated as matter of fact. However, as the newer historical interpretations have shown, conceptualization of race dramatically affected Spanish perceptions of indigenous peoples, ranging from primitive tribes in the Caribbean to the established Aztec and Incan Empires in Central and South America. Antonio Feros (2017) observes two phenomena growing out of these contacts. First, the violent confrontations highlighted the differences between Spanish and indigenous peoples in race and such traits associated with ethnicity as religion, dress, diet, language, and customs (see also Lee, 2011). Second, Feros argues that Spain's population legitimized their identities of "Spanishness" and "whiteness" by forcing the supposedly inferior indigenous peoples into slavery and serfdom (see also Pagden, 2009).

By the early 1600s, several other European nations carved out empires in the western hemisphere. Of these, the British conflicts with the North American Indians offer substantive historiographies relating race and ethnicity to warfare. Peter Silver (1997) and Jill Lepore (1998) reveal how differences in race, religions, ways of life, and brutal tactics among Indians helped unify the British colonists. Those differences took the form of hierarchies, with the Indians being relegated to inferior barbarians. Lepore uses King Philip's War during the 1670s as a tragic example of tensions between Indians and colonists erupting into vicious fighting. She adds an analytical twist, however, by arguing that the colonists needed to justify their brutal actions in moral terms, lest they be labeled as savages like the Indians.

Taking a longer, more holistic approach, Wayne Lee (2011) synthesizes warfare on the British Isles and in the North American colonies. As seen in the title of his book, the terms "barbarians and brothers" refer to white and Indians as belligerents. If one of these belligerents saw a foe as a "brother," then restraints on war were more likely to be observed. Conversely if a belligerent saw a foe as "barbarian," then restraints would be less likely observed, with the result of greater levels of violence. The dynamic played out in Dunmore's War in 1773–1774, as Glenn Williams (2017) demonstrates by tracing the escalation of hostilities between British colonists and the Shawnee tribe near the Ohio River. He asserts the racial animosities among white colonists and Indians alike made the fighting that much more bitter. Williams also gives agency to the Indians, rather than treating them merely as victims of colonial expansion.

British colonial policies regarding the Indians sometimes drove wedges between the white colonists, as in the cases of Bacon's Rebellion in 1676 and the Conestoga Massacre in 1763. James Rice (2013) relates how Nathaniel Bacon led a revolt in rural Virginia because Bacon and his followers—many of whom were escaped or former indentured servants—wanted protection from Indian attacks and more economic opportunities. After driving Virginia's government from the capital, Bacon's rebels turned to fighting a destructive war against Indians, which was justified in part by racial differences. Kevin Kenney (2009) examines another conflict on the Pennsylvanian frontier where settlers of Ulster-Irish ethnicity wanted land previously given to the Conestoga Indians, so they massacred this tribe living on the land. Racial differences with the Indians and ethnic differences between Ulster-Irish and British colonists elevated the violence levels.

When British colonists moved westward beyond the Appalachian Mountains to acquire inexpensive land later in the eighteenth century, Patrick Griffin (2007) reveals how race and citizenship helped bind white settlers together in opposition first to the Indians and later to the British Army. "White" and "citizen" became synonymous and exclusive constructions during the War for Independence. Robert G. Parkinson (2016) compares the Revolutionary rhetoric portraying the Indians as savages with that of African slaves as insurrectionists. For Parkinson, the American Revolutionaries saw the defeat of the Indians and suppression of slave insurrection as a "common cause" imbued by racism. Studies by Sylvia Frey (1992) and Gerald Horne (2014) have demonstrated that the white Revolutionaries had legitimate fears of African slave cooperation with the British Army because the slaves made up about one-fifth of the population of the thirteen colonies. Alan Taylor (2013) finds repeated attempts by British invaders to incite slave revolts during the War of 1812. In all, these historians show that warfare helped sustain racially charged notions of nationalism among the whites in the Revolutionary and Early Republic periods. These feelings mutually reinforced the white American opposition to Indians and African slaves.

Throughout the nineteenth century, the ongoing race- and ethnic-based struggles helped propel American expansion across the continent as part of "manifest destiny." Mark E. Neely Jr. (2007) uses the Mexican War and the Civil War to illustrate this process. White Americans believed in their inherent superiority to the Mexicans, so the United States attacked Mexico in 1846 with the goal of absorbing more territory into the white American republic. Neely then turns to the American Civil War to demonstrate that racism permeated the conflict. Race-based slavery in the South and race-based free labor in the North were the core causes of the conflict. Once the fighting started, African Americans in the Union Army endured heinous treatment on and off battlefields. An anthology edited by Gregory Urwin (2004) examines the Battles of Olustee, Fort Pillow, and Poison Spring, where white Confederates believed their African American enemies were savages and brutally murdered those who surrendered or were wounded.

Several decades before the works by Urwin and Neely, James McPherson (1965) broke historiographical ground by publishing his classic study of African Americans during the Civil War. He looked at all aspects of their experiences, including slave and free, rural and urban, and on the home and battle fronts. In the North, for example, African Americans encountered racial discrimination despite their desire to serve the Union and fight for freedom. In so doing, McPherson gives agency to blacks, albeit severely limited by institutional, legal, and personal racisms.

More recently, Joseph Glatthaar (2000) has focused on the U.S. Colored Troops. He notes that white Union officers commanding the USCT units included everyone from avowed racists to rabid abolitionists, which meant that their African American experiences varied depending on their commanders' racial beliefs. Turning to President Abraham Lincoln's role in opening military service to blacks, Jason David Smith (2013) used the term "military emancipation project" to define the reasons behind his support of the USCT. Even with support from Lincoln and northern abolitionists, African Americans in uniform faced racist assumptions that they were inferior to white soldiers. One example of institutionalized racism occurred in the medical field. According to Margaret Humphreys (2008), Union Army medical doctors conceded that blacks were no different from whites in terms of physiology, yet argued they found no empirical evidence that raised the African Americans to intellectual or moral equality with whites.

Irish and German Americans, many of whom recently immigrated to the United States, served in the Civil War by the hundreds of thousands. Susannah Bruce (2006) details the motivations and experiences of one subset: Catholic Irish Americans in the Union Army. She finds that they initially joined the Union cause in part because of loyalties to their new home nation of the United States. Other Irish Americans fighting for the Union also identified with slaves because they also faced discrimination at the hands of white Protestants. Switching to the Confederate States of America, David T. Gleeson (2013) examines the Irish Americans who lived in the slave states and joined the Confederate military. He believes their motivations for service were localized, spurred by a collective sense of martial valor, and united in the Democratic Party against Protestant nativism.

Most German Americans fought for the northern free states or the border slave states during the Civil War. One vehicle for understanding this ethnic group's involvement is biography. Mary Bobbitt Townsend (2010) writes about Major General Peter J. Osterhaus, one of the most successful German American generals to serve in the Union Army. He came to the United States in the 1850s with military education and with experience on the losing side of the Revolution of 1848 in Prussia. Stephen D. Engle's biography (1993) of Major General Franz Sigel yields much the same analysis of another, albeit not always successful, Prussian officer who also fled Prussia. Both Osterhaus and Sigel brought nationalistic fervor to the United States that helped draw them and many other German immigrants to the Union cause.

The contribution of Indians has received increasing attention in recent years. Clarissa Confer (2006) writes about the 1st and 2nd Cherokee Mounted Rifles that served in the Confederate States Army. The Cherokee tended to favor the Confederate cause, especially after having suffered at the hands of the U.S. government during the Trail of Tears in the 1830s. Switching sides, M. Jane Johansson (2016) examines the role played by the First Indian Home Guard, which included Muscogee Creeks, Seminoles, and African American soldiers. Using the commentary by the white Union Army officer commanding the Home Guard, Johansson brings this mixed-race unit to life as it fought in Arkansas and the Indian Territory.

Essays in Susannah Ural's edited anthology (2010) provide brief overviews of African Americans, Native Americans, Irish Americans, and German Americans. The contributors map the many overlapping yet sometimes contradictory reasons of race or ethnicity that affected why groups chose sides. The essays also reveal how and why the Civil War experiences may have solidified the identities of self and other among minorities.

After the Civil War ended, the U.S. Army's operations shifted to the American West, where racism and ethnocentrism were ubiquitous on the frontier. While perceived racial differences remained entrenched in the U.S. Army, Kevin Adams (2009) finds evidence that ethnic diversity among white Europeans played a less divisive role than earlier in American history. The Frontier Army of the late 1800s became an ethnic melting pot. Adams does, however, note that class consciousness separating the enlisted and officer's ranks proved to be the most disruptive force in the Army.

One case study of race on the frontier focused on African Americans serving in four regiments during the Indian Wars. William and Shirley Leckie (1967) were among the first historians to discuss the hard-won respect for the blacks in the Army, known as the "Buffalo Soldiers." Nevertheless, according to the Leckies, racism in the Army and among its white officers combined with lack of appreciation for their service to create stereotypes about African Americans almost impossible to reverse. Writing more than three decades later, William Dobak and Thomas Phillips (2001) made two points that clarify, if not revise, the Leckies' book. Dobak and Phillips argue that the U.S. Army's black and white regiments received similar treatment, and that the Buffalo Soldiers did not consistently outperform nor underperform when compared to white soldiers.

Biographies of Henry Flipper and Charles Young offer insights into the racial discrimination faced by the first African Americans to enter the U.S. Military Academy and earn commissions as Army officers in 1877 and 1888. Both overcame being born into slavery to gain admittance into the Academy. According to Charles M. Robinson III (2008), Flipper remains best known for his court-martial conviction for "conduct unbecoming" of an officer in 1881. Although this sentence is usually cited as evidence of racism in the Army, Robinson instead argues that Flipper was indeed guilty of making the mistake that ultimately led to his dismissal from the Army. Brian Shellum's two volumes (2006 and 2010) chronicle how Young earned grudging respect at West Point and then advanced through the ranks to become the first black officer to command a Regular Army regiment and to hold the rank of colonel. However, Young encountered racism, such as President Woodrow Wilson's policies that blocked blacks from higher promotions.

In recent years, historians have compared the influence of the racism in the American West with other regions and conflicts. Edward B. Westermann (2016) analyzes the similarities and differences between American manifest destiny on the North American frontier and Adolf Hitler's quest for *Lebensraum* in Eastern Europe. Racist dogmas spurred each conquest that included many atrocities. Westermann argues that, whereas brutal acts by American soldiers occurred on relatively limited scales, the Nazi program undertook the systematic extermination of Jews and other racially inferior groups on a much larger scope in Eastern Europe.

Using a similar comparative approach, James O. Gump (1994) explores the American and British reactions to unexpected battlefield victories by nonwhite Lakota and Cheyenne warriors at Little Bighorn in 1876 and Zulu warriors at Isandhlwana in 1879. After regrouping and defeating these indigenous peoples, the Americans shunted the Indian tribes onto reservations, and the British tried to integrate the Zulus into their empire.

Gump is also one of many authors exploring how and why the Europeans conquered so much of the nonwhite, nonindustrialized world in what became known as the "new imperialism." During the late nineteenth century, the Europeans found an intellection foundation for their conquests in Social Darwinism. This ideology twisted the theory of natural selection espoused by Charles Darwin in 1859. Theorists like Herbert Spencer and William Graham Sumner incorrectly extended Darwin's ideas about biological science to human relations. This resulting Social Darwinism argued that racial and ethnic distinctions helped drive struggles for dominance between white people and other supposedly inferior races and ethnicities (Frederickson, 2002; Brace, 2005).

Social Darwinism justified Europe's new imperialism, such as the British subjugation of the Indian subcontinent in the nineteenth century. The British belief in the Indian people as racially inferior came to a head in the unsuccessful Sepoy Mutiny. The term "sepoy" refers to those Indian soldiers hired by the British East India Company. Since the sepoys chafed at promotion limits and cultural indignities foisted on them by their white employers, many sepoy units mutinied in 1857 only to be crushed by the regular British Army and other Indian units loyal to the British. Although the conventional (i.e., British) interpretation, like Saul David's (2009), tends to marginalize the effects of British racism, others, like Kaushik Roy (2006), have reinserted race as a dynamic factor into military historiography of the Sepoy Mutiny and its aftermath.

According to Mrinalini Sinha (1995), British prejudices in India took other forms, such as categorizing certain ethnic groups as "martial races." Prized for their fighting skills and courage, the Punjabis and Sikhs were deemed to be ideal warriors. Meanwhile, the Bengalis found themselves in the unappealing ethnic category and portrayed as effeminate shopkeepers, beggars, and civil servants within the British racial hierarchy in India.

The United States proved to be no less imperialist or racist than the Europeans. White Americans believed that they could export civilization, Protestant Christianity, and democracy to the nonwhite world during the late nineteenth and early twentieth centuries. After winning the Spanish-American War in 1898 and acquiring Spain's possessions in the western Pacific Ocean, Americans discovered a new frontier to subdue. Historians including Leon Wolff (1961), Michael Hunt (1987), and David J. Silbey (2007) have examined how American forces spent the years 1899–1902 securing control of the Philippines by waging a racially charged war against Filipino insurgents.

The United States also established its hegemony in the Caribbean between 1898 and 1935. Racial and ethnic, together with religious and economic, components affected every American intervention to stabilize Cuba, Haiti, Panama, Honduras, Mexico, Nicaragua, and the Dominican Republic (Colby, 2013). Of these, Haiti emerged as the most exotic place, with its cultural intersections among black, mulatto, and white races, as well as Catholic and Voodoo religions. Mary Renda (2001) explains that the U.S. Marines adopted a paternalist approach in governing: they played the roles of benevolent but stern fathers, and relegated the Haitian to roles as inferior children needing direction from the American father figures.

While the United States and the Europeans tried to maintain control of their far-flung empires, the First World War (1914–1918) started the slow decline of Europe. The four-year conflict caused incredibly high losses of life. Because the British and French suffered several million casualties, their governments turned to their global empires to replace the losses.

In addition to other colonial possessions, the British looked to India as a source of one million soldiers to fight in the First World War. According to Shrabani Basu (2016), Indians of many castes, regions, languages, and religions served under British command on the Western Front and in Africa and Asia. Some soldiers sought opportunities to fight alongside the British, while others from the Indian political elites hoped that supporting the British cause would accrue favor with the imperial government and eventually yield greater autonomy or even self-rule for India. To the British, however, the nonwhite Indians could be thrown into battle because they were expendable.

For their part, the French brought some five hundred thousand *troupes indigènes* from French West Africa to fight on the Western Front. Until recent decades, historical interpretations have portrayed the white French society and its Army as "color-blind" because they treated nonwhite races with relative respect and equality. This egalitarian treatment extended not merely to the West Africans but also to the African American soldiers assigned to the French Army (Barbeau and Henri, 1974; Michel, 1982). However, revisionists Tyler Stovall (1998) and Richard Fogarty (2008) have proven the older interpretations to be more myth than reality. They discovered that the French leaders viewed their colonial forces like the British did—as replacement troops for a badly mauled French Army. Yet, for the West Africans and other French colonial troops, fighting against white Europeans irrevocably changed their visions of themselves. Their service aroused feelings of nationalism and sowed the seeds of decolonization decades later (Glenford Howe, 2002; Das, 2011).

Beginning in the 2000s, American military historians started producing nuanced studies of African Americans and other minorities during the First World War. Chad Williams (2010) shows how the 380,000 African American soldiers personified the black community's attempts to leverage military service as means to attaining more equality as American citizens (see also Barbeau and Henri, 1974). Immigrants of Italian and Irish descent also used military service to gain fuller acceptance in America's society and polity, as described in Richard Slotkin's book (2005) on the ethnic "melting post" in the 77th Division.

The U.S. Army formed two segregated infantry divisions to deploy to France during the First World War. The experiences of men in each unit, however, were altogether different. Frank Roberts (2004) argues that men in the 93rd Division proved their mettle in combat while serving under French command. Of the four regiments in this division, the 369th Infantry Regiment—known as the "Harlem Hellfighters"—has garnered the most historical attention. Coauthors Jeffrey Sammons and John Morrow (2014), as well as Slotkin (2005), have explored why the 369th came to symbolize the heroism of African Americans during the First World War.

Another division segregated by race—the 92nd—also fought in France. Negative historical treatments have criticized some units in this division because they collapsed during the Meuse-Argonne Campaign in late 1918. However, as Robert Ferrell (2011) contends in his revisionist book, the entire division does not deserve the label of failure because many units in the 92nd did perform well under fire. Despite service and sacrifices, veterans of both divisions returned home after the First World War to find that nothing regarding race relations had changed in the United States (see also Slotkin, 2005).

Among other minorities, the historiography of Latino Americans in the U.S. military is sparse in comparison to African Americans in part because, other than Puerto Ricans, Latino Americans never served in segregated units. José Ramírez (2009) is one of the few historians to focus entirely on Mexican Americans, many of whom volunteered to serve in the U.S. military. They acquitted themselves well on the battlefields of the First World War. Nevertheless, they returned home after the war ended and faced ongoing racist policies.

After a brief interlude of uneasy peace, another global conflict started during the 1930s. Most histories of the Second World War appearing between the late 1940s and early 1980s concentrated on battles, leaders, weapons, or strategy (e.g., see Churchill, 1948–1953; Guderian, 1952; Van Creveld, 1982). When mentioning race or ethnicity, these traditional studies tended to do so only in passing without any reflective analysis. However, vitriolic racism and ethnocentrism saturated the Eastern European Front, where Germany clashed with the Soviet Union, and the Asia-Pacific region, where the United States and its allies fought Japan.

War crimes became commonplace during long, bloody campaigns between Germany and the Soviet Union between 1941 and 1945. Histories of the Eastern Front appearing between the 1940s and 1970s tended to posit blame for German racial atrocities on *Waffen-SS* units manned by Nazi Party members. Moreover, some older studies and popular histories have venerated the *Heer* (German Army) for its operational effectiveness while overlooking its racism or collusion in the Holocaust. Books by Heinz Guderian (1952) and John Laffin (1965) have crafted collective hagiographies of the German soldier that appeal to large numbers of readers.

In the 1980s, however, historians began challenging the presumed aloofness of the *Heer* from the Holocaust and other war crimes. Among the most vocal revisionists was Omer Bartov (1985), who argued that the *Heer* engaged in more racist and ethnocentric barbarism as the war dragged on. This shift occurred in part because a generation of younger soldiers had been indoctrinated with Nazism's racist beliefs before they joined the military. The German leadership likewise wanted to deflect attention from the reality of Soviet victories by encouraging their soldiers to commit atrocities.

Stephen Fritz (1995) presents a convincing counterargument to Laffin's sympathetic portrayal of German soldiers. Fritz found that, despite considering themselves to be moral people, many Germans did commit war crimes against Jews and Slavs. The German historian Wolfram Wette (2002) plumbed the depths of prejudices that made German soldiers at all levels and ranks complicit in the Holocaust and other war crimes. Hitler played on prejudices already held as truth in German culture. Geoffrey Megargee (2006) concentrates on the German invasion of the Soviet Union in 1941. He discovered that, even this early in the war, German soldiers routinely murdered Jews and Slavs on the Eastern Front.

During the Second World War, the racially charged ferocity on the Eastern Front was matched by fighting in the Pacific theater. John Dower (1987) broke ground in his comparative study of racism in the United States and Japan. He argues that the Japanese exhibited severely prejudiced assumptions about their enemies. Americans also embraced racial stereotypes that dehumanized the Japanese as the "Yellow Peril." The mutual loathing and paranoia intensified the violence, whether on home fronts (Tchen and Yates, 2014) or battlefields. Craig Cameron (1994) and Peter Schrijvers (2002) concur with Dower that killing surrendered Japanese soldiers, firebombing Japanese cities, and even dropping the atomic bomb became easier to justify because the Japanese were dehumanized in American eyes.

For their part, the Japanese attitudes about Americans played out in their portrayal of the United States as an impure mixed-race nation. Indeed, the Japanese believed their homogenous culture and their *Yamato* race to be superior to all whites and other Asians, which, according to Dower (1987) and Yuki Tanaka (1996), legitimized Japanese mistreatment of enemies. Gerald Horne (2003) makes a unique study of how the Japanese used British imperialism's racist justifications to cast themselves as liberators of Asians under white British control. This symbolism, however, obscured the reality of Japan's merciless brand of imperialism.

Other historians have joined Dower and Horne in documenting Japanese abuse of Asian civilians, such as the rape of women and girls in Nanjing (formerly Nanking) sanctioned by their military commanders. Rape reinforced racism as an ideology, and racism reinforced rape as a weapon. The Japanese exploited millions of Koreans, Chinese, and Filipinos as sex slaves, forced laborers, and biological warfare experimental subjects because the Japanese racial hierarchy made them unworthy of humane treatment (Ienaga, 1979; Tanaka, 1996; Chang, 1997; Yoshimi, 2002; Barenblatt, 2005).

Apart from works on the Eastern Front and the Pacific war, a significant body of literature has analyzed race and ethnicity on the American home front and in the U.S. military establishment. One point of departure from traditional interpretations is Michael C. C. Adams' short overview (1993) of the United States. While conceding that the conflict achieved some admirable goals, he nevertheless systematically dispels myths of the "good war." Adams exposes the dark underside of the conflict. American racism and ethnocentrism may have decreased in some ways, like employment opportunities; but they mutated into different forms of discrimination, such as lower wages, promotion limitations, and short-term employment only for the conflict's duration (see also Ulbrich, 2011).

African Americans during the Second World War have received more attention than any other period in U.S. military history. In his official Army history, Ulysses Lee (1966) recorded problems and contributions of African Americans in segregated units. Neil Wynn (2010) offers a brief overview of the Second World War in which he avers that the wartime experiences of African Americans played significant parts in the longer struggle for civil rights.

Arguably the most recognizable African Americans in the Second World War were the "Tuskegee Airmen" of the U.S. Army Air Force. According to Stanley Sandler (1998), the training of black pilots started as an experiment because racial prejudices caused skepticism among white officers regarding black pilots' skill and intelligence levels. African American pilots of the 332nd Fighter Group slowly garnered respect from their white counterparts because of their skills as fliers and courage during combat. Sandler emphasizes that the 332nd's pilots felt constant pressure to meet higher standards than required of white pilots.

Histories of other African American units include books by Mary Pat Kelly (1995), Melton McLaurin (2009), and Gina DiNicolo (2014). The authors tell similar stories of blacks confronting racial discrimination in American society and its military and later facing danger on battlefields. This struggle came to be known as the "Double V"—victory over racism at home and overseas. Authors have written about black individuals attempting to advance through the ranks, such as biographies of Benjamin O. Davis Jr. by Marvin Fletcher (1989) and Daniel "Chappie" James Jr. by James McGovern (2002).

Despite the valor in combat by black units, historians have found that nine out of ten African Americans in the U.S. military found themselves relegated by discriminatory policies to menial support roles (Ulbrich, 2011). Menial did not equate to meaningless, however, as evinced by the appraisal of the "Red Ball Express" by David Colley (2000). This logistical effort saw six thousand trucks carrying 12,500 tons of supplies race across France to support the U.S. Army every day during late 1944. African American soldiers made up 75 percent of the truck drivers.

As females of color, African American women faced dual discriminations. The starting point for understanding the experiences of black women is Brenda L. Moore (1996). She traces the history of the 6888th Central Postal Battalion as the first all-female, all-black unit in the Women's Army Corps to deploy overseas. The 855 women volunteered because they felt service was their patriotic duty, desired some escape from their civilian lives, wanted to refute racist and sexist assumptions about them by most whites, or some combination of these. Yet, like black men, their wartime contributions failed to translate into advances in civil or women's rights during the immediate postwar years.

Other minorities endured discrimination during the Second World War. Nearly two million Latino Americans, Native Americans, and Jewish Americans served in the U.S. military. Maria Rivas-Rodriguez (2005), Jere Franco (1999), and Deborah Moore (2004) find that these respective groups had to reconcile that paradox that fighting in the war required them to fight for an oppressive government and society. Like African Americans, they wanted their loyal service to justify their calls for equal rights as citizens.

Of all the minorities in the United States, Japanese Americans suffered from the most blatant racism. The attack on Pearl Harbor in December 1941 intensified white Americans' bigoted attitudes regarding people of Japanese descent that were already present in prewar society. Daniel Rogers (1972) established the dominant interpretation of the Japanese American internments; and more recently Greg Robinson (2009) added a comparative study of the events and decisions that led to internments. Suspicions about Japanese American loyalties, according to Rogers and Robinson, prompted a willing President Franklin Roosevelt to issue Executive Order 9066 in February 1942. Although most of the 110,000 Japanese Americans were U.S. citizens, this order classified them as enemy aliens, deprived them of legal rights, and mandated that they be forced to evacuate from their homes on the West Coast to internment camps in very remote areas. Robinson also finds comparable Japanese internment policies in Canada and Latin America.

Despite unlawful and racist treatment, several thousand Japanese Americans volunteered and were conscripted into U.S. military service because they wanted to prove their loyalty to the United States. James McCaffrey (2013) has written an operational history of Japanese Americans during the Second World War, demonstrating that they faced constant harassment from white soldiers while continually defying racist assumptions regarding their patriotism and proving their mettle in tough fighting in Italy. Another book by Takaski Fujitani (2013) compares why otherwise degraded minorities—Japanese Americans and Koreans—served in the United States' and Japan's militaries respectively. In so doing, Fujitani blends racism and nationalism as ideological factors that affected wartime mobilizations.

After the Second World War ended, some attitudes about race changed in the United States. Histories of desegregation of the American armed forces include books by Morris McGregor (1985), Jon Taylor (2012), and Rawn James (2013). They explain how several progressive white leaders, including President Harry Truman and later General Matthew Ridgeway, played critical roles in desegregating the military. Truman's Executive Order 9981 abolished racial discrimination in 1948, but it took Ridgeway's leadership during the Korean War to integrate the Army in practice.

These achievements did not occur without setbacks, however. Older histories have blamed the poor combat performance of the still-segregated African American 24th Infantry Regiment in the Korean War on inherent racial inferiority of African American soldiers. However, as documented by coauthors William Bowers et al. (1997), the 24th performed poorly under fire because it was understrength and inadequately trained and equipped, not due to racially based shortcomings. In fact, most white American units also performed poorly in combat in 1950, so Bowers and his coauthors conclude that the 24th has been unfairly evaluated in older histories.

Beyond the Korean War, the Cold War proved to be fertile ground for studying race and ethnicity. Both intertwined with ideological differences pitting the United States with its allies against the Soviet Union with its allies. Meanwhile, the mostly nonwhite, non-European nations of the Third World found themselves being played as pawns in the global struggle between democracy and communism. Thomas Borstelmann (2003) has noted these dynamics by layering racism onto the history of the Cold War. He offers the provocative analog that race-based decolonization was the twentieth century's version of emancipation of race-based slavery. He also gives agency to Third World peoples by exploring how and why they could play the United States and Soviet Union against one another.

The conflicts in Vietnam and Kenya represent the varied historiographies assessing racism and ethnocentrism during the Cold War. First, racism permeated the American experience in the Vietnam War. James Westheider's book (1997) remains the best secondary source for understanding race relations in the U.S. military. He develops three typologies of racism: personal, perceived, and institutional. These allow for analysis of whether discrimination was intentional or random. Lastly, according to Westheider, the U.S. military tried to drive wedges between African Americans and Vietnamese so that these two groups could not make common cause as oppressed nonwhite groups.

Several years after Westheider's book appeared, Herman Graham III (2003) delved still deeper into psyches of African Americans fighting in Vietnam. Constructions of race and gender overlap in attempts by young black males to prove their masculinity, despite being an oppressed racial minority. The assassination of Martin Luther King Jr. in 1968 caused some African American servicemen to seek inspiration from the militant Black Power movement and define their manhood irrespective of white masculine ideals. However, according to Graham, this widening rift between white and black soldiers did not exist in actual combat, where survival required cooperation across racial lines.

Lawrence Allen Eldridge (2012) explores the attempts by the African American press to resolve the Vietnam War's contradictions. Blacks were being sent to Vietnam ostensibly to stop communist oppression and fight another race also considered inferior, yet they still suffered racism back in the United States. Ultimately, the African American press decided that Vietnam was detrimental to their race because too many blacks went to Vietnam and the conflict's cost deprived Johnson's Great Society of money.

An anthology edited by Douglas Bristol and Heather Marie Stur (2017) includes the most recent historiographical trends on race in the U.S. military. Several essays point to the Vietnam War as a watershed. The contributors depict the U.S. military as a social engineering laboratory where reforms in race relations sometimes evolved more quickly than in civilian society. However, the military could also be slower and more resistant to new ideas at other times.

The effects of racism and ethnocentrism during the Vietnam War went beyond American attitudes and actions. According to anthropologist Gerald C. Hickey (2002), the ethnic Vietnamese majority oppressed an ethnic minority group in the Central Highlands called the Montagnards. Hickey's book is part memoir and part history in that he draws on years of living with the Montagnards and experiencing the Vietnam War through their eyes, yet his analysis also benefits from hindsight. Hickey's perspective cannot help but show favoritism to his study's subjects.

Among many examples of violent decolonization during the Cold War, Kenya stands as a useful case study because this nation's struggle for independence defies conventional historical models. For many decades before the 1950s, native Kenyans served as cheap labor for the white minority landholders and British colonial government. Then in 1952, the Mau Mau Rebellion erupted when a fraction of the Kenyan population revolted against British imperial rule. Among the conventional interpretations, Robert Edgerton (1990) believes that the insurgent Mau Mau rebels fought for political freedom, economic opportunity, and social justice that could occur only if the British were expelled from Kenya. This argument is not unlike a Marxist interpretation, with its fixation on class struggle as the prime mover in history.

However, Wanyabari Maloba (1993) finds no such ideological motivation. Instead, he argues that the Mau Mau Rebellion was more religious in practice and traditional in goals—hardly the rhetoric of Marxist revolution. This is not to say that the rebellion was not vicious. The Mau Mau's insurgent fighters attacked white British people and even murdered their black countrymen if they supported imperial rule. The British Army could likewise be ruthless in its suppression of the insurgency. The Mau Mau rebels never capitalized on underlying problems of race-based divisions of wealth and labor (Baxter, 2012). Quite the contrary, the British defeated the Mau Mau rebels because they drove wedges between the majority of Kenyans and the small fraction of rebel insurgents. The British thus divided the Kenyans into "good" blacks and "bad" blacks, while maintaining their belief that "whiteness" made them superior. The isolated Mau Mau insurgents eventually found themselves with no popular support by 1960, when the rebellion ended (Maloba, 1993). These points undermine Marxist interpretations of the rebellion.

In closing, despite the growing bodies of literature, many historical topics are ripe for research on race and ethnicity vis-à-vis war and society. Military historians can draw analytical techniques from Loring Brace (2005) and George Frederickson (2002) to enrich their understanding of racial theories in modernity. For any inquiry into race or ethnicity in American military history, scholars should start with Bernard Nalty's survey (1986) and Alexander Bielakowski's edited reference work (2013). Both serve as springboards that could lead to innovative scholarship on the Second World War and the Korean War. Although histories of units, branches, and individuals exist, no single scholarly survey of race or gender exists for either conflict. One of the best analytical filters is found in typologies of racism in James Westheider's book (1997). His three typologies of racism—personal, perceived, and institutional—have stood the test of time. These could be extended to ethnocentrism and racism in European and global military histories. More histories, using Michael Hunt (1987) and Thomas Borstelmann (2003) as models, can explore the shifting influences of race and ethnicity on political, economic, strategic, or other more traditional areas of inquiry. Finally, as seen in James Gump (1994), Greg Robinson (2009), Takaski Fujitani (2013), and Edward Westermann (2016), the comparative studies of race and ethnicity in warfare can offer opportunities to open discourses about similarities and differences among Western and non-Western examples. Scholars need not limit themselves comparing race or ethnicity in different times or places, but rather they should also integrate different disciplines, like race and gender studies, into military history.

Bibliography

Adams, Kevin. *Class and Race in the Frontier Army: Military Life in the West, 1870–1890*. Norman: University of Oklahoma Press, 2009.

Adams, Michael C. C. *The Best War Ever: America and World War II*. Baltimore: Johns Hopkins University Press, 1993.

Barbeau, Arthur, and Florette Henri. *The Unknown Soldiers: African-American Troops in World War I*. Philadelphia: Temple University Press, 1974.

Barenblatt, Daniel. *A Plague Upon Humanity: The Hidden History of Japan's Biological Warfare Program*. New York: Harper Perennial, 2005.

Bartov, Omer. *The Eastern Front 1941–1945: German Troops and the Barbarisation of War*. London: St. Anthony's/Macmillan, 1985.

Basu, Shrabani. *For King and Another Country: Indian Soldiers on the Western Front, 1914–1918*. New Delhi: Bloomsbury India, 2016.

Baxter, Peter. *Mau Mau: Kenyan Emergency, 1952–1960*. London: Helion, 2012.

Bielakowski, Alexander M., ed. *Ethnic and Racial Minorities in the U.S. Military: An Encyclopedia*. Santa Barbara, CA: ABC-Clio, 2013.

Borstelmann, Thomas. *The Cold War and the Color Line: American Race Relation in the Global Arena*. Cambridge, MA: Harvard University Press, 2003.

Bowers, William T., William M. Hammond, and George L. MacGarrigle. *Black Soldier/White Army: The 24th Infantry Regiment in Korea*. Washington: U.S. Army Center of Military History, 1997.

Brace, C. Loring. *"Race" Is a Four-Lettered Word: The Genesis of a Concept*. New York: Oxford University Press, 2005.

Bristol, Douglas Walter Jr., and Heather Marie Stur, eds. *Integrating the US Military: Race, Gender, and Sexual Orientation Since World War II*. Baltimore: Johns Hopkins University Press, 2017.

Bruce, Susannah Ural. *The Harp and the Eagle: Irish-American Volunteers and the Union Army, 1861–1865*. New York: New York University Press, 2006.

Cameron, Craig M. *American Samurai: Myth, Imagination, and the Conduct of Battle in the First Marine Division, 1941–1951*. New York: Cambridge University Press, 1994.

Chang, Iris. *The Rape of Nanking: The Forgotten Holocaust of World War II*. New York: Basic Books, 1997.

Churchill, Winston. *The Second World War*. 6 vols. Boston: Houghton Mifflin, 1948–1953.

Colby, James M. *The Business of Empire: United Fruit, Race, and U.S. Expansion in Central America*. Ithaca: Cornell University Press, 2013.

Colley, David P. *Road to Victory: The Untold Story of World War II's Red Ball Express*. Washington: Brassey's, 2000.

Confer, Clarissa. *Cherokee Nation in the Civil War*. Norman: University of Oklahoma Press, 2006.

Crost, Lyn. *Honor by Fire: Japanese Americans at War in Europe and the Pacific*. Novato, CA: Presidio Press, 1994.

Das, Santanu, ed. *Race, Empire and First World War Writing*. Cambridge, UK: Cambridge University Press, 2011.

David, Saul. *The Indian Mutiny*. New York: Penguin Books, 2003.

DiNicolo, Gina M. *The Black Panthers: A Story of Race, War, and Courage—the 761st Tank Battalion in World War II*. Yardley, PA: Westholme, 2014.

Dobam, William A., and Thomas D. Phillips. *The Black Regulars, 1866–1898*. Norman: University of Oklahoma Press, 2001.

Dower, John W. *War Without Mercy: Race and Power in the Pacific War*. New York: Pantheon, 1987.

Edgerton, Robert B. *Mau Mau: An African Crucible*. London: I.B. Taurus, 1990.

Eldridge, Lawrence Allen. *Chronicles of a Two-Front War: Civil Rights and Vietnam in the African American Press*. Columbia: University of Missouri Press, 2012.

Eliav-Feldon, Meriam, et al., eds. *The Origins of Racism in the West*. Cambridge, UK: Cambridge University Press, 2009.

Engle, Stephen D. *Yankee Dutchman: The Life of Franz Sigel*. Fayetteville: University of Arkansas Press, 1993.

Feros, Antonio. *Speaking of Spain: The Evolution of Race and Nation in the Hispanic World*. Cambridge, MA: Harvard University Press, 2017.

Ferrell, Robert. *Unjustly Dishonored: An African American Division in World War I*. Columbia: University of Missouri Press, 2011.

Fletcher, Marvin E. *America's First Black General: Benjamin O. Davis, Sr., 1880–1970*. Lawrence: University Press of Kansas, 1989.

Fogarty, Richard S. *Race and War in France: Colonial Subjects in the French Army, 1914–1918*. Baltimore: Johns Hopkins University Press, 2008.

Franco, Jere B. *Crossing the Pond: The Native American Effort in World War II*. Denton: University of North Texas Press, 1999.

Frederickson, George M. *Racism: A Short History*. Princeton: Princeton University Press, 2002.

Frey, Sylvia R. *Water From the Rock: Black Resistance in Revolutionary War*. Princeton: Princeton University Press, 1992.

Fritz, Stephen G. *Frontsoldaten: The German Soldier in World War II*. Lexington: University Press of Kentucky, 1995.

Fujitani, Takashi. *Race for Empire: Koreans as Japanese and Japanese as American During World War II*. Berkeley: University of California Press, 2013.

Glatthaar, Joseph T. *Forged in Battle: The Civil War Alliance of Black Soldiers and White Officers*. Baton Rouge: Louisiana State University Press, 2000.

Gleeson, David T. *The Green and the Gray: The Irish in the Confederate States of America*. Chapel Hill: University of North Carolina Press, 2013.

Graham, III, Herman. *The Brothers' Vietnam: Black Power, Manhood, and the Military Experience*. Gainesville: University of Florida Press, 2003.

Griffin, Patrick. *American Leviathan: Empire, Nation, and Revolutionary Frontier*. New York: Hill and Wang, 2007.

Guderian, Heinz. *Panzer Leader*. Heidelberg: Kurt Vowinkel Verlag, 1950; Cambridge: De Capo Press, 1952.

Gump, James O. *The Dust Rose Like Smoke: The Subjugation of the Zulu and the Sioux*. Lincoln: University of Nebraska Press, 1994.

Guyatt, Nicholas. *Bind Us Apart: How Enlightened Americans Invented Racial Segregation*. New York: Basic Books, 2016

Hickey, Gerald C. *Window on a War: An Anthropologist in the Vietnam Conflict.* Lubbock: Texas Tech University Press, 2002.

Horne, Gerald. *The Counter-Revolution of 1776: Slave Resistance and the Origins of the United States of America.* New York: New York University Press, 2014.

——. *Race War! White Supremacy and the Japanese Attack on the British Empire.* New York: New York University Press, 2003.

Howe, Glenford. *Race, War, and Nationalism: A Social History of West Indies in the First World War.* Kingston, Jamaica: Ian Randle, 2002.

Humphreys, Margaret. *Intensely Human: The Health of the Black Soldier in the American Civil War.* Baltimore: The Johns Hopkins University Press, 2008

Hunt, Michael H. *Ideology and U.S. Foreign Policy.* New Haven: Yale University Press, 1987.

Ienaga, Saburō. *The Pacific War: 1931–1945.* New York: Pantheon, 1979

James, Rawn, Jr. *The Double V: How Wars, Protest, and Harry Truman Desegregated America's Military.* New York: Bloomsbury, 2013.

Johansson, M. Jane, ed. *Albert C. Ellithorpe, the First Indian Home Guards, and the Civil War on the Trans-Mississippi Frontier.* Baton Rouge: Louisiana University Press, 2016.

Kelley, Mary Pat. *Proudly We Served: The Men of the USS Mason.* Annapolis: Naval Institute Press, 1995.

Kenney, Kevin. *Peaceable Kingdom: The Paxton Boys and the Destruction of William Penn's Holy Experiment.* New York: Oxford University Press, 2009.

Laffin, John. *Jackboot: The Story of the German Soldier.* London: Cassel, 1965.

Leckie, William H., and Shirley A. *Buffalo Soldiers: A Narrative of the Negro Cavalry in the West.* Norman: University of Oklahoma Press, 1967.

Lee, Ulysses. *United States Army in World War II: Special Studies—Employment of Negro Troops.* Washington, DC: Government Printing Office, 1966.

Lee, Wayne E. *Barbarians and Brothers: Anglo-American Warfare, 1500–1865.* Oxford: Oxford University Press, 2011a.

——, ed. *Empires and Indigenes: Intercultural Alliance, Imperial Expansion, and Warfare in the Early Modern World.* New York: New York University Press, 2011b.

Lepore, Jill. *The Name of War: King Philip's War and the Origins of American Identity.* New York: Knopf, 1998.

MacGregor, Morris J. *Integration of the Armed Forces: 1940–1965.* Washington, DC: U.S. Army Center of Military History, 1985.

Maloba, Wunyarabi O. *Mau Mau and Kenya: An Analysis of a Peasant Revolt.* Bloomington: Indiana University Press, 1993.

McCaffrey, James M. *Going for Broke: Japanese Soldiers in the War Against Nazi Germany.* Norman: University of Oklahoma Press, 2013.

McGovern, James. *Black Eagle: General Daniel "Chappie" James, Jr.* Tuscaloosa: University of Alabama Press, 2002.

McLaurin, Melton A. *The Marines of Montford Point: America's First Black Marines.* Chapel Hill: University of North Carolina Press, 2009.

McPherson, James M. *Negro's Civil War: How American Blacks Felt and Acted During the War for the Union.* 1965. New York: Vintage, 2003.

Megargee, Geoffrey P. *War of Annihilation: Combat and Genocide on the Eastern Front, 1941.* Lanham, MD: Rowman and Littlefield, 2006.

Michel, Marc. *L'appel à l'Afrique: Contributions et reactions à l'effort de guerre en AOF, 1914–1919.* [The Call to Africa: Contributions and Reactions to the War Effort in AOF, 1914–1919.] Paris: Publications de la Sorbonne, 1982.

Moore, Brenda L. *To Serve My Country, To Serve My Race: The Story of the Only African-American WACs Stationed Overseas During World War II.* New York: New York University Press, 1996.

Moore, Deborah D. *G.I. Jews: How World War II Changed a Generation.* Cambridge, MA: Harvard University Press, 2004.

Morison, Samuel Eliot. *Admiral of the Ocean Sea: A Life of Christopher Columbus.* 2 vols. Boston: Little, Brown, 1942.

Nalty, Bernard C. *Strength for the Fight: A History of Black Americans in the Military.* New York: Free Press, 1986.

Neely, Mark E., Jr. *The Civil War and the Limits of Destruction.* Cambridge, MA: Harvard University Press, 2007.

Pagden, Anthony. "The Peopling of the New World: Ethnos, Race and Empire in the Early Modern World." In *The Origins of Racism in the West*. Eds. Mariam Eliav-Feldon, et al. Cambridge, UK: Cambridge University Press, 2009.

Parkinson, Robert G. *The Common Cause: Creating Race and Nation in the American Revolution*. Chapel Hill: University of North Carolina Press, 2016.

Ramírez, José A. *To the Line of Fire!: Mexican Texans and World War I*. College Station: Texas A&M University Press, 2009.

Renda, Mary. *Taking Haiti: Military Occupation and the Culture of U.S. Imperialism*. Chapel Hill: University of North Carolina Press, 2001.

Rice, James D. *Tales From a Revolution: Bacon's Rebellion and the Transformation of Early America*. New York: Oxford University Press, 2013.

Rivas-Rodriguez, Maria. *Mexican Americans and World War II*. Austin: University of Texas Press, 2005.

Roberts, Frank E. *American Foreign Legion: Black Soldiers of the 93d in World War One*. Annapolis: U.S. Naval Institute Press, 2004.

Robinson, Charles M., III. *The Fall of a Black Army Officer: Racism and the Myth of Henry O. Flipper*. Norman: University of Oklahoma Press, 2008.

Robinson, Greg. *A Tragedy of Democracy: Japanese Confinement in North America*. New York: Columbia University Press, 2009

Rogers, Daniel. *Concentration Camps USA: Japanese American and World War II*. New York: Henry Holt, 1972.

Roy, Kaushik. *Brown Warriors of the Raj: Recruitment and Mechanics of Command in the Sepoy Army, 1859–1913*. New Delhi, India: Manohar, 2008.

———, ed. *War and Society in Colonial India*. New York: Oxford University Press, 2006.

Sammons, Jeffrey T., and John H. Morrow, Jr. *Harlem's Rattlers and the Great War: The Undaunted 369th Regiment and the African American Quest for Equality*. Lawrence: University Press of Kansas, 2014.

Sandler, Stanley. *Segregated Skies: All-Black Combat Squadrons of World War II*. Washington, DC: Smithsonian Institute Press, 1998.

Schrijvers, Peter. *The GI War Against Japan: American Soldiers in Asia and the Pacific During World War II*. New York: New York University Press, 2002.

Scott, Lawrence P., and William M. Womack, Sr. *Double V: The Civil Rights Struggle of the Tuskegee Airmen*. East Lansing: Michigan State University Press, 1998.

Shellum, Brian G. *Black Cadet in a White Bastion: Charles Young at West Point*. Lincoln: Bison Books, 2006.

———. *Black Officer in a Buffalo Soldier Regiment: The Military Career of Charles Young*. Lincoln: University of Nebraska Press, 2010.

Silbey, David J. *A War for Frontier and Empire: The Philippine-American War, 1899–1902*. New York: Hill and Wang, 2007.

Silver, Peter. *Our Savage Neighbors: How Indian War Transformed Early America*. New York: W.W. Norton, 1997.

Sinha, Mrinalini. *Colonial Masculinity: The "Manly Englishman" and the "Effeminate Bengali" in the Late Nineteenth Century*. Manchester, UK: Manchester University Press, 1995.

Slotkin, Richard. *The Lost Battalions: The Great War and the Crisis of American Nationality*. New York: Henry Holt, 2005.

Smith, John David. *Lincoln and the U.S. Colored Troops*. Carbondale: Southern Illinois University Press, 2013.

Stovall, Tyler. "The Color Line Behind the Lines: Racial Violence in France During the Great War," *American Historical Review* 103 (June 1998): 737–769.

Tanaka, Yuki. *Hidden Horrors: Japanese War Crimes in World War II*. Boulder, CO: Westview Press, 1996.

Taylor, Alan. *The Internal Enemy: Slavery and War in Virginia, 1772–1832*. New York: W.W. Norton, 2013.

Taylor, Jon E. *Freedom to Serve: Truman, Civil Rights, and Executive Order 9981*. New York: Routledge, 2012.

Tchen, John Kuo Wei, and Daniel Yeats, eds. *Yellow Peril! An Archive of Anti-Asian Fear*. New York: Verso, 2014.

Townsend, Mary Bobbitt. *Yankee Warhorse: Major General Peter Osterhaus*. Columbia: University of Missouri Press, 2010.

Ural, Susannah J., ed. *Civil War Citizens: Race, Ethnicity, and Identity in America's Bloodiest Conflict*. New York: New York University Press, 2010.

Urwin, Gregory J. W., ed. *Black Flag Over Dixie: Racial Atrocities and Reprisals in the Civil War*. Carbondale: Southern Illinois University Press, 2004.

Van Creveld, Martin. *Fighting Power: German and U.S. Army Performance, 1939–1945*. Westport, CT: Praeger, 1982.

Westermann, Edward B. *Hitler's Ostkrieg and the Indian Wars: Comparing Genocide and Conquest*. Norman: University of Oklahoma Press, 2016.

Westheider, James. *Fighting on Two Fronts: African Americans and the Vietnam War*. New York: New York University Press, 1997.

Wette, Wolfram. *The Wehrmacht: History, Myth, and Reality*. Frankfurt, Germany: S. Fischer Verlag, 2002. Trans. Deborah Lucas Schneider. Cambridge: Harvard University Press, 2007.

Williams, Chad L. *Torchbearers of Democracy: African American Soldiers in the World War I Era*. Chapel Hill: University of North Carolina Press, 2010.

Williams, Glenn F. *Dunmore's War: The Last Conflict of America's Colonial Era*. Yardley, PA: Westholme, 2017.

Wolff, Leon. *Little Brown Brother: America's Forgotten Bid for Empire Which Cost 250,000 Lives*. Garden City: Doubleday, 1961.

Wynn, Neil A. *The African American Experience During World War II*. Lanham, MD: Rowman and Littlefield, 2010.

Yoshimi, Yoshiaki. *Comfort Women: Sexual Slavery in the Japanese Military During World War II*. New York: Columbia University Press, 2002.

24 Gender and Sexuality

Heather Marie Stur

Thirty years ago, Joan Wallach Scott (1986, 1988) issued a call to scholars to consider gender a "useful category of analysis" for understanding the power relationships that move politics, diplomacy, and war. Significant to Scott's appeal was its implication that the work of social historians incorporating women into historical narratives was not enough. To truly grasp the construction of power, scholars must scrutinize the definitions of masculinity and femininity and examine the involvement of men and women in establishing, perpetuating, and challenging gender ideas. Since then, historians, sociologists, political scientists, and international relations scholars have responded by using gender analysis to understand the social and cultural constructions of masculinity, the obvious and hidden ways in which women have been key actors in what was once considered the male realm of high politics, and the ways in which intimate encounters both enforce and challenge international power relationships. What these scholars have found is that race and sexuality often intersect with gender to establish power hierarchies, so in some ways the three categories cannot be separated. This approach has also expanded scholars' thinking about what constitutes war and diplomacy and who participates in the fighting and negotiating.

Scholars have analyzed gender and sexuality in military history through examinations of men's and women's roles in wartime as well as cultural analyses of gender constructions. Jeffrey Goldstein's seminal *War and Gender* argues that culturally constructed roles for men and women are central to war across time and place (Goldstein, 2001). John Lynn's examination of women and war in early modern Europe illustrates both what women did in war and how gender constructions shaped their experiences (Lynn, 2008). Political scientist and feminist scholar Cynthia Enloe has produced a body of work throughout her career that established the foundation for much of the subsequent work on gender, sexuality, and militarization. Beginning in the early 1980s (Chapkis, 1981), Enloe began examining the impact of war and militarization on women's lives (Enloe, 1988, 1989, 1993). Her entry into this broad subject began with a question about the ways in which military and diplomatic history, as well as political science and international relations, had up to that point portrayed war and other global interactions as the purview of men: "Where are the women?" Enloe has sought to answer this question by exploring a wide range of women's experiences in war throughout the world, from American servicewomen to diplomats' wives, from prostitutes to washerwomen who offer services to American GIs stationed abroad.

Enloe's work on gender, militarization, and international relations has demonstrated that women long have been crucial to the military's image as a powerful masculine institution, even while they are required to occupy subordinate positions in martial hierarchies. Her writings reveal how militaries rely on and sustain gender, from the state-sanctioning of prostitution at the request of an occupying military to the enforcement of specific ideas about masculine sexuality. Hierarchical gender constructions were and continue to be perpetuated on stateside and overseas bases, where military policies have shaped men's and women's sexuality, determined entrepreneurial and women's economic opportunities, regulated public health, and controlled entertainment. Women who signed up for the armed services, Enloe notes, were required to "behave like the gender 'woman'" so as not to disrupt the masculine image of the military. While Enloe's subjects typically are women, her gender analysis of women's experiences illuminates the ways in which definitions of masculinity and femininity work in tandem to create wartime cultures that cast men as protectors, and women as either in need of protection or available to provide care, sustenance, and sex to male troops. Enloe's overarching point is that the very meanings of war and what a soldier is depend on women as "symbols, consumers, workers, and emotional comforters" (Enloe, 1989, xvii).

Jean Bethke Elshtain (1987), another pioneering scholar on the subject of women, gender, and war, has sought to understand the origins of the men as warriors/women as peacemakers dichotomy. Exploring these images from Plato's writings to depictions of late twentieth-century terrorists, Elshtain illustrates that the actions of men and women throughout history have complicated these gender dynamics. That the dichotomy has remained central to wartime cultures, despite the challenges issued by actual human behaviors, should serve as a call to scholars (especially feminists) to continue the work of deconstructing these categories, Elshtain argues. She also challenges scholars to conceptualize new ways of thinking about gender and men's and women's roles in war. Subsequent historians and other scholars have applied the thematic and theoretical approaches that Enloe's and Elshtain's works exemplify to studies of specific wars, areas, and time periods.

Women's experiences in World War II planted the seeds of second-wave feminism, and the body of literature on the conflict includes studies of the gender politics that influenced women in the workplace and shaped depictions of women in popular culture. D'Ann Campbell's work on American women during World War II illustrates how inserting women into narratives and applying a gender analysis have led to a reassessment of a war itself. Campbell (1984, 1993) took a wider view of the era, to include the Great Depression, and found that economic necessity, not patriotic fervor, motivated women to enter the factories during the war. Moreover, neither the entry of women into the labor force nor the creation of the Women's Army Corps (WAC) in 1943 transformed Americans' beliefs about gender roles or who was eligible for military service.

World War II–era gender constructions embedded in the images of "the idealized wife, the promiscuous WAC, the seductive fräulein, the maternal bomber, and the legendary Tokyo Rose" defined American soldier culture, Ann Pfau (2008, 10) has argued in her work on gender and the war. Written on the sides of bombers and the walls of barracks, kept close in photographs and letters, women, both real and imagined, figured prominently in American soldiers' attempts to make sense of their military service overseas in World War II. As Pfau demonstrates, during World War II women reminded men of why they fought and what was waiting back home for them. Women also symbolized danger to homefront domestic comforts and to servicemen themselves. With her study of this aspect of wartime culture, Ann Pfau joins an expanding group of scholars who are shifting the focus of military history away from combat. Pfau's work pushes beyond the "greatest generation" image to provide a more nuanced, and honest, picture of America's World War II servicemen. Acknowledging that most fought willingly, Pfau reveals that during World War II, vague concepts like freedom and democracy did not necessarily motivate American men to pick up arms and answer the government's call to fight. What did inspire them was the thought of women, especially those back home, and the longing for the domestic life waiting for them as husbands and fathers.

Gender and sexual norms also shaped the experiences of American servicewomen during World War II. Leisa Meyer (1997) found that the common perception of WACs was that they were either lesbians or whores who viewed military enlistment as a way into a world where they could act out their sexual deviance. Both critics and supporters relied on long-established discourses about race, gender, and citizenship when building their cases for or against women's incorporation into the Army. Meyer points out that some opponents of bringing women into the Army feared it would upset the long-standing belief that military service is the ultimate measure of masculinity. Others considered the WAC a hideout for promiscuous heterosexual women and for lesbians, endangering traditional values regarding family and domesticity. Allan Berube's seminal work (1990) on gays and lesbians in World War II illustrates the dual battle they fought as they both served the American war effort and struggled against antigay military policies. That U.S. servicemen and women were segregated on posts and steered toward entertainment districts in military towns opened opportunities for sexual freedom, even if sexuality was grounded in heteronormative beliefs about male and female sexual expression, as Beth Bailey and David Farber argued (1994) in their cultural study of Hawaii during World War II. In his study of the development of a gay movement after 1945, John D'Emilio called World War II "a nationwide coming out experience" (1983, 24). Regarding American women's experiences in the war, Campbell concluded that World War II was not the "good war" for everyone.

The intersection of gender and sexuality is central to the historiography of women and World War II. Marilyn Hegarty (2007) and Meghan Winchell (2008) have highlighted the roles of U.S. government agencies, such as the USO, the U.S. Public Health Service, and the Office of War Information, that perpetuated a sexual double standard that normalized male soldiers' heterosexual behavior and demonized women's sexuality. The contradiction in this approach to sexuality led to rumors about the motives of women entertainers in military-sanctioned morale boosting programs, and the branding of women who interacted with U.S. servicemen as whores. At the same time, as Mary Louise Roberts (2014) has revealed in her work on U.S. troops in World War II France, American servicemen abroad purchased the services of prostitutes and perpetrated sexual assaults and rapes. Petra Goedde's (2002) writing on American soldiers and German women in the postwar occupation of Germany demonstrates the role of sex in U.S.-German relations. Work such as Katharine H. S. Moon's (1997) study of prostitution during the Korean War carries the theme beyond World War II, focusing on U.S. and South Korean government policymaking that institutionalized prostitution as part of the relationship between the two nations.

Scholars of Europe in World War II have also produced both social histories of what women did during the war and cultural histories analyzing the ways in which gender shaped men's and women's wartime experiences. Reina Pennington's work on Soviet women pilots (2002), the various writings on British women spies (Escot, 2010; Helm, 2005; Mulley, 2012; Basu, 2007; Kramer, 1996), and studies of women's participation in the French Resistance (Delbo, 1997; Moorehead, 2011; Weitz, 1995) all detail the myriad ways in which European women contributed to the Allied war effort. Other scholars have analyzed the gender ideas that formed the framework in which European men and women participated in the war. These works (Higonnet et al., 1987; Wingfield and Bucur, 2006; Heineman, 1999; Rose, 2004; Nye, 2007) focus on homefront experiences, the conflict's impact on family structures, the use of sexual assault and rape as weapons of war, popular constructions of heroism, and public memories of the war. Central to these approaches is a commitment to scrutinizing gender constructions in order to challenge the notion that categories such as "man" and "male" are normative touchstones off of which "woman" and "female" are defined as different. George Mosse and others have illustrated how entrenched gender constructions have shaped not only the wars themselves but also how subsequent generations remember them (Mosse, 1991; Grayzel, 2012).

Like the extensive scholarship on World War II, the broad interest among historians in the Vietnam War has led to the creation of a vast historiography on the conflict, including studies focused on women, gender, and sexuality. In the Vietnam War historiography, scholars have paid close attention to masculinity and race in their examinations of gender and wartime culture, in addition to exploring women's experiences in the war. At the forefront is Susan Jeffords (1989), whose work explores the impact of the Vietnam War on American cultural representations of masculinity, and postwar efforts to use popular culture to "remasculinize" America in the wake of its military defeat in Vietnam. As Jeffords explains, the Vietnam War consisted of not just battlefields but also "fields of gender," where "enemies are depicted as feminine, wives and mothers and girlfriends are justifications for fighting, and vocabularies are sexually motivated" (Jeffords, 1989, xi). Jeffords analyzes post–Vietnam War literature, television shows, and movies—including Tim O'Brien's writings; *The A Team*; Vietnam War movies, such as *Platoon* and *The Deer Hunter*; and the Rambo character—to show how American conceptions of masculinity changed from the end of the Vietnam War into the 1980s. Jeffords argues that cultural portrayals of masculinity reflected desires and efforts to recover from the emasculating experience of the war. Jerry Lembcke (2010) takes a similar approach as Jeffords but with a different cultural product: "Hanoi Jane," specifically the idea of Jane Fonda as the ultimate traitor. Lembcke argues that Americans have used the Hanoi Jane image to obscure the social activism related to civil rights and women's liberation that Fonda's actual political involvement emphasized.

Kara Dixon Vuic's (2009) study of the Army Nurse Corps (ANC) in Vietnam melds a social history of women and men who served in the ANC with a cultural analysis of the ways in which gender shaped recruits' Vietnam experiences. Vuic positions the ANC as a lens on both Americans' experiences in the Vietnam War and the transformations occurring in U.S. society. By the late 1960s, when the ANC launched a recruitment campaign to answer the need for nurses in Vietnam, "second-wave" feminism had challenged traditional notions of women's work, and opposition to the war meant that the Army could not necessarily rely on appeals to patriotism. The Army courted nurses via Operation Nightingale, an all-out recruitment initiative that sent recruiters throughout the country to schools, hospitals, and conferences, selling the ANC as an avenue to professional equality and respect. In 1967, Congress approved legislation that allowed women, including nurses, to be promoted to the military's highest ranks. The ANC had begun admitting men beginning in 1955, and in 1968, approximately 25 to 30 percent of the nurses stationed in Vietnam were men—even though in the civilian world, men made up only about 1 percent of nurses. Male nurses faced accusations of homosexuality and effeminacy; to counter those assumptions, Army public relations staff emphasized male nurses who were husbands and fathers, as well as those who enjoyed "manly" pursuits, such as sports. The Army also stationed male nurses with combat units and in hospitals that were under increased threat of attack, enforcing the image of a man at war as a soldier first, and then a nurse.

In order to explain the ideas Americans used to make sense of the Vietnam War, Heather Stur (2011) uses the lenses of gender and sexuality to examine the experiences of men and women on the ground in Vietnam, as well as the pervasive images of women and gender that surfaced in policymakers' conversations, State and Defense Department pamphlets, newspapers and magazines, and soldier culture. Analyzing cultural representations of American and Vietnamese women and men, Stur argues that popular ideas about women and gender influenced the ways in which U.S. policymakers, soldiers, and ordinary citizens made sense of and justified the war in Vietnam, interactions between Americans and Vietnamese, and the Cold War world. A cluster of intertwined images reflecting gendered and racialized notions of U.S. power—the "dragon lady," the "girl next door," and the "gentle warrior"—applied domestic cultural norms to international relations, even as those norms were being challenged on the homefront. By confronting the notion that war is an exclusively male realm, and by investigating the imagery designed to perpetuate the illusion that war is men's business, Stur's work builds on the scholarship of Cynthia Enloe, Susan Jeffords, and other predecessors. Although much of the Vietnam War historiography views the conflict from American perspectives, scholars have begun to incorporate Vietnamese voices, including those of women, into the literature. Douglas Bristol and Heather Stur (2017) edited a recent anthology that blends constructions of race, gender, and sexuality in the U.S. military during Vietnam and all of post–World War II America.

Sandra Taylor's (1999) study of South Vietnamese women who fought and otherwise supported the National Liberation Front (NLF) illustrates the impact of the Vietnam War on rural families, especially in the Mekong Delta region. Women's stories straddle the line between the homefront and the battlefront. Women assisted the communist cause in various roles, from regular infantry soldier to the highest level of NLF authority, as exemplified by Nguyen Thi Dinh, a founder of the NLF and high-ranking official in the People's Liberation Armed Forces (PLAF). NLF authorities saw a propaganda value in the women of PLAF, and they disseminated photographs of armed young women walking point and loading artillery to media outlets throughout the world during the Vietnam War.

Yet reality did not always square with the image of gender equality in the liberation forces, as Karen Gottschang-Turner (1998) has found. During and after the war, female soldiers did not have the same access to services and support that their male comrades had, and the war affected women differently, especially regarding fertility and reproduction. For a young woman who enlisted in PLAF or the People's Army of Vietnam (PAVN), they gave their childbearing years to the war; and when it was over, the physical toll it had taken on their bodies meant that some would never become mothers in a culture that values family over all other social units. Taylor's and Gottschang-Turner's important books have provided the structure in which historians must continue to add Vietnamese women's voices, especially the voices of those women who fought for the Saigon government. Their perspectives remain largely absent from the broader historiography of the Vietnam War.

Decolonization, post-1945 wars for independence, and guerrilla movements have offered scholars case studies of women and war, both in terms of women's actual service and propaganda images of "people's wars." Like the women of the NLF, Filipinas served in various military capacities in the Hukbalahap rebellions, first against Japanese occupation during World War II, and then against the U.S.-backed Philippine government (Lanzona, 2009). Although most Huk women performed domestic and secretarial duties—traditional "women's work"—some managed to earn positions of military leadership. Women also joined revolutions in Nicaragua, Chiapas, and El Salvador in the second half of the twentieth century (Kampwirth, 2002). Karen Kampwirth has found that Latin American women had numerous reasons for joining revolutionary movements, from family economic needs to the influence of liberation theology. Challenging the conventional interpretation of these revolutions as primarily peasant movements, Kampwirth reveals that many female revolutionaries came from cities rather than rural villages, and were more likely than their male counterparts to have a high school or college education. The Latin American women of Kampwirth's study are the descendants of groups such as the *soldaderas*, women who played various roles, from camp follower to armed combatant, in the Mexican Revolution (Salas, 1990). In an important comparative study, Cora Ann Presley brings decolonizing Africa into the conversation with her work on Kikuyu women during Kenya's Mau Mau rebellion (Presley, 1992).

Adhering to Joan Scott's reminder that gender is not just about women, scholars have used gender analysis to understand men's experiences in war and diplomacy. Where World War II served as one of the early focal points of historians' work on women and war, pioneering studies of men and masculinity used World War I as the case study (Keegan, 1976; Fussell, 1975; Leed, 1981). Feminist scholars and others who sought to answer Scott's call began broader research into the "man question" in international relations and war (Zalewski and Parpart, 1997). Just as Scott asserted that inserting women's stories into political, military, and diplomatic history was not enough, Zalewski, Parpart, and others set out not to incorporate more men into the story but to illustrate the ways in which ideas about masculinity function in defining power. There is no shortage of men's perspectives on war and international relations, they argue. However, scholars must investigate the ways in which societies construct masculinity, which masculinities become the dominant images and why, and how women and men have challenged prevailing gender norms in war and diplomatic contexts.

Cultural studies of the Vietnam War have illustrated the impact of Cold War–era conceptions of masculinity on the generation of American men who served in Vietnam. As Richard Slotkin (1992, 520) writes, John Wayne symbolized the "perfection of soldierly masculinity" in the 1960s. The John Wayne image, transmitted through generic cowboy and soldier characters in movies and on TV, thus represented U.S. martial prowess along with a broader collection of American virtues, including patriotism, courage, Christian faith, and unremitting dedication to protecting the civilization embodied in the girl next door (Stur, 2011). Growing up in a militarized culture predicated on defending the American way of life, many of the young men who went to Vietnam, as Andrew Huebner has observed, considered John Wayne the embodiment of their "martial dreams" (Huebner, 2008, 250). Born during World War II and coming of age during the nebulous early Cold War, they witnessed empires collapse and new nations emerge in the initial phases of the postcolonial struggles that defined the second half of the twentieth century. The generation of American boys whose fathers had fought against Germany and Japan was encouraged to make sense of these circumstances not by looking ahead but by looking back, beginning with childhood games of playing cowboys and Indians and reenacting World War II (Engelhardt, 1995). For a brief moment in the late 1950s, Fidel Castro also provided a tantalizing image of masculinity—that of a rugged revolutionary—which inspired *Fidelismo* in young, restless American men (Gosse, 1993).

Borrowing the methodologies of postcolonial studies (McClintock, 1995; Stoler, 1995, 1997, 2002, 2006, 2013) that examined European imperialism, U.S. military and diplomatic historians have shown how gender shaped American attitudes and policies toward war since the end of the nineteenth century. The imperative to prove masculinity through war, a central characteristic of American identity, led U.S. politicians to authorize military interventions in Cuba and the Philippines at the turn of the twentieth century (Hoganson, 1998; Belkin, 2012). Paternalism motivated subsequent interventions in the Caribbean, including the U.S. occupation of Haiti (Renda, 2001). Naoko Shibusawa (2006) describes how U.S. military and political leaders, journalists, authors, and Hollywood executives cast Japan as a submissive woman—a "geisha"—during the post–World War II occupation in order to shift American public opinion of Japan, formerly a hated enemy, toward accepting Japan as a Cold War ally.

Efforts to cultivate warm, patriarchal feelings among Americans for Asia continued throughout the 1950s, Christina Klein (2003) has explained, in order to foster support for U.S. involvement in the affairs of allies and potential allies in Asia. Andrew Rotter (2000) shows that beliefs about masculinity and femininity shaped U.S. policymakers' attempts to understand the newly independent India, which they cast as effeminate, and therefore indecisive and weak, based on their assumptions about Indian men. The gendering of power that cast the United States as a global father figure influenced American relations with the Middle East, Melani McAlister (2001) has argued, writing that U.S. policymakers envisioned America presiding over a family of nations, a mind-set that provided a moral justification for U.S. engagement in the region.

Seth Jacobs (2005) traces these cultural approaches to America's early political involvement in Vietnam, where the United States supported Ngo Dinh Diem as leader of South Vietnam. A Catholic and staunch anti-communist, Diem was the "miracle man" for U.S. policymakers in an otherwise untrustworthy, less manly Asian nation, Jacobs writes. Central to these foreign policy approaches was an American self-perception that linked power with heterosexual masculinity, as Robert Dean (2003) and Donald Mrozek (1980) have shown. Dean notes that the "lavender scare" that targeted gays in the State Department and other federal government institutions in the 1950s created a climate in which political weakness was equated with effeminacy and, worse, homosexuality. Dean and Mrozek argue that this mind-set had a profound impact on both Kennedy and Johnson, who assessed U.S. foreign relations in general and American intervention in Vietnam in particular in gendered and sexualized terms. For them, a tough, prepared, aggressive U.S. military must be on hand to rescue nations from the grip of communism.

Some scholars have approached the intersection of gender and war through an analysis of space, looking at conceptions of the homefront and the battlefront as "separate spheres," with all the gender implications that phrase encompasses. Historians of World War I have illustrated how citizens conceptualized the idea of the homefront as a space separate from the battlefield during the conflict. Susan Grayzel's (2012) work on the impact of air warfare on Britain illustrates how air raids destroyed the notion that two fronts existed during the war, forcing governments to recognize that the state was obligated to protect citizens during wartime. In exchange for expecting men to leave their families to fight for national security, Grayzel writes, the state must assume the role of a father figure in protecting a soldier's family while he is away at war. The resulting relationship is one in which we see the intersection of the warfare state and the welfare state. Krisztina Robert's spatial analysis of women's employment in World War I Britain illustrates how ideas about what constituted the domestic and military fronts shaped wartime identities and social roles.

Historians of the U.S. Civil War have also used spatial analysis to explore issues of gender and conflict. Lyde Cullen Sizer's (2011) survey of recent works on women in the U.S. Civil War examines various notions of shifting geographies, including Judith Giesberg's concept of an "alternate wartime geography" (Giesberg, 2009, 10). Sizer explains that Giesberg has called for an examination of the women's movement into new spaces as a result of war. In doing so, Sizer states, they "mapped an alternative wartime geography dictated by the material conditions of war rather than the ideological constraints of gender or the limits of the middle-class imagination" (Sizer, 2011, 537). Sizer notes that Giesberg emphasizes motion; women did not stand still on the shifting Civil War landscape. Sizer points out that Giesberg refers to Edward Ayers' idea of "practice theory" to understand the broader impact of the movement of civilians during the Civil War. Practice theory, Sizer writes, is the notion that "social power is understood through practice, or everyday life, rather than through categories" (Sizer, 2011, 537). As Sizer concludes, "All of these (moving) constituencies forced attention to their plight and their opinions, whether they won their specific battles or not, as the war 'spilled over into everyday spaces'" (543–544). In his study of Confederate military and civilian morale in Mississippi, Alabama, and Tennessee in the final two years of the Civil War, Bradley Clampitt (2011) argues that the homefront and battlefront were inseparable during the U.S. Civil War. "Any military history that neglects the homefront overlooks the fundamental human concerns of home and family, wives, children, and other loved ones, common to most soldiers in any conflict," he contends (Clampitt, 2011, 1). Because concerns related to home and family were so central in the minds of Confederate soldiers, an analysis of the relationship between the homefront and the battlefront offers a means for understanding issues such as soldiers' conceptions of their masculinity and masculine responsibility.

American Revolutionary War scholar Holly Mayer's (1996) work on camp followers offers a framework for understanding the particular experiences of women who were part of a "military community," a social unit in which families, merchants, and other camp followers "live and work with the military and accept, willingly or not, its governance of their affairs"

(Mayer, 1996, 1). In this world, the "way of life reinforced female dependence at the same time it demanded female fortitude and, occasionally, supported female initiative" (124). In Mayer's study, a woman's most patriotic act was to send her father, brother, husband, or sons to war, and in doing so, "subordinate her needs to those of the nation" (125).

Existing in other war and national contexts, such as the *cantinieres* and *vivandieres* who supported French troops from the eighteenth century to World War I (Cardoza, 2010; Lynn, 2008), women have been closer to the front lines than popular images of war have shown. The image of the camp follower has shifted across various wartime cultures, from the families and domestics of the American Revolutionary and Civil War eras to the prostitutes of Korean War "camptowns" (Yuh, 2002). The fluidity of the camp follower definition points to the changing conceptions of gender, sexuality, and militarization, depending on time and place.

Historians and other scholars of war and militarization have responded with conviction to Joan Scott's assertion that gender analysis is a valuable tool for understanding war, politics, and diplomacy. Yet there is more work to be done related to gender, sexuality, and war, both in terms of covering specific wars and offering new perspectives on various conflicts. Writers have devoted much attention to World War II, but we still know very little about masculinity and military cultures during that war or the impact of the war on homefront conceptions of masculinity. Examining both of the world wars through the lenses of gender and sexuality, but from non-Western perspectives, could tell us something about international power relationships that transcended the wars. More analysis of the roles of Muslim women in armed forces and the gender constructions of conservative Muslim societies is crucial to understanding conflict in the Middle East and southwestern Asia. Can a gender analysis help us understand more deeply why images of Vietnamese "long-haired warriors" and *Fidelismo* had so much international appeal in the 1950s, 1960s, and 1970s? Why do twenty-first-century young Americans wear t-shirts bearing Che Guevara's likeness?

Pop culture depictions of servicemen and women provide clear starting points at which to begin a gender analysis. In 2007, the men's magazine *Maxim* published a feature about women of the Israel Defense Forces (IDF), and the text introducing the photographs read, "They're drop-dead gorgeous and can take apart an Uzi in seconds. Are the women of the Israeli Defense Forces the world's sexiest soldiers?" ("The Chosen Ones"). What could a gender and sexuality analysis of this piece tell us about Americans' views of Israel and Israeli servicewomen? Are they sexy in spite of their ability to take apart an Uzi quickly, or because of it? What can the piece tell us about the servicewomen who agreed to be photographed and interviewed? Are they agents making decisions about how the media portrays them, or are they victims of a hypersexualized American culture? Should U.S. policymakers look to the gender integration of the IDF as a model for American forces to follow (Van Creveld, 2000; Swirski and Safir, 1993; Lomsky-Feder and Ben-Ari, 2000)? Given that the gender conventions that inform military cultures are typically constructed on the homefront, applying a gender analysis to a war can tell us as much about the related civilian world as it does about the military. During Operation Desert Shield and Desert Storm, the first major conflicts in which America's gender-integrated military fought, some of the most vocal opponents of the deployment of women were civilians. Servicewomen have been highly visible in America's twenty-first-century wars, from the Jessica Lynch controversy to the work of female engagement teams and Marine Corps Lionesses. The Pentagon's 2015 decision to open combat specialties to women launched renewed debates over whether servicewomen are physically capable of serving in combat (MacKenzie, 2015). War concentrates and makes visible the issues of gender and sexuality that inform the policymaking and public opinion about the conflict.

Bibliography

Bailey, Beth, and David Farber. *The First Strange Place: Race and Sex in World War II Hawaii*. Baltimore: Johns Hopkins University Press, 1994.

Basu, Shrabani. *Spy Princess: The Life of Noor Inayat Kahn*. Medford, OR: Omega, 2007.

Belkin, Aaron. *Bring Me Men: Military Masculinity and the Benign Façade of American Empire, 1898–2001*. New York: Oxford University Press, 2012.

Berube, Allan. *Coming Out Under Fire: The History of Gay Men and Women in World War II*. New York: Free Press, 1990.

Bethke-Elshtain, Jean. *Women and War*. New York: Basic Books, 1987.

Bristol, Douglas Walter Jr., and Heather Marie Stur, eds. *Integrating the US Military: Race, Gender, and Sexual Orientation Since World War II*. Baltimore: Johns Hopkins University Press, 2017.

Campbell, D'Ann. *Women at War With America: Private Lives in a Patriotic Era*. Cambridge, MA: Harvard University Press, 1984.

———. "Women in Combat: The World War II Experience in the United States, Great Britain, Germany, and the Soviet Union," *The Journal of Military History* 57, no. 2 (April 1993): 301–323.

Cardoza, Thomas. *Intrepid Women: Cantinieres and Vivandieres of the French Army*. Bloomington: Indiana University Press, 2010.

Chapkis, Wendy, ed. *Women in Militaries*. Washington, DC: Institute for Policy Studies, 1981.

"The Chosen Ones: Israeli Defense Forces," *Maxim*, September 13, 2007, <www.maxim.com/women/chosen-ones-israeli-defense-forces>. Accessed May 26, 2016.

Clampitt, Bradley. *The Confederate Heartland: Military and Civilian Morale in the Western Confederacy*. Baton Rouge: Louisiana State University Press, 2011.

Dean, Robert. *Imperial Brotherhood: Gender and the Making of Cold War Foreign Policy*. Amherst: University of Massachusetts Press, 2003.

Delbo, Charlotte. *Convoy to Auschwitz: Women of the French Resistance*. Trans. Carol Cosman. Boston: Northeastern University Press, 1997.

D'Emilio, John. *Sexual Politics, Sexual Communities: The Making of a Homosexual Minority in the United States, 1940–1970*. Chicago: University of Chicago Press, 1983.

Engelhardt, Tom. *The End of Victory Culture: Cold War America and the Disillusioning of a Generation*. New York: Basic Books, 1995.

Enloe, Cynthia. *Bananas, Beaches, and Bases: Making Feminist Sense of International Politics*. Berkeley: University of California Press, 1989.

———. *Does Khaki Become You? The Militarization of Women's Lives*. Oakland, CA: Pandora Press, 1988.

———. *The Morning After: Sexual Politics at the End of the Cold War*. Berkeley: University of California Press, 1993.

Escot, Beryl E. *The Heroines of SOE: F Section: Britain's Secret Women in France*. Stroud: The History Press, 2010.

Fussell, Paul. *The Great War and Modern Memory*. New York: Oxford University Press, 1975.

Giesberg, Judith. *Army at Home: Women and the Civil War on the Northern Home Front*. Chapel Hill: University of North Carolina Press, 2009.

Goedde, Petra. *GIs and Germans: Culture, Gender, and Foreign Relations, 1945–1949*. New Haven, CT: Yale University Press, 2002.

Goldstein, Jeffrey. *War and Gender: How Gender Shapes the War System and Vice Versa*. Cambridge, UK: Cambridge University Press, 2001.

Gosse, Van. *Where the Boys Are: Cuba, Cold War America, and the Making of a New Left*. New York: Verso, 1993.

Gottschang-Turner, Karen. *Even the Women Must Fight: Memories of War From North Vietnam*. Hoboken, NJ: Wiley, 1998.

Grayzel, Susan. *At Home and Under Fire: Air Raids and Culture in Britain From the Great War to the Blitz*. New York: Cambridge University Press, 2012.

Hegarty, Marilyn. *Victory Girls, Khaki-Wackies, and Patriotutes: The Regulation of Female Sexuality During World War II*. New York: New York University Press, 2007.

Heineman, Elizabeth D. *What Difference Does a Husband Make? Women and Marital Status in Nazi and Postwar Germany*. Berkeley: University of California Press, 1999.

Helm, Sarah. *A Life in Secrets: Vera Atkins and the Missing Agents of WWII*. New York: Nan A. Talese, 2005.

Higonnet, Margaret R., et. al., eds. *Behind the Lines: Gender and the Two World Wars*. New Haven, CT: Yale University Press, 1987.

Hoganson, Kristin. *Fighting for American Manhood: How Gender Politics Provoked the Spanish-American and Philippine-American Wars*. New Haven, CT: Yale University Press, 1998.

Huebner, Andrew. *The Warrior Image: Soldiers in American Culture From the Second World War to the Vietnam Era*. Chapel Hill: University of North Carolina Press, 2008.

Jacobs, Seth. *America's Miracle Man in Vietnam: Ngo Dinh Diem, Religion, Race, and U.S. Intervention in Southeast Asia*. Durham, NC: Duke University Press, 2005.

Jeffords, Susan. *The Remasculinization of America: Gender and the Vietnam War*. Bloomington: Indiana University Press, 1989.

Kampwirth, Karen. *Women and Guerrilla Movements: Nicaragua, El Salvador, Chiapas, Cuba*. State College: Penn State University Press, 2002.

Keegan, John. *The Face of Battle: A Study of Agincourt, Waterloo, and the Somme*. New York: Viking 1976.

Klein, Christina. *Cold War Orientalism: Asia in the Middlebrow Imagination, 1945–1961*. Berkeley: University of California Press, 2003.

Kramer, Rita. *Flames in the Field: The Story of Four SOE Agents in Occupied France*. Reprint. London: Penguin, 1996.

Lanzona, Vina. *Amazons of the Huk Rebellion: Gender, Sex, and Revolution in the Philippines*. Madison: University of Wisconsin Press, 2009.

Leed, Eric. *No Man's Land: Combat and Identity in World War I*. New York: Cambridge University Press, 1981.

Lembcke, Jerry. *Hanoi Jane: War, Sex, and Fantasies of Betrayal*. Amherst: University of Massachusetts Press, 2010.

Lomsky-Feder, Edna, and Eyal Ben-Ari. *The Military and Militarism in Israeli Society*. Albany: SUNY Press, 2000.

Lynn, John. *Women, Armies, and Warfare in Early Modern Europe*. Cambridge, UK: Cambridge University Press, 2008.

MacKenzie, Megan. *Beyond the Band of Brothers: The U.S. Military and the Myth That Women Can't Fight*. Cambridge, UK: Cambridge University Press, 2015.

Mayer, Holly. *Belonging to the Army: Camp Followers and Community During the American Revolution*. Columbia: University of South Carolina Press, 1996.

McAlister, Melani. *Epic Encounters: Culture, Media, and U.S. Interests in the Middle East Since 1945*. Berkeley: University of California Press, 2001.

McClintock, Anne. *Imperial Leather: Race, Gender, and Sexuality in the Colonial Context*. London: Routledge, 1995.

Meyer, Leisa. *Creating GI Jane: Sexuality and Power in the Women's Army Corps During World War II*. New York: Columbia University Press, 1997.

Moon, Katharine H. S. *Sex Among Allies: Military Prostitution in U.S.-Korea Relations*. New York: Columbia University Press, 1997.

Moorehead, Caroline. *A Train in Winter: An Extraordinary Story of Women, Friendship, and Resistance in Occupied France*. New York: HarperCollins, 2011.

Mosse, George. *Fallen Soldiers: Reshaping the Memory of the World Wars*. Oxford: Oxford University Press, 1991.

Mrozek, Donald. "The Cult and Ritual of Toughness in Cold War America." In *Rituals and Ceremonies in Popular Culture*. Ed. Ray B. Browne. Bowling Green, OH: Bowling Green University Popular Press, 1980.

Mulley, Clare. *The Spy Who Loved*. London: Macmillan, 2012.

Nye, Robert. "Western Masculinities in War and Peace," *The American Historical Review* 112 (April 2007): 417–438.

Pennington, Reina. *Wings, Women, and War: Soviet Airwomen in World War II Combat*. Lawrence: University Press of Kansas, 2002.

Pfau, Ann. *Miss Yourlovin: GIs, Gender, and Domesticity During World War II*. New York: Columbia University Press, Gutenberg-e, 2008.

Presley, Cora Ann. *Kikuyu Women, the Mau Mau Rebellion, and Social Change in Kenya*. Boulder, CO: Westview Press, 1992.

Renda, Mary. *Taking Haiti: Military Occupation and the Culture of U.S. Imperialism, 1915–1940*. Chapel Hill: University of North Carolina Press, 2001.

Roberts, Mary Louise. *What Soldiers Do: Sex and the American GI in World War II France*. Chicago: University of Chicago Press, 2014.

Rose, Sonya O. *Which People's War? National Identity and Citizenship in Wartime Britain, 1939–1945*. New York: Oxford University Press, 2004.

Rotter, Andrew. *Comrades at Odds: The United States and India, 1947–1964*. Ithaca, NY: Cornell University Press, 2000.

Salas, Elizabeth. *Soldaderas in the Mexican Military: Myth and History*. Austin: University of Texas Press, 1990.

Scott, Joan Wallach. *Gender and the Politics of History*. New York: Columbia University Press, 1988.

———. "Gender: A Useful Category of Historical Analysis," *American Historical Review* 91, no. 5 (December 1986): 1053–1075.

Shibusawa, Naoko. *America's Geisha Ally: Reimagining the Japanese Enemy*. Cambridge, MA: Harvard University Press, 2006.

Sizer, Lyde Cullen. "Mapping the Spaces of Women's Civil War History," *The Journal of the Civil War Era* 1, no. 4 (2011): 536–548.

Slotkin, Richard. *Gunfighter Nation: The Myth of the Frontier in Twentieth-Century America*. New York: Antheum, 1992.

Stoler, Ann Laura. *Carnal Knowledge and Imperial Power: Race and the Intimate in Colonial Rule*. Berkeley: University of California Press, 2002.

———. *Haunted by Empire: Geographies of Intimacy in North American History*. Durham, NC: Duke University Press, 2006.

———, ed. *Imperial Debris: Of Ruins and Ruination*. Durham, NC: Duke University Press, 2013.

———. *Race and the Education of Desire: Foucault's History of Sexuality and the Colonial Order of Things*. Durham, NC: Duke University Press, 1995.

———. *Tensions of Empire: Colonial Cultures in a Bourgeois World*. Berkeley: University of California Press, 1997.

Stur, Heather Marie. *Beyond Combat: Women and Gender in the Vietnam War Era*. New York: Cambridge University Press, 2011.

Swirski, Barbara, and Marilyn P. Safir. *Calling the Equality Bluff: Women in Israel*. New York: Teachers College Press, 1993.

Taylor, Sandra C. *Vietnamese Women at War: Fighting for Ho Chi Minh and the Revolution*. Lawrence: University Press of Kansas, 1999.

Van Creveld, Martin. "Armed but Not Dangerous: Women in the Israeli Military," *War in History* 7 (January 2000): 82–98.

Vuic, Kara Dixon. *Officer, Nurse, Woman: The Army Nurse Corps in the Vietnam War*. Baltimore: Johns Hopkins University Press, 2009.

Weitz, Margaret Collins. *Sisters in the Resistance: How Women Fought to Free France, 1940–1945*. Hoboken, NJ: Wiley, 1995.

Winchell, Meghan. *Good Girls, Good Food, Good Fun: The Story of USO Hostesses During World War II*. Chapel Hill: University of North Carolina Press, 2008.

Wingfield, Nancy Meriwether, and Maria Bucur. *Gender and War in Twentieth-Century Eastern Europe*. Bloomington: Indiana University Press, 2006.

Yuh, Je-Yeon. *Beyond the Shadow of Camptown: Korean Military Brides in America*. New York: New York University Press, 2002.

Zalewski, Marysia, and Jane Parpart. *The "Man" Question in International Relations*. Boulder, CO: Westview Press, 1997.

25 Militarism and Nationalism

Ingo Trauschweizer

The Second World War saw the cataclysmic culmination of a lethal blend: nationalism and militarism propelled the criminal ideologies of Germany, Italy, and Japan in a war of conquest and extermination. The Allies withstood the initial onslaught, mobilized their resources, and defeated the atavistic challengers to the liberal world order. With the collapse of Germany and Japan came the end of militarism in those countries, and while the Soviet Union, herself heavily militarized, presented a serious challenge to the liberal democracies, the last great power conflict resulted in the end of history, the defeat and disappearance of political ideology as a motivating factor in human affairs. That is the history of the twentieth century read in hindsight and rendered in black and white. A closer reading suggests militarism and nationalism—both cultural forms of modernity—were not exclusive to dictatorships, and they remain alive, if perhaps in more benign expressions, in the global and regional powers of our time.

What is militarism, what is nationalism? The literature suggests ambiguity: nationalism is sometimes discussed as a political framework imposed by elites and at other times as an idea that grew more organically; militarism is more commonly depicted as an organizational practice, but sometimes appears as a bellicose ideology. Historian John Breuilly (2013) explains that interpretations of nationalism shifted in the course of the twentieth century, from seeing it as the outgrowth of the nation-state and an ideology imposed by elites to control the masses, to discussions of language and ethnicity as well as culture and polity that suggest a more organic growth process of nationalism in the modern age. Most scholars see the dawn of modern nationalism in the American and French Revolutions, with antecedents in early modern northwestern Europe, but Azar Gat (2013) asserts that a longer view reveals an evolution from the beginnings of civilization in which the Age of Revolutions serves as an emphatic punctuation. Breuilly represents the scholarly mainstream when he concludes

> we can distinguish between nationalism as ideas, politics, and sentiments . . . however, only with the formation of territorial states with large populations and pressures to extend participation beyond privileged elites is there a systematic shift of the national idea to be inclusive and participatory and political.
>
> (2013, 13–14)

Militarism is primarily associated with Germany and Japan, but the term itself has remained diffuse. The most frequent uses fall in the categories of military domination of civilian society and politics or aggressive foreign policy and propensity to war (Trauschweizer, 2012). Most inclusively, sociologist Stanislav Andreski (1968) posits militarism comes in different types that range from militancy (an aggressive foreign policy) to militarization (state, society, economy, and culture following military structures or principles) and militocracy (preponderance of military over civilian leaders). When coupled with "the adulation of military virtues," these three forms could evolve into "a clear case of militarism" (184–186).

Most scholars depict militarism and nationalism as European creations sprung on the world in the nineteenth and twentieth centuries as Western empires and commercial networks expanded (Vagts, 1959; Gellner, 1983). Their kinship first seemed particularly evident in the radical phase of the French Revolution. Both played a prominent role in the unifications of Italy and Germany and in the modernization of Japan. From there it seemed a straight line to the maelstrom of the world wars, events that were both caused and propelled by militarism and nationalism in ever more radicalized forms. Looking beyond the classic examples of France, Germany, and Japan, scholars have begun serious discussion of how militarism and nationalism contributed to global empires, like that of Great Britain, or how they interacted in multiethnic ones, like those of Austria-Hungary, Russia and the Soviet Union, or the Ottoman Empire. There are particularly promising lines of inquiry in world history. Did the spread of the European nation-state and of militarism project Western ideologies that non-Western societies had to adopt or reject, or did leaders in nineteenth-century Khedival Egypt, Meiji Japan, Qing China, Latin America, and elsewhere adapt European practices such as the bureaucratic state and the modern military deliberately and selectively? Did the United States—born in war, remade in the Civil War—develop an anti-militarist national culture? This essay explores the literature on these questions and aims to show that a global perspective may help us move beyond stereotypes.

War and armed forces are the connective tissue between militarism and nationalism. But war also serves as a pillar on which rests the modern nation-state. Barbara Ehrenreich (1997) observes that social scientists in the Cold War era created a sanitized depiction of the nation, imagined without the passions roused by war (viz. Anderson, 1983). Ehrenreich advances an intriguing argument about nationalism and war worship as a quasi-religion in Nazi Germany and Imperial Japan, and detects a somewhat lighter blend of patriotism and war remembrance and celebration in the United States after the Second World War. She sees nationalism and war as almost indissolubly linked throughout modern history. Ehrenreich's argument represents an emotional expansion on Charles Tilly's recognition that war made the state and the state made war. Tilly and Ehrenreich, in turn, serve as a foundation for contemporary interpretations that see war—and militarization—as the central engine in the evolution of civilization and the emergence of the modern state (Gat, 2006; Morris, 2014).

The growth of the nation-state idea and the militarization of culture accelerated after 1763, when the modern world order took shape. Britain's victory in the Seven Years' War, and the debt she incurred to finance operations from North America to India, had world-historical consequences. Regardless of whether one agrees with Fred Anderson's (2000) revisionist argument that the American Revolution evolved from the war itself, or the more conventional reasoning that Britain's policies in the 1760s and 1770s enraged and incited conservative or radical revolutionaries, there is little doubt about the causal relationship of the Seven Years' War and the American Revolution. Equally, France's debt (and defeat), renewed war (and more debt) in the 1770s, and a decaying sociopolitical system helped bring about the French Revolution in 1789. The war also boosted the Industrial Revolution, under way in Britain in the mid-eighteenth century, in at least two areas: increased demand of the military sector during the war and British control of commerce in the Atlantic and the Indian Ocean thereafter. It was this revolution in geopolitics that consolidated the shift of wealth and power from east to west and ultimately led the nations, states, and empires of the world to alter political structures, harness resources from around the globe, and control mass societies (Darwin, 2008). Measured by military prowess, industrialization, and empire building, Germany, Russia, the United States, and Japan emerged as Britain's main new competitors in the course of the nineteenth century, while France remained a power with global ambitions. These great powers stood among the most heavily militarized states and societies of the early twentieth century.

During the French Revolution military force was an essential ingredient in the transition from bureaucratic absolutism of the ancien régime to popular sovereignty. The *levée en masse* of 1793, at a moment of desperate need to save the revolution and the nation from foreign invasion, and subsequent conscription laws after 1798, profoundly altered the relationship of state and society. In the words of the soldier-author of the *Marseillaise*, the rousing, bloodthirsty, nationalist anthem of 1792, the glory days had arrived and the children of the fatherland were to rise up against all enemies. Yet, the revolution was not uniformly welcomed and at times had to be propelled from Paris to the countryside by force of arms and representatives on mission, those revolutionary commissars who stirred up the people and ensured sufficient support for the new order. In extreme cases, such as in the Vendée region of western France, royalists, counterrevolutionaries, or simply protesters against conscription were suppressed by a display of great force and brutality (Bell, 2007). During the revolution and under Napoleon Bonaparte, France experienced near perpetual war and the army became the embodiment of the nation. In post-Napoleonic Europe military institutions simultaneously came to play a smaller role in the states' growing bureaucracies and a greater one in the militarization of society and the formation of national identity, most obviously so in Germany after 1871. But by the late nineteenth century, conscript armies served as schools of their respective nations all across continental Europe (Frevert, 2004; Jansen, 2004).

A wider lens suggests these were not simply European developments. Historian Lloyd Kramer (2011) depicts the similar ways in which American and European national identities and political cultures were formed. American nationalism may have been more defined by a political nation-state and European nationalisms more by cultural and ethnic communities, but the results were surprisingly similar. George Washington, military leader of the American Revolution and first president of the United States, believed that only a respectable army could win independence and demonstrate political maturity of the new nation. Washington was wrong, in the sense that victory in the War of Independence ultimately rested more on local militia forces and French allies than on victories won by his Continental Army. Yet, like most of the other founding fathers, he considered the European model of state building as normative and recognized that military institutions, war, and the state were closely aligned. Political scientist Charles Tilly (1990) provides the mantra for the modern warfare state, yet he also concludes there was not simply one size or type of state; instead, there was a spectrum from concentrated coercion—most notably Russia and Brandenburg-Prussia—to states such as the Dutch Republic and Venice that were built on concentrated capital. Britain, France, and, one surmises, the United States took their positions in between these extremes.

Elsewhere in the Americas, we can observe the correlation of war, violent upheaval, and militarized state formation in events ranging from the Haitian Revolution, when slave armies rose against their masters and fended off foreign intervention by Britain, Spain, and France, to the wars of independence in Spain's colonies. Wars of independence and violent revolutions are also civil wars, and what emerged were socioeconomically weak and politically factious nation-states (McFarlane, 2013). Most of the new states were dominated by elites of European descent even though the majority of the people were of indigenous or mixed races. Very few countries escaped the clutches of military dictators or strongmen ruling with the backing of arms. The shadow cast by Argentina's Juan Manuel de Rosas or Mexico's Antonio López de Santa Anna clouded the history of the Americas for much of the nineteenth century. Historian John Lynch (1992) argues that *caudillos* like Venezuela's José Antonio Páez facilitated the emergence of national identity by linking central government to nationalist sentiment in order to make it acceptable to the people. Rosas, on the other hand, depended on the balance of power between regions, prevented the integration of a nation-state, and stunted Argentina's development until the mid-nineteenth century. Anthony McFarlane (2013) concludes, "wars had swept away the old order; in their wake they left leaders and groups who had found in violence and militarism an efficient means to remake their societies, to exercise political command, and to impose their demands" (423). Modern political culture in Latin America thus rested on an unsteady foundation.

Latin American militarism primarily took the form of militocracy and did not result in frequent aggressive wars. Armies in South and Central America have fought in relatively few interstate conflicts. They are better understood as institutions built for domestic struggles and a foundation of political power (Andreski, 1968). In favoring political loyalty over military effectiveness, these armies were comparable to professional forces of Europe's monarchies after the Napoleonic Wars. Yet, securing the power of *caudillos*, families, or political parties did not lead to strong state institutions. Sociologist Miguel Angel Centeno (2002) concludes that state formation in Latin America remained incomplete. He argues there is a causal relationship between limited war and limited states. This has contributed to socioeconomic inequality and to the state's inability to assert a monopoly of violence. Latin American militarism and nationalism are different from the European ideal type in that they did not correlate with the emergence of strong nation-states but remained personal or factional realms. Put in more general terms, Latin American nation-states emerged slowly and met the modernization challenge of the nineteenth century incompletely.

Egypt experienced modernization after the attempt by Napoleon Bonaparte to seize the Levant for France. Through the first half of the nineteenth century, Mehmed Ali, the governor of the Ottoman province, employed aspects of European-style state building in an attempt to gain autonomy. As his power grew, Mehmed Ali replaced relationships of the mediated state with more modern structures (Fahmy, 1997). This included centralizing tax administration instead of tax farming, increasing revenue streams by investing in industries and cash crops (especially cotton), controlling trade through Red Sea ports, and building an empire in Sudan. Militarism was central to Mehmed Ali's venture; nationalism was not. By the 1820s he had introduced military conscription and planned for a modern navy, and his armies soon fought in Syria, contesting Constantinople's control of the Levant until 1841. Facing intervention from Britain, Russia, Prussia, and Austria, Mehmed Ali withdrew from Syria, but he had gained the guarantee of dynastic inheritance from the sultan. His sons and grandson continued to expand Egypt's African empire and invested heavily in the Suez Canal. Modernization seemed to progress, but Egypt's debt burden proved too much when cotton prices declined after the American Civil War and the Depression of 1873 struck. Britain intervened to take control of the Egyptian treasury and, following a failed revolt from within the army against growing foreign influence, Egypt came under British occupation, though it nominally remained part of the Ottoman Empire until the First World War. Yet, if modernization did not result in independence, it had yielded a modern state and integrated Egypt into the world economy—and Mehmed Ali's dynasty remained in Cairo until the early 1950s.

As Egypt's modernization project stalled, Japan's was gathering momentum and China was in decline. Both East Asian empires were forced in the mid-nineteenth century to reconsider their relationship with the Western world. For China, the defeat in two Opium Wars (1839–1842 and 1856–1860), coupled with the Taiping Civil War that devastated the country's southeastern region in the 1850s and 1860s, was a shock to the system. The foreigners in the imperial palace in Beijing, the Manchu Qing dynasty, held on to a semblance of power until 1911, but the Chinese state weakened, foreign empires carved out their spheres of control, and foreign ideas undermined traditional society. Reform-minded officials attempted selective modernization of the economy, primarily in an attempt to build a stronger army and navy, but this did not translate into recovery of political or military strength for the emperor, and it elevated provincial leaders at the expense of the center. The late nineteenth century saw devolution of control and a shift from state bureaucracy to a system of personalized leadership; this held promise for the empire at first, but it depended on Qing loyalists leading the new armies. Li Hongzhang offers a good example. Li recruited an army in Anhui province on orders from Zeng Guofan, himself a scholar-bureaucrat turned general in the Taiping War (Platt, 2012). After he had helped defeat the rebel armies, Li built China's first modern arsenal (at Shanghai in 1867) and its first Prussian-style military academy (at Tianjin in 1885). Li's investment in modern armed forces and infrastructure proved insufficient in the war against Japan in 1894–1895. Outsiders' incursions and the stultification of the system had weakened the state, and the next generation of military-political leaders, exemplified by Yuan Shikai, shed their loyalty to the Qing. They became the warlords that ravaged China in the first half of the twentieth century (McCord, 1993).

Japan, in stark contrast, managed to hold off foreign powers and defeat China and, more stunningly, Russia in 1904–1905. Much had changed since the arrival of U.S. Navy Commodore Matthew C. Perry in 1853. Perry, we are told, forced the opening of the island's economy to the West and caused a revolution. The events of the Meiji Restoration that ended a civil war were more complex, as reform-minded officials within the Tokugawa Shogunate had already begun to look toward Europe for ways to modernize Japan. The restoration of the emperor—albeit more symbol than leader of Japan—was the result of factional infighting among the elite following the ossification of social, political, and economic structures after two and a half centuries of Tokugawa rule. In the new political system, the armed forces and industries feeding Japan's military were at the heart of the modernization push. Generals like Yamagata Aritomo and Katsura Taro emerged as the nation's political leaders in the late nineteenth and early twentieth centuries; until the Second World War military leadership and civilian leadership were closely intertwined. Yet, recent scholarship has also shown that the Japanese Army did not hold a monolithic view on empire, nation, strategy, and politics (Lone, 2000; Drea, 2009). In the short term, Japan rose as a great power by virtue of her victories over China and Russia. Japan's modernizers had taken the lesson: strength of nation and state, at the dawn of the twentieth century, was measured in military might and empire building.

That lesson was shared by two new nation-states in Europe and by the United States. Historian Thomas Bender (2006), though not considering militarism explicitly, notes that one could observe a global trend in the mid-nineteenth century: the development of national economies and centralization of the nation-state. But there were different trajectories. Abraham Lincoln's philosophy closely linked freedom, nation, and republic, leading to liberal and progressive nationalism; Otto von Bismarck's policies conjoined force and the nation, which resulted in militarism and illiberal nationalism. Anticipating Bismarck's path, Italy emerged from wars in the mid-nineteenth century as an independent nation-state, with visions of an empire in the Mediterranean and Africa that persisted until the downfall of Benito Mussolini's fascist regime. Even more dramatically, Germany was shaped in wars with Denmark (1864), Austria (1866), and France (1870–1871). And Lincoln reimagined the foundations of the American nation-state in the Civil War (1861–1865), most directly in his Gettysburg Address, which stressed the providential nature of the United States and projected the idea of a government for and by a free people to the world.

As long as Bismarck governed, Germany's imperial adventures were limited and militarism held in check, but that changed under the brash young emperor Wilhelm II. Isabel Hull (2005) concludes German militarism at the dawn of the twentieth century had not pervaded civilian society, but manifested in the extremism of an army insulated from civilian oversight. Institutional ethos and public adulation of the army explain the atrocities in colonial wars in Africa and China as well as in occupied areas during the world wars. Michael Geyer (1986) also notes the separation of military men—managers of violence—from civilian and political spheres, which led to a focus on operations, disregard for the political aims of strategy, and propensity to embrace violence as an aesthetic. But German militarism was not singular. Within the army two views competed: one focused on controlling society while another favored aggressive war (Förster, 1985). Germans also encountered sharp critiques of militarism, particularly from the increasingly popular political left (Stargardt, 1994). In the United States after the Civil War, the consolidation of a continental empire gave way to globalism with commercial interests across the Atlantic and Pacific Ocean. This required a modern navy, which in turn helped bring about a military-industrial complex and enabled the maritime empire that grew from the Spanish-American War (Epstein, 2014). The United States was in good company: in Britain, too, a military-industrial complex had arisen from naval expansion (McNeill, 1982), and this fed into what historian David Edgerton (2005) termed "liberal militarism" throughout much of the twentieth century.

In public memory wars played a distinct role as nation-builders. Examples include American commemoration of wars from the revolution through the twentieth century (Piehler, 1995) and the rendering of the liberation of Europe in World War II, with the D-Day landings in Normandy at the heart of the narrative, as a pillar of American national identity (Dolski, 2016; Dolski et al., 2014); Germany after 1871, where the middle class romanticized the Wars of Unification and glorified the role of the army (Becker, 2001); and Japan after the victories over China and Russia. Naoko Shimazu (2009) shows Japanese society was an active yet often fickle partner of the state. She concludes a blend of modern means and traditional values pervaded Japanese memory culture and notes subtle differences between official and popular nationalism. The armed forces crafted their heroic accounts with emphasis on the modern, industrialized, and disciplined, while in popular culture traditions such as the warrior code came to play a powerful part in celebrating the nation's heroes. This grew more powerful as personal memories faded and "war commemoration took on cultural and commercial dimensions" (12), partly to support an aggressive foreign policy and renewed warfare in China. We should not assume German or Japanese elites defined the militarism of the twentieth century by themselves; the people played a significant role in the volatile mix of traditions and modernity in the Japanese and German empires.

How widespread was militarism on the eve of the First World War? James Sheehan (2008) suggests that it appeared more prominent in hindsight because its proponents, concerned about rising pacifism, stated their arguments increasingly vocally. Heather Jones (2014) adds that war culture in Europe arose from the Moroccan and eastern Mediterranean crises of 1911. In Germany, the supposedly enthusiastic popular response to the outbreak of war in 1914 has been rendered more complex by studies that depict anxiety in urban and rural areas away from the capital cities (Verhey, 2000; Ziemann, 2007), a point also noted for Russia by David Stone (2015), who depicts initial enthusiasm in major cities and a much wider range of emotional responses to the war elsewhere. For Japan, Stewart Lone (2011) finds that people in the early-twentieth-century countryside did not uncritically embrace the power of the military. People in modern nation-states, as Alon Confino (1997) shows for Germany, had multiple layers of identity from the national to the local.

We should not limit ourselves to the challengers of world order when considering the depth of militarism and nationalism. One need not agree with the extent of Lenin's critique of imperialism to recognize Britain's position at the wheel of global trade. And Britain, herself pervaded by a militarized and heroic culture of empire that affected political parties and voters across the spectrum (Johnson, 2013), went into league with Russia, Tilly's most coercive state, and with France, a nation desiring revenge against Germany and a society that did not lag far behind her neighbor's militarism (Vogel, 1997; Chrastil, 2010). Even the multiethnic empires of Austria-Hungary (Déak, 1992) and the Ottoman Turks relied on a heavy dose of militarization, partly due to fear of Russian expansion. Competition between empires and a rigid alliance system bred fear and mistrust. At the centennial of World War I, historians continue to debate what led to war in the summer of 1914. German militarism remains among the most prominent explanations, but Russian and French aggressive diplomacy has also been noted. Then there is the reeling Ottoman Empire, and the Austrians who made fatal decisions in July, misjudging the spiral of escalation triggered by war with Serbia (Watson, 2014; McMeekin, 2015). The war that ensued deepened nationalism and militarism among the victors and vanquished; it was to be most pervasive among those who lost (Germany), started anew (Soviet Union), or were disappointed by their share of the spoils (Italy and Japan).

Chief among the lessons of the war was the need to mobilize soldiers, society, and economy on an unprecedented scale. The totalizing of the war, what William McNeill (1982) calls managerial metamorphosis, favored empires with control of maritime communications and access to global resources. American steel, Argentine beef, soldiers from Africa and India, and Chinese workers helped Britain and France survive. For leaders of the Soviet Union, once it had become apparent that the world revolution was delayed, the quest for autarky and security required radical collectivization and industrialization. Red militarism resulted from Josef Stalin's presumption of inevitable war against capitalist aggressors. Some historians suggest the Red Army's ability to hold and repulse the German invaders justified Stalin's policies (Barber and Harrison, 1991), but the violence and terror of the 1930s had weakened the state, nearly decapitated the army, and frayed social cohesion (Stone, 2000; Miner, 2016). Soviet soldiers fought for a wide range of reasons, from patriotism to fear of Stalinist terror, but many were persuaded by the extreme brutality of the Germans. And while the Soviet economy was able to churn out tanks, the war effort depended increasingly on American lend-lease aid, and the Soviet Union hung on by a bare thread (Miner, 2018).

Germany and Japan also recognized the need to attain autarky. Historian Adam Tooze (2006) considers strains on food supplies in both countries a major cause of the Second World War. The perceived need to gain resources facilitated the emergence of militaristic and genocidal policies of the Nazis. Nationalist and socialist sentiment and ideology, and militarism, mattered greatly, but did not by themselves explain the nature and scope of the war. Historians often seek structure in chaos, and nationalism, militarism, and imperialism serve as a plausible explanation for the upheavals of the first half of the twentieth century. Yet, as Tooze (2014) cautions in his sweeping study of international order after World War I, we should not forget about contingency. To wit, the interwar years appear as mere pause in a thirty years' war in Winston Churchill's memoirs of World War II, but a sense of foreboding did not appear in his writings at the end of the 1920s. Tooze (2014) places the United States at the center of a beleaguered liberal world order, and the Great Depression stands as watershed: without it, the democratization and multiparty system of the Taisho period in Japan (1912–1926) may well have extended into the reign of Emperor Hirohito, and Adolf Hitler may not have attained the unshackled power in Germany that allowed him to launch a war of conquest and extermination. Considering the culture of violence that enabled war and the Holocaust, Omer Bartov (2000), building on the argument of sociologist Zygmunt Bauman, concludes war, genocide, and modern identity were closely linked. This implies modern societies and the modern state itself carry the seeds of destruction that men of violence could nurse and grow.

Once the United States joined the wars in East Asia and Europe, the balance of resources favored the Allies. But, contrary to the argument that winning the war was a matter of brute force, ingenuity at all levels, military and political organization, and careful diplomacy played a significant role in ensuring the victory of Allied arms (Stoler, 2000; Murray and Millett, 2001). In the process, expedient structures in state, society, economy, and science became permanent fixtures. When did the United States become a warfare state? Scholars have pointed to the beginnings of the Cold War, the National Security Act of 1947, and the cultural shift of the Cold War era (Sherry, 1995; Hogan, 1998; Stuart, 2008); the mobilization effort and mixed military economy of the Second World War, particularly since the Franklin D. Roosevelt administration employed the New Deal philosophy of an interventionist state (Sparrow, 2011; Thorpe, 2014); the transformation of understandings of citizenship from voluntarism and group rights to individuals' rights and a sense of enforced and selective obligation during and after the First World War (Capozzola, 2008); or the mixed military economy in the Union during the Civil War and subsequent managerial culture in American business (Wilson, 2006). Long gone were the days of a weak American state, although recent literature suggests that even the weakness of the state in the nineteenth century has been overstated (Novak, 2008).

Scholars and critics who fear the entrenchment of American empire and militarism seek the foundations of both in the twentieth century, anywhere from the aftermath of the Spanish–American War or Woodrow Wilson's worldview to the warfare state of the Second World War and the Cold War (Johnson, 2004; Bacevich, 2005). There is another explanation for American militarism, however, driven less by elites and the power of the state and more by the agency of a people relying upon themselves, in local communities, for defense and expansion (Grenier, 2005; Lee, 2011). This led to a culture that combined the warrior spirit of the frontier with a political ethos of self-reliance, and it propelled alternative tracks of state building, wherein the European model served as superstructure, but the people long retained premodern functions of warfare and local autonomy that put pressure on the state. We may consider militarization, the process by which militarism arises and deepens, as not simply top-down or elite-driven, and democracies or republics as not immune (Gillis, 1989; Caverley, 2014). The debate over a new American militarism (Bacevich, 2005) and an ongoing process of militarization (Kohn, 2009) thus has deeper roots even though the locus of sovereignty shifted; once resting in the people, in the twentieth century it was assumed by the state. The main challenge today consists of recognizing a new framework: perpetual war, which should no longer be treated as a state of exception, has shifted how we interpret rights and threatens to pervert the Constitution (Dudziak, 2012).

One has to be careful not to overstate the case and allow for distinctions between militarization and fully developed garrison states. To face the realities of the Cold War, political scientist Harold Lasswell revisited his garrison state thesis in 1962. Twenty-five years earlier he had argued Japan represented a novel threat stemming from close integration of state, military, and nation. In 1941, he clarified the garrison state, characterized by militarization of technology and the rise of specialists on violence in politics and society, was not yet fully realized, even by Nazi Germany. In the Cold War era, the Soviet bloc had developed into a close approximation of a garrison state that cut across international borders. But Lasswell also noted structures trumped ideology, reiterated the importance of military elites in politics, and held out hope that the complete grasp of military elites on world society could be prevented (Lasswell, 1997). His contemporaries in the United States had begun to fear permanent mobilization. C. Wright Mills wrote of power elites in 1956, in 1973 Arthur Schlesinger Jr. pointed at the emergence of an imperial presidency, and in between President Dwight D. Eisenhower warned of the undue influence of the military-industrial complex in his 1961 farewell address. The United States did not become a garrison state (Friedberg, 2000), even though the military-industrial complex could not be dismantled as long as the Cold War confrontation lasted. But where Michael Sherry (1995) was optimistic about the reversal of a culture of militarism, Andrew Bacevich (2005) notes militarization of U.S. foreign policy, economy, and political culture has persisted into the twenty-first century.

In the Communist world, the Soviet Union relied on its military-industrial complex, and militarism was an important feature in Mao Zedong's continuous revolution in China.

Communist modernization required domestic programs, from land reform to forced-draft industrialization and the Cultural Revolution, and Mao used conflicts to rally the nation behind the Communist Party and its great helmsman (Chen Jian, 2001). Yet the army, always closely connected to Chinese society, could serve as a power center, and Mao faced opposition from Marshal Peng Dehuai over the agrarian policies of the Great Leap Forward and from Marshal Lin Biao over rapprochement with the United States in the early 1970s (Li, 2007). Since Deng Xiaoping and his successors have built a blend of Communist Party rule and public-private economy, China's wealth and power have increased; but her foreign policy and state system remain heavily militarized. But perhaps Vietnam is the most poignant example for a successful Communist militarist state born in war. Following Ho Chi Minh's September 1945 declaration of independence, the new regime mobilized society and economy for protracted war against the French and their Vietnamese allies. Initial attempts to build a volunteer army gave way to conscription and a total war effort (Goscha, 2011). While North Vietnam experienced years of peace, revolution, and industrialization in the 1950s and 1960s, the survival of the revolution was threatened and communist party leaders responded with appeals to militarism and nationalism, and they built a garrison state to meet the pressing challenges of reunification and war against the United States (Nguyen, 2012).

Among pro-Western militocracies, South Korea's rise from poverty to prosperity under General Park Chung Hee in the 1960s and 1970s stands out as an example of authoritarian development (Brazinsky, 2009). In many instances, decolonization led to a brief moment of hope for democracy followed by descent into military rule. Examples range from Joseph Mobutu's Zaire to Ne Win's Burma (or Myanmar), General Suharto's Indonesia, and the Philippines under Ferdinand Marcos' rule by martial law. In Latin America, too, the nineteenth-century experience with *caudillos* extended to the recent past. Prominent twentieth-century cases include Nicaragua under the Somoza family (until the late 1970s), Guatemala for much of the Cold War era, Augusto Pinochet's Chile (1973–1990), and military dictatorships in Argentina and Brazil. And Egypt bears witness to the political power of modern military institutions. Perhaps the current presidency of General Abdel Fatah el-Sisi is not an obvious case of militarism—and perhaps the same could be said for earlier military-officers-turned-presidents, from Gamal Abdel Nasser to Hosni Mubarak—but Egypt's political fate has been controlled by the military more than by political parties since the monarchy was overthrown by a group of officers in 1952.

In Western Europe militarism and nationalism declined throughout the Cold War, and Europeans came to consider prosperity and social welfare the main purpose of the state. Historian James Sheehan (2008) notes this was an indirect result of living under the protection of the United States. The transatlantic alliance facilitated economic integration and lowered readiness among Europeans to pay for defense. The exceptions were Britain and France, which waged their long twilight struggles at the end of empire. Britain, in particular, maintained her warfare state longer than others in Western Europe (Edgerton, 2005). But ultimately, empires receded and militarism and nationalism were tempered. West Germany, which emerged from the ruins of the Third Reich, built its identity on stability and prosperity, not nationalism and militarism. The catastrophic defeat of 1945 had destroyed militarism as an acceptable cultural phenomenon—leading to significant domestic opposition to West German rearmament in the 1950s—and it had crippled nationalism for the time being. Even after the Cold War, as a new German nationalism appeared, the culture of antimilitarism persisted (Berger, 1998). Japan saw a similar dynamic as the people, facing absolute defeat, starvation, and the trauma of the atomic bombings of Hiroshima and Nagasaki, embraced a new order (Dower, 1999). It took catastrophic defeat to uproot militarism, but national identity could be rebuilt on different pillars.

We may conclude, optimistically, that the disappearance of militarism and the waning of nationalism in Europe and Japan suggest future world history will be shaped less by wars and violence or by repressive rule of elites or the military. Or we may conclude, pessimistically, that the geographic diffusion of militarism and nationalism ensures they will remain a defining factor in human affairs. Above all, we should remember that militarism takes different shapes and is not always an elite-driven concept intended to control society or wage war. Germany, for example, experienced more than one form of militarism (within the army and within society). But even the United States has experienced militarism. And neither nation experienced the kind of military dictatorships that took root in all too many countries in the nineteenth and twentieth centuries. Whether pessimists or optimists, by expanding our perspective to a global consideration of militarism and nationalism we may appreciate more fully the destructive facets of modernity.

Bibliography

Anderson, Benedict. *Imagined Communities: Reflections on the Origin and Spread of Nationalism*. London: Verso, 1983.

Anderson, Fred. *Crucible of War: The Seven Years' War and the Fate of Empire in British North America, 1754–1766*. New York: Knopf, 2000.

Andreski, Stanislav. *Military Organization and Society*. Second and enlarged edition. Berkeley: University of California Press, 1968.

Bacevich, Andrew J. *The New American Militarism: How Americans Are Seduced by War*. New York: Oxford University Press, 2005.

Barber, John, and Mark Harrison. *The Soviet Home Front: A Social and Economic History of the USSR in World War II, 1941–1945*. New York: Longman, 1991.

Bartov, Omer. *Mirrors of Destruction: War, Genocide, and Modern Identity*. Oxford: Oxford University Press, 2000.

Becker, Frank. *Bilder von Krieg und Nation: Die Einigungskriege in der bürgerlichen Öffentlichkeit Deutschlands 1864–1913*. [Images of War and Nation: the Wars of Unification and the German Bourgeois Public.] Munich: R. Oldenbourg Verlag, 2001.

Bell, David A. *The First Total War: Napoleon's Europe and the Birth of Warfare as We Know It*. New York: Houghton Mifflin Harcourt, 2007.

Bender, Thomas. *A Nation Among Nations: America's Place in World History*. New York: Hill and Wang, 2006.

Berger, Thomas U. *Cultures of Antimilitarism: National Security in Germany and Japan*. Baltimore, MD: The Johns Hopkins University Press, 1998.

Brazinsky, Gregg. *Nation Building in South Korea: Koreans, Americans, and the Making of Democracy*. Chapel Hill: University of North Carolina Press, 2009.

Breuilly, John, ed. *The Oxford Handbook of the History of Nationalism*. Oxford: Oxford University Press, 2013.

Capozzola, Christopher. *Uncle Sam Wants You: World War I and the Making of the Modern American Citizen*. Oxford: Oxford University Press, 2008.

Caverley, Jonathan D. *Democratic Militarism: Voting, Wealth, and War*. Cambridge, UK: Cambridge University Press, 2014.

Centeno, Miguel Angel. *Blood and Debt: War and the Nation-State in Latin America*. University Park: The Pennsylvania State University Press, 2002.

Chen Jian. *Mao's China and the Cold War*. Chapel Hill: University of North Carolina Press, 2001.

Chrastil, Rachel. *Organizing for War: France, 1870–1914*. Baton Rouge: Louisiana State University Press, 2010.

Confino, Alon. *The Nation as a Local Metaphor: Württemberg, Imperial Germany, and National Memory, 1871–1918*. Chapel Hill: The University of North Carolina Press, 1997.

Darwin, John. *After Tamerlane: The Global History of Empire Since 1405*. London: Bloomsbury, 2008.

Deák, István. *Beyond Nationalism: Social and Political History of the Habsburg Officer Corps, 1848–1918*. New York: Oxford University Press, 1992.

Dolski, Michael R. *D-Day Remembered: The Normandy Landings in American Collective Memory*. Knoxville: The University of Tennessee Press, 2016.

Dolski, Michael R., et al. *D-Day in History and Memory: The Normandy Landings in International Remembrance and Commemoration*. Denton: University of North Texas, 2014.

Dower, John. *Embracing Defeat: Japan in the Wake of World War II*. New York: W. W. Norton, 1999.

Drea, Edward J. *Japan's Imperial Army: Its Rise and Fall, 1853–1945*. Lawrence: University Press of Kansas, 2009.

Dudziak, Mary. *War Time: An Idea, Its History, Its Consequences*. New York: Oxford University Press, 2012.

Edgerton, David. *Warfare State: Britain, 1920–1970*. Cambridge, UK: Cambridge University Press, 2005.

Ehrenreich, Barbara. *Blood Rites: Origins and History of the Passions of War*. New York: Metropolitan Books, 1997.

Epstein, Katherine C. *Torpedo: Inventing the Military-Industrial Complex in the United States and Great Britain*. Cambridge, MA: Harvard University Press, 2014.

Fahmy, Khaled. *All the Pasha's Men: Mehmed Ali, His Army and the Making of Modern Egypt*. Cambridge, UK: Cambridge University Press, 1997.

Förster, Stig. *Der doppelte Militarismus: Die Deutsche Heeresrüstungspolitik zwischen Status-Quo Sicherung und Aggression 1890–1913*. [Dual Militarism: German Armaments Policy Between Status Quo and Aggression.] Stuttgart: Franz Steiner Verlag, 1985.

Frevert, Ute. *A Nation in Barracks: Modern Germany, Military Conscription and Civil Society*. Oxford: Berg, 2004.

Friedberg, Aaron L. *In the Shadow of the Garrison State: America's Anti-Statism and Its Cold War Grand Strategy*. Princeton, NJ: Princeton University Press, 2000.

Gat, Azar. *War in Human Civilization*. Oxford: Oxford University Press, 2006.

Gat, Azar, with Alexander Yakobson. *Nations: The Long History and Deep Roots of Political Ethnicity and Nationalism*. Cambridge, UK: Cambridge University Press, 2013.

Gellner, Ernest. *Nations and Nationalism*. Ithaca: Cornell University Press, 1983.

Geyer, Michael. "German Strategy in the Age of Machine Warfare, 1914–1945." In *Makers of Modern Strategy: From Machiavelli to the Nuclear Age*. Ed. Peter Paret. Princeton: Princeton University Press, 1986.

Gillis, John R., ed. *The Militarization of the Western World*. New Brunswick, NJ: Rutgers University Press, 1989.

Goscha, Christopher. *Vietnam: Un État née de la Guerre, 1945–1954*. [Vietnam: A State Born of War, 1945–1954.] Paris: Armand Collin, 2011.

Grenier, John. *The First Way of War: American War Making on the Frontier, 1607–1814*. Cambridge, UK: Cambridge University Press, 2005.

Hogan, Michael J. *A Cross of Iron: Harry S. Truman and the Origins of the National Security State, 1945–1954*. Cambridge, UK: Cambridge University Press, 1998.

Hull, Isabel V. *Absolute Destruction: Military Culture and the Practices of War in Imperial Germany*. Ithaca, NY: Cornell University Press, 2005.

Jansen, Christian, ed. *Der Büger als Soldat: Die Militarisierung europäischer Gesellschaften im langen 19. Jahrhundert: Ein internationaler Vergleich*. [The Citizen as Soldier: Militarization of European Societies in the Long Nineteenth Century: An International Comparison.] Essen: Klartext Verlag, 2004.

Johnson, Chalmers. *The Sorrows of Empire: Militarism, Secrecy, and the End of the Republic*. New York: Henry Holt, 2004.

Johnson, Matthew. *Militarism and the British Left, 1902–1914*. Houndmills: Palgrave Macmillan, 2013.

Jones, Heather. "The German Empire." In *Empires at War, 1911–1923*. Eds. Robert Gerwarth and Erez Manela. Oxford: Oxford University Press, 2014.

Kohn, Richard H. "The Danger of Militarization in an Endless 'War on Terrorism,'" *Journal of Military History* 73 (January 2009):177–208.

Kramer, Lloyd. *Nationalism in Europe and America: Politics, Cultures, and Identities Since 1775*. Chapel Hill: University of North Carolina Press, 2011.

Lasswell, Harold. *Essays on the Garrison State*, ed. Jay Stanley. New Brunswick: Transaction, 1997.

Lee, Wayne. *Barbarians and Brothers: Anglo-American Warfare, 1500–1865*. New York: Oxford University Press, 2011.

Li, Xiaobing. *A History of the Modern Chinese Army*. Lexington: University Press of Kentucky, 2007.

Lone, Stewart. *Army, Empire and Politics in Meiji Japan: The Three Careers of General Katsura Taro*. New York: St. Martin's Press, 2000.

―――. *Provincial Life and the Military in Imperial Japan: The Phantom Samurai.* London: Routledge, 2011.

Lynch, John. *Caudillos in Spanish America, 1800–1850.* Oxford: Oxford University Press, 1992.

McCord, Edward A. *The Power of the Gun: The Emergence of Modern Chinese Warlordism.* Berkeley: University of California Press, 1993.

McFarlane, Anthony. *War and Independence in Spanish America.* New York: Routledge, 2013.

McMeekin, Sean. *The Ottoman Endgame: War, Revolution, and the Making of the Modern Middle East, 1908–1923.* New York: Penguin, 2015.

McNeill, William H. *The Pursuit of Power: Technology, Armed Force and Society Since A.D. 1000.* Chicago: University of Chicago Press, 1982.

Mills, C. Wright. *The Power Elite.* New York: Oxford University Press, 1956.

Miner, Steven Merritt. *The Furies Unleashed: The Soviet Peoples at War, 1941–1945.* New York: Simon and Schuster, 2018.

Morris, Ian. *War! What Is It Good For? Conflict and the Progress of Civilization From Primates to Robots.* New York: Farrar, Straus, Giroux, 2014.

Murray, Williamson, and Allan R. Millett. *A War to Be Won: Fighting the Second World War.* Cambridge, MA: Harvard University Press, 2001.

Nguyen, Lien-Hang T. *Hanoi's War: An International History of the War for Peace in Vietnam.* Chapel Hill: University of North Carolina Press, 2012.

Novak, William J. "The Myth of the Weak American State," *American Historical Review* 113, no. 3 (June 2008): 752–772.

Piehler, G. Kurt. *Remembering War the American Way.* Washington, DC: Smithsonian Institution Press, 1995.

Platt, Stephen R. *Autumn in the Heavenly Kingdom: China, the West, and the Epic Story of the Taiping Civil War.* New York: Knopf, 2012.

Schlesinger, Arthur M., Jr. *The Imperial Presidency.* Boston: Houghton Mifflin, 1973.

Sheehan, James. *Where Have All the Soldiers Gone? The Transformation of Modern Europe.* Boston: Houghton Mifflin, 2008.

Sherry, Michael S. *In the Shadow of War: The United States Since the 1930s.* New Haven, CT: Yale University Press, 1995.

Shimazu, Naoko. *Japanese Society at War: Death, Memory, and the Russo-Japanese War.* Cambridge, UK: Cambridge University Press, 2009.

Snyder, Timothy. *Bloodlands: Europe Between Hitler and Stalin.* New York: Basic Books, 2010.

Sparrow, James T. *Warfare State: World War II Americans and the Age of Big Government.* New York: Oxford University Press, 2011.

Stargardt, Nicholas. *The German Idea of Militarism: Radical and Socialist Critics, 1866–1945.* Cambridge, UK: Cambridge University Press, 1994.

Stoler, Mark R. *Allies and Adversaries: The Joint Chiefs of Staff, the Grand Alliance, and U.S. Strategy in World War II.* Chapel Hill: University of North Carolina Press, 2000.

Stone, David R. *Hammer and Rifle: The Militarization of the Soviet Union, 1926–1933.* Lawrence: University Press of Kansas, 2000.

―――. *The Russian Army in the Great War: The Eastern Front, 1914–1917.* Lawrence: University Press of Kansas, 2015.

Stuart, Douglas T. *Creating the National Security State: A History of the Law That Transformed America.* Princeton, NJ: Princeton University Press, 2008.

Thorpe, Rebecca U. *The American Warfare State: The Domestic Politics of Military Spending.* Chicago: University of Chicago Press, 2014.

Tilly, Charles. *Coercion, Capital, and European States, AD 990–1990.* Cambridge, MA: Basil Blackwell, 1990.

Tooze, Adam. *The Deluge: The Great War and the Remaking of Global Order, 1916–1931.* London: Allen Lane, 2014.

―――. *Wages of Destruction: The Making and Breaking of the Nazi Economy.* London: Allen Lane, 2006.

Trauschweizer, Ingo. "On Militarism," *The Journal of Military History* 76, no. 2 (April 2012): 507–543.

Vagts, Alfred. *A History of Militarism: Romance and Realities of a Profession.* Revised edition. New York: Meridian Books, 1959.

Verhey, Jeffrey. *The Spirit of 1914: Militarism, Myth and Mobilization in Germany.* New York: Cambridge University Press, 2000.

Vogel, Jakob. *Nationen im Gleichschritt: der Kult der "Nation der Waffen" in Deutschland und Frankreich, 1871–1914.* [Nations in Lockstep: The Cult of the Nation in Arms in Germany and France.] Göttingen: Vandenhoeck & Ruprecht, 1997.

Watson, Alexander. *Ring of Steel: Germany and Austria-Hungary in World War I, 1914–1918.* New York: Basic Books, 2014.

Wilson, Mark R. *The Business of Civil War: Military Mobilization and the State, 1861–1865.* Baltimore, MD: The Johns Hopkins University Press, 2006.

Ziemann, Benjamin. *War Experiences in Rural Germany, 1914–1923.* Trans. Alex Skinner. Oxford: Berg, 2007.

26 Memory and Memorialization

Michael Dolski

Wars linger long after the cessation of formal hostilities. From economic dislocation, or the trauma inflicted on civilians and combatants, to national debates on course and conduct of conflicts, and hero worship of martial predecessors, the presence of the military past is often a palpable reality. People and collectives draw on the past to shape identity. War, as a traumatic event, often serves as a defining feature in personal and national remembrance. Individuals and groups literally re-member the past to make sense of the present and anticipate the future. The past, then, is of both individual and collective concern. Historians studying group efforts to solidify common understandings of the past have wrapped these activities into concepts such as collective memory and memorialization. The implications are that the past is malleable rather than constant, that narrative choices reveal perspectives held by storytellers, and groups struggling for political or social power often employ renderings of the past to validate claims for support.

The study of memory and memorialization offers a unique window through which to view the operating mechanics of societies. The relationship between memory and identity has intrigued scholars across multiple disciplines for some time. Individuals like Ernest Renan, Friedrich Nietzsche, and Sigmund Freud, to name several, each explored through the lens of memory the manners in which people and societies engage with the past. In the 1920s, French sociologist Maurice Halbwachs began a systematic review of the group dynamics involved in remembrance (Halbwachs, 1992). Halbwachs asserted that memories of the past represented reconstructions rather than accurate recordings. The interactions between individuals and their larger social groups heavily shaped this reconstruction process. Individuals helped to define yet at the same time were subject to attempts by elements within their society to assert visions of the past, frequently for present political purpose. Social context and societal pressures, in this argument, often determined how people thought about and remembered the past.

Throughout the mid-twentieth century scholars drew on these foundations to explore myths, often in a national context. Joseph Campbell, Roland Barthes, and Richard Slotkin, for instance, each considered the process of myth formation and the influence of myths on societies. Surveys such as theirs suggested ways whereby groups create common understandings of the past and transmit those views throughout their societies, and across generations.

The explorations of memory, identity, trauma, and myth-making forged the way for the study of collective remembrance of military events. By the 1960s and 1970s, a small group of academics began probing more concertedly the ways that societies and their constituents commemorate wars. Bruce Rosenberg (1974), for instance, described the American inclination to remember martial defeats. Taking the loss of George Armstrong Custer's force at the Battle of Little Bighorn as the framework, Rosenberg discerned an American myth-making tradition that extolled heroic defeats as exemplars of national fortitude. Paul Fussell, on the other hand, explored British efforts to turn victory in the First World War into a later-day defeat (Fussell, 1975). His analysis focused on literary tropes prevalent in works produced by authors from privileged backgrounds in the decade following the conflict. Fussell identified an ironic slant that represented the gulf between initial glamorized expectations and subsequent dire perceptions of the war after its conclusion.

Differing in methodological and thematic approaches, these early analyses were sporadic in nature, predating the 1990s "memory boom" that caught up academics and societies alike (Evans and Lunn, 1997; Winter, 2006). A number of factors drove this boom, but one important contributor for American authors was the national obsession with aging war veterans, particularly those of the Second World War. The public devoted increasing attention to the supposed last opportunities for capturing the testimony of the "greatest generation," as they were likened (Brokaw, 1998). Veneration carried dark undertones, however, as hapless officials at the Smithsonian Institution discovered when advancing critical commentary on the popular Second World War in the 1990s. While preparing a display on the *Enola Gay*, the plane that dropped the atomic bomb on Hiroshima, museum officials sought to depict the city's devastation and the lives torn apart as a result. Veterans and their families felt such a move cast aspersions on the American war effort. They characterized it as a venal attempt to equate Japanese "victims" with the supposed real victims of the war—the American people who were treacherously attacked in 1941 (Linenthal and Engelhardt, 1996). Scholars, partially due to such controversies, turned to the modes and meanings of this collective remembrance effort with greater frequency and sophistication.

Halbwachs (1992) identified a duality in approach by stating "the individual remembers by placing himself in the perspective of the group, but one may also affirm that the memory of the group realizes and manifests itself in individual memories" (40). Individual recall is impossible to divorce from this social context. In the modern era, nationalism has pervaded that social context and, as a result, it shapes the processes of war memory and memorialization. As "imagined communities," nations emerge from collectives that draw on different cultural ties, such as shared understandings of the past, to bind together disparate citizens (Anderson, 1991). Since modern wars can represent the largest and most traumatic endeavors for nations, they occupy a pronounced place in commemorative practice. Common understandings of major wars can validate the nation either through tales of victory (and perhaps conquest), which reassure and embolden, or through defeat, from which lessons can emanate for improvement. Wars, understandably, form a divisive touchstone for debates about domestic political choices, apportionment of resources, the role of the military in society, and a nation's place in the international setting.

While authors diverge over terminology, methodology, assessment of causal power, points of emphasis, and topics of coverage, the field of memory and military affairs is now well established and rapidly expanding. In an attempt to convey some of the broader trends and themes in this vibrant area of study, what follows is a topical review of works focused on remembrance of several specific conflicts. Scholarly analyses of war and memory often gravitate toward those conflicts that draw significant popular attention. For the United States, three wars in particular demonstrate this dynamic. Two of them, the Civil War and the Vietnam War, constitute something of an American soul-searching through memory and memorialization. The Second World War has attracted increasing American scholarship using memory as a lens. It appears ahead, however, to show also the ways in which convergent national studies and international comparative studies add to the overall discussion.

The mediated, present-oriented nature of collective remembrance activities explains why certain conflicts draw more attention than others. The following consideration of works by no means suggests that other American wars remain uncovered in the historiography of memory, nor does it indicate that only or primarily American remembrance activities attract the interest of historians. Rather, the case studies demonstrate the animated discussion surrounding the commemoration of several conflicts that continue to shape American identity. The significance to the present manifests itself in an outpouring of popular commemorative activity and scholarly analyses of that memory work.

Although national context provides shape for remembrance processes, the evolution of commemorative practice and the meanings people ascribe to wars are inherently convoluted, iterative, mutable, and often contested (Kammen, 1991). Scholars have struggled over the dynamic interplay between a group's reconstruction of the past and an individual's memory (actual or contrived) of the same. Furthermore, nations remain nebulous constructs of social units that are more or less gelled together (Trouillot, 1995; Sharma and Gupta, 2006). National amorphousness indicates that differences can and usually do materialize within the collective interpretation of the military past. Nations often continue fighting their wars through remembrance activities. The phenomenon of the re-membered past helps explain why memory and military affairs provide such interesting veins for scholastic mining efforts. Collective struggles to shape, define, and reconstruct the past shed revealing light on how a society operates, the ways it perceives itself, and which groups successfully employ cultural power (Piehler, 1995). Choosing which wars to commemorate, for instance, reveals how a society (at least its mainstream) prefers to think of itself.

Some scholars, for instance, have analyzed the American Revolution and its war in American identity and remembrance practices (Purcell, 2002; Lengel, 2011). But the memory and memorialization of other, more recent conflicts draw far more sustained attention. The selection demonstrates in part the tremendous and enduring public fixation on certain events. It also highlights the magnetic pull of divisive scuffles over their interpretation. Ascription of meaning often remains fiercely contested because divergent interpretations of these wars relate to something intrinsic about America as a nation in the past and present.

Thus explained, the historiographical attention to the American Civil War makes sense. The innumerable books, films, monuments, memorials, parks, cemeteries, songs, games, speeches, and reenactments devoted to the conflict indicate that this history is alive in the present. The perennial firestorms that erupt over its larger meaning also warrant scholarly attention, as they suggest something about society itself, a point historians have increasingly focused on over the past several decades.

The divisions that spurred the war and fractured the nation continued after the conclusion of hostilities in 1865. Almost immediately, southern sympathizers began to explain away the defeat as a valiant effort in a worthy cause, but one that was undone by the sheer weight of (often ineptly applied) northern material wealth. This interpretation evolved into a "Lost Cause" myth that celebrated southern honor and martial prowess despite the Confederacy's ultimate demise (Foster, 1987; Gallagher, 2000). Although politically freighted and soothing to those who suffered defeat, this line of interpretation ultimately proved appealing across the North-South divide. It fostered a fixation on battles and the thrill of supposedly honorable combat, while pushing social and political questions to the wayside. Certain battles attracted scholarly and public attention. Gettysburg stood foremost since it evoked the apex of the Lost Cause, supposedly representing the turning point of the war (Reardon, 1998). Part of the explanation for the enduring focus on the battles, and this battle in particular, stemmed from "living history" enthusiasts who routinely dressed in period garb and reenacted moments of camp life or battle (Horowitz, 1998). The draw to certain places, such as Gettysburg, even led to rampant commercialization, thus commodifying memory in a particularly American way (Linenthal, 1991; Weeks, 2003). Battle history, then, presented something that different groups could rally around, either to celebrate victory or to mourn defeat, in no small measure because it sidestepped the thornier causes and consequences of the conflict.

By the 2000s, a shift in emphasis emerged as historians began exploring in greater depth the link among race, memory, and the war. David Blight's *Race and Reunion* (2001) is an important example. Blight indicated that over time white northerners and southerners used the honorable fight narrative to reknit society at the expense of attending to racial justice. Slavery as a cause for war fell into oblivion, the rights of emancipated slaves quickly ebbed, and by the turn of the century the country seemed united in a celebration of (white) military prowess in magnanimous victory or noble defeat. Blight used the fiftieth anniversary of the Gettysburg battle as his capstone for this reconciliationist narrative (see also Blight, 2002; Brundage, 2005). This thrust sought to restore slavery and racism as central features in the story of the Civil War and of America. Arguments in this vein emanated out of and contributed to the great "culture wars" of the 1990s and beyond, which included caustic debates over the use (and abuse) of the American past (Linenthal and Engelhardt, 1996). Scholars since have continued to assess the role of race in the Civil War and its memory, adding new dimensions in the process.

In a demonstration of the present influencing interpretations of the past, recent wars have helped shift scholarly attention to questions of suffering and bereavement over loss following the Civil War (Faust, 2008; Blair, 2004). These studies show the emotive power of loss and its trans-generational effects on individuals, families, and small communities. Individual and group identities, thus, become much more clearly linked to the commemorative focus on this war. Further indicating presentist influences on scholarship, the literature has also devoted more attention to the ways that returning veterans sought (and continue to seek) to define their personal experiences and, in the process, shape collective remembrance of their wars (Harris, 2014).

The Civil War remains hotly contested even 150 years after its conclusion. The recent sesquicentennial observances of the war occurred along with successful efforts to further delegitimize the Confederate flag. Yet the country has also experienced bitter outcries, protests, and violence prompted by racially charged police activity. As many of these Civil War studies attempt to explain, the interpretation of the war depends in large part on contemporary social context (Blight, 2001). Lingering tensions over the meaning of the war constitute an accurate reflection of the wartime era as much as they speak to racial politics and the polarization of our current period (Goldfield, 2013).

The Civil War's persistent popularity explains the focus of historians on its commemoration. Other wars, however, attract minimal public attention and, often as a result, fewer scholarly analyses of their remembrance. The First World War has fascinated those focused on, for example, British remembrance practices (Fussell, 1975; Lloyd, 1998; Watson, 2004). Other works have included additional belligerents, such as Germany (Mosse, 1990), a combination of European nations in comparative review (Winter, 1995; Goebel, 2007), or non-Europeans, such as Australians (Thompson, 1994). American remembrance of the war, on the other hand, has only recently earned sustained scholarly attention, other than perhaps through the field of literary analysis (Hynes, 1997).

Similar to a tendency with Civil War historiography, U.S.-focused scholarship on First World War commemoration has probed official efforts to repatriate soldiers' remains (Budreau, 2009). Scholars have also explored the commemorative role of veterans, changing interpretations of the war and its varying presence in mainstream remembrance (Snell, 2008; Trout, 2012). From the American perspective, however, the historiography on First World War remembrance remains as pronounced as the war in the collective consciousness, which is to say not that much. The relatively recent death of the last American veteran of the war certainly put a final boundary on some form of First World War commemoration (Rubin, 2013). Yet the vibrancy of Civil War remembrance indicates that direct participants in the events are not the sine qua non of collective memory. The relative absence of First World War remembrance indicates that Americans find minimal ties from that war to the present.

Instructive, then, is the fixation on the Second World War, both in American society and, as a result, by scholars interested in the workings of memory and memorialization. Coined the "Good War" by Studs Terkel's oral history collection of the same name (1984), the second war provided a far more popular tale than the first. This book represented the explosion of popular interest in oral history and witness testimony that continues to this day. Terkel tried to capture the incongruity of glorifying war with his title, but the appellation aptly characterized popular sentiments of that conflict. Americans felt that they had engaged in a great crusade to liberate the world from evil oppression. Unlike its predecessor, this war resulted in decisive defeat of the enemy and, in the process (as commemorated), pulled America out of depression and thrust it into a leadership role in the world that endures today (Adams, 1994).

Works on this war have demonstrated an intriguing diversity of methodology and perspective. Some have analyzed popular cultural representations of the war, such as in music and movies (Beidler, 1998). Others have focused on particular battles and the ways in which Americans reconstruct them through commemorative practice (Rosenberg, 2003; Burrell, 2006; Dolski, 2016). These analyses of specific moments provide a great deal of clarity on the nuanced interpretation of an enormous, and enormously significant, event in the American experience. While a tendency persists to make broad-based claims about the entire conflict in national remembrance (Bodnar, 2010), the widely varying experience of the Pacific and European wars—and the divergent interpretations of each—demonstrates that differentiated study is more appropriate.

Americans prefer a glorifying hue in narratives of this war. This slant developed over several interpretations, from an initially less bloody version (glorious because great success required limited sacrifice) to an increasingly grisly take (glorious because America sacrificed so much of value—namely, its own citizens). It is also an interpretation that is widely protected by much of society. For instance, the *Enola Gay* controversy mentioned earlier fed into larger arguments about the "cultural wars" of the 1990s. Through the combination of the two controversies, mainstream media and the average public seemingly rejected as so much academic sophistry attempts to question the actions of America's "greatest generation." The intent of Smithsonian officials to place the *Enola Gay* near pictures of Japanese suffering undercut America's proclaimed morality by suggesting racially tinged obliteration of others lay at the heart of even this most popular war (Linenthal and Engelhardt, 1996).

The particular directions of American-centric remembrance studies stand out more starkly when compared with the evaluations of other countries' commemorative practices. For example, reflecting the American approach, studies of British Second World War remembrance demonstrate a similar emphasis on the "good" features of the war, such as saving humanity from implacable evil. The shining moment for the British, however, also carried negative connotations. Imperial declension followed the war along with subordination to American interests and even political, military, or economic control (Connelly, 2004; Edwards, 2015).

German remembrance of the Second World War has attracted numerous scholars and studies. It is somewhat difficult to segregate the war from the genocidal actions of the Nazi regime, as the two were indeed deeply intertwined. The imbrication of this war and this genocide suggests that some of the vast literature on Holocaust remembrance is worth consideration. In both a reflection of the oral history fervor of war studies, as well as a challenge of its claims for representativeness, Holocaust literature has probed the conflicted ways victims recall their experiences (Langer, 1991; Agamben, 1999; Jacobs, 2010). Some works focus instead on the politics surrounding creation of memorials and their varied interpretations once established (Young, 1993). Several scholars have even dissected the American obsession with the Holocaust, as evidenced by constant reference in film, books, and speeches. Placing a museum to the Holocaust in the heart of the national capital, alongside other national shrines and museums, appropriates an event that, after all, happened to other people in distant countries. Yet, as a partial explanation of the "good war" phenomenon, Americans tend to use the Holocaust as a morality tale to validate the past and to foster acceptance of globalist policies in the present (Flanzbaum, 1999; Novick, 1999).

German society has struggled to develop a useable version of this past (Maier, 1988). Evolving conceptions of Germany's actions in the war were fed by international developments, such as the Cold War and anti-militarist attitudes in the wake of Vietnam, as well as domestic pressures like tension over Germany's fracturing and generational turmoil (Herf, 1997; Bartov, 2003). Regarding the latter point, by the 1960s students and foreign scholarship fomented a critical understanding of the Nazi regime and crimes propagated on its behalf. Commemorative practice, however, continued to cordon off guilt to the supposedly few Nazis or the regime proper while alleging that most Germans remained detached from the war and the Holocaust. Popular views also maintained that the military fought a noble albeit losing battle. Mimicking the American "Lost Cause" myth, this argument overlooked the ideology espoused by the armed forces and the many war crimes perpetrated as a result (not to mention the pass given to larger society). After German reunification, society's tendency to sanitize the war lingered even when confronted by details of the Holocaust it enabled. In the 1990s, a powerful traveling exhibit pointedly depicted the crimes of the *Wehrmacht*, which elicited a widespread though brief and ultimately shallow public debate on the war (Bartov, 2003; Wette, 2006). Despite critical efforts to push wider acceptance of these terrible realities, German society maintains a complicated relationship with the wartime past (Evans, 2015).

The utility of studying each society's commemorative struggle with the Second World War should not mask the benefit of comparative analysis. The unique features of each national context stand out more profoundly when contrasted with versions of the same events advanced by other nations. This more comparative dimension is a relatively recent development in Second World War remembrance historiography. Some scholars have reviewed aspects of the war, such as key battles, while incorporating the perspectives of multiple countries to show similarities and differences (Dolski et al., 2014). For instance, the D-Day battles acquire a different meaning in the context of America ("good war" salvation of others), Germany (noble defeat preceding its own liberation), and France (destruction by others, even while celebrating French military performance in the battles). Pushing beyond the Western-centric focus offered earlier, others have argued for regional analyses of the Pacific war's remembrance by its varied participants (Jager and Mitter, 2007; Murray, 2016).

Matching recent Civil War trends, scholars have argued that race presents a significant influence on American memory and memorialization of the Second World War (Rosenberg, 2003). The "good war" narrative often elides the greater brutality of the Pacific war (for Americans), the varied sense of cultural difference between European or Asian opponents, and the commemorative fixation on saving Europe as opposed to destroying Asia. Race, in fact, pervades the remembrance of America's twentieth-century wars in Asia. Echoing the discussion on the First World War earlier, the Korean War has retained a relatively minor role in American popular consciousness and historical analyses of collective remembrance. The uncertain and deflating conclusion to the war plays some role in its popular neglect, which has resulted in the frequently used "forgotten war" designation (Edwards, 2000). Nevertheless, a more concerted review of its course and conduct would raise troubling questions about racism, rampant destruction, and noncombatant casualties, as well as enduring harmful aftereffects, like military-related prostitution (Cho, 2008). Perhaps there is something more willful about American forgetfulness regarding this war. Predictably, these murkier issues fall to the wayside in the few moments of American mass reflection on the Korean War.

Whereas mainstream America tends to avoid commemorating the Korean War, it has remained painfully enthralled in seemingly interminable debates over the Vietnam War. The soul-searching and recriminations reveal continuing debates in American society about the role (and trustworthiness) of government, proper use of military force, the limits of dissenting citizenship, and the potential militarization of society (Bacevich, 2005). Sensibly, scholars have subjected the commemorative politics and practices associated with this war to repeated and in-depth analyses. In many ways, the Vietnam War remembrance literature indicates intriguing trends in the historiography of war memory and memorialization. Since the end of the war in the mid-1970s (even before), popular American engagement with the Vietnam War has obscured many of the questions about the outbreak, conduct, and course of the conflict, instead wrapping up the experience as an American tragedy. A haunting fixation on American losses or what Americans did to one another (as opposed to the Vietnamese or others) shaped the public discourse, especially since the 1980s.

Scholarly reviews of this conflict's powerful presence have taken many different approaches. One work considered the development of the war in film and television shows (Anderegg, 1991). These productions often employed tropes about violent veterans disrupting society upon their return, which served as a screen memory (literally and figuratively) for debates about what America did to its soldiers and the Vietnamese people. Other scholars noted the mixture of metaphors and the uncanny ways that guilt or anger over Vietnam percolated throughout American society, such as with the discourse about AIDS (Sturken, 1997). Prevalent myths received scholarly treatment as well. Different books considered such inflammatory topics as allegations of Americans spitting upon their returning soldiers (Lembcke, 1998) or the apotheosis of Jane Fonda as a figure of intense hatred by conservative society (Lembcke, 2010). Recently, a turn to other cultural productions, such as the popular music of the era, has revealed an interesting approach to considering how people think about, remember, and even at times fantasize about the past (Bradley, 2015).

Traditional analyses of memory and memorialization often include consideration of monuments. None stands more iconic in American remembrance of the Vietnam War than does the Vietnam Veterans Memorial Wall in Washington, DC. Some have deftly described its politicized design and building process (Allen, 2009). Others have explored different ways in which Americans make sense of the war through this monument. For instance, one ritualized activity there is the invocation of the dead through hand scribbling the Wall's embossed names of American casualties onto a piece of paper. Visitors also frequently leave items and notes explaining their significance. Taking scribbled names and leaving meaningful votives are actions that reveal the highly personalized engagement with the war's many meanings that this site affords (Sturken, 1997; Hass, 1998). Beyond that monument, others have turned to an exploration of the thousands of memorials dedicated to the war (primarily to the American veterans) spread throughout the country (Hagopian, 2009). These works probe American struggles over interpretation of the Vietnam War as a futile endeavor, tragic defeat, heroic albeit stunted victory, or an indictment against capitalized militarism. These acidic debates continue in large part because they speak to American identity and America's role in the world, both consequential particularly in a new season of divisive wars abroad (Appy, 2015).

This intensive self-reflection does come at a cost, though, as alluded to earlier with the Second World War historiography. What is often missing in these probing analyses is consideration of the fluid interaction between memories or remembrance practices of different countries. The relative dearth of comparative remembrance studies is understandable in light of the difficulty of discussing a "national" memory with any degree of coherence. Nevertheless, Vietnam War memory historiography shows interesting trends to account for this gap. Some scholars have turned increased attention to Vietnamese remembrance practices tied to the "American War" (to distinguish it among their string of twentieth-century conflicts). Vietnamese cultural practices, for instance, include using ghost stories that seek to place the overwhelming death and destruction in a familiar context (Kwon, 2006; Schwenkel, 2009). Alternatively, this scholarship provides apt opportunities for comparative work by describing the "other" side of remembrance tied to heavily studied events, like the My Lai massacre (Kwon, 2008). Chong explored Vietnamese official efforts to exploit the notorious picture of a young Kim Phuc running in pain after suffering through a napalm attack on her village. In the process, Chong also recounted some of the American, Soviet, Canadian, and even Cuban reactions. Even more ambitious, recent efforts have placed the war in a regional context and thereby analyzed the remembrance of traditionally overlooked groups, such as the Hmong, the Lao, and even the South Koreans who fought as U.S. allies in South Vietnam (Nguyen, 2016).

These last few works present an opportunity to discuss the interdisciplinary opportunities for studies of war memory and memorialization. Many of the works cited earlier emanate from other disciplines (e.g., Nguyen representing English, and Kwon standing for anthropology). The interdisciplinary possibilities of this field become readily apparent when reviewing some of the major journals or outlets for memory studies, such as *H-Memory* or *History & Memory*. Recent academic conferences, including annual meetings of the Society for Military History, have also focused attention on remembrance. Historians, and other interested parties, would be remiss to pass over the fruitful methodological approaches presented in these settings.

Historiographical gaps are the lifeblood of new scholarship; although history and memory studies have exploded over the past two decades, there remain many interesting topics for consideration. Much of the extant scholarship offers unique ways to look at societies struggling to define themselves through recalling the past. Yet another approach open to scholars is the consideration of an active process of forgetting or silencing the past (Trouillot, 1995; Ricoeur, 2004). As some scholars have rightly reminded us, the absence of sustained attention on historically relevant subjects is also worthy of exploration. For instance, Americans have tended to pay more attention to the flare-up "hot" wars in the mid- to late twentieth century at the expense of considering the contours and meanings of the larger Cold War context (Wiener, 2012). In the effort, then, each constituent conflict becomes decontextualized, which empowers American claims to innocent victimhood or noble intentions with military interventions abroad.

The tension between remembering and forgetting is, as Halbwachs (1992) and others have indicated, hardly free from influence. Particularly in the American context, jockeying by different groups to assert emotional or political power often expands into the realm of memory politics (Bodnar, 1992; Kammen, 1991; Piehler, 1995). The vast American "marketplace of ideas" affords many groups the opportunity to assert their claims to the past. This effort is often conducted to gain social purchase and perhaps elicit public support or policy changes (Foner, 1998). Recurrent themes in American commemorative politics include this struggle for social power, which has depended in large part on the democratization of memory over the past few decades (Kammen, 1991). New actors appear in the stories told about the past (not just the "grand figures" of the past), but at the same time new actors participate in the discussions of that past. The stories told, often echoing cultural divides of the present period, tend to splinter into traditional positive remembrance of the past or revisionist critique of (somehow falsified) collective memories (Dolski, 2016).

Throughout all of this memory work, the stress between present and past remains palpable. As Halbwachs (1992) rightly noted, individuals remember, whereas the social context influences how and what they recall. The differentiation of wars deserving consistent attention and those fit for oblivion serves to flesh out characteristics of these social settings.

Furthermore, the imperative to honor predecessors while drawing on (a positive) past to validate the (potentially positive) present does tend to obscure historical events just as much as it helps keep them alive (Dolski, 2016). Yet historians should remember that collective memory usually forms the "history" that mass society engages with on a frequent basis. To understand that important point helps to shed light on the significance of the tortuous relationship between race and American war memory for the three conflicts considered earlier.

American society in particular seems prone to engage with historical issues in several distinctive ways. Literary treatments of warfare, particularly those penned by veterans, attract steady public interest. These sources carry profound influence on widespread understandings of the events (Hynes, 1997). Film also offers a prominent avenue for mass society to learn from or experience history. Thus, film studies are serious contributors to the discussion on American remembrance—particularly regarding warfare, as it remains a popular and prominent topic in film and television (Rollins and O'Connor, 2008). The American approach to remembering war usually includes a significant degree of commodification (Nguyen, 2016). Selling "memories" occurs either through kitsch transactions or via mass industrialized production, like Hollywood filmmaking or even the sale of something like the Volkswagen Beetle (Evans, 2015). Consumerism, a particular hallmark of American society, permeates its attempt to memorialize the martial past (Groot, 2009).

One other American idiosyncrasy with war memory appears in the treatment of the soldier dead (Sledge, 2005). The United States government likely expends more than the rest of the world combined in its attempt to recover, identify, and return to families the remains of missing service personnel from past conflicts. Earlier wars, such as the Civil War (Faust, 2008) and the First World War (Budreau, 2009), witnessed an intense American effort to recover and transfer home remains, often to a far greater degree than exhibited by other belligerents. The current iteration of this effort largely grew out of debates over official treachery and individual (American) victimhood in Vietnam. Yet this activity has since expanded to encompass other conflicts as well (Allen, 2009). The remains issue serves as an excellent liminal zone contrasting official/national memories with those of the families affected. The government, of course, prefers to extol its intent to "leave no man behind," whereas the families manifest a far greater diversity of opinions, ranging from happy acceptance to sharp rebuke (too little, too late), to outright rejection of this effort. The very fact that the United States devotes so much attention to dealing with its soldier dead suggests the significance of martial duty in the American collective consciousness. It also shows an intriguing attempt to sanitize the wars of the past by glorifying the (silent) dead and using their recovery as an example of official largess (Wagner, 2015). When considered in relation to critiques of the wars in question, say the Korean War (Cho, 2008), the multivalent intentions involved in shaping the past for collective consumption become readily apparent.

This review of the historiography on war memory and memorialization indicates that the field is energetic, extensive, and continuing to expand at a dramatic pace. Popular outlets focus a great deal of attention on the commemoration of wars past, and, as a result, scholars have increased the amount of coverage on military remembrance. Consider the decision by academic presses, such as Oxford University or the University of Tennessee, to foster new book series on this topic, which demonstrates its enduring salability. The foregoing chief case studies serve only as indicators of the overall field. One could readily encounter titles on other conflicts, those focused on societies other than America, or within the burgeoning subcategory of nationally comparative studies.

The market presence of these works does much to indicate the extreme popularity of the "memory boom." Scholars and the public remain deeply interested in the remembrance of military affairs. Anniversaries of major wars or their key events tend to provoke increased public and popular attention, which in turn elicits new scholarly treatments of the phenomenon. Take, for instance, the recent smattering of books on American Civil War remembrance timed near the sesquicentennial anniversaries of the war (2011–2015) or the few recent works on American memorialization of the First World War right as the hundredth anniversary years kicked off (2014–2018). Beyond the magnetism of milestone anniversaries, new conflicts tend to draw renewed interest in commemorations of previous military affairs. As one example, consider the extremely provocative and highly emotional discussion of the Vietnam War, and remembrance of its events and its (American) soldiers, in light of the recent wars against terrorism (Simpson, 2006).

What these points suggest is that people will continue to find value in the attempt to shape society's views on wars past. Differences over identity politics or the use of military force in the world today often involve the use of war memory and memorialization by contending groups. Claims for social justice or access to resources also involve people drawing on the martial past to advance a cause or rebut the arguments of others. Regarding resources, consider the public debate over reparations payments to African Americans to compensate for the lingering impact of slavery and recent commemorative debates over the Civil War. Obviously, people discern value in drawing on changing interpretations of the past for present, often quite political, purposes. With that reality bound to continue, the market for studies on war remembrance remains open for new entries.

Bibliography

Adams, Michael C. C. *The Best War Ever: America and World War II*. Baltimore: Johns Hopkins University Press, 1994.

Agamben, Giorgio. *Remnants of Auschwitz: The Witness and the Archive*. Trans. by Daniel Heller-Roazen. New York: Zone Books, 1999.

Allen, Michael J. *Until the Last Man Comes Home: POWs, MIAs, and the Unending Vietnam War*. Chapel Hill: University of North Carolina Press, 2009.

Anderegg, Michael, ed. *Inventing Vietnam: The War in Film and Television*. Philadelphia: Temple University Press, 1991.

Anderson, Benedict. *Imagined Communities: Reflections on the Origin and Spread of Nationalism*. Rev. ed. New York: Verso, 1991.

Appy, Christian G. *American Reckoning: The Vietnam War and Our National Identity*. New York: Viking, 2015.

Bacevich, Andrew. *The New American Militarism: How Americans Are Seduced by War*. New York: Oxford University Press, 2005.

Bartov, Omer. *Germany's War and the Holocaust: Disputed Memories*. Ithaca: Cornell University Press, 2003.

Beidler, Philip D. *The Good War's Greatest Hits: World War II and American Remembering*. Athens: The University of Georgia Press, 1998.

Blair, William A. *Cities of the Dead: Contesting the Memory of the Civil War in the South, 1865–1914*. Chapel Hill: University of North Carolina Press, 2004.

Blight, David W. *Beyond the Battlefield: Race, Memory, and the American Civil War*. Amherst: University of Massachusetts Press, 2002.

———. *Race and Reunion: The Civil War in American Memory*. Cambridge, MA: Harvard University Press, 2001.

Bodnar, John E. *The "Good War" in American Memory*. Baltimore: The Johns Hopkins University Press, 2010.

———. *Remaking America: Public Memory, Commemoration, and Patriotism in the Twentieth Century*. Princeton: Princeton University Press, 1992.

Bradley, Doug. *We Gotta Get Out of This Place: The Soundtrack of the Vietnam War*. Amherst: University of Massachusetts Press, 2015.

Brokaw, Tom. *The Greatest Generation*. New York: Random House, 1998.

Brundage, W. Fitzhugh. *The Southern Past: A Clash of Race and Memory*. Cambridge, MA: Belknap Press of Harvard University Press, 2005.

Budreau, Lisa M. *Bodies of War: World War I and the Politics of Commemoration in America, 1919–1933*. New York: New York University Press, 2009.

Burrell, Robert S. *The Ghosts of Iwo Jima*. College Station: Texas A&M University Press, 2006.

Cho, Grace M. *Haunting the Korean Diaspora: Shame, Secrecy, and the Forgotten War*. Minneapolis: University of Minnesota Press, 2008.

Chong, Denise. *The Girl in the Picture: The Story of Kim Phuc, the Photograph, and the Vietnam War*. New York: Penguin, 1999.

Connelly, Mark. *We Can Take It!: Britain and the Memory of the Second World War*. New York: Pearson, 2004.

Dolski, Michael R. *D-Day Remembered: The Normandy Landings in American Collective Memory*. Knoxville: University of Tennessee Press, 2016.

Dolski, Michael R., et al., eds. *D-Day in History and Memory: The Normandy Landings in International Remembrance and Commemoration*. Denton: University of North Texas Press, 2014.

Edwards, Paul M. *To Acknowledge a War: The Korean War in American Memory*. Westport, CT: Greenwood Press, 2000.

Edwards, Sam. *Allies in Memory: World War II and the Politics of Transatlantic Commemoration, c. 1941–2001*. New York: Cambridge University Press, 2015.

Evans, Martin, and Ken Lunn, eds. *War and Memory in the Twentieth Century*. New York: Berg, 1997.

Evans, Richard J. *The Third Reich in History and Memory*. New York: Oxford University Press, 2015.

Faust, Drew Gilpin. *This Republic of Suffering: Death and the American Civil War*. New York: Knopf, 2008.

Flanzbaum, Hilene, ed. *The Americanization of the Holocaust*. Baltimore: Johns Hopkins University Press, 1999.

Foner, Eric. *The Story of American Freedom*. New York: W. W. Norton, 1998.

Foster, Gaines M. *Ghosts of the Confederacy: Defeat, the Lost Cause, and the Emergence of the New South, 1865–1913*. New York: Oxford University Press, 1987.

Fussell, Paul. *The Great War and Modern Memory*. New York: Oxford University Press, 1975.

Gallagher, Gary W., and Alan T. Nolan, eds. *The Myth of the Lost Cause and Civil War History*. Bloomington: Indiana University Press, 2000.

Goebel, Stefan. *The Great War and Medieval Memory: War, Remembrance and Medievalism in Britain and Germany, 1914–1940*. New York: Cambridge University Press, 2007.

Goldfield, David. *Still Fighting the Civil War: The American South and Southern History*. Baton Rouge: Louisiana State University Press, 2013.

Groot, Jerome de. *Consuming History: Historians and Heritage in Contemporary Popular Culture*. New York: Routledge, 2009.

Hagopian, Patrick. *The Vietnam War in American Memory: Veterans, Memorials, and the Politics of Healing*. Amherst: University of Massachusetts Press, 2009.

Halbwachs, Maurice. *On Collective Memory*. Ed. and trans. Lewis Coser. Chicago: University of Chicago Press, 1992.

Harris, Keith. *Across the Bloody Chasm: The Culture of Commemoration Among Civil War Veterans*. Baton Rouge: Louisiana State University Press, 2014.

Hass, Kristin Ann. *Carried to the Wall: American Memory and the Vietnam Veterans Memorial*. Berkeley: University of California Press, 1998.

Herf, Jeffrey. *Divided Memory: The Nazi Past in the Two Germanys*. Cambridge, MA: Harvard University Press, 1997.

Horowitz, Tony. *Confederates in the Attic: Dispatches From the Unfinished Civil War*. New York: Pantheon, 1998.

Hynes, Samuel. *The Soldiers' Tale: Bearing Witness to Modern War*. New York: Viking, 1997.

Jacobs, Janet. *Memorializing the Holocaust: Gender, Genocide and Collective Memory*. London: I.B. Tauris, 2010.

Jager, Sheila Miyoshi, and Rana Mitter, eds. *Ruptured Histories: War, Memory, and the Post-Cold War in Asia*. Cambridge, MA: Harvard University Press, 2007.

Kammen, Michael G. *Mystic Chords of Memory: The Transformation of Tradition in American Culture*. New York: Vintage Books, 1991.

Kwon, Heonik. *After the Massacre: Commemoration and Consolation in Ha My and My Lai*. Los Angeles: University of California Press, 2006.

———. *Ghosts of War in Vietnam*. New York: Cambridge University Press, 2008.

Langer, Lawrence L. *Holocaust Testimonies: The Ruins of Memory*. New Haven, CT: Yale University Press, 1991.

Lembcke, Jerry. *Hanoi Jane: War, Sex, and Fantasies of Betrayal*. Amherst: University of Massachusetts Press, 2010.

———. *The Spitting Image: Myth, Memory, and the Legacy of Vietnam*. New York: New York University Press, 1998.

Lengel, Edward G. *Inventing George Washington: America's Founder, in Myth and Memory*. New York: Harper-Collins, 2011.

Linenthal, Edward, and Tom Engelhardt, eds. *History Wars: The Enola Gay and Other Battles for the American Past*. New York: Metropolitan Books, 1996.

Linenthal, Edward Tabor. *Sacred Ground: Americans and Their Battlefields*. Urbana: University of Illinois Press, 1991.

Lloyd, David William. *Battlefield Tourism: Pilgrimage and the Commemoration of the Great War in Britain, Australia and Canada, 1919–1939*. New York: Bloomsbury Academic, 1998.

Maier, Charles. *The Unmasterable Past: History, Holocaust, and German National Identity*. Cambridge, MA: Harvard University Press, 1988.

Mosse, George. *Fallen Soldiers: Reshaping the Memory of the World Wars*. New York: Oxford University Press, 1990.

Murray, Stephen C. *The Battle Over Peleliu: Islander, Japanese, and American Memories of War*. Tuscaloosa: University of Alabama Press, 2016.

Nguyen, Viet Thanh. *Nothing Ever Dies: Vietnam and the Memory of War*. Cambridge, MA: Harvard University Press, 2016.

Novick, Peter. *The Holocaust in American Life*. New York: Houghton Mifflin, 1999.

Piehler, G. Kurt. *Remembering War the American Way*. 1995; reprint, Washington, DC: Smithsonian Books, 2004.

Purcell, Sarah. *Sealed With Blood: War, Sacrifice, and Memory in Revolutionary America*. Philadelphia: University of Pennsylvania Press, 2002.

Reardon, Carol. *Pickett's Charge in History and Memory*. Chapel Hill: University of North Carolina Press, 1997.

Ricoeur, Paul. *Memory, History, Forgetting*. Trans. Kathleen Blamey and David Pellauer. Chicago: University of Chicago Press, 2004.

Rollins, Peter C., and John E. O'Connor. *Why We Fought: America's Wars in Film and History*. Lexington: The University Press of Kentucky, 2008.

Rosenberg, Bruce. *Custer and the Epic of Defeat*. University Park: Pennsylvania State University Press, 1974.

Rosenberg, Emily S. *A Date Which Will Live: Pearl Harbor in American Memory*. Durham, NC: Duke University Press, 2003.

Rubin, Richard. *The Last of the Doughboys: The Forgotten Generation and Their Forgotten War*. New York: Houghton Mifflin Harcourt, 2013.

Schwenkel, Christina. *The American War in Contemporary Vietnam: Transnational Remembrance and Representation*. Bloomington: Indiana University Press, 2009.

Sharma, Aradhana, and Akila Gupta, eds. *The Anthropology of the State: A Reader*. New York: Wiley-Blackwell, 2006.

Simpson, David. *9/11: The Culture of Commemoration*: Chicago: University of Chicago Press, 2006.

Sledge, Michael. *Soldier Dead: How We Recover, Identify, Bury, and Honor Our Military Fallen*. New York: Cambridge University Press, 2005.

Snell, Mark A. *Unknown Soldiers: The American Expeditionary Forces in Memory and Remembrance*. Kent: Kent State University Press, 2008.

Sturken, Marita. *Tangled Memories: The Vietnam War, The AIDS Epidemic, and the Politics of Remembering*. Berkeley: University of California Press, 1997.

Terkel, Studs. *"The Good War": An Oral History of World War II*. New York: The New Press, 1984.

Thompson, Alistair. *Anzac Memories: Living With the Legend*. New York: Oxford University Press, 1994.

Trouillot, Michel-Rolph. *Silencing the Past: Power and the Production of History*. Boston: Beacon Press, 1995.

Trout, Steven. *On the Battlefield of Memory: The First World War and American Remembrance, 1919–1941*. Tuscaloosa: University of Alabama Press, 2012.

Wagner, Sarah. "A Curious Trade: The Recovery and Repatriation of Vietnam MIAs," *Comparative Studies in Society and History* 57, no. 1 (2015): 161–190.

Watson, Janet S. K. *Fighting Different Wars: Experience, Memory, and the First World War in Britain*. New York: Cambridge University Press, 2004.

Weeks, Jim. *Gettysburg: Memory, Market and an American Shrine*. Princeton: Princeton University Press, 2003.

Wette, Wolfram. *The Wehrmacht: History, Myth, Reality*. Cambridge, MA: Harvard University Press, 2006.

Wiener, Jon. *How We Forgot the Cold War: A Historical Journey Across America*. Los Angeles: University of California Press, 2012.

Winter, Jay. *Remembering War: The Great War Between Memory and History in the Twentieth Century*. New Haven: Yale University Press, 2006.

———. *Sites of Memory, Sites of Mourning: The Great War in European Cultural History*. New York: Cambridge University Press, 1995.

Young, James. *The Texture of Memory: Holocaust Memorials and Meaning*. New Haven, CT: Yale University Press, 1993.

27 War and Culture

Matthew S. Muehlbauer

The preceding chapters in this volume have addressed numerous aspects of the war and society field. But an additional topic that requires some commentary is war and culture. It is problematic, in that culture is difficult to define, particularly in a manner that facilitates consistent methodological and analytical approaches. Culture is about meaning, how it is constructed, expressed, and disseminated. It is manifested in various types of media and art, ideas and concepts, but also in action and speech. It is shared, and helps to define, communities and networks of various scales, and in varying contexts. Discrete cultures, for example, supposedly exist from the supranational level down to particular regions, municipalities, and neighborhoods; for particular ethnic, racial, class, and gender categories; for specific vocations and professions; and for those who adopt distinctive lifestyles, or endure unique circumstances. In this regard, culture and identity are crucially interlinked. (For interested readers, separate chapters in this volume address the intersection of war and gender, and war and race, written by Heather Stur, and by David Ulbrich and Alexander Bielakowski, respectively.)

Moreover, culture both reflects and stems from social interactions, which in turn are varied and occur in diverse contexts. In this regard, culture can also be contested. Such challenges often echo clashes over power, resources, or a group's demands for greater recognition. Whether accepted or disputed, cultural meanings exist within a broader social context, and studies of the former require the latter receive some attention.

This chapter will regard cultural studies as those whose primary focus is analyzing the creation and propagation of meaning, even if such also address the social relations and constructions that facilitate it. More specifically, the discussion will highlight works that examine the role played by war in this phenomenon, in particular those that delve into the creation of expectations for conduct in warfare.

Before doing so, it bears noting that not all attempts to apply "culture" to warfare have been welcomed by scholars. Most infamous in this regard is Victor David Hanson's *Carnage and Culture: Landmark Battles in the Rise of Western Power* (2001). Hanson claims that over the past 2,500 years, Western combatants have been more militarily successful and effective than those elsewhere because of their culture, identifying specific facets of it, such as "civic militarism," capitalism, individual freedom, reason and rational inquiry, and a preference for brutal, decision-seeking combat. To demonstrate the military and cultural superiority of the "Western" way of war, he selects nine battles between Western and non-Western opponents from the ancient to the modern era, including Salamis (supposedly illustrating the West's emphasis on individual freedom), Lepanto (the power of capitalism), and Rorke's Drift (combat discipline).

Carnage and Culture has received no shortage of criticism. Hanson's focus on various battles scattered around the globe, over the course of two and a half millennia, is not sufficient to prove continuity. He examines particular characteristics of Western culture only in the context of a specific battle, without considering how they might have changed across time and space, or if they have occurred in non-Western societies.

Among other points John A. Lynn, one of Hanson's fiercest critics, notes European societies of the late medieval and early modern eras did not exemplify civic militarism, nor did armies demonstrate a preference for decisive battles. He also contests Hanson's assertions about the distinctiveness of Western culture, and the degree to which Western forces have been victorious over non-Western ones (Lynn, 2003). Another problem is that the book promotes a simplistic dichotomy between "Western" and "Eastern" modes of warfare, one that indulges rather than challenges stereotypes. Jeremy Black observes that the idea of a "Western" way of war serves a contemporary purpose, offering some conservative commentators a means to challenge "political correctness and multiculturalism" while also "encouraging American interventionism" (2004, 57–58).

Hanson was not the first to advance the notion of a "way of war." Black expresses skepticism of cultural approaches to conflict because some echo previous writings claiming to identify such distinct "ways" for a given country (Black, 2004a; 2004b). The epitome here is the notion of an American way of war, propounded by Russell F. Weigley (1973). Weigley claimed the United States preferred short, resource-intensive wars in which conventional operations sought to overwhelm an opponent and limit casualties, with chief examples being the Union effort in the American Civil War, and the Second World War. Prior to the former, he claimed that limited resources had forced the country to employ strategies of attrition, but thereafter U.S. leaders preferred potentially quicker and more destructive strategies of annihilation. Writing at the end of the Vietnam War, Weigley's book sought in part to explain why U.S. prowess in conventional operations—so successful in the Second World War—had failed to achieve similar results in the Vietnam War. His answer was that the Cold War had ushered in a departure from previously successful strategies, in that American leaders could no longer achieve "decisive" victories at a reasonable cost.

The claim that the United States had abandoned a traditional, previously successful strategy echoed Sir Basil Henry Liddell Hart's earlier notion of a British "way in warfare" (1932), one he claimed emphasized landing small professional forces on the periphery of Europe, and one that Britain had supposedly forsook when it committed a mass army to the Continent in World War I. Weigley's analysis added a cultural component in his assertion that Americans preferred short, intense wars (though he treated it as axiomatic), and it dominated interpretations of U.S. military history for the next quarter century. Even then, it continued to influence arguments of American warfighting in social science and defense literature (Gray, 2006). But as with Hanson (and Liddell Hart), it came under serious critique. Among other points, critics noted that American leaders had often chosen not to resort to overwhelming use of force, and rather had adapted to circumstances, even after the Civil War. In fact, over the course of its entire history, the types of conflicts the United States had engaged in demonstrated variety and adaptability (Linn, 2002; Echevarria, 2014).

Weigley wrote at a time when work on cultural theory and perspectives had not yet gained much currency among historians of war. Hanson could not make that claim with *Carnage and Culture*. Rather, over the past few decades various historians have borrowed concepts from other disciplines to pursue more refined and analytically rigorous approaches to examining war and culture. Some have been drawn to anthropologist Clifford Geertz's notion of culture as "fiction"—not as something untrue but rather as something created to impart meaning (1973). Those seeking to explain changes in the conduct of war are often attracted to the notion of "paradigm shift" promulgated by Thomas S. Kuhn in 1962 (2012). A philosopher of science, Kuhn asserted that scientific inquiry occurred within a paradigm comprising ideas, methodologies, experimental procedures, and so forth. Such a conceptual regime shapes the nature of research by both identifying avenues of inquiry and obscuring those possibilities that cannot be comprehended with the paradigm. Yet, when experimentation yields enough results beyond what it can explain, the result is a paradigm shift wherein a new regime of concepts and experimental approaches becomes predominant.

These do not exhaust the list of ideas historians have employed in their methodologies. But one that students and scholars of conflict should examine is the model that John A. Lynn developed to specifically address the interaction between war and culture, which draws upon earlier scholarship. In his *Battle: A History of Combat and Culture* (2003), he proffers eight case studies of how culture shapes the ways in which warfare can be conceived and prosecuted. A number of these stem from early modern Europe, in which he specializes, but also include ancient and medieval examples, as well as ones from colonial India, World War II, and modern Egypt. Lynn observes that "the essential value of using a cultural approach in military history is . . . in distinguishing the mental from the material." He thus focuses on "what could be termed 'conceptual culture,' that is values, beliefs, assumptions, expectations, preconceptions, and the like" (xix–xx). Moreover, Lynn distinguishes between this conceptual culture, which he deems the "discourse" of war, and the reality of the war.

Lynn claims that approaches to war stem from interactions between this discourse and actual experience. A given society will wage war according to its prevailing discourse; but when real developments occur that question or possibly invalidate it, the result is a modification in the discourse. Sometimes the difference between discourse and reality is so great that a society might create a "perfected" form of war, approximating an idealized form of conflict (e.g., medieval tournaments), or an "alternative" discourse that allows for actions more violent or excessive than what would be accepted by the orthodox one (e.g., if an enemy is depicted as "barbaric"). Lynn observes that a society, given its various component groups, can possess multiple discourses of war. That said, he notes that "not all segments of society are equal," and that the views of elites "usually weigh more heavily" in perspectives on and conceptions of warfare (333). Professionalized military forces also generally have their own discourse on war.

Other scholars have scrutinized the relationship between war and culture through the lens of discourse. The most powerful of these discern how the wars have produced long-term and potent cultural consequences. In his seminal *The Great War and Modern Memory* (1975), Paul Fussell examined the writings of combat veterans, arguing that the dissonance between expectations and the reality of combat in the World War I helped create the modern sense of irony. Jill Lepore examines an early American case, observing that in the aftermath of King Philip's War, colonists wrote about it in a way that bolstered their sense of English identity—including depicting native warriors as "savage" (1998). More recently, Peter Silver has tied native attacks in Pennsylvania's backcountry during the Seven Years' War to political pamphleteering in the colony, which helped create the idea of race (2008).

Whereas these authors have specifically examined written works, Peter Paret has scrutinized depictions of warfare in European art from the sixteenth to the early twentieth century. He notes how these have varied not just over time, and with evolving artistic movements, but also by the social class of the intended audience (1997). More recently, Frans de Bruyn and Shaun Regan have edited an anthology that examines various cultural aspects of the Seven Years' War, including treatments in literature and fine art, as well as depictions of identity (2014). Drew Gilpin Faust has scrutinized how the American Civil War upended social conventions and perspectives on death (2008). Her book is related to what has become the most prominent academic intersection of culture and war in the past generation, the history of memory, particularly research that addresses the relationship between war and public memory. Whereas some works explore the processes that have shaped how particular events are recalled in popular culture (Reardon, 1997), others address the question of memorialization (Winter, 1995). Readers interested in this literature should consult Michael Dolski's chapter in this volume, which assesses memory and memorialization in more detail (see also Dolski, 2016).

But as implied by his book's title, Lynn's framework is geared toward considering how culture shapes approaches to combat, and how the latter change in reaction to experience. As with some of the immediately aforementioned books, he is not the first to do. Rather, his model synthesizes various conceptual approaches and acknowledges prior fruitful methodologies, which can be subsumed within it. These include investigations into how a discourse of combat forms and evolves over time, and how one might become dominant over alternatives; as will be seen, multiple and rival discourses can exist within the same military institution. Another examines how preparations for and conduct in combat reflect a specific cultural context. A third methodology assesses how various institutions and other social actors mediate the interaction between discourse and reality, which is of particular salience for studies of specific armed forces and their cultures. The remainder of the chapter will discuss a far-from-exhaustive set of books that exemplify these approaches (including a few that adopt a more tightly focused "ways of war" approach), published both before and after Lynn's *Battle*.

Consider ancient history. Despite the controversy stirred by *Carnage and Culture*, Hanson had previously expounded upon the nature and experience of hoplite combat, noting the broader social context and cultural practices that surrounded it in his work actually entitled *The Western Way of War* (1989). Hoplites were generally not full-time soldiers, but rather free men whose wealth and status enabled them to bear arms on behalf of their polis or city-state. Hence the phalanxes of heavy infantry fielded by opposing poleis sought to limit conflict to a single, brutal battle, thus constraining the degree of death and destruction. Moreover, various rituals evolved to reinforce each side's acceptance of the battle's outcome as final, such as the exchange of bodies and erection of monuments. Hanson claimed that the legacy of hoplite battle is a Western bias toward "decisive" conflict, generally relying on infantry assault. In contrast to his focus on the preliminaries, particulars, and immediate consequences of combat in this work, J. E. Lendon gives more attention to discourse. More particularly, he examined epic, myth, and other cultural items whose provenance lay in the past, and how they provided archetypes of proper conduct in combat. Lendon compares how Classical Greeks and Romans acted in accordance with that discourse over time, noting how the latter sought to emphasize the martial values of *disciplina* and *virtus* (2005).

To date, ancient societies in the Mediterranean and Near East have received most academic treatment. But in recent years, scholars have given more attention to Asian texts (Lynn, 2003). Much work has been devoted to translating Chinese military "classics" compiled during the Han Dynasty (206 BCE—220 CE), of which Sunzi's *The Art of War* is the most famous. Among other points, the latter is well known for its pragmatic perspective regarding war, wherein military force should be conserved and applied to objectives that would best preserve and enhance the authority of the state. However, when it was originally written in the Warring States period (circa 480–475 BCE to 221 BCE), it may not have reflected prevailing notions about the use of force, but rather sought to challenge widespread chivalric ideals in which nobles regarded war as a means to demonstrate personal courage and martial prowess (Meyer, 2012). Relying upon work by Richard W. Kaeuper and others, Lynn similarly explored the discrepancy between medieval European chivalric ideals and the violence and terror knights propagated among common folk in operations known as *chevauchées*. For him such incongruity led to the creation of an idealized form of combat in the form of the tournament (Kaeuper, 1999; Lynn, 2003).

The debate over a "military revolution" often impinges upon broader discussions of war in the early modern period, a topic reviewed in the chapters written by Aaron Graham for this anthology. Of greater import for studying cultural approaches to warfare, however, is the related but distinct phenomenon of evolving separate civilian and military institutions (at least in Europe). Observing how large standing forces emerged as an alternative to mercenary units and temporary militia or levee formations, historians have illustrated the customs, practices, and beliefs that began to distinguish them from civilian entities, and how the privilege of command was mostly reserved for men of noble birth for whom military service was no longer an obligation (Corvisier, 1979; Hale, 1998; Howard, 1976). With regards to discourse, Hew Strachan (1983) has noted that discussions of topics such as tactics and "universal" principles of war served to justify the independence and autonomy of military institutions and professionals in European history.

Some have examined how these developments occurred in particular countries. In seventeenth-century France, for example, the monarchy manipulated aristocratic values, particularly honor, not simply to entice nobles to become officers but also to persuade them to help fund the costs of maintaining their commands (Lynn, 1997). In Prussia, the cooperation between the state and *junker* nobility in maintaining the latter's dominance in civil society and the army has been a staple of European historiography (e.g., Craig, 1964). But in one of his first works Peter Paret highlighted the impact of this relationship upon military organization and tactics. In particular, late-eighteenth-century noble officers expected to command units of close-ordered line infantry immediately responsive to their direct orders, while disparaging the need for light infantrymen who fought more autonomously in loose, open-ordered formations (1966).

The growing division between civilian and military spheres fed the growth of military thought and writings, discourse that in turn reflected cultural and intellectual developments in Europe. The two broader topics here are the Enlightenment and the reaction to it in light of the French Revolutionary and Napoleonic Wars. With respect to military thought, these produced what has been dubbed the Military Enlightenment and Military Romanticism (Lynn, 2003; Gat, 1989). The seventeenth-century French engineer Sébastien Le Prestre de Vauban has long been regarded as the primary example of how Enlightenment principles of science, reason, and control were applied to military pursuits, specifically fortress design and siegecraft. Janis Langins (2004), however, has examined how France's Royal Corps of Engineering sought to preserve these principles in the face of external challenges a century later, and the consequences for their relevance in the wars that followed the outbreak of the French Revolution. Regarding the latter, David Bell (2007) has argued that intellectual developments during the Enlightenment, and political ones during the first years of the Revolution, created the conditions by which France abandoned the relatively restrained pursuit of war prior to 1792 and engaged in the "first total war." Moreover, this conceptual change was accompanied by an acceleration in the distinction between civilian and military spheres, and growing social participation in the latter, as he demonstrates with his analysis of writings throughout the period.

Of course the experience of the Napoleonic Wars produced the most renowned thinker on the subject of war, the Prussian officer Carl von Clausewitz. Peter Paret's comprehensive biography (1985) sought to identify the cultural and intellectual influences upon his writings, leading up to the renowned, though unfinished, *On War*. While numerous works have sought to explore and analyze that tome in recent years, students interested in the cultural context that shaped Clausewitz's work, including Enlightenment writers and those (like Clausewitz) who challenged their ideas, should engage with the work of Azar Gat, particularly his first book (1989). Gat has since expanded his analysis of military thought to address the evolution of ideas in the later nineteenth and twentieth centuries (2001). Antulio J. Echevarria II has also written a good introduction to *On War* (2007), and Jon Tetsuro Sumida (2008) has offered another interpretation of it.

As they began to create distinct, permanent military institutions, Europeans also began exploring and expanding beyond their continent, ushering in an era of regular and ongoing interactions between peoples originating (later, claiming origins) from different parts of the globe. Not surprisingly, culture has become a prominent motive for analyzing these relationships, including those that encompass warfare. Arguably the most prescient work for understanding their impact on the conduct of war is that of Wayne Lee. He notes (2011a) that, in the absence of restraints, violence in war escalates to become "frightful." Lee identifies four factors that affect such restraints: a polity's capacity to mobilize force; its ability to control it; political calculation; and culture. Whereas usually seen as damping violence, cultural norms sometimes encourage retaliation or aggressive action. That said, Lee notes that antagonists who share a common culture can more easily comprehend and observe restraints on applications of force. Those who do not, however, lack the "grammar" that would enable them to recognize such limits, which can facilitate frightfulness in war. The latter is not just a problem among foes, though, for military cooperation among culturally dissimilar allies has also been a challenge. Lee edited a volume that scrutinizes examples of this phenomenon in a variety of places during the early modern period (2011b). One popular topic in this regard is how the British East India Company fought effectively with sepoy units of native Indians (Roy, 2011). Lee has also employed these ideas in a recent survey of global military history (2016).

As with Lynn's *Battle*, Lee's framework both builds off prior scholarship and helps refine approaches for future research. Fields that have been particularly fertile for contrasting different cultural perspectives on war, and how the experience of fighting a culturally dissimilar foe shapes consequent approaches to warfare, are Native American and early American studies. For example, Daniel Richter revealed that a chief purpose of Iroquois warfare had been to acquire captives to replace the dead, but that in the seventeenth century such "mourning war" was undermined by epidemics and the expansion of conflict for trade purposes (1992).

Many writers have contrasted native combat, which relied upon stealthy movement, with the firepower-intensive, close-order tactics familiar to European colonists and settlers. Some argue that over time, native warfare became more lethal due to acquisition of European weapons (Richter, 1992) and, in the case of King Philip's War, from having previously witnessed the carnage wrought by colonists against native foes in the Pequot War (Malone, 1991). Lee, though, contrasts the experience of the Tuscarora, who adopted Western-style fortifications, with the Cherokees, who eschewed them in favor of flight when faced with invasion (2004). Others have noted that native peoples became dependent upon European trade and goods, especially gunpowder, to maintain their fighting effectiveness (Starkey, 1998).

Differences between native and Western tactics enabled warriors to avoid battle if those so chose, which they usually did (unless settlers fought with native allies). As a result, early English settlers often escalated violence and attacked native fields, villages, and noncombatants (Hirsch, 1988; Karr, 1998). John Grenier (2005) asserts that settlers adopted this attitude in what became the "first" American way of war, employed exclusively against native populations up into the nineteenth century—one of a number of books that have sought to revamp Weigley's approach. As such, it offers an example of what Lynn would regard as an "alternative" discourse of war, one that better reflected the circumstances of white settlers in the North American borderlands than prevailing European notions that had developed to encompass conflict between forces fielded and maintained by dynastic states.

Steven C. Eames (2011) delves deeply into the society that supported the colonists' form of borderland warfare. Eames argues that settlers fought a type of warfare germane to their frontier circumstances, one that borrowed from both European and native approaches to conflict. He also notes that military service was regarded as a male civic duty in the region, with the issue being how to delimit such service in ways that enabled men to fulfill other obligations to family and community. Eames does not just examine the way New England men fought and how they prepared for war. He addresses how military service was regarded in the larger community, noting that militia systems served administrative and recruiting functions and as a means to organize local defense, rather than as active combat formations.

The role of militia is crucial for understanding early U.S. military policy, which rejected European nations' reliance upon large, standing armed forces for national defense. Instead, policy makers agreed that state militia establishments would provide manpower for any major conflict, while establishing a small, peacetime army to guard the frontiers, garrison coastal fortifications, and provide the nucleus for an expanded army in a large war (Kohn, 1975). Recently Rick Herrera has illuminated the martial aspects of early U.S. culture (2015). Herrera argues that military service was a crucial aspect of republicanism for the generations of men reaching adulthood between independence and the secession crisis (though many states did not enforce mandatory militia training as time went on). The display of martial virtue in service to the nation during periods of crisis was a defining aspect of citizenship, the touchstone here provided by the American Revolution.

Regarding the nineteenth and twentieth centuries, many monographs have explored the culture of particular military institutions and organizations. Those addressing Britain often do so in the context of assessing the challenges presented by World War I. Tim Travers (1987), for example, argues that British Army operations on the Western Front reflected the persistence of prewar attitudes among its top commanders. These included anti-modernism and anti-intellectualism, and class biases that denigrated common soldiers but championed officers as "gentlemen." Similarly, British commanders believed battles should be structured and organized, that morale and discipline would counter the increased effectiveness of firepower, and gave little thought to how new weapons might require organizational or conceptual change. When combined with the fixation on offensive operations so prominent in the prewar and early war years, the results would prove devastating on a scale heretofore unknown in combat (Van Evera, 1984).

Andrew Gordon offers an insightful parallel with his examination of the Royal Navy in the years leading up to and including the Battle of Jutland (1996). Whereas Travers employs Kuhn's notion of paradigm shift, Gordon utilizes Norman Dixon's distinction between "authoritarian" and "autocratic" military personalities, specifically arguing that the Royal Navy had too many of the former—who responded to routine and hierarchy and had flourished during the peacetime years of the late nineteenth century—and too few of the latter, who spurned routine and embraced initiative. Relying upon biographical presentations and analysis of prewar issues, such as efforts to develop decentralized tactics, Gordon demonstrates conflict between discourses of war among two distinct groups of British naval officers, and compares their performance during the Jutland operation.

Among Continental states, Germany has received the most attention with respect to the military history of modern Europe. Robert Citino (2005) has illustrated that Prussian and, later, German armies consistently emphasized rapid operational movement of large units (*bewegungskrieg*) for the purpose of engaging and defeating enemy forces so quickly and decisively as to preclude the possibility of long, drawn-out conflict—which in turn reflected Prussia's exposed position in Central Europe and its limited resources. This work is arguably the best attempt to delineate a broader "way of war," because Citino analyzes the specific temporal, geographic, and sociopolitical circumstances that contributed to its persistence—as well as how reliance upon it failed when utilized in the different strategic circumstances of the world wars. Citino touches on how *bewegungskrieg* cultivated certain traits, such as a flexible command system and initiative and aggression among officers, while suppressing others, like logistical and intelligence capabilities.

Isabel V. Hull, however, directly addresses the institutional culture of the Imperial German Army (2005) during 1871–1918. Her argument powerfully demonstrates how many of the same traits identified by Citino facilitated German officers developing risky and, regarding noncombatants, brutal policies. That, in turn, stemmed from their predilection to employ increasingly extreme solutions and excessive levels of violence when confronted with intractable problems or challenges. Her creative and wide-ranging analysis scrutinizes historical cases ranging from the war against the Herero in Southwest Africa to idiosyncrasies arising from the creation of the German Empire, to various examples from the World War I. But Hull also employs social science theory to note that organizational culture can create solutions to problems that may be at odds with overtly expressed goals (and thus "irrational"), but that are nonetheless part of a coherent cognitive framework for its members. In short, Hull examines how the German Army's discourse of war propelled it toward excessive violence. Conversely, other scholarship examines how the linkages among state, society, and the armed forces in the nineteenth and twentieth centuries reflected discourses on the nature of military service and civic obligation in Prussia, and later Germany (Frevert, 2004).

Scholars have also examined discourses within U.S. military institutions. Prominent here is Brian Linn (2007), who, similar to Gordon, has identified three distinct groups of officers within the U.S. Army over the past two centuries. He dubs these Guardians, who have emphasized homeland defense, deterrence, and engineering and technological approaches to military policy; Heroes, who have emphasized martial spirit and adaptability, with an emphasis on combat and operations (and the only one to explicitly address the issues of unconventional war); and Managers, who have focused on the capacity to mobilize and deploy resources for large-scale conflict. Linn argues that the interaction of these three groups' perspectives has shaped the U.S. Army's stance toward war and military preparation over time, with all three having selectively chosen examples from prior experience when arguing for current and future policies. He also notes that all three groups have generally shared a disdain for civilian opinion, and have often assumed a lack of popular support. In contrast to Linn, another attempt to address U.S. Army discourse has highlighted French influences upon its tactical doctrine during the nineteenth and into the twentieth century (Bonura, 2012).

A recent, innovative work by Jörg Muth (2011) compares the "command cultures" of the U.S. and the German armies, or the institutional means by which they developed leadership qualities among military cadets and early-career officers, from the late 1800s to World War II. He concludes that, with the short-lived exception of the U.S. Army's Infantry School (when commanded by George C. Marshall), the German institutions were far superior to their American equivalents. German instructors were far more engaged with their students not just as teachers but also as mentors and role models, and were selected for both military experience and subject knowledge. Conversely, American faculty often lacked both practical experience and academic knowledge of the fields they were assigned to teach. Not surprisingly, pedagogy of the German schools was more modern and intellectually stimulating.

With respect to the impact of World War II, Andrew Krepinevich (1986) has scrutinized the U.S. Army's discourse of war with respect to fighting the Vietnam War, which emphasized fire-intensive conventional operations. Conversely, Army leadership downplayed and deflected efforts to develop counterinsurgency and pacification efforts. Adrian Lewis has taken a broader approach, producing a survey of U.S. military history since 1945 that addresses cultural assumptions about war (2012). Broadly, he asserts that the more limited wars during and after the Cold War, the end of conscription, and other changes were at odds with American conceptions about when and how to fight wars, and have generated commitments that are ultimately unsustainable (an argument that reworks Weigley's thesis).

As implied by this discussion, most research that assesses the culture of military institutions tends to focus on Western armies. But among those that, like Gordon's, address sea forces, some demonstrate how preexisting conceptions of combat channeled the development of naval technology in the late nineteenth and early twentieth centuries. For example, Robert L. O'Connell (1991) asserts that American naval officers' fixation on dreadnoughts—with their similarity to sailing-era ships of the line—reflected emotive and psychological needs, reinforcing a sense of order and familiarity in a period of rapid change, as opposed to rational analysis of contemporary developments. As a result, they relegated more revolutionary technologies, such as the torpedo, submarine, and airplane, to subsidiary, defensive roles, reserving offensive action solely for battleships.

Whereas that book illuminates how preexisting discourses of war shaped naval officers' reactions to new technologies, Gian Gentile demonstrates the converse regarding American air power (2000). He argues that the *U.S. Strategic Bombing Survey*, published in the years immediately following World War II, promoted the view that air forces should focus on bombing strategic targets that impeded the enemy's ability to wage war. Although his analysis indicates the underlying data did not consistently demonstrate the effectiveness of this approach, the claim helped establish an independent U.S. Air Force. In the process, the *Bombing Survey* established a discourse that made strategic bombing the sine qua non of American air power ever since.

The foregoing discussion also demonstrates that most work on war and culture in English addresses Western cases. In addition to the exceptions noted earlier, some intriguing studies have addressed Japan. Oleg Benesch examines how the idea of Bushido developed during the Meiji restoration and thereafter, and its relationship to militarism and Japan's military engagements abroad (2014). Edward Drea traces the evolution of the Japanese Army in this period (2009), while David C. Evans and Mark R. Peattie scrutinized the Japanese navy in an earlier work (1997). Echoing some of O'Connell's points, the latter argued that Japanese naval officers before World War II assumed they would fight a short war won with a single decisive battle, and hence concentrated on constructing large capital ships while consigning other vessels and weapons systems to secondary roles.

Similar to examining the culture of particular military institutions has been work on "strategic culture." It is supposedly the culture that encompasses a particular country's strategic decision-making community, rather than just a single institution. Though popular for a time in policy circles, critics have noted the idea lacks conceptual clarity and has been invoked with little analytic rigor; one deemed it a reworking of previous "way of war" approaches (Black, 2004a; Echevarria, 2014). A recent work that addresses both notions is C. Christine Fair's work on the Pakistani Army (2014). Fair argues that the army dominates the state as well as Pakistani politics, and sees its mission as challenging India's influence and protecting the country's Islamic identity. In this regard, outright victory over India is not as important as constantly maintaining the means to challenge it. As with some other recent works, such as Citino's, Fair demonstrates that research examining a "way of war" or "strategic culture" can encompass rigorous, critical investigations of their topics, and do so when they identify and analyze specific contextual factors. This said, due to the Pakistani Army's political and policy dominance, Fair's work is essentially a treatment of a specific institutional culture, similar to many works discussed earlier.

Unlike those authors, though, Fair is not a historian but a political scientist. Hence in addition to its focus on a non-Western institution, her work offers an example that addresses the subject of war and culture from a different discipline. Sociologist Miguel Angel Centeno has produced another work in this regard (2002). Rather than focus on a particular armed force, he considers Latin America broadly since the demise of the Spanish Empire. Centeno argues that the relationships between war, the state, and nationalism have evolved differently from the European case and as described by models developed by scholars such as Charles Tilly (1990). In Latin America, most wars did not occur between states, but rather within them; war did not offer a means to strengthen states and centralized power, but weakened them; and nationalism never acquired the broader cultural resonance that it achieved for a time in Europe. Religious scholar Kelly Denton-Borhaug (2011) offers a different example of an alternative disciplinary approach. Scrutinizing recent American culture and popular discourse, she asserts Christian notions of redemption and salvation have been exploited to promote the notion of sacrifice in war, which in turn has supported military policies.

To conclude, scholarship that examines the interaction of war and culture encompasses a disparate range of methodologies, a set of which (memory) has coalesced into a distinct subfield, and is addressed earlier in this volume. The others range from examinations of war's impact on cultural forms, to investigations of the meanings attached to war and military service in broader society, to how cultures evolve within specific military institutions. Readers have been advised to consider John Lynn's model, which offers a framework for addressing these multiple approaches. As with other fields of war and society studies, those considering culture would benefit from greater attention to non-Western cases, as well as more comparative analyses, of which Lee and Muth offer notable examples. A few works have been mentioned that demonstrate the insights other academic disciplines offer for the topic, and future research would also benefit from interdisciplinary efforts. In the meantime, the lack of standardized methodologies among scholars of war and culture can be considered a strength, for it is a field that invites creative research approaches—so long as they are pursued with analytic rigor.

Bibliography

Bell, David A. *The First Total War: Napoleon's Europe and the Birth of Warfare as We Know It*. New York: Houghton Mifflin, 2007.

Benesch, Oleg. *Inventing the Way of the Samurai: Nationalism, Internationalism, and Bushido in Modern Japan*. New York: Oxford, 2014.

Black, Jeremy. "Determinisms and Other Issues," *Journal of Military History* 68 (2004a): 1217–1232.

———. *Rethinking Military History*. New York: Routledge, 2004b.

Bonura, Michael A. *Under the Shadow of Napoleon: French influence on the American Way of Warfare From the War of 1812 to the Outbreak of WWII*. New York: New York University Press, 2012.

Centeno, Miguel Angel. *Blood and Debt: War and the Nation-State in Latin America*. University Park: Pennsylvania State University Press, 2002.

Citino, Robert M. *The German Way of War: From the Thirty Years' War to the Third Reich*. Lawrence: University Press of Kansas, 2005.

Corvisier, André. *Armies and Societies in Europe, 1494–1789*. Trans. Abigail T. Siddal. Bloomington: University of Indiana Press, 1979.

Craig, Gordon A. *The Politics of the Prussian Army 1640–1945*. New York: Oxford University Press, 1964.

De Bruyn, Frans, and Shaun Regan, eds. *The Culture of the Seven Years' War: Empire, Identity, and the Arts in the Eighteenth-Century Atlantic World*. Toronto: University of Toronto Press, 2014.

Denton-Borhaug, Kelly. *U.S. War-Culture, Sacrifice, and Salvation*. Oakville, CT: Equinox, 2011; New York: Routledge, 2014.

Dolski, Michael R. *D-Day Remembered: The Normandy Landings in American Collective Memory*. Knoxville: University of Tennessee Press, 2016.

Drea, Edward J. *Japan's Imperial Army: Its Rise and Fall, 1853–1945*. Lawrence: University Press of Kansas, 2009.

Eames, Steven C. *Rustic Warriors: Warfare and the Provincial Soldier on the New England Frontier, 1689–1748*. New York: New York University Press, 2011.

Echevarria, Antulio J., II. *Clausewitz & Contemporary War*. New York: Oxford University Press, 2007.

———. *Reconsidering the American Way of War: U.S. Military Practice From the Revolution to Afghanistan*. Washington, DC: Georgetown University Press, 2014.

Evans, David C., and Peattie, Mark R. *Kaigun: Strategy, Tactics, and Technology in the Imperial Japanese Navy, 1887–1941.* Annapolis: Naval Institute Press, 1997.

Fair, C. Christine. *Fighting to the End: The Pakistan Army's Way of War.* New York: Oxford University Press, 2014.

Faust, Drew Gilpin. *This Republic of Suffering: Death and the American Civil War.* New York: Knopf, 2008.

Frevert, Ute. *A Nation in Barracks: Modern Germany, Military Conscription and Civil Society.* New York: Berg, 2004.

Fussell, Paul. *The Great War and Modern Memory.* New York: Oxford University Press, 1975.

Gat, Azar. *A History of Military Thought: From the Enlightenment to the Cold War.* New York: Oxford University Press, 2001.

———. *The Origins of Military Thought: From the Enlightenment to Clausewitz.* New York: Oxford University Press, 1989.

Geertz, Clifford. *The Interpretation of Cultures: Selected Essays.* New York: Basic Books, 1973.

Gentile, Gian. *How Effective Is Strategic Bombing?: Lessons Learned From World War II and Kosovo.* New York: New York University Press, 2001.

Gordon, Andrew. *The Rules of the Game: Jutland and the British Naval Command.* Annapolis: Naval Institute Press, 1996.

Gray, Colin S. *Irregular Enemies and the Essence of Strategy: Can the American Way of War Adapt?* Carlisle, PA: U.S. Army War College, 2006.

Grenier, John. *The First Way of War: American War Making on the Frontier.* New York: Cambridge University Press, 2005.

Hale, J.R. *War & Society in Renaissance Europe 1450–1620.* Revised Edition. Montreal: McGill-Queen's University Press, 1998.

Hanson, Victor Davis. *Carnage and Culture: Landmark Battles in the Rise of Western Power.* New York: Doubleday, 2001.

———. *The Western Way of War: Infantry Battle in Ancient Greece.* Second Edition. Berkeley: University of California Press, 2000.

Herrera, Ricardo A. *For Liberty and the Republic: The American Citizen as Soldier, 1775–1861.* New York: New York University Press, 2015.

Hirsch, Adam J. "The Collision of Military Cultures in Seventeenth-Century New England," *Journal of American History* 74 (1988): 1187–1212.

Howard, Michael. *War in European History.* New York: Oxford University Press, 1976.

Hull, Isabel V. *Absolute Destruction: Military Culture and the Practices of War in Imperial Germany.* Ithaca: Cornell University Press, 2005.

Kaeuper, Richard W. *Chivalry & Violence in Medieval Europe.* New York: Oxford University Press, 1999.

Karr, Ronald Dale. "'Why Should You Be So Furious?' The Violence of the Pequot War," *Journal of American History* 85 (1998): 876–909.

Kohn, Richard. *Eagle and Sword: The Federalists and the Creation of the Military Establishment in America, 1783–1802.* New York: The Free Press, 1975.

Krepinevich, Andrew F., Jr. *The Army and Vietnam.* Baltimore: The Johns Hopkins University Press, 1986.

Kuhn, Thomas S. *The Structure of Scientific Revolutions.* Fourth Edition. Chicago: University of Chicago Press, 2012.

Langins, Janis. *Conserving the Enlightenment: French Military Engineering From Vauban to the Revolution.* Cambridge, MA: The MIT Press, 2004.

Lee, Wayne E. *Barbarians and Brothers: Anglo-American Warfare, 1500–1865.* New York: Oxford University Press, 2011a.

———, ed. *Empires and Indigenes: Intercultural Alliance, Imperial Expansion, and Warfare in the Early Modern World.* New York: New York University Press, 2011b.

———. "Fortify, Fight, or Flee: Tuscarora and Cherokee Defensive Warfare and Military Culture Adaptation," *Journal of Military History* 68 (2004): 713–770.

———. *Waging War: Conflict, Culture, and Innovation in World History.* New York: Oxford University Press, 2016.

Lendon, J. E. *Soldiers and Ghosts: A History of Battle in Classical Antiquity.* New Haven: Yale University Press, 2005.

Lepore, Jill. *The Name of War: King Philip's War and the Origins of American Identity.* New York: Knopf, 1998.

Lewis, Adrian R. *The American Culture of War: The History of U.S. Military Force From World War II to Operation Enduring Freedom.* Second Edition. New York: Routledge, 2012.

Liddell, Hart, and Basil Henry. *The British Way in Warfare.* London: Faber & Faber, 1932.

Linn, Brian M. "The American Way of War Revisited," *Journal of Military History* 66 (2002): 501–530.

———. *The Echo of Battle: The Army's Way of War.* Cambridge, MA: Harvard University Press, 2007.

Lynn, John A. *Battle: A History of Combat and Culture.* Boulder: Westview Press, 2003. Revised edition, 2004.

———. *Giant of the Grand Siecle: The French Army, 1610–1715.* New York: Cambridge University Press, 1997.

Malone, Patrick. *The Skulking Way of War: Technology and Tactics Among the New England Indians.* Plymouth: Plimoth Plantation, 1991.

Meyer, Andrew Seth, tr. *The Dao of the Military: Liu An's Art of War.* New York: Columbia University Press, 2012.

Muehlbauer, Matthew S. "Holy War and Just War in Early New England, 1630–1655," *Journal of Military History* 81 (July 2017): 667–692.

Muth, Jörg. *Command Culture: Officer Education in the U.S. Army and the German Armed Forces, 1901–1940, and the Consequences for World War II.* Denton: University of North Texas Press, 2011.

O'Connell, Robert L. *Sacred Vessels: The Cult of the Battleship and the Rise of the U.S. Navy.* New York: Oxford University Press, 1991.

Paret, Peter. *Clausewitz and the State: The Man, His Theories, and His Times.* Princeton: Princeton University Press, 1985.

———. *Imagined Battles: Reflections of War in European Art.* Chapel Hill: University of North Carolina Press, 1997.

———. *Yorck and the Era of Prussian Reform, 1807–1815.* Princeton: Princeton University Press, 1966.

Reardon, Carol. *Pickett's Charge in History & Memory.* Chapel Hill: The University of North Carolina Press, 1997.

Reid, Brian Holden. "The British Way in Warfare: Liddell Hart's Idea and Its Legacy," *RUSI Journal* 156, no. 6 (2011): 70–76.

Richter, Daniel K. *The Ordeal of the Longhouse: The Peoples of the Iroquois League in the Era of European Colonization.* Chapel Hill: University of North Carolina Press, 1992.

Roy, Kaushik. *War, Culture and Society in Early Modern South Asia, 1740–1849.* New York: Routledge, 2011.

Silver, Peter. *Our Savage Neighbors: How Indian War Transformed Early America.* New York: W.W. Norton, 2008.

Starkey, Armstrong. *European and Native American Warfare, 1675–1815.* Norman: University of Oklahoma Press, 1998.

Strachan, Hew. *European Armies and the Conduct of War.* Boston: George Allen & Urwin, 1983.

Sumida, Jon Tetsuro. *Decoding Clausewitz: A New Approach to On War.* Lawrence: University of Kansas, 2008.

Tilly, Charles. *Coercion, Capital, and European States, AD 990–1990.* Cambridge, MA: Basil Blackwell, 1990.

Travers, Tim. *The Killing Ground: The British Army, the Western Front and the Emergence of Modern Warfare 1900–1918.* London: Allen & Unwin, 1987.

Van Evera, Stephen. "The Cult of the Offensive and the Origins of the First World War," *International Security* 9 (1984): 58–107.

Weigley, Russell F. *The American Way of War: A History of United States Military Strategy and Policy.* Bloomington: Indiana University Press, 1973.

Winter, Jay. *Sites of Memory, Sites of Mourning: The Great War in European Cultural History.* New York: Cambridge: 1995.

Index